MW00837275

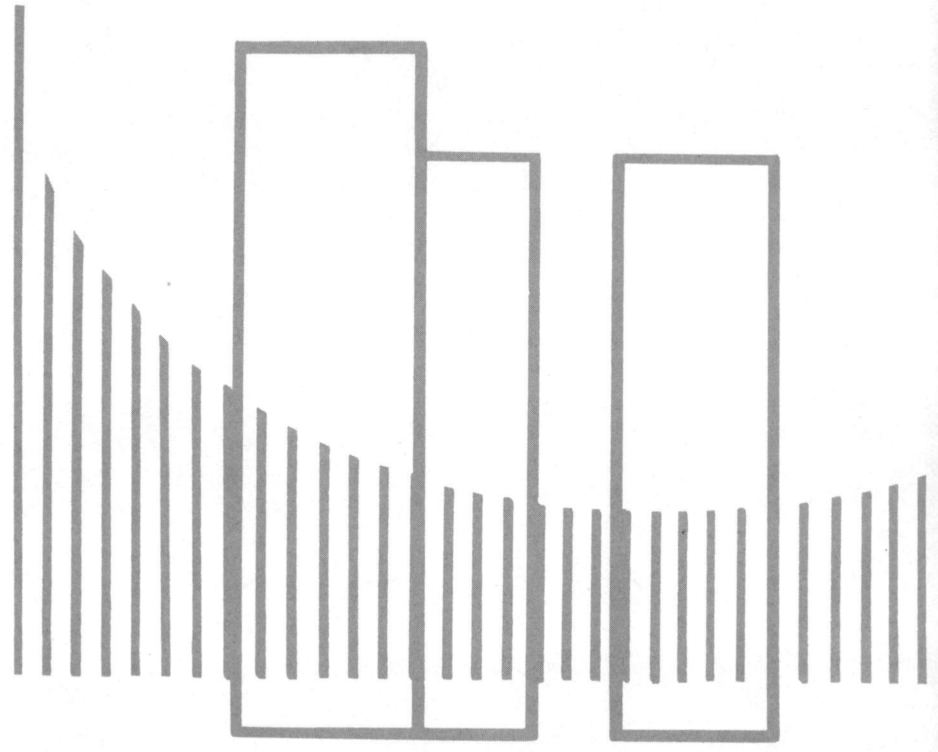

GUILLERMO OWEN

Associate Professor of Mathematical Sciences, Rice University

M. EVANS MUNROE

Professor of Mathematics, University of New Hampshire

W. B. SAUNDERS COMPANY

Philadelphia • London • Toronto • 1971

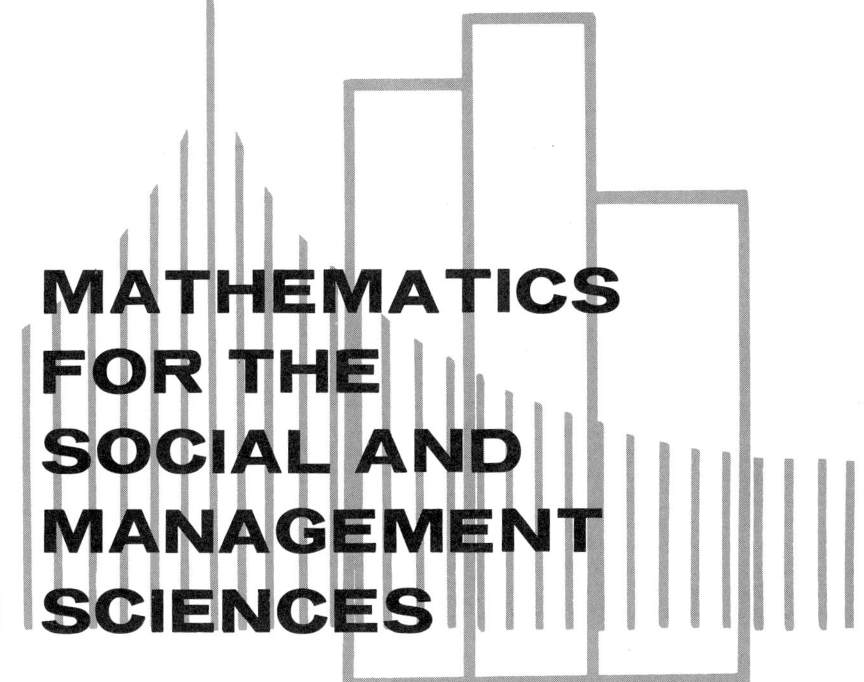

Finite Mathematics and Calculus

W. B. Saunders Company: West Washington Square
Philadelphia, Pa. 19105

12 Dyott Street
London, WC1A 1DB

1835 Yonge Street
Toronto 7, Ontario

SBN 0-7216-7040-7

Mathematics for the Social and Management Sciences: *Finite Mathematics and Calculus*

 Press of W. B. Saunders Company. Library of Congress catalog card number 74-151681.

Print No.: 9 8 7 6 5 4 3 2 1

PREFACE

One of the principal developments in mathematics during the past fifty years has been the dramatic increase in its applications to the social and management sciences. After centuries in which mathematics had been studied, either for itself, or as a handmaiden to the physical sciences, a broad new field of applications has become evident. Subjects such as operations research, statistical analysis, computer systems design, and input-output analysis are fundamentally mathematical in nature, and clearly oriented towards management, economics, and the social sciences. This orientation has been well understood by both industry and education, the former requiring, and the latter providing, much greater quantitative training for the student of social and management science.

This book was written for the purpose of providing the beginning student in the social and management sciences with the basic elements of mathematics. We have, therefore, endeavored to present those fundamental mathematical methods which are useful in these sciences.

In order to motivate the reader, we have at all times introduced practical problems. It is our belief that he will learn more quickly if he sees the value of possible applications. Let us therefore look at the book's contents from the point of view of applications.

The first chapter covers the elements of set theory, with specific reference to counting problems. The related subject of mathematical logic is introduced, and its use in the design of a computer is discussed.

Chapter II discusses analytic geometry, with references to linear equations and inequalities, and systems of these. Such topics as fixed and variable costs, and break-even analysis are treated. Chapter III continues the study of analytic geometry, discussing the conic sections and other non-linear (exponential and logarithmic) curves. Applications include supply and demand curves, product transformation curves, and compound interest.

Chapter IV deals with the basic elements of finite probability theory, using the set-theoretic concepts of Chapter I. The Bayes formula, and sampling from finite populations, are discussed, as is also the criterion of optimizing expected values.

Chapter V deals with the differential calculus. Our main purpose here is to show how the differential calculus can be used to describe marginal effects, and to compute maximal and minimal values of functions. The discussion includes multidimensional optimization problems, and we reach this "advanced" stage in one chapter by limiting the discussion of calculus formulas to those functions most used in the social and management sciences. In Chapter VI, dealing with the integral calculus, our desire is to show how this can be used to obtain cumulative effects.

Chapter VII introduces the reader to vectors and matrices, as aids in the study of linear systems. Applications include a simple Leontief input-output model, and a representation of socio-psychological relations by matrices. Chapter VIII continues the study of linear models, and develops linear programming as a tool for allocation analysis. Transportations problems and shadow prices are discussed.

Finally, in Chapter IX we return to probability theory, using methods developed in previous chapters. Such important distributions as the Poisson and normal are discussed, with applications to economics and psychology. Finite Markov chains are also discussed, with some reference to waiting lines.

In general, the logical relationships among the various chapters are as shown in the diagram:

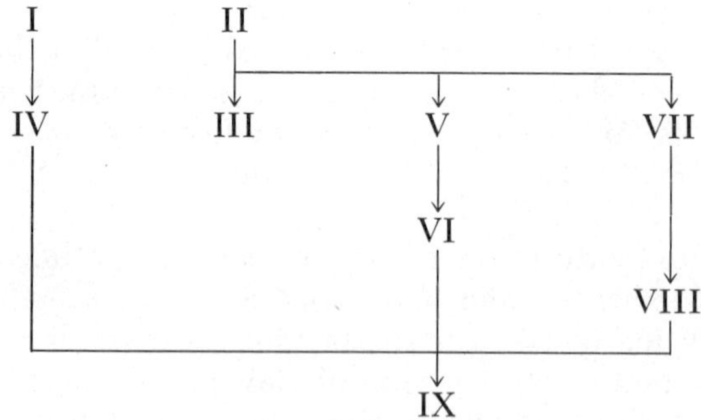

It is a pleasure to give thanks to all those who have helped us with the writing of this book. Professor Howard Frisinger of Colorado State University gave many valuable suggestions, as did Professor William H. Caldwell of the University of South Carolina. Our publisher, W. B. Saunders Company, gave us considerable assistance, financial and technical, in the preparation of the manuscript. Finally, our wives bore patiently with our work.

G. O.
M. E. M.

CONTENTS

Chapter I

SETS, FUNCTIONS, AND LOGIC 1

1. Sets 1
2. Operations on Sets 6
3. Relations and Functions 11
4. Counting Techniques 19
5. Permutations and Combinations 23
6. Propositions and Connectives 30
7. Truth-Tables 32
8. The Conditional Connectives 36
9. Applications to Circuits 40
10. Binary Arithmetic and the Construction of Computers . 46

Chapter II

ANALYTIC GEOMETRY 53

1. The Cartesian Plane 53
2. Graphs and Equations 56
3. Linear Equations 60
4. Inequalities 71
5. Systems of Equations and Inequalities 80
6. Higher Dimensions 99

Chapter III

NON-LINEAR CURVES 113

1. The Conic Sections 113
2. Classification of Conics 121
3. The Circle 124
4. The Ellipse 128
5. The Parabola 132
6. The Hyperbola 137
7. Systems of Equations 147
8. The Exponential and Logarithmic Functions 157

Chapter IV

THE THEORY OF PROBABILITY 169

1. Probabilities 169
2. Probability Spaces 172
3. Conditional Probability. Bayes' Formula 177
4. Compound Experiments 180
5. Repetition of Simple Experiments: The Binomial Distribution 191
6. Drawings With and Without Replacement 194
7. Random Variables 199
8. Expected Values. Means and Variances 206
9. Rules for Computing the Mean and Variance 209
10. Two Important Theorems 213

Chapter V

DIFFERENTIAL CALCULUS 220

1. Derivatives 220
2. Power Functions 227
3. Differentials 232
4. The Chain Rule 235
5. Products and Quotients 241
6. Higher Order Derivatives 246
7. Maxima and Minima 251
8. Exponentials 261
9. Logarithms 266
10. The Trigonometric Functions 272
11. Partial Derivatives 278
12. Lagrange Multipliers 287

Chapter VI

INTEGRAL CALCULUS 301

1. Area Under a Curve 301
2. The Fundamental Theorem 307

3. Inverse of the Chain Rule 314
4. Integration by Substitution 320
5. Integration by Parts 327
6. Applications 334
7. Improper Integrals 341
8. Numerical Integration 348

Chapter VII

VECTORS AND MATRICES 354

1. The Idea of Abstraction 354
2. Addition and Scalar Multiplication 356
3. Scalar Products of Vectors 364
4. Matrix Multiplication 368
5. Inverse Matrices 374
6. Row Operations and the Solution of Systems of Linear Equations 387
7. Solution of General $m \times n$ Systems of Equations 395
8. Computation of the Inverse Matrix 405

Chapter VIII

LINEAR PROGRAMMING 413

1. Linear Programs 413
2. The Simplex Algorithm: Slack Variables 422
3. The Simplex Tableau 426
4. The Simplex Algorithm: Objectives 431
5. The Simplex Algorithm: Choice of Pivots 433
6. The Simplex Algorithm: Rules for Stage I 438
7. The Simplex Algorithm: Proof of Convergence 441
8. Equation Constraints 447
9. Degeneracy Procedures 450
10. Some Practical Comments 454
11. Duality 456
12. Transportation Problems 470
13. Assignment Problems 479

Chapter IX

ADVANCED PROBABILITY 489

1. Discrete Probability Spaces 489
2. Discrete Random Variables 493
3. Markov Chains 500
4. Regular and Absorbing Markov Chains 504
5. Continuous Random Variables 511

6. Transformation of Continuous Random Variables 520
7. The Normal Distribution 524

APPENDIX

1. The Solution of Equations 535
2. The Principle of Induction 542
3. Exponents and Logarithms 544
4. The Summation Symbol 550
5. Table of Common Logarithms 554
6. Table of Natural (Naperian) Logarithms 556
7. Table of Exponential Values 558
8. The Normal Distribution 559

ANSWERS TO SELECTED PROBLEMS 561

INDEX 593

CHAPTER I

SETS, FUNCTIONS, AND LOGIC

1. SETS

We shall give in this chapter a brief outline of the elementary theory of sets. It is quite likely that the reader has already studied this subject; nevertheless, the notions that we shall cover here are so fundamental to mathematics as a whole that such an outline is worthwhile, if only to refresh the reader's memory.

A *set,* as the word is used in mathematics, is an undefined concept. The general idea is that a set contains elements—though there is one, the empty set, that has no elements. In fact, a set is determined by its elements: two sets are equal if they have the same elements. (This is in contrast to the possibility that two different societies might have exactly the same members—but then a society is somewhat more than just the set of its members.) We shall use braces to denote a set (defined by its members). Thus,

$$S = \{a,b,c,d\}$$

which should be read as "S is the set whose elements are $a,b,c,$ and d." Often, a set is defined in terms of a particular property. In such a case, the notation is, say,

$$\{x \mid x \text{ has property } P\}$$

to represent "the set of all x that have the property P." Thus,

$$\{x \mid 0 \leq x < 5\}$$

is the set of all non-negative numbers smaller than 5, while

$$\left\{ y \,\middle|\, \begin{array}{l} y \text{ is a girl} \\ y \text{ has blue eyes} \end{array} \right\}$$

is the set of all blue-eyed girls.

We have, therefore, two methods of defining a set. One is the *roster* method, in which all the elements of the set are enumerated; the other is the *property* method, in which some *characteristic* property of the elements is given. (Note: a property is characteristic of the elements of a set if every element of the set, and no others, has this property.) As an example, the set of integers between 1 and 5 can be defined by the roster method:

$$A = \{1,2,3,4,5\}$$

or by the property method:

$$A = \left\{ n \,\middle|\, \begin{array}{l} n \text{ is an integer} \\ n \text{ is between 1 and 5, inclusive} \end{array} \right\}$$

In a similar manner, suppose that Jones, Roberts, Smith, and Thompson are the members of the board of directors of the ABC Manufacturing Corporation. In this case, we can give a set D by the roster method as

$$D = \{\text{Jones, Roberts, Smith, Thompson}\}$$

or by the property methods as

$$D = \{x \mid x \text{ is a member of the board of directors of ABC Corporation}\}.$$

Closely related to the idea of a set is the idea of belonging: an element belongs (or does not belong) to a set. We use the notation

$$x \in S$$

to signify that x is an element of set S. For instance, if R is the set of real numbers, the symbols

$$5 \in R$$

represent the (true) statement that 5 is a real number. A line drawn through the set element sign represents the negation of this relation. If Q is the set of rational numbers, then

$$\sqrt{2} \notin Q$$

represents the statement that the number 2 does not have a rational square root.

Similarly, in the example of the directors of the ABC Corporation, we will have

$$\text{Smith} \in D$$

and

$$\text{Lewis} \notin D$$

since Lewis is not a member of the board.

The two fundamental notions of set and belonging can be used to define further relations. Thus:

I.1.1 Definition. We say S is a *subset* of T, with notation

$$S \subset T$$

if every element of S is also an element of T (Figure I.1.1).

Thus, if S is the set of girls, and T is the set of human beings, we have

$$S \subset T$$

to represent the fact that all girls are human.

I.1.2 Definition. We say $S = T$ if $S \subset T$ and $T \subset S$.

We define two sets as equal if they have the same elements. If, for instance, S is the set of human beings, while T is the set of featherless bipeds (Plato's definition of human beings), we will (if Plato is right) find that $S = T$. We see that two sets may be actually equal even if defined by very different means.

There are, in general, two sets that are of great importance. One of these, the *universal set*, is not a well-defined notion; it must generally be referred to the context. If we are dealing with analysis of

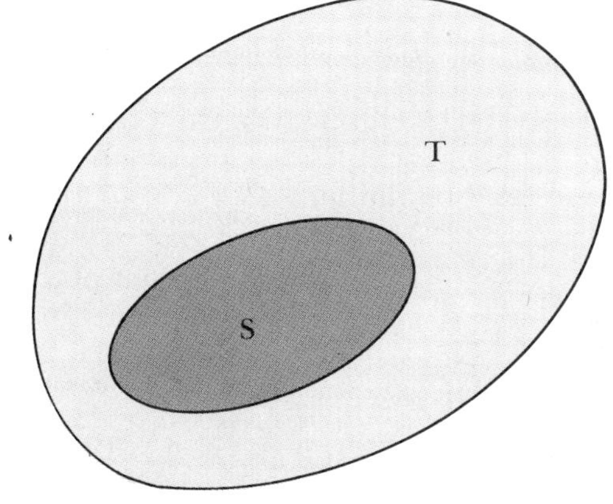

FIGURE I.1.1 S is a subset of T:$S \subset T$.

real numbers, the universal set will be the set of all real numbers. If we are dealing with human beings, the universal set may be the set of all men. If we are dealing with plane geometry, the universal set will probably be the set of all points in the plane. In general, in the absence of any other statement, the universal set, for a given context, will be the set of all elements that are relevant to the context.

Another very important set is the *empty set*, $\varnothing$. This is a set that has no elements. As such, it is clearly unique; moreover, it is easy to see that, for any S, $\varnothing \subset S$.

It is of importance to notice that, in Definitions I.1.1 and I.1.2, no importance whatsoever is given to the order in which the elements of a set may be given. Thus we can write, for the sets A and B above,

$$A = \{1,3,5,2,4\}$$

or

$$B = \{\text{Smith, Thompson, Roberts, Jones}\}$$

without in any way altering the sets. We can, moreover, repeat the elements of a set as often as desired. Thus we have

$$A = \{1,1,2,4,5,2,3,4,3,2,1\}$$

The reader should satisfy himself that, according to Definitions I.1.1 and I.1.2, this set is indeed equal to $\{1,2,3,4,5\}$.

I.1.3 Example. A baseball manager has four outfielders, Root, Brown, Gordon, and Ernest. These can be assigned to any of the three outfield positions, except that Brown is a slow runner and so cannot play center field. There are then 18 possible outfields, shown in Table I.1.1, as O1 through O18. The set of all these outfields will be our universal set, U.

From this universal set, there are many ways in which we can take subsets. For example, we can let A be the set

$$A = \{Oi \mid \text{Root plays}\}$$

which can also be given by the roster method:

$$A = \{O1, O2, O3, O4, O5, O6, O7, O9, O11, O12, O13, O15, O16, O17\}$$

and let B be the set

$$B = \{Oi \mid \text{Root plays right field}\}$$

or

$$B = \{O7, O9, O13, O17\}$$

It is easy to see that $B \subset A$.

TABLE I.1.1

Outfield	Left Field	Center Field	Right Field
O1	Root	Gordon	Ernest
O2	Root	Gordon	Brown
O3	Root	Ernest	Gordon
O4	Root	Ernest	Brown
O5	Brown	Root	Gordon
O6	Brown	Root	Ernest
O7	Brown	Gordon	Root
O8	Brown	Gordon	Ernest
O9	Brown	Ernest	Root
O10	Brown	Ernest	Gordon
O11	Gordon	Root	Ernest
O12	Gordon	Root	Brown
O13	Gordon	Ernest	Root
O14	Gordon	Ernest	Brown
O15	Ernest	Root	Gordon
O16	Ernest	Root	Brown
O17	Ernest	Gordon	Root
O18	Ernest	Gordon	Brown

PROBLEMS ON SETS AND FUNCTIONS

1. From the universal set of Example I.1.3, give (by the roster method) the following sets:

(a) $A_1 = \{Oi \mid \text{Gordon plays center field}\}$

(b) $A_2 = \{Oi \mid \text{Gordon plays, but Root does not play left field}\}$

(c) $A_3 = \{Oi \mid \text{Ernest plays, and Root plays center field}\}$

(d) $A_4 = \{Oi \mid \text{Brown plays center field and Gordon plays right field}\}$

(e) $A_5 = \{Oi \mid \text{Either Gordon or Brown plays right field}\}$.

2. Which of the following sets of real numbers are equal?

(a) $\{x \mid x^2 - 2x = 0\}$

(b) $\{0, 1\}$

(c) $\{0\}$

(d) $\{0, 2\}$

(e) $\varnothing$

(f) $\{y \mid y^3 - 2y^2 = 0\}$

(g) $\{x \mid x^2 + 1 = 0\}$

(h) $\{2, 0\}$

(i) $\{x \mid x^2 - 2x = 0 \text{ and } x^2 - x = 0\}$

2. OPERATIONS ON SETS

In the preceding section we discussed the elementary concepts of sets and elements. In this section we shall study methods by which new sets can be obtained from others. We shall use diagrams, known as *Venn diagrams*, to illustrate these concepts. Figure I.2.1 shows two sets, S and T.

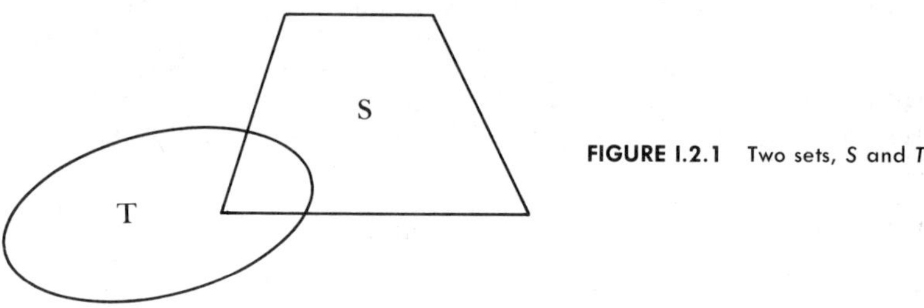

FIGURE I.2.1 Two sets, S and T.

I.2.1 Definition. The *union* of two sets, S and T, is the set of all elements which belong to either S or T (or possibly both). The notation for the union of S and T is $S \cup T$ (Fig. I.2.2).

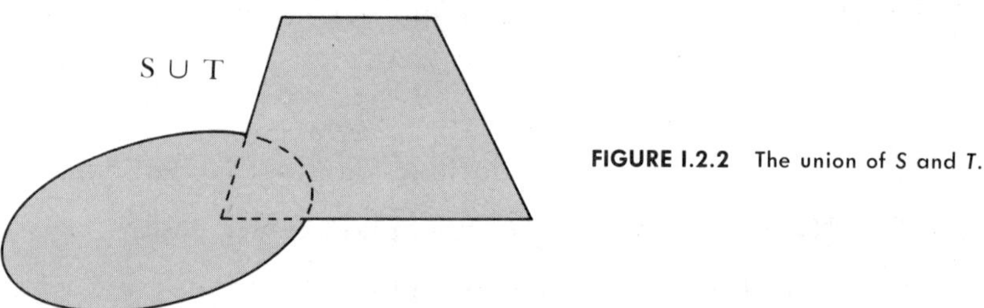

FIGURE I.2.2 The union of S and T.

If the sets S and T are given by the roster method, we can give their union by the simple expedient of listing first the elements of S, then those of T. Some elements may then be repeated, but we saw above that this is really unimportant. Thus, for the set of Example I.1.3, let C be the subset

$$C = \{\text{Oi} \mid \text{Gordon plays left field}\}$$

Then

$$C = \{\text{O11, O12, O13, O14}\}$$

If B is as given above, then

$$B \cup C = \{\text{O7, O9, O13, O17, O11, O12, O13, O14}\}$$

We see that the element O13 appears twice; we may, if we wish, delete it once:

$$B \cup C = \{O7, O9, O13, O17, O11, O12, O14\}$$

Another important operation on sets is the *intersection* operation:

I.2.2 Definition. The *intersection* of two sets, S and T, is the set of all elements of S which are also elements of T. The notation for the intersection of S and T is $S \cap T$ (Fig. I.2.3).

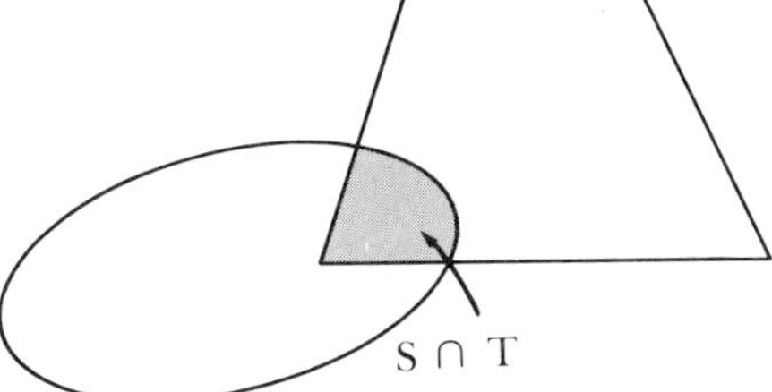

FIGURE I.2.3 The intersection of S and T.

Returning to Example I.1.3, we will have

$$B \cap C = \{O13\}$$

so that $B \cap C$ contains only one element, i.e., the outfield in which Root plays right field, while Gordon plays left field, with Ernest in center.

In the example of the board of directors of ABC Corporation (p. 2), let us assume that the company has three large stockholders: Jones, Thompson, and Knowles. Let E be the set of large stockholders of ABC Corp., i.e.,

$$E = \{\text{Jones, Thompson, Knowles}\}$$

Then we see that

$$D \cup E = \{\text{Jones, Thompson, Smith, Roberts, Knowles}\}$$

is the set of all people who are either large stockholders or members of the board of directors, while

$$D \cap E = \{\text{Jones, Thompson}\}$$

is the set of large stockholders who are also members of the board of directors.

I.2.3 Definition. The *complement* of a set S is the set of all elements of the universal set (whichever it may be in the context)

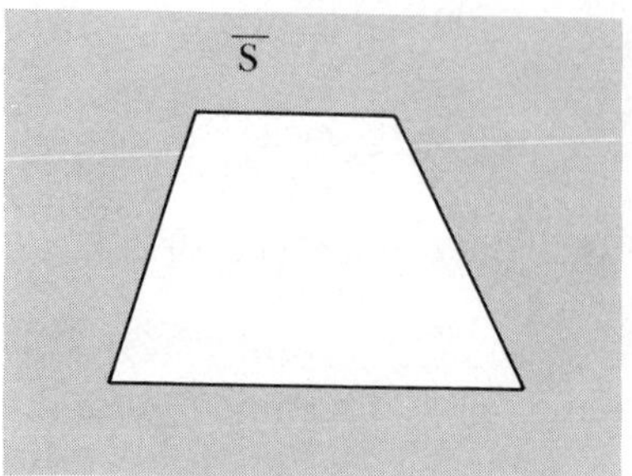

FIGURE I.2.4 The complement of S.

that do not belong to S. The notation for the complement of S is $\overline{S}$ (Fig. I.2.4).

In Example I.1.3, the set $\overline{A}$ consists of those outfields in which Root does not play:

$$\overline{A} = \{O8, O10, O14, O18\}$$

while $\overline{B}$ would consist of those in which he does not play right field.

In the example of ABC Corporation, it is not clear how complements should be defined, because we have not given a universal set. Thus, $\mathcal{U}$ may here be the set of all humans, in which case $\overline{D}$ would consist of all humans who are not on the board of directors of ABC Corp.; $\mathcal{U}$ may also be the set of all people who have ever heard of ABC Corp., or any one of many other sets. These considerations, and others, lead to the following definition:

I.2.4 Definition. The relative complement of S in T is the set of elements of T that are not elements of S. Its notation is $T-S$ (Fig. I.2.5).

To give an example, let us assume that S is the set of all girls, while T is the set of blue-eyed people. The universal set, in this context, is the set of all human beings. Then $S \cup T$ is the set of all human beings who either are girls or have blue eyes. $S \cap T$ is the set of all blue-eyed girls. $\overline{S}$ is the set of human beings who are not girls, i.e., of men, adult women, and boys, while $\overline{T}$ is the set of human beings whose eyes are not blue. Finally, $S-T$ is the set of girls with eyes that are not blue, while $T-S$ is the set of blue-eyed men, women, and boys.

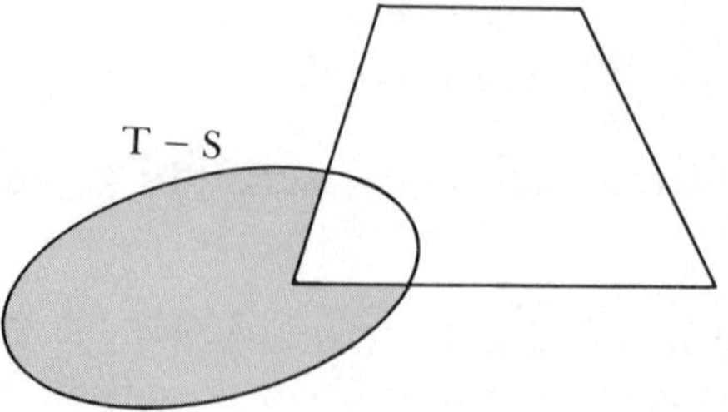

FIGURE I.2.5 The relative complement of S in T.

Returning to Example I.1.3, we see that $A-B$ will consist of outfields in which Root plays either center field or left field; $B-C$, of those in which Root plays right field but Gordon does not play left field, and so on:

$$A-B=\{O1, O2, O3, O4, O5, O6, O11, O12, O15, O16\}$$

$$B-C=\{O7, O9, O17\}$$

In the example of ABC Corporation, we see that $D-E$ consists of those members of the board of directors who are not large stockholders:

$$D-E=\{\text{Roberts, Smith}\}$$

while $E-D$ consists of the large stockholders who are not members of the board of directors:

$$E-D=\{\text{Knowles}\}.$$

The several operations which we have defined satisfy certain rules. Thus, returning to Example I.1.3, we note that $A \cup B=A$, while $A \cap B=B$. This may seem somewhat surprising, until we notice that $B \subset A$. The question, then, is whether this always happens. We state the following:

1.2.1 If $S \subset T$, then $S \cup T=T$

1.2.2 If $S \subset T$, then $S \cap T=S$

It is easy to see that (1.2.1) and (1.2.2) are true. In effect, we note (for 1.2.1), that, if x belongs to either S or T, it must (since $S \subset T$) belong to T. Thus $S \cup T \subset T$. On the other hand, it is clear that in all cases $T \subset S \cup T$. Therefore $S \cup T=T$. Similar reasoning proves (1.2.2). These are not, of course, the only laws satisfied by our operations. Others are given below:

1.2.3 $$S \cup T=T \cup S$$

1.2.4 $$S \cap T=T \cap S$$

1.2.5 $$(R \cup S) \cup T=R \cup (S \cup T)$$

1.2.6 $$(R \cap S) \cap T=R \cap (S \cap T)$$

1.2.7 $$R \cap (S \cup T)=(R \cap S) \cup (R \cap T)$$

1.2.8 $$R \cup (S \cap T)=(R \cup S) \cap (R \cup T)$$

1.2.9 $$S-T=S \cap \overline{T}$$

1.2.10 $$\overline{(S \cup T)}=\overline{S} \cap \overline{T}$$

1.2.11 $$\overline{(S \cap T)} = \overline{S} \cup \overline{T}$$

1.2.12 $$S \cup \overline{S} = \mathcal{U}$$

1.2.13 $$S \cap \overline{S} = \varnothing$$

These laws are all easily proved, and so we shall not stop to do so, but leave them rather as exercises for the reader. Laws (1.2.3) and (1.2.4) are known as the *commutative* laws; (1.2.5) and (1.2.6) are the *associative* laws, and (1.2.7) and (1.2.8) are the *distributive* laws.

The commutative and associative laws (1.2.3) through (1.2.6) are extremely important; they mean, in effect, that if we perform the same operation (either $\cup$ or $\cap$) on many sets, the order in which these sets are taken is of no importance. (But note that if both operations are used, then order is important; see, for example, the distributive laws.) Thus, we can talk about the union or the intersection of a collection of sets. If, for example, we are given sets $S_1, S_2, \ldots, S_{10}$, we will use the notation

$$\bigcup_{i=1}^{10} S_i \quad \text{or} \quad \bigcap_{i=1}^{10} S_i$$

to denote the sets

$$S_1 \cup S_2 \cup \ldots . \cup S_{10}$$

or

$$S_1 \cap S_2 \cap \ldots . \cap S_{10}$$

respectively. If we write

$$\mathcal{S} = \{S_1, S_2, \ldots, S_{10}\}$$

to denote the collection of these sets, the notation $\cup(\mathcal{S})$, $\cap(\mathcal{S})$ is also in common use.

PROBLEMS ON OPERATIONS ON SETS

1. Prove the laws (1.2.3)-(1.2.13).
2. Consider the sets A_1, A_2, A_3, A_4, A_5, given in Exercise 1, page 5. Form the following sets:

(a) $A_1 \cup A_3$

(b) $(A_1 \cup A_2) \cap A_4$

(c) $A_2 \cap A_4 \cap A_5$

(d) $\overline{A}_2 \cup A_3$

(e) $(A_1 - A_3) \cup A_5$

(f) $(A_1 \cap A_3) \cup (\bar{A}_1 - A_4)$

(g) $A_5 \cap (A_1 \cup A_2)$

(h) $A_5 \cup (A_1 \cap A_3)$

3. Five hundred people were classified by age, sex, and political affiliation, with the results given below:

	Men 34 or under	Men over 35	Women 34 or under	Women over 35
Democrats	47	52	68	73
Republicans	51	53	81	75

Let the universal set, $\mathscr{U}$, consist of these 500 people. Let W be the set of women, Y the set of young people (34 or under), and D the set of Democrats. Give the number of elements in each of the following sets:

(a) $W \cap Y$

(b) $W \cup Y$

(c) $D \cup (Y \cap W)$

(d) $D - Y$

(e) $\bar{Y} - W$

(f) $\overline{(D \cup Y)}$

(g) $Y - (D \cap \bar{W})$

(h) $(\bar{W} \cap \bar{Y}) \cup W \cup Y$

3. RELATIONS AND FUNCTIONS

Another very important operation on sets is the *Cartesian product.* The notion of a Cartesian product is defined in terms of an *ordered pair.* Very briefly, if a and b are any elements of any two sets (or possibly of the same set), we can put them together to form the ordered pair (a,b). This pair is *ordered* in the sense that (a,b) and (b,a) are not equal unless, of course, $a = b$.

We define equality for ordered pairs by:

$$(a,b) = (c,d) \qquad \text{if and only if} \qquad a = b \qquad \text{and} \qquad c = d$$

Two ordered pairs are equal if, and only if, corresponding terms are equal. The notion of an ordered pair should not, indeed, be new to the reader. The development of plane (two-dimensional) analytic geometry in Chapter II is based, precisely, on ordered pairs. (In effect, each point in the plane corresponds to an ordered pair of real numbers, and it is well known that the order is important. For example, the points (0,1) and (1,0) are distinct.)

More generally, we can talk about ordered triples (a,b,c), ordered quadruples (a,b,c,d), and, generally, ordered n-tuples. These are all things that we will study in Chapter II.

We are now able to define the Cartesian product of two sets:

I.3.1 Definition. The *Cartesian product* of two sets A and B, denoted by $A \times B$, is the set of all ordered pairs (a,b), where $a \in A$ and $b \in B$.

I.3.2 Example. Let $A = \{1,2,3\}$, and let $B = \{x,y\}$. Then $A \times B$ is the set

$$A \times B = \{(1,x),(2,x),(3,x),(1,y),(2,y),(3,y)\}.$$

As another example, we can let C be the set of all men, while D is the set of all women. Then $C \times D$ is defined by

$$C \times D = \{(x,y) \mid x \text{ is a man, } y \text{ is a woman}\}$$

and

$$C \times C = \{(x,y) \mid x \text{ and } y \text{ are men}\}$$

while, if R is the set of all real numbers,

$$C \times R = \{(x,y) \mid x \text{ is a man, } y \text{ is a real number}\}.$$

A question naturally arises as to why we should be interested in sets such as $C \times R$, above. What possible good can it do us to consider all ordered pairs consisting of a man and a number? The answer is that, in general, we will be interested only in some such pairs. Thus, a telephone directory will give us a list of certain pairs of men and numbers, and these can be of considerable use. This leads us to the idea of a relation.

I.3.3 Definition. A *relation* from a set A to a set B is a subset of the Cartesian product $A \times B$.

If $\mathscr{R}$ is a relation from A to B, and $(a,b) \in \mathscr{R}$, we usually write $a\mathscr{R}b$. Some examples are:

I.3.4 Example. Let A be the set of all men, and B, the set of all women. Let $\mathscr{R}$ be the relation of marriage: we will have $(a,b) \in \mathscr{R}$ or

$a\mathscr{R}b$, if (and only if) a is married to b. Thus, $\mathscr{R}$ consists of all married couples.

I.3.5 Example. Let X be the universal set of Example I.1.3, and let Y be the set of the four outfielders:

$$Y = \{\text{Root, Gordon, Ernest, Brown}\}.$$

We can define a relation $\mathscr{R}$ as follows: $x\mathscr{R}y$ if outfielder y is used in any of the three positions in outfield x. Then (O1, Root), (O1, Gordon), and (O12, Gordon), *among others*, are elements of $\mathscr{R}$. On the other hand, (O2, Ernest) is *not* an element of $\mathscr{R}$. We would write

$$\text{O1 } \mathscr{R} \text{ Root}$$

and

$$\text{O2 } \not\mathscr{R} \text{ Ernest.}$$

Again, let D be the set of directors of ABC Corporation, and let Y be the set of integers. Then we can define a relation, $\mathscr{K}$, from D to Y by $d\mathscr{K}y$ if director d was on the board of directors in the year y. Assume that Smith has been on the board every year since 1963, but not before; then the pairs (Smith, 1963), (Smith, 1964), . . . , (Smith, 1970) all belong to $\mathscr{K}$. On the other hand, (Smith, 1962) does not belong to $\mathscr{K}$.

We see from Example I.3.4 that one element in the Set A may be related to several elements in the set B, (as in the case of a man with several wives) or to none of the elements of B (as in the case of a bachelor). If, however, we restrict A to be, not the set of all men, but only the set of all married men in a monogamous country, then each element of A is related to exactly one element of B. We would say, in this case, that, if $(a,b) \in \mathscr{R}$, then b is *the* wife of a. Each man, in this restricted set, has a unique, well-defined wife. It is this idea that gives rise to the concept of a function.

I.3.6 Definition. A *function* from a set A into a set B is a relation f from A to B such that, for each $a \in A$, there is a *unique* $b \in B$ with $(a,b) \in f$.

A function is a very special type of relation. For any $a \in A$, we let $f(a)$ be that element of B such that $(a,f(a)) \in f$. Precisely because of its uniqueness, the element a determines $f(a)$, and we often think of a function as a rule that assigns an element of B to each element of A. This element, $f(a)$, is known as the *image* of a (under the function f).

In general, there are two ways to think of a function. One interpretation is simply that a function is a collection of ordered pairs, satisfying a specific condition (two different pairs must have different first elements). This interpretation is best exemplified by tabulated data. Most tables can be thought of as representing functions. Thus, a table of populations of cities is a set of ordered pairs; in each pair, the first element is a city, while the second element is a number (the city's population). A table of gross daily receipts in some business is also a set of ordered pairs; here, the first element is a day, while the second element is a number (gross receipts, in dollars, for that day). A table of national capitals is a set of ordered pairs, the first elements being countries, while the second elements are cities (their capitals). Mathematical tables are also representations of functions, and the reader can doubtless think of many other such examples.

The second interpretation of a function is a *mapping*, i.e., a rule which assigns, to each element of the set A, some element of the set B. The name mapping is, itself, most suggestive: a map, as usually understood, is a representation of part of the earth's surface on a piece of paper, and we can think of the mapping as a rule which assigns, to each point on some part of the earth, a corresponding point (its image) on a piece of paper. In this case, it is of course impossible to give a complete listing of every point (on the earth) and its image (on the paper), but a good cartographer can explain the rule by which the correspondence is obtained. In effect, this rule determines the function, and (by extension) we often say that the rule *is* the function.

If f is a function from A into B, then A is called the *domain* of f. The set

$$f(A) = \{b \mid b = f(a) \text{ for some } a \in A\}$$

is called the *range* of f. It is also called the *image* of A under f. Clearly, $f(A) \subset B$.

For example, let the function f have as domain the set of real numbers, and let it be defined by $f(x) = x^2$. Then the range is the set of non-negative real numbers. Similarly, let the domain of g be the set of human beings, and let $g(x) =$ the father of x. Then the range of g is the set of all men who have ever fathered a child.

In Example I.1.3, we can define a function by letting $f(\text{Oi})$ be the centerfielder in outfield Oi. Then

$$\begin{aligned} f(\text{O1}) &= \text{Gordon} \\ f(\text{O7}) &= \text{Gordon} \\ f(\text{O11}) &= \text{Root} \end{aligned}$$

and so on. In this case, the domain is the set of all 18 outfields; the image or range is the set {Gordon, Root, Ernest} of the three possible centerfielders.

In the example of ABC Corporation, we can define a function f by letting $f(x)$ be the year in which director x first joined the board of directors. In this case the domain of the function is D; the range will be the set of years in which these men joined the board. Thus, if

Jones joined in 1969, Roberts in 1957, and Smith and Thompson both in 1963, we will have the set {1957, 1963, 1969} as range.

We will, of course, be most interested in functions whose domains and ranges are both subsets of the real numbers. In the example $f(x) = x^2$, mentioned above, the domain given is the set of real numbers, while the range is a subset of the reals. This is quite common in mathematics, since, after all, most of our mathematics will deal with numbers. Since functions are so often closely related to numbers, the question quite naturally arises as to whether the arithmetic operations can also be carried out on functions. Is it meaningful, for instance, to talk about the sum or product of two functions?

In what follows, we shall assume that the functions we deal with all have their ranges in the real numbers (i.e., the ranges are subsets of the set of all real numbers). We wish to define the sum of two functions. Note that, if f and g are functions, $f+g$ is not intrinsically defined. Other definitions could be devised, but the following has proved to be useful and is generally adopted. If f and g are functions, $f+g$ is the function consisting of all ordered pairs $(a, b+c)$, where (a,b) is in f, and (a,c) is in g. In the examples below, f and g are represented in tabular form (i.e., as tables which list all pairs belonging to the functions); so is $f+g$:

f		g		$f+g$	
−5	−1	−5	−1	−5	−2
−1	4	−1	3	−1	7
0	5	0	−2	0	3
2	−1				
3	−2	3	0	3	−2
		4	4		

As can be seen, the general idea is to "match up" equal entries in the two domains, and add corresponding entries in the ranges. Note that, unless both $f(x)$ and $g(x)$ exist, $(f+g)(x)$ does not exist. In other words, the domain of $f+g$ is the intersection of the domains of f and g.

Subtraction, multiplication and division are defined in a similar manner. The reader should formulate precise definitions. Examples:

f		g		$f-g$		fg		f/g	
−5	−1	−5	−1	−5	0	−5	1	−5	1
−1	4	−1	3	−1	1	−1	12	−1	4/3
0	5	0	−2	0	7	0	−10	0	−5/2
2	−1								
3	−2	3	0	3	−2	3	0		
		4	4						

Note that division by zero is not defined; where it is indicated, that ordered pair is deleted from f/g.

1.3.7 Example. Let X be the set of all men, and let f and g be functions from the set X to the set of real numbers, where $f(x)$ is the assets of individual x (in dollars) while $g(x)$ is his liabilities. Then the function $f-g$ will define a man's net worth: $(f-g)(x)= f(x)-g(x)$.

As another example, let g and r be functions from the set of all countries into the reals, $g(x)$ being country x's gross national product for 1970 in its local currency, while $r(x)$ is the rate of exchange, in dollars per unit of x's currency. Then the product function, gr, defined by $gr(x)=g(x)r(x)$, would give country x's gross national product for 1970 in U.S. dollars.

An alternative definition of function addition is the following (the reader should satisfy himself that this is equivalent to the definition given above):

1.3.8 Definition. Let f and g be real-valued functions with domains S and T respectively. Then the *sum* of f and g is the function h with domain $S \cap T$, defined by $h(x)=f(x)+g(x)$. We write $h=f+g$.

In a similar way, it is possible to define the functions $f-g$ and fg. The quotient function f/g is defined similarly, but only for values of x such that $g(x) \neq 0$. Finally, if c is a constant, the function cf is similarly defined.

There is another important operation on functions, known as the *composition* operation. We can understand it best, perhaps, by an example.

Let X be the set of all married men in a monogamous country. If we wish to know the name of some man's father-in-law, there are two usual ways to do it. One is to ask directly the name of x's father-in-law. This will be a well-defined function, which we may denote $g(x)$. The other method is to ask the name of x's wife, and then find the name of her father. If we let $w(x)$ be x's wife, and $f(y)$ be y's father, then $f(w(x))$ is also x's father-in-law. We have here three functions: the wife function, w; the father function, f; and the father-in-law function, g. It is clear that they are connected by the equation

$$g(x)=f(w(x))$$

or, in words, "the father-in-law of x is the father of the wife of x." We will say, here, that g is the composition of f and w, and write

$$g=f \circ w.$$

Note that $f \circ w$ and $w \circ f$ are not the same; $f \circ w$ is the father-in-law function, whereas, if we consider $w \circ f$, we see that it should be defined as

$$w \circ f(x)=w(f(x))$$

which is the wife of x's father (i.e., x's mother or stepmother) and not x's father-in-law.

We have, then, the notion of a composite function: we define it in terms of our two interpretations of a function. If we think of a function as a set of ordered pairs, then the composite function

$$f \circ g,$$

read "f circle g," is defined as the set of all ordered pairs (a,c) such that, for some b, (a,b) is in g, and (b,c) is in f. The following example shows how this is done:

g		f		$f \circ g$	
3	−2	−5	−1		
−5	−1	−1	3	−5	3
2	−1	0	−2	2	3
		3	0		
−1	4	4	4	−1	4
0	5				

Informally, pair off second entries of g with equal first entries of f; then take the corresponding first entries of g and second entries of f to form the composite function.

If, on the other hand, we think of a function as a mapping, the idea of composition is very easily understood. Indeed, the function g "maps" a point a into its image, $g(a)$. The function f, in turn, maps $g(a)$ into the point $f(g(a))$. The composite function is simply the result of carrying out the mapping g, followed by the mapping f. Thus, $f \circ g$ takes the point a to its "twice-removed" image, $f(g(a))$. The following diagram shows this idea schematically:

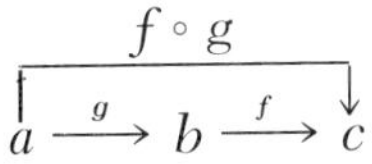

$$\overbrace{a \xrightarrow{g} b \xrightarrow{f} c}^{f \circ g}$$

The following definition is given, for the sake of reference:

1.3.9 Definition. Let g be a function from a set A into a set B, and let g be a function from B into a set C. Then the composition of f and g is the function h, from A into C, defined by

$$h(x) = f(g(x))$$

We have, of course, $h = f \circ g$.

1.3.10 Example. Let f and g be defined by $f(x) = x^2$, and $g(x) = x - 1$. (In each case, the domain is the set of all real numbers.) Then

$$(f+g)(x) = x^2 + x - 1$$

$$(f-g)(x) = x^2 - x + 1$$

$$(fg)(x) = x^3 - x^2$$

$$(f/g)(x) = \frac{x^2}{x-1} \quad \text{where } x \neq 1$$

$$f \circ g(x) = (x-1)^2 = x^2 - 2x + 1$$

$$g \circ f(x) = x^2 - 1$$

PROBLEMS ON RELATIONS AND FUNCTIONS

1. Let X be the set of all states of the United States. For each x in X, let $f(x)$ be the population of state x (in the 1970 census) and let $g(x)$ be the area of state x in square miles. Then f and g are functions. What is the function f/g?

2. Let $X = \{a, b, c, d\}$, and let $Y = \{a, x, y\}$. What is $X \times Y$?

3. Let f be the function consisting of all pairs (x,y), where x is a city, and y is the population of that city in 1970. Let g be the function consisting of all pairs (a,b), where a is a state of the United States, and b is the capital of that state. What is the composite function $f \circ g$? What is $g \circ f$ (if anything)?

4. For any set A, let I_A be a function from A to A, defined by

$$I_A(x) = x \text{ for all } x \in A$$

Then I_A is called the identity function on A. Show that, if f is any function from A into B, then

$$f \circ I_A = f$$

whereas, if g is any function from C into A, then

$$I_A \circ g = g$$

5. Let X be the set of all outfields discussed in Example I.1.3, and let the function g, on X, be defined by

$$x = g(\text{Oi}) \quad \text{if } x \text{ is the left fielder in outfield Oi.}$$

Give an explicit listing of all elements of g.

6. Let f, g, and h be functions with domain R (the set of all reals) defined by

$$f(x) = x^2 - x$$
$$g(x) = 3x + 4$$
$$h(x) = \sqrt[3]{x}$$

Define:

(a) $f + h$

(b) g/f

(c) $f \circ h$

(d) $f - 2g + h \circ f$

(e) $(f + g) \circ (h - g)$

(f) $(h - g) \circ (f + g)$

4. COUNTING TECHNIQUES

In mathematics, it is often very important to know the number of ways in which something can be done. The resulting counting process may in some cases be trivial, but in others will require considerable ingenuity. We give first a definition:

1.4.1 Definition. Let A be a set. Then $N(A)$ is the number of elements in A.

Thus $N(\varnothing) = 0$, $N(\{1, 3, 5\}) = 3$, and, if R is the set of all real numbers, $N(R) = \infty$ (infinity). We shall be interested here in finite sets, although most of what we say will be valid also for infinite sets.

We shall use two fundamental principles in counting; we give the first here:

1.4.1 If $A \cap B = \varnothing$, then $N(A \cup B) = N(A) + N(B)$

Thus, in the special case of *disjoint* sets, (i.e., sets with empty intersection) the number of elements in the union is the sum of the number of elements in each set. As an example, let A be the set of male students in a class, and B the set of female students. Then $A \cup B$ is the set of all students in the class. It is clear that A and B are disjoint. Then, the number of students is equal to the number of male students, plus the number of female students.

Suppose, now, that A, B, and C are *pairwise disjoint* sets (any two have an empty intersection). Then, by (1.2.8),

$$(A \cup B) \cap C = (A \cap C) \cup (B \cap C) = \varnothing \cup \varnothing = \varnothing$$

so that $A \cup B$ and C are disjoint. Therefore,

$$N(A \cup B \cup C) = N(A \cup B) + N(C)$$

But A and B are disjoint, and so $N(A \cup B) = N(A) + N(B)$. It follows that

$$N(A \cup B \cup C) = N(A) + N(B) + N(C)$$

and, in general, if $A_1, A_2, \ldots, A_k$ are pairwise disjoint sets, then

$$N(A_1 \cup A_2 \cup \ldots \cup A_k) = N(A_1) + N(A_2) + \ldots + N(A_k) \tag{1.4.2}$$

If, for example, Mr. Jones, the pet store owner, has 8 puppies, 7 kittens, 11 rabbits, 6 goldfish, and 3 parrots (and no other pets) in stock, then he has a total of $8 + 7 + 11 + 6 + 3 = 35$ pets in stock.

An important corollary of (1.4.1) is given by

I.4.2 Theorem.

$$N(A - B) = N(A) - N(A \cap B)$$

This follows directly from the fact that $A \cap B$ and $A - B$ are disjoint sets with $(A - B) \cup (A \cap B) = A$.

Principle (1.4.1) and its extension (I.4.2) are valid for disjoint sets. When the given sets are not disjoint, however, the situation becomes more complicated. The following is an important result.

I.4.3 Theorem. For any two sets A and B,

$$N(A \cup B) = N(A) + N(B) - N(A \cap B) \tag{1.4.3}$$

The truth of I.4.3 can best be seen by looking at Figure I.4.1, in which the union $A \cup B$ is dissected into the three disjoint sets $A - B$, $B - A$, and $A \cap B$. Then

$$N(A \cup B) = N(A - B) + N(B - A) + N(A \cap B)$$

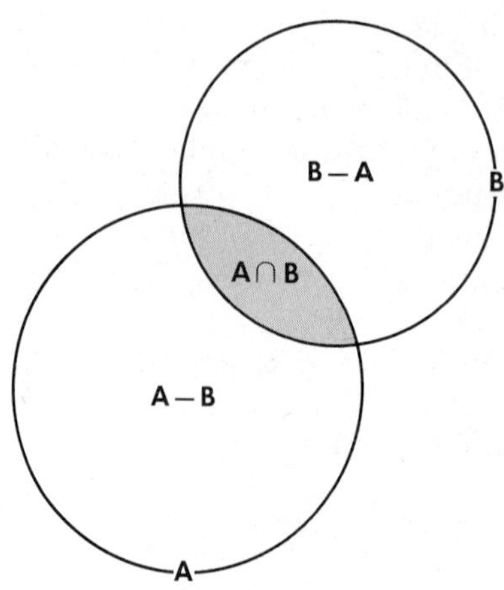

FIGURE I.4.1 Dissection of $A \cup B$ into the disjoint sets $A - B$, $A \cap B$, and $B - A$.

Using I.4.2, this gives us

$$N(A \cup B) = N(A) - N(A \cap B) + N(B) - N(B \cap A) + N(A \cap B),$$

and this can be simplified to give equation (1.4.3). Another way to see the reasonableness of (1.4.3) is to see that, if we simply add the number of elements in A to that in B, we will count the elements of $A \cap B$ twice. We must therefore subtract this number to get $N(A \cup B)$.

I.4.4 Example. In the freshman class at State College, there are 121 students. Of these, 37 take mathematics, 48 take biology, and 16 take both mathematics and biology. How many take neither math nor biology?

We let A be the set of freshmen taking math, and B, those that take biology. Then $N(A) = 37$, $N(B) = 48$, and $N(A \cap B) = 16$. Therefore,

$$N(A \cup B) = 37 + 48 - 16 = 69$$

Let X be the set of all freshmen. Then

$$N(X - (A \cup B)) = N(X) - N(X \cap (A \cup B))$$

but $A \cup B \subset X$, and so $X \cap (A \cup B) = A \cup B$. Therefore,

$$N(X - (A \cup B)) = 121 - 69 = 52$$

and there are 52 freshmen who take neither mathematics nor biology.

I.4.5 Example. In Mathematics 101, all the girls have either blond hair or blue eyes. If 7 girls have blond hair, 5 have blue eyes, and 3 have both blond hair and blue eyes, how many girls are there in all?

Clearly, there are $7 + 5 - 3 = 9$ girls in this class.

For more than two sets, the analysis is similar, but will, naturally, give more complicated formulas. For three sets, Figure I.4.2 explains the situation. If we form the expression

$$N(A) + N(B) + N(C) - N(A \cap B) - N(B \cap C) - N(A \cap C)$$

we see that the elements in the triple intersection $A \cap B \cap C$ have been added three times and subtracted three times as well. Hence they have not been counted at all, and must be added once again. The correct formula is

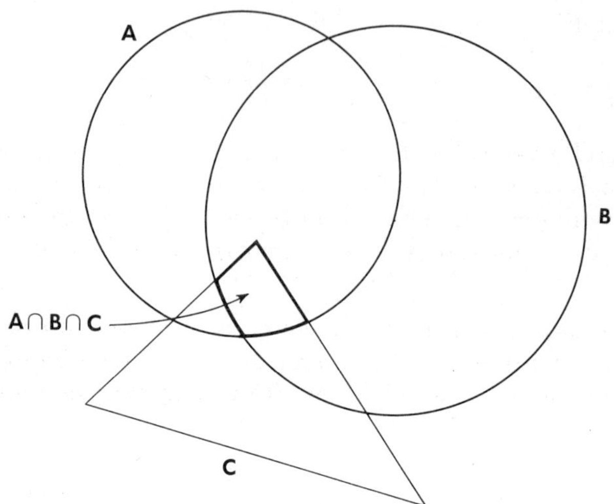

FIGURE I.4.2 The area $A \cap B \cap C$ belongs to all of $A \cap B$, $A \cap C$, and $B \cap C$.

1.4.4 $$N(A \cup B \cup C) = N(A) + N(B) + N(C) - N(A \cap B) - N(A \cap C) - N(B \cap C) + N(A \cap B \cap C)$$

The formula (1.4.4) can also be obtained by treating $A \cup B \cup C$ as the union of the two sets $A \cup B$ and C. For four or more sets, the general idea is to add the number of elements in each set, subtract those in the intersections of two sets, add those in the intersections of three sets, subtract those in the intersections of four sets, and so on. Clearly, the formula becomes extremely lengthy for more than three sets (it will contain 15 terms for four sets, 31 terms for five sets, and so on).

1.4.6 Example. The machine parts produced by a certain factory are tested for three defects, A, B, and C. It is found that 27 have defect A, 31 have B, 17 have C, 8 have both A and B, 5 have both B and C, 6 have both A and C, and 2 have all three defects. How many defective parts are there?

We have here

$$N(A \cup B \cup C) = 27 + 31 + 17 - 8 - 5 - 6 + 2 = 58$$

so there are 58 defective parts.

PROBLEMS ON COUNTING TECHNIQUES

1. In a certain boys' club, every member owns at least one cat or dog. If 20 boys own dogs, 17 own cats, and 7 own both dogs and cats, how many boys are there in the club?

2. A poll of 220 families reveals that 120 read the *Post*, 75 read the *Times*, and 87 read the *Chronicle*. Moreover, 38 read both *Post* and *Times*, 33 read *Post* and *Chronicle*, and 21 read *Times* and *Chronicle*, while 11 read all three papers. How many of the families read none of these papers?

3. At County Community College, every freshman must take one of mathematics, English, or biology. A study of the class lists shows that 108 take English, 81 take math, and 63 take biology. It also shows that 57 take both math and English, 21 take both English and biology, and 8 take both math and biology. Finally, 2 take all three subjects. How many freshmen are there, in all, at this college?

5. PERMUTATIONS AND COMBINATIONS

The second fundamental principle to be used in counting is the following:

1.5.1 If there are m ways to perform act A, and if, for each of these, there are n ways to perform act B, then there are mn ways to perform the *two* acts, A and B.

We obtain from this principle the following:

I.5.1 Theorem. For sets A and B,

1.5.2 $$N(A \times B) = N(A)N(B).$$

This follows clearly from (1.5.1). In effect, the elements of $A \times B$ are ordered pairs. The first term can be chosen in $N(A)$ different ways, and the second, in $N(B)$ different ways.

As an example, suppose that a congressional committee consists of 8 Democrats and 6 Republicans. Then there are $8 \times 6 = 48$ ways to choose a subcommittee consisting of one Democrat and one Republican.

It is easy to see that (1.5.2) generalizes to

1.5.3 $$N(A_1 \times A_2 \times \ldots \times A_k) = N(A_1)N(A_2) \ldots N(A_k)$$

Thus, suppose that a ball can be classified by color, weight, and material. If there are 5 possible colors, 4 possible weights, and 7 possible materials, then there are $5 \times 4 \times 7 = 140$ possible combinations of these.

I.5.2 Example. We consider a more complicated problem. If a set A has n elements, how many subsets (including A itself and the empty set $\varnothing$) will it have?

Let us assume that the elements of A have been labeled $a_1, a_2, \ldots, a_n$. Then, for any $S \subset A$, we can define an ordered n-tuple $(x_1, x_2, \ldots, x_n)$ by

$$x_i = \begin{cases} 1 & \text{if } a_i \in S \\ 0 & \text{if } a_i \notin S \end{cases}$$

for $i = 1, 2, \ldots, n$. It is clear that there is one such n-tuple for each $S \subset A$, and conversely, so that the number of n-tuples is the same as the number of subsets. Now, the set of such n-tuples is clearly the Cartesian product

$$\{0,1\} \times \{0,1\} \times \ldots \times \{0,1\}$$

and must therefore have $2 \times 2 \times \ldots \times 2 = 2^n$ elements. Thus A has 2^n subsets. For this reason, the symbol 2^A is used to denote the set of all subsets of A.

As an example, let $S = \{a,b,c\}$. Then $n = 3$, and we see that S must have $2^3 = 8$ subsets. These can, indeed, be enumerated:

$$\begin{array}{cccc} \varnothing & \{b\} & \{c\} & \{b,c\} \\ \{a\} & \{a,b\} & \{a,c\} & \{a,b,c\} \end{array}$$

It may be checked that there are no other subsets.

Another important counting problem deals with the different permutations of a finite set. Briefly speaking, a permutation is nothing other than an ordering of the set.

Suppose S has n elements. If we wish to order these somehow, we must first choose the first element of the ordering. This can clearly be done in n different ways. The second element must then be chosen; this can be done in $n-1$ ways (since it cannot be the same as the first). Then there are $n-2$ choices for the third element, $n-3$ for the fourth, and so on. We find, by (1.5.3), that there are

$$n(n-1)(n-2) \ldots 2 \times 1$$

different ways to order these n elements. This number is usually denoted by the symbol $n!$, and called "n factorial." In general, we define

1.5.4 $$n! = 1 \times 2 \times 3 \ldots (n-1)n$$

whenever n is a positive integer. Thus,

$$\begin{aligned} 1! &= 1 \\ 2! &= 1 \times 2 = 2 \\ 3! &= 1 \times 2 \times 3 = 6 \\ 4! &= 1 \times 2 \times 3 \times 4 = 24 \end{aligned}$$

and so on. We also define

$$0! = 1$$

While this may not seem an intuitively proper definition, it has great practical advantages, among them the fact that the recurrence relation

1.5.5 $$(n+1)! = n!(n+1)$$

is then satisfied for $n = 0$.

1.5.3 Example. A class has 10 students, who are ordered according to average grades. How many possible orderings are there (discounting the possibility of ties)?

The number of orderings is clearly equal to the number of permutations of the 10 students, or 10! This is

$$10! = 10 \cdot 9 \cdot 8 \cdot 7 \cdot 6 \cdot 5 \cdot 4 \cdot 3 \cdot 2 \cdot 1 = 3{,}628{,}800$$

1.5.4 Example. A baseball team consists of nine men. These must be put into a batting order. How many different batting orders are possible? How many are possible if we assume that the pitcher must bat ninth?

In the first case, the number of possible batting orders is

$$9! = 9 \cdot 8 \cdot 7 \cdot 6 \cdot 5 \cdot 4 \cdot 3 \cdot 2 \cdot 1 = 362{,}880$$

If we assume that the pitcher must bat ninth, then we are clearly interested only in the permutations of the remaining eight players. The number of these is

$$8! = 8 \cdot 7 \cdot 6 \cdot 5 \cdot 4 \cdot 3 \cdot 2 \cdot 1 = 40{,}320$$

1.5.5 Example. How many different pairs of batting orders can two baseball teams (of nine men each) form, if we assume that, in both cases, the pitcher must bat ninth?

In this case, each team has 8!, or 40,320 batting orders. The number of different pairs of batting orders will, by (1.5.3), be equal to 40,320 times 40,320, or 1,625,702,400.

In some cases, it is required to know, not all the possible orderings of a set, but, rather, the number of ways in which the first k elements of an ordering may be chosen. For example, a baseball club has 25 men on its roster. The number of different teams that its manager can field is equal only to the possible orderings of 9 men (from among the 25); it does not matter how the other 16 sit on the bench.

For a problem such as this, the first element in the ordering may be chosen in n different ways; the second element may be chosen in $n-1$ ways; the third in $n-2$ ways; and so on down to the kth, which may be chosen in $n - k + 1$ ways. We obtain the number

$$P_{n,k} = n(n-1)\ldots(n-k+1)$$

or, more concisely,

1.5.6 $$P_{n,k} = \frac{n!}{(n-k)!}$$

as the number of *permutations of k elements from a set with n elements.*

As an example, let $S = \{a,b,c,d\}$ so that $n = 4$, and let us look for the permutations of two elements from S. By (4.2.4), we have

$$P_{4,2} = \frac{4!}{(4-2)!} = \frac{4!}{2!} = 12$$

possible permutations. We can, in fact, enumerate them:

$$\begin{array}{cccc} (a,b) & (b,a) & (c,a) & (d,a) \\ (a,c) & (b,c) & (c,b) & (d,b) \\ (a,d) & (b,d) & (c,d) & (d,c) \end{array}$$

Note that, for permutations, *order* is important: the pair (a,c) and the pair (c,a) are listed separately. In the concrete example of a baseball team, it is clear that order is important, since interchanging two of the players will, in effect, change the team: if the pitcher moves to center field, while the center fielder pitches, it is hardly the same team any more.

1.5.6 Example. An international affairs club must choose three members to be chairmen of its committees on diplomatic relations, trade relations, and military relations. Assuming that the club has 25 members, and that no member can be chairman of more than one of the committees, in how many ways can this be done?

Here we want the number of permutations of 3 elements from a set with 25 elements. This is the number $P_{25,3}$, which is

$$P_{25,3} = 25 \cdot 24 \cdot 23 = 13{,}800$$

1.5.7 Example. A baseball team has 12 outfielders. How many different outfields (of 3 men each) can it field?

Again we want the permutations of 3 elements, this time, from a set with 12 elements. This is $P_{12,3}$, or $12 \cdot 11 \cdot 10 = 1320$.

1.5.8 Example. A psychologist is studying the relative strengths of certain personality traits in married couples. He considers seven different traits for the husband, and nine different traits for the wife. He then lists the four strongest traits in the husband and the six strongest traits in the wife (all in order of strength) and calls this the couple's profile. How many possible profiles are there?

For the husband, we must consider the number of permutations of four elements from a set of seven; this is $P_{7,4}$, or $7 \cdot 6 \cdot 5 \cdot 4 = 840$. For the wife, it is $P_{9,6}$, or $9 \cdot 8 \cdot 7 \cdot 6 \cdot 5 \cdot 4 = 60{,}480$. Since there are 840 possibilities for the husband's part of the profile and 60,480 for the wife's part, we conclude that there are $840 \cdot 60{,}480$, or 50,803,200 possible profiles.

As we mentioned earlier, in permutations, order is important. In some cases, however, this need not be so. If the 9 men selected from 25 are chosen, not to play baseball, but to attend a course on the fine points of the game, the order in which they appear is not important. A similar consideration holds in dealing a hand for bridge: the only thing that matters is *which* 13 cards are received—not the order in which they are received. In cases such as these, we are dealing with *combinations*. Briefly speaking, a *combination of k elements from a set S is a subset of S with exactly k elements.*

Let us look now for the number of combinations of k elements from a set with n elements. We know that there are $P_{n,k}$ permutations, but the number $C_{n,k}$ of combinations is generally smaller, since the same combination will give rise to several different permutations. (If, for example, $k = n$, then $P_{n,n} = n!$, but there is only one combination, namely the set itself.) The best way to compute $C_{n,k}$ is to consider the choice of a permutation as a two-step process. In the first step, a combination (a subset S with k elements) is chosen; in the second, the elements of S are ordered, to give a permutation.

According to the principle (1.5.1), we must now have

$$P_{n,k} = C_{n,k}(k!)$$

as there are $k!$ ways to order each of the combinations. This means that

$$C_{n,k} = \frac{P_{n,k}}{k!}$$

or

1.5.7 $$C_{n,k} = \frac{n(n-1)\ldots(n-k+1)}{1 \times 2 \times 3 \ldots k}$$

is the number of combinations. This is also sometimes written in the form

1.5.8 $$\binom{n}{k} = \frac{n!}{k!(n-k)!}$$

$C_{n,k}$ and $\binom{n}{k}$ are equivalent expressions; the latter is more commonly used.

As an example, the number of possible poker hands is given by the expression

$$\binom{52}{5} = \frac{52 \times 51 \times 50 \times 49 \times 48}{1 \times 2 \times 3 \times 4 \times 5}$$

which, simplified, gives a grand total of 2,598,960 poker hands (for draw or five-card stud; it is of course different for six- or seven-card variations of the game).

1.5.9 Example. A congressional committee consists of 12 Democrats and 9 Republicans. A subcommittee must be chosen, containing 3 Democrats and 2 Republicans. In how many ways can this be done?

Clearly, the order in which the subcommittee members are chosen does not alter the subcommittee. We are interested here in combinations. The number of ways in which the 3 Democrats can be chosen is

$$\binom{12}{3} = \frac{12 \times 11 \times 10}{3 \times 2 \times 1} = 220$$

while the number of ways in which the 2 Republicans can be chosen is

$$\binom{9}{2} = \frac{9 \times 8}{2 \times 1} = 36$$

The total number of ways in which the subcommittee can be chosen is 220 times 36, or 7920.

1.5.10 Example. There are 26 houses in a small town. A poll taker is to visit 7 of them. (The order in which he does it is not important.) In how many different ways can the 7 houses be chosen?

Clearly, we want combinations here. The answer is

$$\binom{26}{7} = \frac{26 \times 25 \times 24 \times 23 \times 22 \times 21 \times 20}{7 \times 6 \times 5 \times 4 \times 3 \times 2 \times 1}$$

or 657,800 different ways.

1.5.11 Example. The town of Example IV.2.8 consists of a north side, with 16 houses, and a south side, with 10 houses. The poll taker is told to visit 4 houses in the north side, and 3 in the south side. In how many ways can this be done?

The 4 houses in the north side can be chosen in $\binom{16}{4}$, or 1820 different ways. The 3 in the south side can be chosen in $\binom{10}{3}$, or 120 ways. In all, there are 1820×120, or 218,400 ways to do this.

PROBLEMS ON PERMUTATIONS AND COMBINATIONS

1. There are 7 men and 12 women in a group of people. In how many ways is it possible to choose a couple (1 man and 1 woman) from the group?

2. A classroom has 10 chairs. There are 7 children in the class. In how many different ways is it possible to assign seats to the children?

3. A Senate committee has 9 Democrats and 7 Republicans. A subcommittee of 3 Democrats and 2 Republicans must be chosen. In how many ways is it possible to do this?

4. Using the data of problem 3, in how many ways is it possible to choose the subcommittee if Senator A, a Democrat, and Senator B, a Republican, refuse to serve on the subcommittee together? In how many ways if it is Senator A and Senator C, both Democrats, who refuse to serve together? (Hint: in each case, it may be easier to compute the number of subcommittees that would have both the hostile senators and subtract this from the total number of possible subcommittees.)

5. A product must be checked to guarantee that it meets 6 different specifications. For each item produced, a card must be filled out, stating which of the 6 specifications is not met. In how many essentially different ways can a card be filled out? (Two cards are filled out in essentially different ways if there is at least one specification that is not met according to one card, but is met according to the other.)

6. A psychiatrist checks a person for 6 personality traits, and lists the 3 strongest on a card, in order of strength. In how many different ways can this card be filled out?

7. Six lines are drawn in a plane. What is the greatest possible number of points of intersection of these 6 lines?

8. A test consists of 5 multiple-choice questions, with 4 possible answers to each, and 10 true-false questions. Assuming that a student

answers all questions, in how many possible ways can he complete the answer sheet?

9. Prove the identity

$$\binom{n+1}{k} = \binom{n}{k} + \binom{n}{k-1}$$

Give a heuristic interpretation of this in terms of combinations.

10. Prove the identity

$$2^n = \binom{n}{0} + \binom{n}{1} + \cdots + \binom{n}{n}$$

by giving heuristic interpretations of the right and left sides of this equation.

6. PROPOSITIONS AND CONNECTIVES

Closely related to the theory of sets is the mathematical theory of logic. In general, it deals with the process of forming compound statements and with finding the *truth-value* of these statements (i.e., whether they are true or false), given the truth-values of the simpler statements that go into the compound.

In essence, this problem was first treated by Aristotle (384-322 B.C.). The mathematical formulation is, however, much later in date, being mainly the work of George Boole (1815-1864). This formulation, while it closely follows Aristotle's reasoning, has the advantage of mathematical strictness, which allows us to see whether the truth of one statement follows logically from another or is merely accidental.

In this section, as well as sections 7 and 8, we shall develop the mathematical theory of symbolic logic. We shall apply this, in sections 9 and 10, to the theory of computer circuits, to see how an elementary computer may be constructed.

Briefly, mathematical logic considers a *proposition* (i.e., a declarative sentence) and assigns to it a *truth-value* (true or false). It then compounds this proposition (possibly with others) and, by logical rules, assigns truth-values to the compounds.

1.6.1 Example. Let us consider the proposition, "I am happy." We shall denote this by the letter, p. Then, p has a truth-value T (true) if the speaker is happy, and a truth-value F (false) if the speaker is unhappy.

Consider now the proposition, "I am not happy." This is known as the *negation* of p, and is represented by the symbols $\sim p$ (the symbol $\sim$ represents negation). The proposition $\sim p$ has the value F whenever p has value T, and it has value T whenever p has value F.

Consider, next, the proposition, "The sun is shining." We shall denote this by q. Again, it may have truth value T or F. We may also form the proposition $\sim q$, "The sun is not shining." The relationship of $\sim q$ to q is exactly that of $\sim p$ to p.

We may also connect the two propositions, generally by the connecting words "and" or "or" (although English usage sometimes requires a different conjunction). In this way we may form the proposition:

"I am happy *and* the sun is shining"

known as the *conjunction* of p and q, and denoted by $p \wedge q$. Now, when will this proposition, $p \wedge q$, be true? Our common experience with such sentences tells us that it will be true if, and only if, both p and q are true. Thus $p \wedge q$ has value T if both p and q have value T. It has value F if either p or q (or possibly both) has value F.

Consider, next, the weaker proposition:

"I am happy, *or* the sun is shining."

We shall denote this by $p \vee q$, and call it the *disjunction* of p and q. When will this be true? Again, common usage tells us that this will be true if either p or q is true. It will be false if both p and q are false. If both p and q are true, then the situation is slightly more complicated, as common usage admits two interpretations.

Under one interpretation, known as *strict* (or *exclusive*) disjunction, the compound is false if both parts are true. We see this in such statements as "You can have your cake or you can eat it," or "The dinner comes with soup or salad." In each of these cases the qualifying phrase "but not both" is clearly understood.

The second interpretation, known as *loose* (or *inclusive*) disjunction, says that the compound is true if both parts are true. Thus, the proposition "You can take math, or you can take Latin" allows for the possibility that the student might take both mathematics and Latin. It is this interpretation that we shall accept: $p \vee q$ has value T if p, or q, or both, have value T. It has value F if both p and q have value F.

It is, of course, possible to form more complicated propositions by using more than one of these connectives. The proposition

"I am happy and the sun is not shining"

would be denoted by $p \wedge \sim q$, while

"I am happy and the sun is shining, or I am not happy"

would be denoted by $(p \wedge q) \vee \sim p$. (The parentheses are used to group $p \wedge q$ together). The reader may be interested in forming other compounds.

These, then, are the most elementary connectives. We may summarize their properties in the form of definitions.

I.6.2 Definition. The *negation* of p, denoted $\sim p$, is true whenever p is false and false whenever p is true.

I.6.3 Definition. The *conjunction* of p and q, denoted $p \wedge q$, is true whenever both p and q are true. It is false otherwise.

I.6.4 Definition. The *disjunction* of p and q, denoted $p \vee q$, is true whenever either p or q, or both, are true. It is false otherwise.

PROBLEMS ON PROPOSITIONS AND CONNECTIVES

1. Find the simple parts in each of the following compound propositions; and write the compounds in symbolic form.

(a) The grass is green, and the weather is hot, but it has not rained.

(b) The people are happy, although the weather is bad.

(c) Either the nation is prosperous, or there is no war.

(d) We are together, and all are good friends.

(e) We are not arming, nor will the enemy attack us.

2. Let p be the proposition "We are singing," let q be "He is playing the piano," and let r be "They are dancing." Form the following compounds:

(a) $p \vee q$

(b) $p \wedge (\sim q \vee r)$

(c) $\sim (p \wedge q) \vee \sim r$

(d) $(p \vee (\sim p \vee q)) \wedge \sim r$

(e) $((p \vee r) \wedge (p \wedge \sim q)) \vee (\sim p \wedge r)$

7. TRUTH-TABLES

For the most elementary compounds, such as those just described, it is very easy to determine the truth-value of the compound from the truth-values of its elements. Definitions I.6.2, I.6.3, and I.6.4 give it to us immediately. For more complicated compounds, however, the result is not so obvious. The reader may, for instance, ask himself what the truth-value of the proposition

$$(p \vee (\sim p \wedge q)) \vee \sim q$$

will be, given that p is false and q is true. It is to deal with such problems that *truth-tables* were developed.

A *truth-table* is, briefly, a short table giving the truth-values of a compound proposition, for any possible combination of truth-values of its elementary components. The truth-table for $\sim p$ is shown in Table I.7.1,

TABLE I.7.1

p	$\sim p$
T	F
F	T

while those for $p \vee q$ and $p \wedge q$ are given in Tables I.7.2 and I.7.3.

TABLE I.7.2

p	q	$p \vee q$
T	T	T
T	F	T
F	T	T
F	F	F

TABLE I.7.3

p	q	$p \wedge q$
T	T	T
T	F	F
F	T	F
F	F	F

For more complicated compounds, a truth-table is generally constructed in a very systematic manner. First, the truth-values of the simple components are considered, then those of the slightly more complicated ones, and so on, step by step, until the truth-table of the desired compound is obtained.

I.7.1 Example. Construct the truth-table of $(p \vee (\sim p \wedge q)) \vee \sim q$.

In our first step, we put the truth-values of each of the simple propositions, p and q, obtaining the table shown as Table I.7.4.

TABLE I.7.4

p	q
T	T
T	F
F	T
F	F

Next, we compute the truth-values of the simplest compounds: $\sim p$ and $\sim q$. We have, then, Table I.7.5.

TABLE I.7.5

p	q	$\sim p$	$\sim q$
T	T	F	F
T	F	F	T
F	T	T	F
F	F	T	T

It is now possible to compute the truth-values of the compound $(\sim p \wedge q)$. These appear in Table I.7.6.

TABLE I.7.6

p	q	$\sim p$	$\sim q$	$(\sim p \wedge q)$
T	T	F	F	F
T	F	F	T	F
F	T	T	F	T
F	F	T	T	F

Next, we introduce the truth-values of the longer compound, $(p \vee (\sim p \wedge q))$. We see these in Table I.7.7.

TABLE I.7.7

p	q	$\sim p$	$\sim q$	$(\sim p \wedge q)$	$p \vee (\sim p \wedge q)$
T	T	F	F	F	T
T	F	F	T	F	T
F	T	T	F	T	T
F	F	T	T	F	F

Finally, we can compute the truth-values for the entire compound. These can be seen in Table I.7.8.

TABLE I.7.8

p	q	$\sim p$	$\sim q$	$(\sim p \wedge q)$	$p \vee (\sim p \wedge q)$	$(p \vee (\sim p \wedge q)) \vee \sim q$
T	T	F	F	F	T	T
T	F	F	T	F	T	T
F	T	T	F	T	T	T
F	F	T	T	F	F	T

It may be seen from Table I.7.8 that this particular compound will be true in all cases. Such a proposition is known as a *tautology*.

I.7.2 Example. Construct the truth-table for the proposition $(p \wedge q) \vee \sim r$.

In this example, there are eight possibilities, since each of the three simple propositions p, q, and r can have two values. As in the previous example, we start by listing all these possibilities:

TABLE I.7.9

p	q	r
T	T	T
T	T	F
T	F	T
T	F	F
F	T	T
F	T	F
F	F	T
F	F	F

Next, we introduce truth-values for the simpler compounds, $(p \wedge q)$, and $\sim r$, as shown in Table I.7.10.

TABLE I.7.10

p	q	r	$p \wedge q$	$\sim r$
T	T	T	T	F
T	T	F	T	T
T	F	T	F	F
T	F	F	F	T
F	T	T	F	F
F	T	F	F	T
F	F	T	F	F
F	F	F	F	T

Finally, we compute truth-values for the entire compound. These are shown in Table I.7.11.

TABLE I.7.11

p	q	r	$p \wedge q$	$\sim r$	$(p \wedge q) \vee \sim r$
T	T	T	T	F	T
T	T	F	T	T	T
T	F	T	F	F	F
T	F	F	F	T	T
F	T	T	F	F	F
F	T	F	F	T	T
F	F	T	F	F	F
F	F	F	F	T	T

36 PROBLEMS ON TRUTH-TABLES

1. Construct the truth-tables for the following compounds:

(a) $(p \wedge q) \vee (\sim p \wedge \sim q)$

(b) $(\sim p \vee \sim q) \wedge (p \vee \sim q)$

(c) $((p \vee q) \wedge (\sim p \vee \sim r)) \vee (p \wedge r)$

(d) $(p \vee \sim q) \wedge (\sim q \vee r)$

(e) $(p \wedge q) \wedge (\sim p \vee \sim q)$

2. Show that the following pairs of compounds are equivalent:

(a) $\sim(p \vee q)$ and $\sim p \wedge \sim q$

(b) $\sim(p \wedge q)$ and $\sim p \vee \sim q$

(c) $(p \vee \sim q) \vee (\sim p \wedge q)$ and $p \vee \sim p$

(d) $p \wedge \sim p$ and $(p \wedge q) \wedge (\sim p \vee \sim q)$

8. THE CONDITIONAL CONNECTIVES

We have, so far, studied the three elementary connectives, $\sim$, $\wedge$, and $\vee$. They are the simplest connectives, but there are others of great importance. These other connectives can be defined, either in terms of the elementary connectives or by their truth-tables. We choose the latter method.

The first connective we consider is the *conditional.* An example of this can be seen in the sentence, "If it rains, we will postpone the picnic." Symbolically, we would write this in the form

$$p \Rightarrow q$$

where p represents "it rains" and q is "we will postpone the picnic." p is called the *hypothesis,* and q the *consequence,* of the conditional.

Let us analyze this proposition. Under which circumstances will it be true? In the first place, it is clear that, if both p and q are true (i.e., it rains, and we postpone the picnic), then the conditional is also true. If both p and q are false (it does not rain, and we do not postpone the picnic), then the conditional will also be true.

Suppose, now, that p is false, but q is true: it does not rain, but we postpone the picnic. Does this make the conditional $p \Rightarrow q$ false? The answer is no, since, after all, $p \Rightarrow q$ states only what will happen in case of rain: we reserve the right to postpone the picnic even if it does not rain. Thus, we will agree that the conditional $p \Rightarrow q$ is true when p is false and q is true.

Finally, suppose that p is true, but q is false: it rains, but we do not postpone the picnic. In this case, it should be clear that the conditional is false.

We conclude, then, that the conditional $p \Rightarrow q$ is true if p is false, or if q is true, or both. It is false if p is true, but q false. We state this in the form of a definition.

I.8.1 Definition. The truth-values of the conditional, $p \Rightarrow q$, are given by Table I.8.1.

TABLE I.8.1

p	q	$p \Rightarrow q$
T	T	T
T	F	F
F	T	T
F	F	T

It should be made quite clear here that the conditional, $p \Rightarrow q$, does not imply, in general, any logical connection between its *hypothesis*, p, and its *consequence*, q. A sentence such as "If dogs can talk, then the moon is made of green cheese," is true, not because of any connection between dogs' loquacity and the substance of the moon, but, quite simply, because the hypothesis ("If dogs can talk") is false. It should be noticed that this property of the conditional is sometimes used for rhetorical effects: "If that man knows what he's talking about, then my name is Napoleon." The point generally is that, since the consequence is obviously false, the hypothesis must be patently untrue. Sometimes, the conditional is used to imply the truth of the consequence by using an obvious hypothesis, even though there is no logical connection: "If he is rich, he is also honest." The man's riches have not made him honest, but the more obvious attribute (wealth) is used rhetorically to affirm the less obvious quality (honesty).

In speaking, the proposition $p \Rightarrow q$ is quite often stated "p implies q". Again, we repeat that the implication need not be a logical implication.

The reader should beware therefore of inferring any logical connection between consequence and hypothesis. For our purposes, we should remember that the truth of the conditional depends merely on that of its parts, as given by Definition I.8.1.

We note also, that the truth-tables for the two compounds, $p \Rightarrow q$, and $(\sim p) \vee q$, are identical. Logically, we can see why this should be so: one will be true whenever the other is. It would have been possible to define the conditional precisely in this manner, as shown in Table I.8.2.

TABLE I.8.2

p	q	$\sim p$	$(\sim p) \vee q$
T	T	F	T
T	F	F	F
F	T	T	T
F	F	T	T

Another important connective is the bi-conditional. We define this also in terms of its truth-table.

I.8.2 Definitions. The truth-values of the bi-conditional, $p \Leftrightarrow q$, are given by Table I.8.3.

TABLE I.8.3

p	q	$p \Leftrightarrow q$
T	T	T
T	F	F
F	T	F
F	F	T

By definition, $p \Leftrightarrow q$ is true when p and q are both true or both false. It is false otherwise. We could have defined it by saying that it is identical with the compound

$$(p \wedge q) \vee (\sim p \wedge \sim q)$$

as it is easy to see that this has the same truth-table. It is more interesting, however, to point out that the compound

$$(p \Rightarrow q) \wedge (q \Rightarrow p)$$

has the same truth-table as $p \Leftrightarrow q$ (Table I.8.4).

TABLE I.8.4

p	q	$p \Rightarrow q$	$q \Rightarrow p$	$(p \Rightarrow q) \wedge (q \Rightarrow p)$
T	T	T	T	T
T	F	F	T	F
F	T	T	F	F
F	F	T	T	T

This shows the relation between the conditional and the bi-conditional connectives, and indeed, the bi-conditional is generally presented in this manner. If p is "I will come," and q is "It does not rain," then $p \Leftrightarrow q$ is generally expressed in the form "I will come if and only if it does not rain," corresponding to the compound

$(p \Rightarrow q) \wedge (q \Rightarrow p)$. It could also be expressed as "Either I will come and it will not rain, or I will not come and it will rain," corresponding to the compound $(p \wedge q) \vee (\sim p \wedge \sim q)$. The two statements are equivalent (i.e., they will be true under the same circumstances), but the first form is much more concise and effective.

Like the conditional, the bi-conditional is sometimes used even when there is no logical connection between the parts; rhetorically, however, this is not nearly so common.

1.8.3 Example. Construct the truth-table for $(p \Rightarrow q) \Rightarrow (q \Rightarrow p)$.

TABLE 1.8.5

p	q	$p \Rightarrow q$	$q \Rightarrow p$	$(p \Rightarrow q) \Rightarrow (q \Rightarrow p)$
T	T	T	T	T
T	F	F	T	T
F	T	T	F	F
F	F	T	T	T

1.8.4 Example. Construct the truth-table for $(p \vee q) \Rightarrow (p \Leftrightarrow q)$.

TABLE 1.8.6

p	q	$p \vee q$	$p \Leftrightarrow q$	$(p \vee q) \Rightarrow (p \Leftrightarrow q)$
T	T	T	T	T
T	F	T	F	F
F	T	T	F	F
F	F	F	T	T

PROBLEMS IN CONDITIONAL CONNECTIVES

1. Construct the truth-tables for the following compounds:

(a) $(p \Rightarrow q) \wedge (q \Rightarrow p)$

(b) $(p \wedge q) \Leftrightarrow (p \vee q)$

(c) $(p \wedge \sim q) \Rightarrow (\sim p \vee q)$

(d) $p \Rightarrow (q \wedge \sim q)$

(e) $((p \vee q) \Rightarrow (q \vee r)) \Leftrightarrow (p \Rightarrow r)$

(f) $(p \Rightarrow q) \Leftrightarrow (\sim p \Rightarrow \sim q)$

(g) $(p \vee \sim q) \Rightarrow p$

(h) $p \Rightarrow (p \vee q)$

(i) $((p \Rightarrow q) \vee (q \Rightarrow r)) \Leftrightarrow (r \Rightarrow p)$

(j) $(p \Leftrightarrow q) \Rightarrow (p \Rightarrow q)$

2. Assume that the following are true: (a) it is raining; (b) today is not Tuesday; (c) I am tired. Which of the following compounds are true?

[Note: the conjunctions "but", "although", have the same logical meaning as the conjunction "and".]

(d) I am tired if and only if either today is Tuesday or it is raining.

(e) Either I am tired, or I am tired although it is Tuesday.

(f) Either, it is Tuesday if it is raining, or, I am tired although it is not Tuesday.

(g) It is raining if today is Tuesday, and, either I am tired or today is not Tuesday.

(h) Either it is raining or today is Tuesday, but I am tired if and only if today is not Tuesday.

9. APPLICATIONS TO CIRCUITS

As an application of the symbolic logic discussed in the last three sections, we will consider here a theory of electrical circuits. The general idea is to build a complex circuit, with many switches; the question then is whether current flows between two terminals (equivalently, whether current will flow through an indicator lamp).

Consider first a simple circuit, with one switch, as shown in Figure I.9.1. The circuit consists of a voltage source (say, a dry cell), a lamp, and a switch P. Current will flow, lighting the lamp, if and only if the switch P is closed. Thus we can think of the lamp as an indicator for the statement

$$p : \text{"the switch } P \text{ is closed."}$$

(Equivalently, we could let the switch correspond to some other

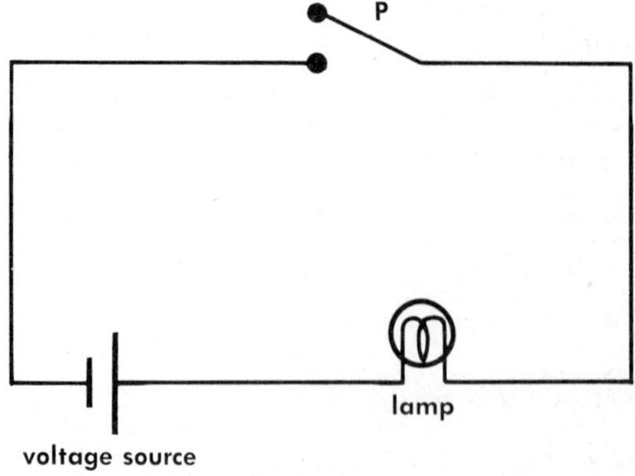

FIGURE I.9.1 A simple circuit.

proposition, and close it if and only if the proposition is true. In this case the lamp will be an indicator for the given proposition.)

Let us suppose, next, that we want to put two switches, P and Q, in a circuit. There are, generally speaking, two ways to do this: they may be placed in *series*, or in *parallel.*

Consider first the case of two switches in *series* (Fig. I.9.2). It is easy to see that current can flow only if both P and Q are closed. Then, if we let p be the proposition "the switch P is closed," and let q be "the switch Q is closed," we see that current will flow if both p and q are true. In other words, the lamp will be an indicator for the *conjunction* $p \wedge q$.

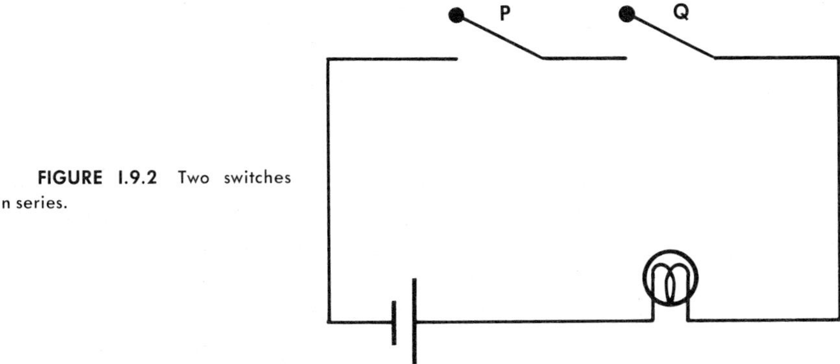

FIGURE I.9.2 Two switches in series.

We look next at the case of two switches in *parallel* (Fig. I.9.3). In this case, current will flow if either one of P and Q (or both) is closed. We conclude that the lamp will be an indicator for the *disjunction* $p \vee q$.

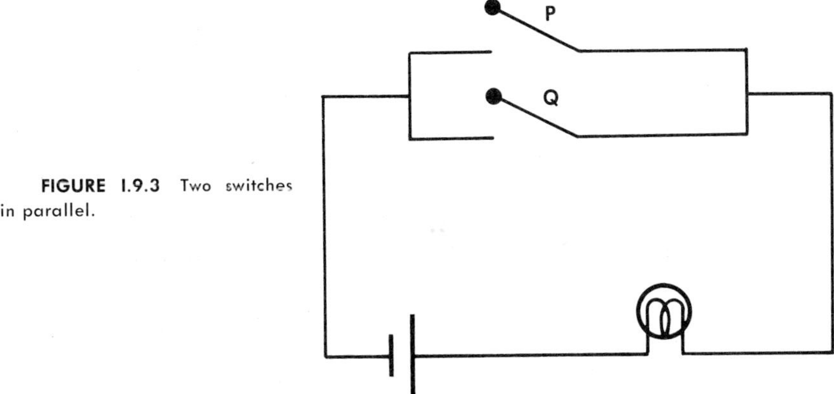

FIGURE I.9.3 Two switches in parallel.

We see then that the two fundamental logical connectives, $\wedge$ and $\vee$, can be represented by the fundamental processes of putting switches in series or in parallel. More complicated circuits can of course be formed; Figure I.9.4 shows how a circuit with four switches P, Q, R, and S can be used to represent the compound proposition $p \vee (q \wedge (r \vee s))$.

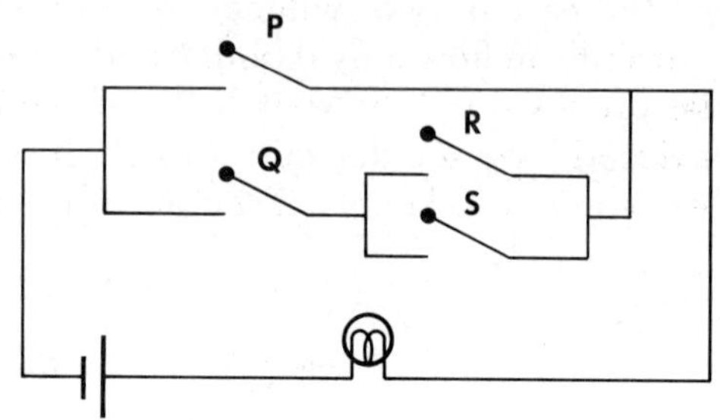

FIGURE I.9.4 Circuit for the compound $p \vee (q \wedge (r \vee s))$.

Up to this point, we have considered only circuits with independent switches; i.e., closing or opening one of the switches has no effect on the remaining switches in the circuit. There is no reason, however, that this should be so. It is possible to connect two switches in such a way that they will always be either both open, or both closed simultaneously; i.e., closing either one of them will cause the other one to close. When two (or more) switches are connected in this way, we shall identify them by using the same letter to represent them. In Figure I.9.5, the two switches labeled P are of this type, as are also the two switches labeled Q. The figure thus represents the proposition $(p \vee q) \wedge (p \vee r) \wedge q$.

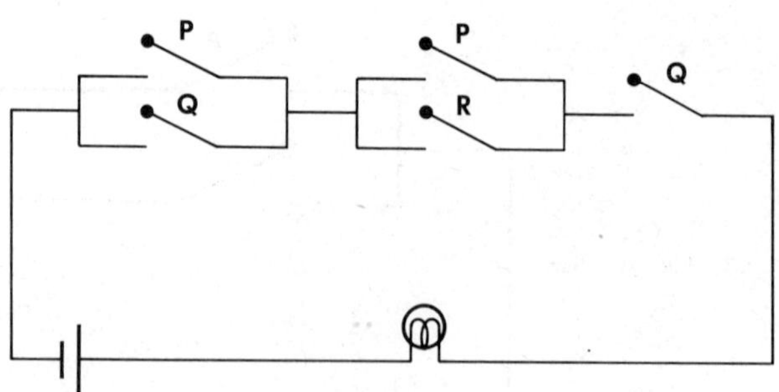

FIGURE I.9.5 Circuit for $(p \vee q) \wedge (p \vee r) \wedge q$.

It is also possible to connect two switches in such a way that closing one of them will cause the other one to open, and conversely. When this happens, we will label such a pair of switches P and P' respectively, Q and Q' respectively, or some other such pair of letters. In Figure I.9.6, P and P' are such a pair of switches; P' is open when-

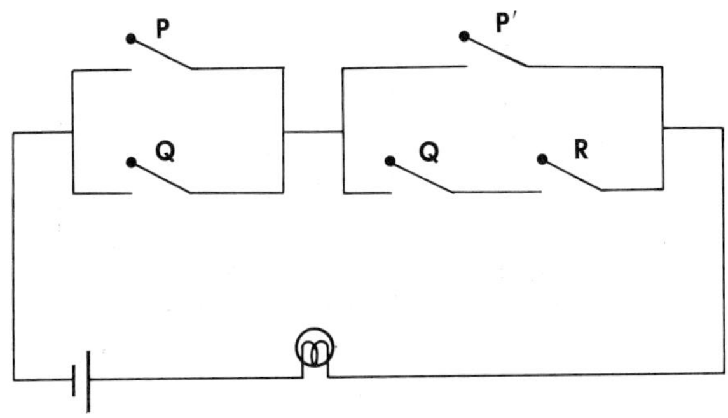

FIGURE I.9.6 Circuit for $(p \vee q) \wedge (\sim p \vee (q \wedge r))$.

ever P is closed, and conversely. Now, if the closing of P corresponds to the proposition

$$p : \text{“}P \text{ is closed,”}$$

we see that the closing of P' corresponds to the negation

$$\sim p : \text{“}P \text{ is not closed.”}$$

This enables us to introduce the negation connective into our calculus of propositions. We see, then, that Figure I.9.6 corresponds to the compound

$$(p \vee q) \wedge (\sim p \vee (q \wedge r)).$$

We see thus how electrical circuits can be used to represent compound propositions made with the three elementary connectives $\sim$, $\wedge$, and $\vee$. We need only remember that elements in parallel correspond to disjunction, and elements in series correspond to conjunction. This gives us a method for constructing the truth-table of any such compound: simply form the corresponding circuit, and, for each combination of open and closed switches, see whether the indicator lamp goes on.

I.9.1 Example. Construct a circuit for the proposition

$$((p \vee q) \wedge (\sim p \vee r)) \vee (q \wedge r).$$

Figure I.9.7 shows the circuit. The general idea is to construct branches for each of the simpler compounds, and then put these together in series or in parallel, according to whether the connective is $\wedge$ or $\vee$. Thus $p \vee q$ is obtained by putting P and Q in parallel, while $\sim p \vee r$ is obtained by putting P' and R in parallel. These two connections are then put in series to give $(p \vee q) \wedge (\sim p \vee r)$. In turn, $q \wedge r$ is obtained by putting Q and R in series. The final arrangement places $q \wedge r$ in parallel with the previous connection.

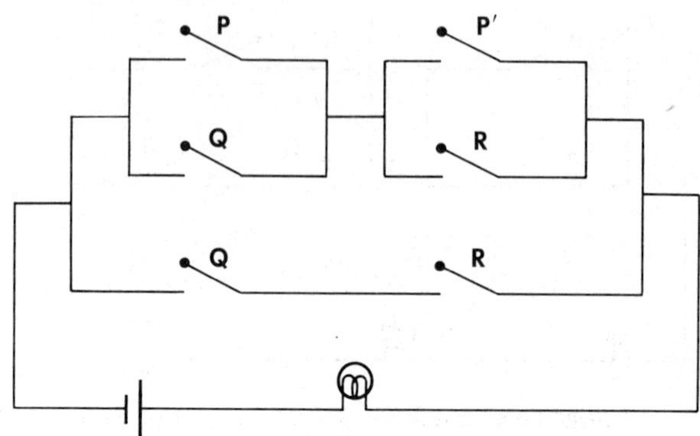

FIGURE I.9.7 Solution of Example I.9.1.

Just as the circuits can be used to calculate truth-tables, it is sometimes desired to do the reverse, and obtain a circuit which will have a specified truth-table. This is done by constructing, for each combination of the elements which will make the compound true, an elementary connection. These elementary connections are then placed in parallel. An example will make this clear.

I.9.2 **Example.** Construct a circuit for the compound $p \Rightarrow q$. We know that the compound is true for the following combinations: p true, q true; p false, q true; and p false, q false. We therefore construct connections for $p \wedge q$ (P and Q in series); $\sim p \wedge q$ (P' and Q in series); and $\sim p \wedge \sim q$ (P' and Q' in series). These are then placed in parallel, as shown in Figure I.9.8.

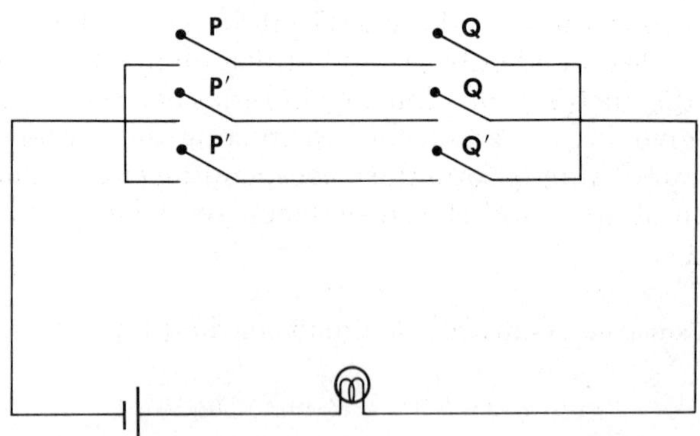

FIGURE I.9.8 A possible circuit for $p \Rightarrow q$.

Of course, this is not the simplest possible circuit for $p \Rightarrow q$. We know that this compound is equivalent to $\sim p \vee q$ (in the sense that they have the same truth-table), so that an equivalent circuit could have been made by putting P' in parallel with Q. However, the circuit of Figure I.9.8 is obtained by a method which is always applicable, and therefore preferable.

I.9.3 Example. Most electric circuits have one single switch which causes a light to go on whenever it is closed, and off whenever the switch is open. Some circuits, however, have two independent switches, such that changing the position of either switch will cause the light to go on (if it is off) or off (if it is on). This is accomplished by putting one connection (P and Q in series) in parallel with a second (P' and Q' in series), as shown in Figure I.9.9. Construct a circuit with three independent switches, P, Q, and R, such that changing the position of any one of them will cause the light to go on (if off) or off (if it is already on).

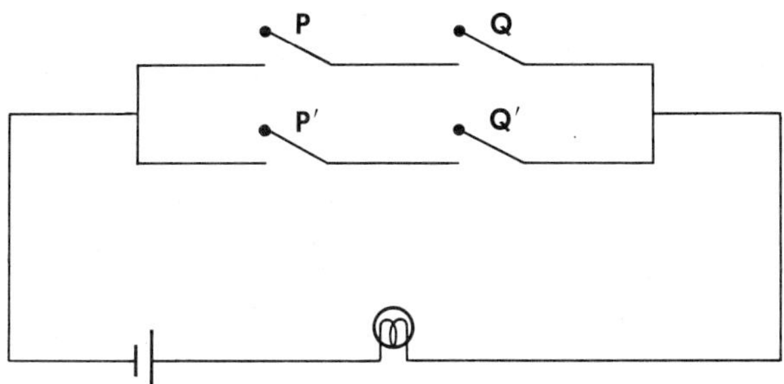

FIGURE I.9.9 A double-switch lighting circuit.

Let us assume that the light is on if all three switches are closed. Opening any one of them will cause the light to go off, and so the light is off if exactly two of the switches are closed. Opening one more switch turns the light back on, and so the light is on if exactly one of the three is closed. Finally, the light will be off if all three switches are open. What we want corresponds, therefore, to a compound which will be true if all of p, q, and r are true, or if exactly one of them is true. It can therefore be expressed in the form

$$(p \wedge q \wedge r) \vee (p \wedge \sim q \wedge \sim r) \vee (\sim p \wedge q \wedge \sim r) \vee (\sim p \wedge \sim q \wedge r).$$

We therefore want four connections in parallel, one for each of $p \wedge q \wedge r$, $p \wedge \sim q \wedge \sim r$, $\sim p \wedge q \wedge \sim r$, and $\sim p \wedge \sim q \wedge r$. Figure I.9.10 shows how this is done.

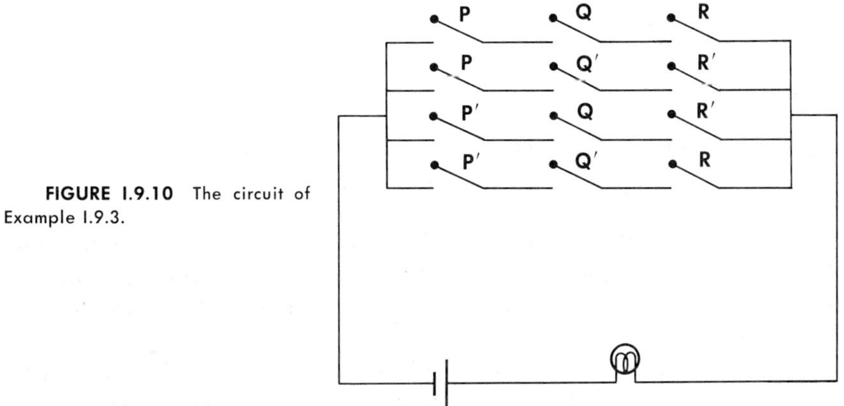

FIGURE I.9.10 The circuit of Example I.9.3.

46 PROBLEMS ON CIRCUITS

1. Construct circuits for the following compounds:

(a) $(p \Rightarrow q) \Rightarrow (q \vee p)$

(b) $((p \wedge q) \vee (p \wedge r)) \wedge (\sim p \vee \sim r)$

(c) $p \Leftrightarrow q$

(d) $(\sim p \wedge r) \Rightarrow (p \vee r)$

(e) $((p \vee q) \wedge (p \vee \sim q) \Rightarrow p$

2. Give a compound statement which represents the circuit below:

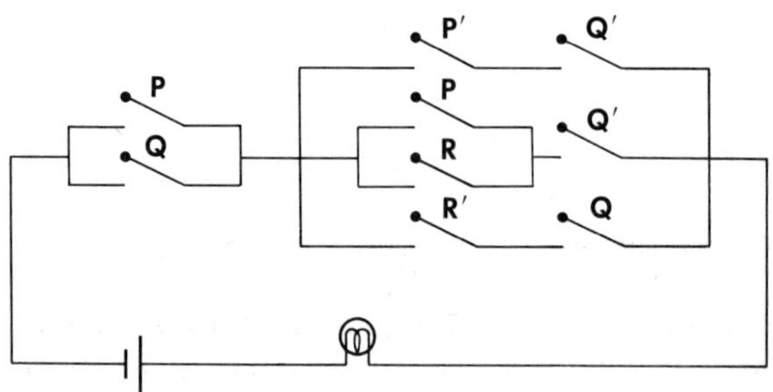

3. The board of directors of a company has three members. The three will vote on decisions, and any two constitute a majority. To obtain a secret vote, each director is provided with a switch. Then, if any two of them (or all three) close their switches, a light will go on. How should this circuit be constructed?

4. Consider the same problem as in Exercise 3 above, with the difference that the board of directors now has four members; one of these (the chairman) has two votes.

5. Design a system similar to that of Example I.9.3, but with four switches instead of three.

10. BINARY ARITHMETIC AND THE CONSTRUCTION OF COMPUTERS

We will, in this section, see how our knowledge of circuits can be used to help design an automatic (logical) computer. Prior to this, however, we must discuss a new notational system—the *binary notation*—for numbers.

As the reader is doubtless well aware, our notational system for numbers is based on the number 10. This is best seen when we consider how simple multiplication or division by 10 is – merely a matter of adding a zero or shifting a decimal point. Because of this, our system is known as the *decimal* system.

Our system is also known as *decimal positional notation* because the meaning of a digit will vary according to its position. Thus, 18 and 81 are different numbers because the digits, 8 and 1, have different meanings in the two numbers.

If we look at a numeral such as 5382, we see that it represents the number 5-thousand, 3-hundred, 8-tens, and 2, or

$$5382 = 5 \times 10^3 + 3 \times 10^2 + 8 \times 10^1 + 2 \times 10^0.$$

(The reader should remember that $10^1 = 10$, and $10^0 = 1$. Those not familiar with exponents should read Appendix 3 in this book.)

This discussion explains how our notation depends on the base 10. The reason that our system is based on 10 lies in the fact that most humans have 10 fingers. This is, however, only a historical accident; if men had 8 fingers, an *octal* system (i.e., one based on the number 8) would probably have developed instead, and would be considered the "natural" system today. In that case, the numeral 5214 would represent the number

$$5 \times 8^3 + 2 \times 8^2 + 1 \times 8 + 4$$

or 2700 (two thousand seven hundred). In other words, the numeral "5214" in octal system and the numeral "2700" in decimal system represent the same number.

In general, any integer greater than 1 can be used as a base for a system of notation. (Some people have even suggested using negative integers as base, a suggestion which we will not consider here.) The trick for changing a number into a new notation is to divide by the base. The quotient, if larger than the base, is divided once again, and so on until a quotient smaller than the base is obtained. Then the remainders obtained, taken in reverse order, will be the digits for the notation. An example will make this clear.

1.10.1 **Example.** Express the decimal numeral 2345 in *hexal* (base 6) notation.

We have here the following sequence of divisions:

$$\begin{aligned} 2345 &= 390 \times 6 + 5 \\ 390 &= 65 \times 6 + 0 \\ 65 &= 10 \times 6 + 5 \\ 10 &= 1 \times 6 + 4 \\ 1 &= 0 \times 6 + 1. \end{aligned}$$

In each line, the left-hand term (dividend) is the quotient from the previous line. The remainders, in reverse order, give the base 6 notation for this number, which is 14505. The reader may check that this is so by seeing that

$$1 \times 6^4 + 4 \times 6^3 + 5 \times 6^2 + 0 \times 6 + 5 \times 1$$

is indeed equal to (the decimal numeral) 2345. Thus the hexal 14505 is equal to the decimal 2345.

1.10.2 Example. Change the hexal numeral 13142 to *binary* (base 2) notation.

We first change from hexal to decimal notation. This is not strictly necessary, but we do so because we are more used to decimal calculations.

$$1 \times 6^4 + 3 \times 6^3 + 1 \times 6^2 + 4 \times 6 + 2 \times 1 = 2006$$

so that hexal 13142 is equal to decimal 2006. Now,

$$\begin{aligned}
2006 &= 1003 \times 2 + 0 \\
1003 &= 501 \times 2 + 1 \\
501 &= 250 \times 2 + 1 \\
250 &= 125 \times 2 + 0 \\
125 &= 62 \times 2 + 1 \\
62 &= 31 \times 2 + 0 \\
31 &= 15 \times 2 + 1 \\
15 &= 7 \times 2 + 1 \\
7 &= 3 \times 2 + 1 \\
3 &= 1 \times 2 + 1 \\
1 &= 0 \times 2 + 1.
\end{aligned}$$

Thus the binary notation for this number (obtained by taking the remainders in reverse order) is

$$11111010110.$$

Again, the reader may check this statement. The reader may also want to satisfy himself that the procedure followed in these two examples will always work.

We know that the use of 10 as base for our notation requires the use of 10 digits, 0, 1, 2, . . ., 9. Generally, the use of the integer n as base will require n distinct digits, 0, 1, . . ., $n-1$. In particular, the use of 2 as base (binary notation) will require only the digits 0 and 1. This is extremely important, because a computer can normally recognize only two possibilities for a circuit: either a current passes, or it does not. Thus binary notation is used by computers: the digit "1" corresponds to current passing through a circuit, while "0" means that no current passes.

Let us suppose that we wish to build an automatic computer. We must teach it to add and multiply (i.e., we must build circuits which will add and multiply). As binary notation is to be used, we must first construct addition and multiplication tables. These are given as Table I.10.1.

TABLE I.10.1

+	0	1
0	0	1
1	1	10

×	0	1
0	0	0
1	0	1

As can be seen, these are extremely simple. In general, they are identical with decimal tables, except that, in binary notation, $1+1=10$ (since "10" represents the decimal number 2).

Addition and multiplication of numbers of more than one digit (in binary notation) are carried out, more or less, as decimal addition and multiplication, with Table I.10.1 to help.

I.10.3 Example. Add the two binary numerals "1001" and "1011." We write

$$\begin{array}{r} 1\,0\,0\,1 \\ \underline{1\,0\,1\,1} \\ 1\,0\,1\,0\,0. \end{array}$$

Note that we have "carried" the digit 1 from the first column $(1+1=10)$ to the second column, and again from the second to the third, and from the fourth to the fifth. In decimal notation the addition would be "$9+11=20$."

I.10.4 Example. Multiply the binary numerals "1001" and "1011." We have here

$$\begin{array}{r} 1\,0\,0\,1 \\ \underline{1\,0\,1\,1} \\ 1\,0\,0\,1 \\ 1\,0\,0\,1 \\ 0 \\ \underline{1\,0\,0\,1} \\ 1\,1\,0\,0\,0\,1\,1. \end{array}$$

Again, note that the numbers are arranged exactly as they would be for long multiplication in decimal notation. The reader may check by translating to decimal notation.

Now that we know how to add and multiply binary numerals,

we must construct a computer which will do this. As mentioned above, a computer represents each number by a series of 0's and 1's, called *bits* (or digits; a bit is nothing other than a *binary digit*). The general idea is that bits are given to the computer as input; these will cause certain switches to open and others to close, and, as a result, current will flow through certain circuits. This gives new bits as the computer's output, which may of course be used as input in further operations.

I.10.5 *Example.* We shall first construct a circuit for multiplication. As we saw above, it is sufficient to multiply one bit at a time from each number; these will then be combined by addition if necessary.

We must, then, multiply two bits. From Table I.10.1, we see that the product of two bits will be 0 unless both bits are 1. Thus it suffices to put two switches, corresponding to the two bits, in series. Current will then flow only if both bits are 1—that is, if the product bit is 1.

I.10.6 *Example.* Let us now consider the problem of constructing a circuit for addition. As before, we need only consider adding two bits together—one from each number—but the situation is complicated by the fact that "carrying" is possible, so that there is now a third input bit, the "carry" bit, to be considered. (Note that carrying does not occur in multiplication.) There will also be two output bits, the "sum" bit, and the "carry" bit.

Let us call the three input bits p, q, and r. The two output bits, "c" (carry) and "a" (sum), are then functions of these, as shown in Table I.10.2.

TABLE I.10.2

p	q	r	c	a
1	1	1	1	1
1	1	0	1	0
1	0	1	1	0
1	0	0	0	1
0	1	1	1	0
0	1	0	0	1
0	0	1	0	1
0	0	0	0	0

We must construct a circuit for each of "a" and "c." For "a," we see that $a = 1$ if all three, or exactly one, of the bits p, q, and r is 1. Thus the circuit for "a" is exactly the same as for Example I.9.3. For "c," we see that $c = 1$ if at least two of the bits p, q, and r are 1. The circuit for c is therefore the same as that considered in problem 3, page 46. Figure I.10.1 shows a possible arrangement of the two circuits. Simpler circuits exist.

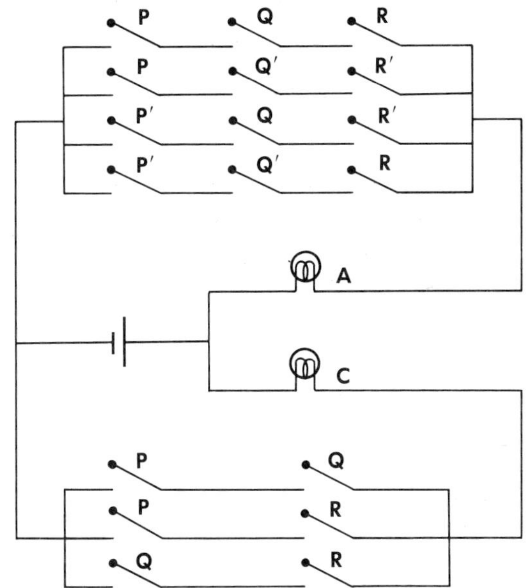

FIGURE I.10.1 A circuit for addition, with "A" and "C" lamps.

PROBLEMS ON BINARY ARITHMETIC AND COMPUTERS

1. Write the decimal numeral 3521 in octal notation.

2. What decimal numeral does the octal numeral 3521 represent?

3. Write the following decimal numerals in binary form: 215; 731; 658; 1350.

4. Write the following binary numerals in decimal form: 101101; 110101; 111110; 10001.

5. Construct a circuit to form the difference of two bits. Note that there will be two output bits: the difference can be 0 or 1, and it can be positive or negative.

6. Construct a circuit which will choose the larger of two bits, p and q. There must be three output bits: one to give the value of this number; and two others to state whether it is p, or q, or a tie.

7. Binary numerals are generally quite long, so, for manual storing, recourse is generally had to the following procedure. The digits of the numeral are split into groups of three (starting from the right) and each of these is written as a decimal digit. Thus the numeral 11101010 would be split in the form (11)(101)(010) and rewritten 352. This is known as *decimal-modified binary form.* Write the following binary numerals in decimal-modified binary form: 11010110; 1011110101; 10110101.

8. Show that the decimal-modified binary form discussed in exercise 7 above is nothing other than octal notation. (Hint: this is due to the fact that $8 = 2^3$.)

CHAPTER II

ANALYTIC GEOMETRY

1. THE CARTESIAN PLANE

Among the intellectual achievements of the Greeks, their mathematics, and especially their geometry, must be given a place of honor. Thales, Pythagoras, Apollonius, Archimedes—to name but a few—have lent their names to important geometric ideas, while Euclid's *Elements* remains one of the fundamental works of classical antiquity.

Yet for all their excellence in geometry, the Greeks often found themselves baffled by problems that could be solved today by a schoolboy. The trouble lay, not in a lack of ingenuity, but rather in the fact that the ancients lacked one of the fundamental tools of modern mathematics.

It remained to René Descartes (1596–1650) to discover this tool. Descartes is primarily known for his philosophic works—the *Discourse on Method* and the *Meditations on Prime Philosophy.* While no one can deny the extent of his influence on modern philosophy, we would yet venture to say that his mathematical work will prove the more important and enduring.

Descartes' great discovery can best be appreciated if we consider the differences between the two mathematical sciences of geometry and algebra. Geometry deals with the relations between points and lines, while algebra deals with numbers. It is generally easy, when dealing with numbers, to decide what should be done with them; because this is not true of points and lines, algebraic problems have always been easier to solve than geometric problems. Descartes' discovery was, precisely, that it is generally possible to solve geometric problems by algebraic means. He showed that the set of all points on a line has a structure identical to that of the set of real numbers.

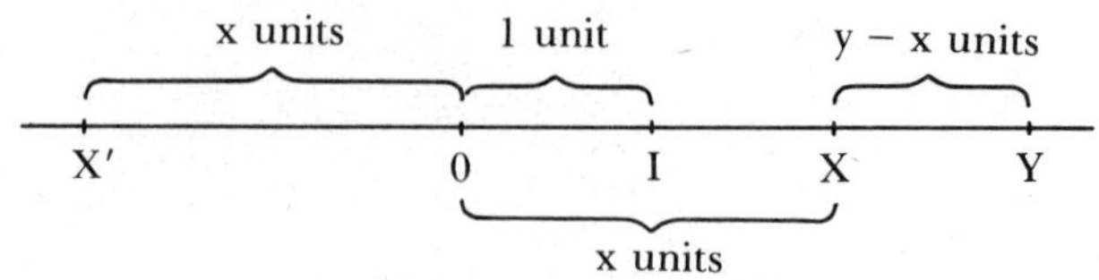

FIGURE II.1.1 The Cartesian line.

Consider, for example, a line (Figure II.1.1). On this line, two points may be taken arbitrarily (the only condition being that they be distinct points), and labeled O and I, respectively. If X is any other point on the line, we associate with X the number

2.1.1
$$x = \frac{\overline{OX}}{\overline{OI}}$$

which is the ratio of the lengths of the two line-segments, $\overline{OX}$ and $\overline{OI}$, with the stipulation that this ratio will be positive if X lies on the same side of O as I (so that $\overline{OX}$ and $\overline{OI}$ have the same direction), while it will be negative if X and I lie on opposite sides of O (so that $\overline{OX}$ and $\overline{OI}$ have opposite directions). The number x is called the *coordinate* of the point X. It is easy to see that the coordinate, x, is merely the distance $\overline{OX}$, measured in units of size $\overline{OI}$, so that the point O, called the *origin*, will have coordinate 0, while the point I will have coordinate 1.

We see, then, that each point on the line can be assigned a number, its coordinate. Conversely, given any positive number x, there are two points X and X' whose distance from the origin is equal to x units. One of these is on the same side of the origin as I and corresponds to x; the other is on the other side of the origin and will correspond to the number $-x$. In this way, each number corresponds to a unique point on the line. We have thus established a one-to-one correspondence between points on the line and real numbers. What is more, the structures of the two systems, in terms of operations that can be performed, are similar. To give an example, the distance $\overline{XY}$ between two points can be expressed in terms of their coordinates by

2.1.2
$$\overline{XY} = y - x$$

so that the geometric relation of distance reduces to the arithmetic operation of subtraction.

Dealing with the geometry of the plane, we find, however, requires a slightly more complicated procedure. It is, in fact, possible to give a one-to-one correspondence between the set of points in the plane and the set of real numbers, but this correspondence is not natural and does not preserve the structure of the system. Instead of assigning a number to each point, then, we assign a pair of numbers.

Consider, in the plane, two straight lines intersecting at a point O. The angle of intersection is not important, but for the sake of convenience, it is best to assume that they intersect at right angles (Figure II.1.2).

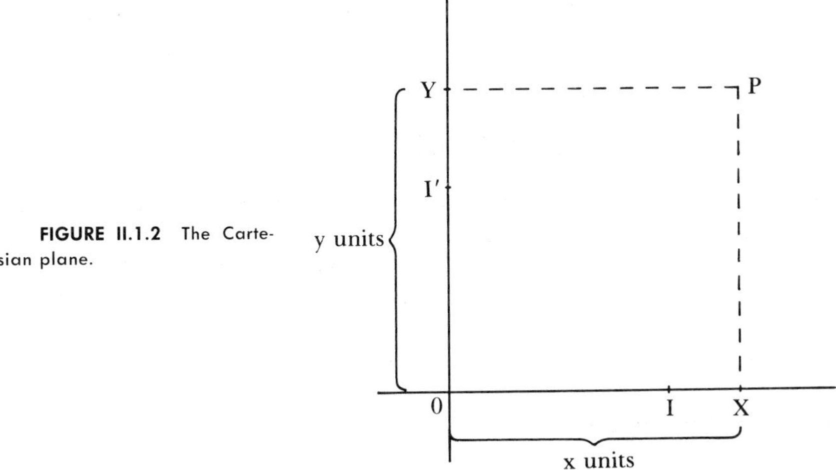

FIGURE II.1.2 The Cartesian plane.

On each of these lines, points I and I' are taken. Although the only need is that they be distinct from O, for convenience they are usually taken to be equidistant from O, and I' is usually obtained from I by a counterclockwise rotation through one right angle.

Consider, now, a point, P, in this plane. Through P, we may draw lines PX and PY, parallel to OI' and OI, respectively. The line PX meets OI at the point X, while PY meets OI' at Y. As in the one-dimensional case, just mentioned, we write

2.1.3
$$x = \frac{\overline{OX}}{\overline{OI}}$$

and

2.1.4
$$y = \frac{\overline{OY}}{\overline{OI'}}$$

where, once again, we agree to let the ratio of two line-segments be positive if they have the same direction, and negative if they have opposite directions.

The two numbers, x and y, given by (2.1.3) and (2.1.4) respectively, are called the *coordinates* of the point P. The number x is generally called the *abscissa*, and y the *ordinate*, of P. The lines OI

and OI' are called the coordinate *axes*; OI is the x-axis, and OI' is the y-axis.

We have thus assigned, to each point in space, a pair of real numbers (x,y), its coordinates. It may be pointed out that, since $OXPY$ is a rectangle, we must have $\overline{OX}=\overline{YP}$, while $\overline{OY}=\overline{XP}$. Thus x and y will be the perpendicular distances of the point P from the y-axis and x-axis, respectively, in terms of the common unit $\overline{OI}$ or $\overline{OI'}$.

2. GRAPHS AND EQUATIONS

We proceed now to study some of the advantages derived from the analytic treatment of geometry. Let us suppose that we are given a relation between two unknowns (or variables), x and y. This relation might be in the form of an equation, say,

2.2.1 $$y = x^2 + 3x$$

or of a word problem,

2.2.2 "x is not smaller than y, but not larger than twice y"

or in many other possible forms. There are necessarily certain pairs of values (x,y) for which the given relation (2.2.1) or (2.2.2) will be true and others for which it will not be true. If we consider the pairs of numbers for which the relation is true, we may plot the position of the corresponding points (i.e., the points having these pairs as their coordinates) on a coordinate plane. The set of these points is called the *graph* of the relation. Figures II.2.1 and II.2.2 show, respectively,

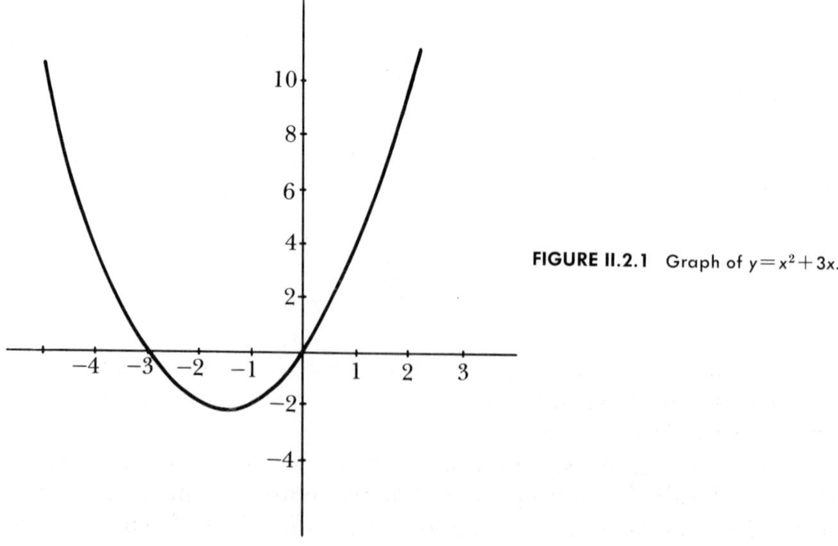

FIGURE II.2.1 Graph of $y=x^2+3x$.

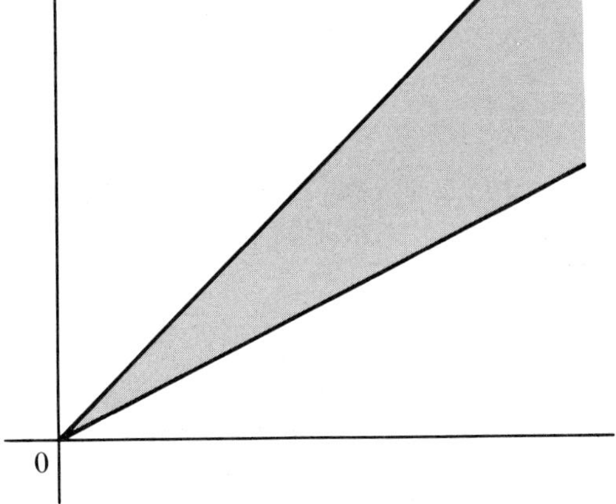

FIGURE II.2.2 Graph of the relation 2.2.2.

the graphs of the relations (2.2.1) and (2.2.2). For the equation (2.2.1), the graph is a curve (in this case, a *parabola*). As for the relation (2.2.2), we find that its graph consists of all the points inside a wedge with its angle at the origin.

Conversely, if we are given a curve in the plane, it is often possible to find a numerical relation that is satisfied by the coordinates of the points on the curve and by no others. If an equation, this relation is said to be the equation of the curve; (2.2.1) is the equation of the parabola shown in Figure II.2.1.

Relations between points can also be expressed analytically by this means. Consider, for example, two points, P and Q (Figure II.2.3), with coordinates (x_1,y_1) and (x_2,y_2) respectively. If we let R have the same ordinate as P, and the same abscissa as Q, we see that the lines

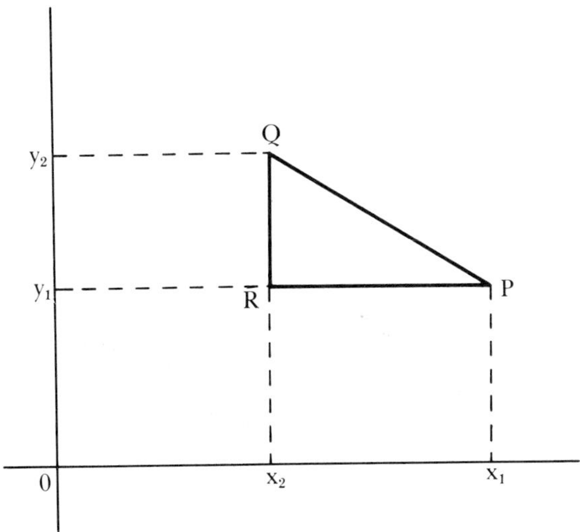

$\overline{PQ}^2 = RPR^2 + \overline{RQ}^2 = (x_1 - x_2)^2 + (y_2 - y_1)^2.$

FIGURE II.2.3

PR and QR, being parallel to the two coordinate axes, must intersect at right angles, and thus, by Pythagoras' Theorem,

2.2.3 $$\overline{PQ^2} = \overline{PR^2} + \overline{QR^2}$$

We note, now, that $\overline{PR} = \overline{X_1X_2}$ and $\overline{RQ} = \overline{Y_1Y_2}$. Replacing these values in (1.2.3), introducing the coordinates, and taking square roots, we obtain

2.2.4 $$\overline{PQ} = \sqrt{(x_2 - x_1)^2 + (y_2 - y_1)^2}$$

the formula for *distance* between two points in the plane.

Consider, finally, the line PQ. It makes an angle, θ, with the x-axis. To find this angle, we note that it is the same as the angle between PQ and PR (since PR is parallel to the x-axis). From elementary trigonometry, we know that

$$\tan\theta = \frac{\overline{RQ}}{\overline{PR}}$$

Now, the tangent of the angle is called the *slope* of PQ (see Figure II.2.4). If we let m represent this slope, and introduce coordinates, we have

2.2.5 $$m = \frac{y_2 - y_1}{x_2 - x_1}$$

as the formula for slope. Although the slope is given in terms of the coordinates of P and Q, it is a property of the line PQ, and the formula (2.2.5) will give the same value if we substitute the coordinates of any two points on PQ. (Sometimes the slope is defined as a property of the two points, P and Q; it is then necessary—though easy—to prove that the slope is constant along a line.)

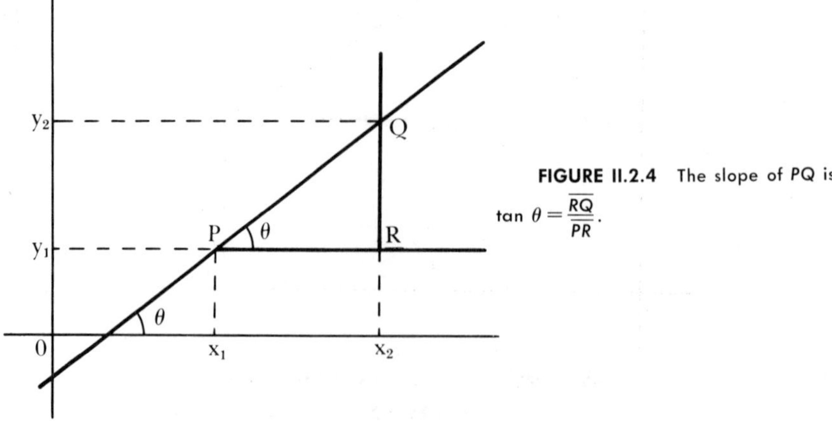

FIGURE II.2.4 The slope of PQ is $\tan\theta = \frac{\overline{RQ}}{\overline{PR}}$.

We point out that if a line is horizontal (parallel to the x-axis) its slope must be 0. In fact, points on such a line must have the same ordinate—and the numerator in (2.2.5) vanishes. On the other hand, if two points lie on a vertical line (i.e., parallel to the y-axis), then the denominator in (2.2.5) will vanish. In this case the slope does not exist, although it is sometimes said that the line has infinite slope.

We are now in a position to prove a few theorems concerning lines and their slopes.

II.2.1 Theorem. Two lines are parallel if and only if both have the same slope (or no slope at all).

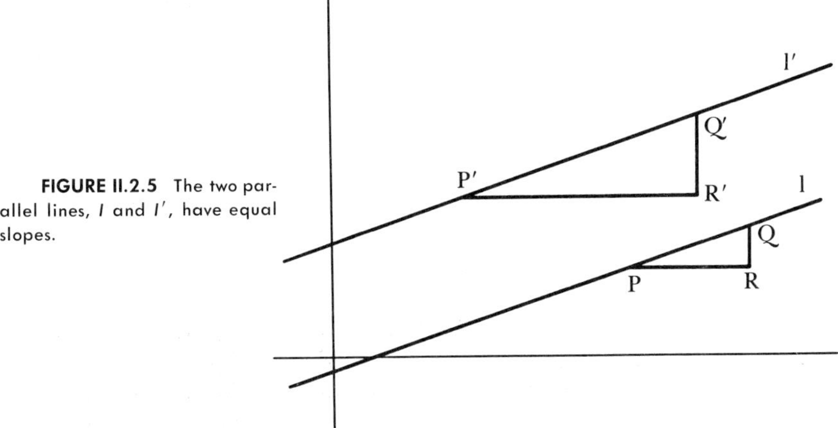

FIGURE II.2.5 The two parallel lines, l and l', have equal slopes.

Proof (see Figure II.2.5). On the lines l and l', take two pairs of points P,Q on l and P', Q' on l'. If l and l' are parallel, the two triangles PQR and $P'Q'R'$ are similar (having corresponding sides parallel) and so

2.2.6
$$\frac{\overline{RQ}}{\overline{PR}} = \frac{\overline{R'Q'}}{\overline{P'R'}}$$

Conversely, if (2.2.6) holds, the two triangles PQR and $P'Q'R'$ are similar. Since PR and $P'R'$ are parallel, this means that PQ and $P'Q'$ are also parallel. This covers the case in which l and l' both have slopes. If the slopes do not exist, then both lines are parallel to the y-axis, and hence to each other.

II.2.2 Corollary. Through a given point P there passes one and only one line with a given slope m.

Proof. This follows from Theorem II.2.1 and the well-known Euclidean fact that through P there passes exactly one line parallel to the line with slope m.

II.2.3 Theorem. Let l and l' be lines with slopes m and m', respectively. Then l and l' are perpendicular if and only if $mm' = -1$.

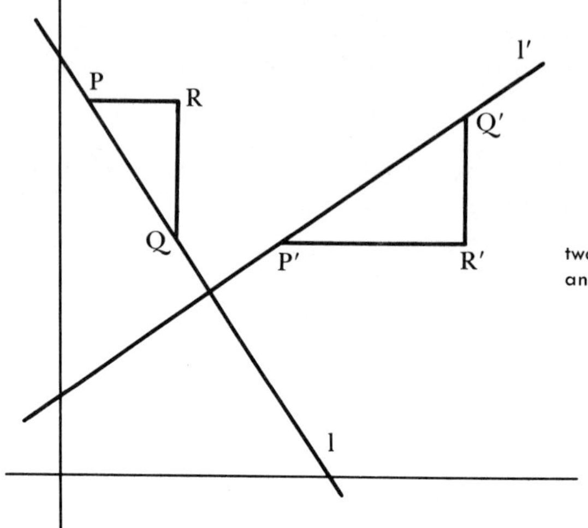

FIGURE II.2.6 The slopes of the two mutually perpendicular lines, l and l', are negative reciprocals.

Proof (see Figure II.2.6). Suppose l and l' are perpendicular. Let P,Q lie on l, while P',Q' lie on l'. Then PR and $R'Q'$ are perpendicular, RQ and $P'R'$ are perpendicular, and the triangles PQR and $Q'P'R'$ have equal angles and are therefore similar. This means that the ratios of corresponding sides are equal, if we disregard the difference in directions. It is not difficult to see, however, that the two slopes must have opposite signs. Thus we have

$$m = \frac{\overline{RQ}}{\overline{PR}} = -\frac{\overline{P'R'}}{\overline{R'Q'}} = -\frac{1}{m'}$$

and so $mm' = -1$. Conversely, if $mm' = -1$, then l' must be perpendicular to l, since there must be a perpendicular to l through P', and this perpendicular must have slope m'; l' is the only line through P' with slope m'.

3. LINEAR EQUATIONS

We now consider the equation of a line. The fundamental property of a straight line, from the point of view of analytic geometry, is that its slope is constant; we will use this property to develop the equation.

Consider, then, a line l passing through a point $P = (x_1, y_1)$ with slope m. (Corollary II.2.2 tells us there is exactly one such line.)

If $Q = (x,y)$ is any other point on this line, we know we must have

2.3.1 $$\frac{y - y_1}{x - x_1} = m$$

or, equivalently,

$$y - y_1 = m(x - x_1)$$

2.3.2 $$y = mx + y_1 - mx_1$$

as the equation of the line l.

Sometimes we are given two of the points on the line, rather than a point and the slope. In such cases we must first find the slope m, by using equation (2.2.5), and then substitute this value in (2.3.1) or (2.3.2).

Suppose that we are given two points, $P_1 = (x_1, y_1)$ and $P_2 = (x_2, y_2)$. The equation for the line passing through these two points, from (2.2.5) and (2.3.1), is seen to be

2.3.3 $$\frac{y - y_1}{x - x_1} = \frac{y_2 - y_1}{x_2 - x_1}$$

This equation is sometimes called the *two-point formula,* while (2.3.1) is the *point-slope formula.*

II.3.1 Example. Find the equation of the line passing through (5,−1) with the slope $m = 2$.

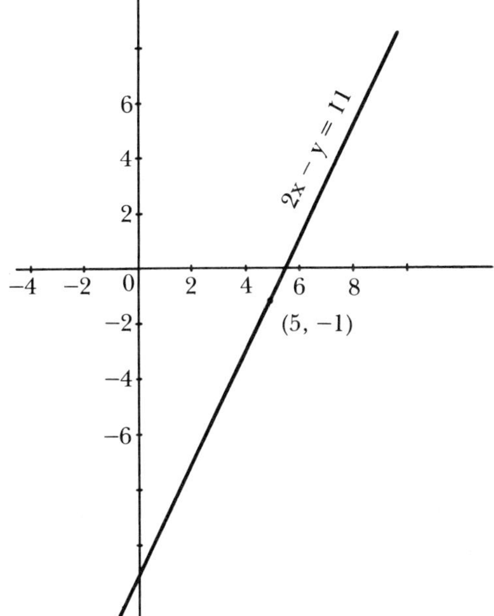

FIGURE II.3.1 Example II.3.1.

In this case, $x_1 = 5$, and $y_1 = -1$. Therefore, from (2.3.1)

$$\frac{y - (-1)}{x - 5} = 2$$

Multiplying through by the denominator, we obtain

$$y + 1 = 2x - 10$$

and, finally,

$$y = 2x - 11$$

or, equivalently,

$$2x - y = 11$$

II.3.2 Example. Find the equation of the line passing through (2,3) and $(-1,4)$.

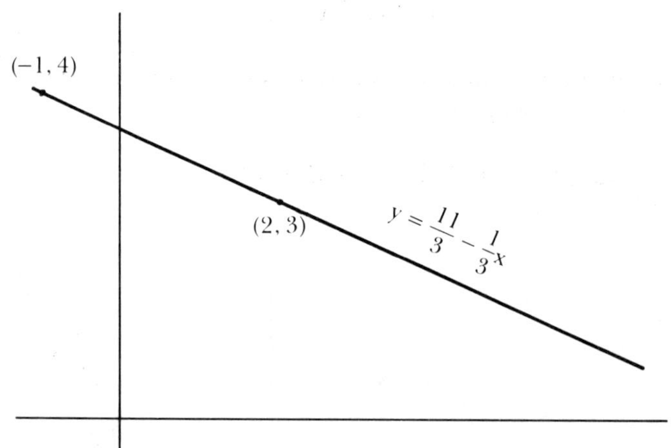

FIGURE II.3.2 Example II.3.2.

In this case, we must use the two-point formula (1.3.3). We have $x_1 = 2$, $y_1 = 3$, $x_2 = -1$, $y_2 = 4$, and so

$$\frac{y - 3}{x - 2} = \frac{4 - 3}{-1 - 2}$$

which, after simplification of the right-hand side, gives us

$$\frac{y - 3}{x - 2} = -\frac{1}{3}$$

or, proceeding as before,

$$y - 3 = \frac{2 - x}{3}$$

$$y = \frac{11}{3} - \frac{1}{3}x$$

which is the desired equation.

The formulas (2.3.1) and (2.3.3) fail when the desired line is parallel to the y-axis. In fact, we mentioned earlier that such a line does not have a slope (i.e., the slope does not exist). Equation (1.3.1) cannot be used since, indeed, it involves the slope. Moreover, two points on such a line are at an equal distance from the y-axis and thus have the same abscissa. This means that the denominator in the right-hand side of (2.3.3) vanishes; this formula, therefore, cannot be used.

Still we would like to have an equation for this sort of line. We can obtain it if we remember that on such a line all points have the same abscissa; if we let a be the common abscissa of all these points, the equation of this line will be

2.3.4 $$x = a$$

II.3.3 Example. Find the equation of the line passing through (3,5) and (3,−2).

We see that these two points have the same abscissa, so that the line is parallel to the y-axis. Thus by (2.3.4), the equation of the line must be

$$x = 3$$

Looking at the solutions of 2.3.1 to 2.3.3, we see that in each case the equations of the straight lines may be rearranged to obtain an equation of the form

2.3.5 $$ax + by = c$$

in which a, b, and c are constants. In fact, we may see from (2.3.2) and (2.3.4) that every line can be given an equation of the form (2.3.5). Such an equation, in which the variables appear only as first-degree terms (i.e., no higher powers or radicals appear, and no variables ever appear in a denominator) is called a *first-degree equation.*

We show, now, that every equation of the form (2.3.5), subject only to the condition that a and b cannot both be zero, is the equation of a straight line. Suppose we have an equation (2.3.5)

$$ax + by = c$$

If we assume that b is not zero, we can subtract the term ax from both sides and divide by b, thus "solving" for y:

2.3.6 $$y = \frac{c}{b} - \frac{a}{b}x$$

Now, assume $P = (x_1,y_1)$ and $Q = (x_2,y_2)$ are distinct points satisfying the equation (2.3.5). We see from (2.3.6) that x_1 and x_2 cannot be equal, since this would make y_1 and y_2 equal as well, meaning that the two points were not distinct. Since x_1 and x_2 are distinct, we can calculate the slope of PQ; it is, by (2.2.5) and (2.3.6),

$$\frac{y_2 - y_1}{x_2 - x_1} = \frac{\frac{c}{b} - \frac{a}{b}x_2 - \frac{c}{b} + \frac{a}{b}x_1}{x_2 - x_1}$$

which, on simplification, gives us

2.3.7 $$m = -\frac{a}{b}$$

The slope, m, of PQ, therefore, is constant for any two points satisfying (2.3.5). This is a characteristic property of straight lines: the graph of the equation must be a straight line; its slope, incidentally, is given by (2.3.7).

Suppose, now, that $b = 0$. Since we have assumed a and b cannot both be zero, it follows that a is different from zero. The equation (2.3.5) will now have the form

$$ax = c$$

and division by a gives us

$$x = \frac{c}{a}$$

so that the equation is satisfied by all points with the abscissa c/a; its graph is the line parallel to the y-axis at a distance c/a from it.

It is apparent that, in any case, the graph of (2.3.5) will be a line. For this reason, first-degree equations are also called *linear equations.*

II.3.4 ***Example.*** Find the equation of the line passing through (1,–3) parallel to the line $x + 2y = 7$.

By Theorem II.2.1 the desired line must have the same slope as $x+2y=7$. Formula (2.3.7) gives us the slope; since $a=1$ and $b=2$,

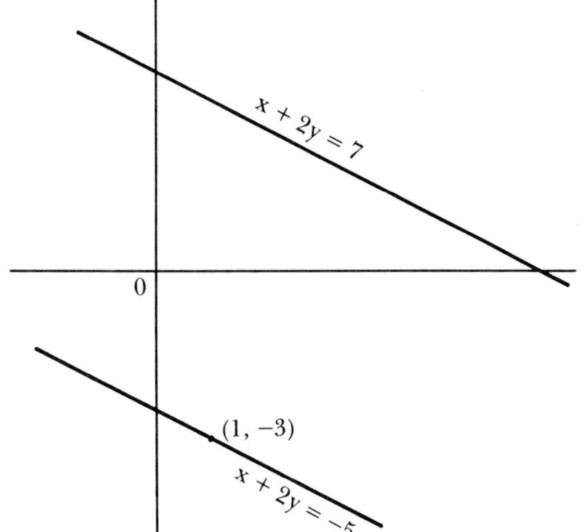

FIGURE II.3.3 Example II.3.4.

we have

$$m = -\frac{1}{2}$$

and the desired equation is

$$\frac{y-(-3)}{x-1} = -\frac{1}{2}$$

which reduces to

$$y = \frac{-1}{2}x - \frac{5}{2}$$

or, equivalently,

$$x + 2y = -5$$

We note that, for these two lines, we have obtained equations in which the constants a and b (called the *coefficients* of x and y, respectively) are equal to 1 and 2, respectively. This is because the lines are parallel: (2.3.7) gives us the slope in terms of these coefficients only; thus, if the coefficients of x and y are the same in two linear equations, the two lines represented must have equal slopes and be parallel (unless they are the same). The converse of this is not quite true: if two lines are parallel, the coefficients in their equations need not be equal, although it is always possible to rewrite their equations in such a way that the coefficients will be the same (because one line may be represented by several equations).

II.3.5 Example. Find the line passing through (3,1) parallel to the line $2x + 3y = 7$.

We saw earlier that any line with equation $2x + 3y = c$ will be parallel to $2x + 3y = 7$ (unless $c = 7$, in which case the two lines are identical). We need only find a line with such an equation passing through (3,1). It is easy to see that this happens if we set $c = 2 \cdot 3 + 3 \cdot 1 = 9$; the desired line has the equation

$$2x + 3y = 9$$

II.3.6 Example. Find the equation of the line passing through (1,2) perpendicular to $x - 2y = 5$.

By (2.3.7), the line $x - 2y = 5$ has slope $m = 1/2$. According to Theorem II.2.3, the desired line must have slope $m' = -2$. Introducing this slope and the coordinates of the given point into (2.3.1), we have

$$\frac{y-2}{x-1} = -2$$

which simplifies to

$$2x + y = 4$$

II.3.7 Example. Find the equation of the line through (−1,3) perpendicular to the line $y = 6$.

The line $y = 6$ can be seen to have slope $m = 0$. Hence the condition $mm' = -1$ cannot be satisfied, and it follows that the slope of the desired line does not exist, i.e., it must be parallel to the y-axis. (This can also be seen from the fact that $y = 6$ is parallel to the x-axis.) Hence the equation we want must be of the form $x = c$. Since the line passes through (−1,3), the equation must be

$$x = -1$$

II.3.8 Example. Find the equation of the line through (2,−2) perpendicular to the line $x = 2$.

In this case, since the line we are given does not have a slope, Theorem II.2.3 does not apply. We see, however, that the line is

$$\frac{y-(-2)}{x-2} = 0$$

This reduces to

$$y = -2$$

which is the desired equation.

In drawing the graph of a linear equation, it is sufficient to know two of the points. Since the graph is a straight line, this determines the line that can then be drawn by using a straight-edge. Drawing the graph is quite simple if two of the points are easy to obtain. Now, a linear equation such as (2.3.5) contains three arbitrary constants (parameters) that determine whether a given point lies on the line. We must generally take these three parameters, a, b, c, into consideration when plotting a point on the line.

On the other hand, there are cases in which only two of them need be taken into consideration. Suppose that we are looking for a point on the line (2.3.5) with ordinate equal to zero. Putting the value $y=0$ into the equation, we obtain

$$ax = c$$

which, assuming a is not zero, reduces to

2.3.8 $$x_0 = \frac{c}{a}$$

which is the abscissa of the point of intersection of (1.3.5) with the x-axis, and is called the *x-intercept* of the equation.

Similarly, the ordinate of the point of intersection of (1.3.5) with the y-axis will be given by

2.3.9 $$y_0 = \frac{c}{b}$$

if b is not equal to zero. The value (2.3.9) is called the *y-intercept*.

We obtain, in this manner, two points $(x_0, 0)$ and $(0, y_0)$ on the line. Each of these depends only on two of the three parameters a, b, and c, and is, therefore, comparatively easy to find.

Two things might conceivably go wrong. One is that the intercepts need not both exist. In fact, if a line is parallel to one of the coordinate axes, it cannot intersect that axis. A line parallel to the y-axis will not have a y-intercept; one parallel to the x-axis will not have an x-intercept. Another possibility is that the two intercepts will both exist, but give us only one point. This happens when the line passes through the origin; in this instance, both intercepts are zero, and so the line intersects both axes at a single point. Since a single point does not determine a line, it follows that we cannot draw a line from its intercepts in these cases. This is really not a serious problem, but it does lessen the possibility of giving a systematic procedure for plotting lines.

We may be given a line in terms of its two intercepts. Each of these intercepts gives us a point so that, in the usual case (when the line does not pass through the origin), the two intercepts determine a line. If the two intercepts are x_0 and y_0, we know that the line passes through $(x_0,0)$ and $(0,y_0)$ and application of the two-point formula yields

$$\frac{y}{x - x_0} = \frac{y_0}{-x_0}$$

which reduces to

$$y_0 x + x_0 y = x_0 y_0 \tag{2.3.10}$$

Formula (2.3.10) is known as the two-intercept formula. Note that if $x_0 = y_0 = 0$, (2.3.10) reduces to the identity $0 = 0$.

Another manner in which a line may be given is by one of the intercepts (generally the y-intercept) and the slope. In fact, if the slope is m and the y-intercept is y_0, we have, from (2.3.1),

$$\frac{y - y_0}{x} = m$$

which reduces to

$$y = mx + y_0 \tag{2.3.11}$$

This is known as the slope-intercept formula.

II.3.9 Example. Find the equation of the line with x-intercept 3 and y-intercept -1.

Using (2.3.10), we obtain immediately

$$-x + 3y = -3$$

as the desired equation.

II.3.10 Example. Find the intercepts of the line $3x + 4y = 12$.

In this case we apply (2.3.8) and (2.3.9), obtaining $x_0 = 4$ and $y_0 = 3$ as the two intercepts.

II.3.11 Example. Find the intercepts of the line $3x = 6$.

Here the x-intercept is easily found to be $x_0 = 2$. The y-intercept, however, does not exist. (Some people would say that the y-intercept is infinite.)

II.3.12 Example. Find the equation of the line with y-intercept $y_0 = -4$ and slope $m = 2$.

Applying (2.3.11), we obtain $y = 2x - 4$ as the desired equation.

II.3.13 Example. Find the intercepts of the line $3x-2y=0$.

For this line, we find that both intercepts are equal to zero, i.e., the line passes through the origin. If we wish to plot the line, we need another point; this may be obtained by solving for y, which gives us

$$y=\frac{3}{2}x$$

Then (1,3/2) or (2,3) will be points on the line.

Fixed and Variable Costs. In making a product, there will normally be some *fixed costs* (e.g., capital equipment or overhead) which will be the same whatever the level of production may be, and other costs, called *variable costs* (e.g., raw materials) which depend directly on the amount of production. Quite often, these variable costs are assumed to be directly proportional to production. Thus, we may be given fixed costs of $\$f$, and variable costs of $\$v$ per unit produced. It is then easy to see that, if x units are produced, the costs, y, are

$$y=f+vx$$

Comparing with formula (2.3.11), we see that f is the y-intercept, and v is the slope of the line relating costs to production.

II.3.14 Example. A steel manufacturer has fixed costs of \$10,000, and variable costs of \$50 per ton of steel. What are his total costs for producing 500 T of steel?

Clearly, we have here

$$y=10{,}000+50x.$$

Since $x=500$, this gives $y=\$35{,}000$.

PROBLEMS ON LINEAR EQUATIONS

1. Find the equations of the lines described as follows:

(a) Passing through (0,1) and (2,5).

(b) Passing through (1,3) and (2,4).

(c) Passing through (1,–2) and (0,0).

(d) Passing through (–1,–2) and (–1,5).

(e) Passing through (1,4) and (6,4).

(f) Passing through (2,5) with slope 1/2.

(g) Passing through (3,6) with slope 2.

(h) Passing through (5,1) with slope 0.
(i) Passing through (2,–3) with slope 1.
(j) Passing through (5,1) with slope 1/4.
(k) Passing through (3,2) and parallel to $2x + 4y = 6$.
(l) Passing through (0,0) and parallel to $x = 5$.
(m) Passing through (1,4) and parallel to $x - 2y = 7$.
(n) Passing through (2,–2) and parallel to $y = x + 2$.
(o) Passing through (3,0) and perpendicular to $x + y = 5$.
(p) Passing through (1,4) and perpendicular to $3x - y = 2$.
(q) Passing through (6,2) and perpendicular to $x + 2y = 4$.
(r) Passing through (2,1) and perpendicular to $2x - y = 10$.
(s) With x-intercept 3 and y-intercept -2.
(t) With x-intercept -1 and y-intercept -4.
(u) With y-intercept 6 and slope 3.
(v) With y-intercept 5 and slope -1.
(w) With y-intercept 2 and slope 0.
(x) With x-intercept -3 and parallel to $2x + y = 3$.
(y) With y-intercept 4 and perpendicular to $x - 2y = 2$.

2. Find the slopes, x-intercepts, and y-intercepts of the following lines; draw the lines on graph paper.

(a) $x + 3y = 6$
(b) $3x - 2y = 5$
(c) $x + 4y = 12$
(d) $-x + 2y = -3$
(e) $x - 4y = 16$

3. Find the distances between the following pairs of points:

(a) (5,1) and (6,1)
(b) (3,1) and (2,4)
(c) (1,4) and (3,1)
(d) (2,6) and (1,1)
(e) (1,4) and (–2,–4)

4. The three points (5,1), (2,4), and (1,0) determine a triangle. Find:

(a) The mid-points of the three sides of this triangle.

(b) The equations of the three medians of this triangle (a median is the line joining one vertex to the mid-point of the opposite side).

(c) The equations of the three altitudes of the triangle (an altitude is the line through one vertex and perpendicular to the opposite side).

5. At a price of \$3 per unit, a firm will supply 10,000 towels per month; at \$5 per unit, it will supply 20,000 per month. Assuming linearity, find the equation relating supply (in towels per month) to price (in \$ per unit).

6. A tool manufacturer finds that, at \$1 per wrench, he can sell 5000 wrenches per month, but, at \$1.50 each, he can only sell 3000 per month. Assuming linearity, find the equation relating demand to price.

7. A manufacturer has fixed costs of \$8000, plus variable costs of \$30 per unit. Find the equation relating costs to production.

8. A miner finds that it costs him \$1200 to produce 6 tons of ore, and \$1500 to produce 12 T. Assuming linearity, find the fixed and variable costs.

4. INEQUALITIES

Up to this point, we have dealt throughout with equations; we shall deal now with the somewhat more general concept of inequalities.

The characteristic symbol of an equation is, of course, the equals sign =. We know its meaning, of course: if two expressions are connected by this sign, they are equal. We propose to deal now, with certain other signs; these are

$$\neq, <, \leq, >, \geq$$

The first sign, $\neq$, is simply an equals sign crossed out, and represents, therefore, the negation of equality. That is, the statement "$a \neq b$" is to be read as "a is different from b." Thus, "$3 \neq 2$" is a true statement, while "$5 \neq 5$" is a false statement.

The second sign, $<$, is to be read as "is less than." Thus, "$a < b$" is read "a is less than b." Since this is defined to mean that the difference $b - a$ is a positive number, the statements

$$2 < 5, -5 < 1, 0 < 3$$

are all true (note for instance that $1 - (-5) = 6$, which is positive), while the statements

$$3 < -7, 2 < 2$$

are both false.

The sign $\leq$ is to be read as "is less than or equal to." We define it by saying that the statement "$a \leq b$" is true whenever either of the two statements "$a = b$" and "$a < b$" is true. Thus the statements

$$\begin{aligned} 0 &\leq 3 \\ 2 &\leq 2 \\ -5 &\leq 1 \end{aligned}$$

are all true, while the statement

$$3 \leq -7$$

is false.

The sign $>$ is read as "is greater than." It can best be defined by saying that the two statements "$a > b$" and "$b < a$" are equivalent, and therefore

$$\begin{aligned} 5 &> 1 \\ -1 &> -8 \\ 2 &> 0 \end{aligned}$$

are all true, while

$$\begin{aligned} 6 &> 6 \\ 5 &> 10 \end{aligned}$$

are both false.

Finally, the sign $\geq$ should be read as "is greater than or equal to." We define it by saying that "$a \geq b$" and "$b \leq a$" are equivalent statements. Examples of true statements are

$$\begin{aligned} 6 &\geq 2 \\ -1 &\geq -1 \\ 3 &\geq 0 \end{aligned}$$

while the statements

$$\begin{aligned} -5 &\geq 1 \\ 3 &\geq 6 \end{aligned}$$

are both false.

In high school algebra, we are taught that there are many things that may be done with equations. One may add (or subtract) the same number to both sides of an equation; one may multiply both sides of an equation by the same number, or divide both sides by the same number (so long as this number is not equal to zero). Equations may also be added together, giving rise to new equations.

Inequalities are not equations, so we must be careful about treating them in the same manner. Nevertheless, with care, some of these same things may be done with inequalities.

We have seen that the inequalities are defined in terms of positive numbers. For instance,

2.4.1 $a < b$ means $b - a$ is positive.

2.4.2 $a \leq b$ means $b - a$ is positive or zero.

2.4.3 $a > b$ means $a - b$ is positive.

2.4.4 $a \geq b$ means $a - b$ is positive or zero.

It follows that, if we want to know which operations may legitimately be performed on an inequality, we must study the properties of positive numbers (as contrasted with negative numbers). These properties are simply as follows:

2.4.5 The sum of two positive numbers is positive.

2.4.6 The product of two positive numbers is positive.

2.4.7 The product of a positive and a negative number is negative.

With these properties established, we can look once again at the inequalities. First, we see that the same number can be added or subtracted to both sides of an inequality. Suppose that we know $a > b$. This means that $a - b$ is a positive number. For any number c, now, we have

$$(a + c) - (b + c) = a - b$$

so that $(a + c) - (b + c)$ is positive, and so

$$a + c > b + c$$

Thus, the inequality is preserved by the addition (or subtraction) of any number to both sides of the inequality. This would still be true if the inequality were of a different type.

Suppose, next, that we have two inequalities of the same type, say

$$\begin{aligned} a &> b \\ c &> d \end{aligned}$$

This tells us that both $a - b$ and $c - d$ are positive. It means that $(a + c) - (b + d)$, as the sum of the two positive numbers $a - b$ and $c - d$, is also positive, and thus

$$a + c > b + d$$

It is apparent that two inequalities, if they are of the same type, can be added together. Note, on the other hand, that if two inequalities have opposite sense, such as $a > b$ and $c < d$, it is not possible to add them. Nor is it possible to subtract inequalities such as $a > b$ and $c > d$. In fact, we have no guarantee that the difference of two positive numbers will be positive.

Finally, suppose we are given the inequality $a > b$, which, we repeat, means that $a - b$ is positive. If c is a positive number, then $ca - cb = c(a - b)$ is the product of two positive numbers, and so is positive. Hence we have

$$ca > cb$$

and the inequality is preserved. If, on the other hand, c is negative, then $ca - cb$ is negative, and $cb - ca$ is positive, so

$$ca < cb$$

and the sense of the inequality is reversed. If $c = 0$, we find that $ca = cb$, and neither of the "strict" inequalities, $ca > cb$, or $ca < cb$, will hold; however, both the "loose" inequalities, $ca \geq cb$ and $ca \leq cb$, will hold.

We have only considered, here, inequalities of the type $a > b$. It may be seen, however, that similar considerations hold also for inequalities of the types $a < b$, $a \geq b$, and $a \leq b$. We thus obtain the following rules for the treatment of inequalities:

2.4.8 Addition of the same number to both sides of the inequality preserves the inequality.

2.4.9 Two inequalities may be added if they are of the same type, giving an inequality of the same type.

2.4.10 Multiplication of both sides of an inequality by the same positive number preserves the inequality.

2.4.11 Multiplication of both sides of an inequality by the same negative number *reverses* the inequality.

2.4.12 In rules (2.4.10) and (2.4.11), if the inequality is "loose" (i.e., of type $\geq$ or $\leq$), the words "positive" and "negative" may be replaced by "non-negative" and "non-positive" respectively.

We shall, in the future, deal mainly with loose inequalities of the form $a \leq b$ or $a \geq b$. The reason for this is that physical, practical constraints are generally of this form; the amount of goods produced by a factory can be positive or zero, but not negative, which leads to a constraint of the form $x \geq 0$, or the total number of hours worked by an employee cannot (in certain cases) be more than 8 hours, which leads to a constraint of the form $t \leq 8$. In both cases we see that the loose inequalities appear naturally.

Break-even Analysis. In the previous section we considered a production model with fixed and variable costs, represented by a linear equation. Suppose now that the product is to be sold on the open market; with a free market, the revenue will be proportional to the amount sold. If the entire product x is sold, revenue will be

$$r = px$$

where p is the unit price. Now, the firm makes a profit if revenue exceeds costs, that is, if

$$px \geq f + vx$$

where, as above, f and v are fixed costs and variable unit costs, respectively. Using (2.4.8), we have

$$(p - v)x \geq f$$

and, if $p > v$, use of (2.4.10) gives

2.4.13 $$x \geq \frac{f}{p - v}$$

Thus, the firm makes a profit if (2.4.13) is satisfied, and the amount $f/(p - v)$ is known as the *break-even point.* (Note, however, that if $p \leq v$, the firm can never make a profit.)

II.4.1 **Example.** Suppose that the steel-making company of Example II.3.13 can sell its steel at \$75 per T. What is its break-even point?

We have here $p = 75$, $v = 50$, $f = 10{,}000$. Then the break-even point is

$$x = \frac{10{,}000}{75 - 50} = 400$$

so that production must be over 400 T to make a profit.

We will now study the graph of a linear inequality (in two variables). We can do this best by considering an example first. Let us take the inequality

$$3x + 2y \leq 6$$

which differs from a linear equation only in that the inequality symbol appears rather than the equals sign. Following the rules (2.4.8) to (2.4.12), we can solve for y in terms of x: first, we add $-3x$ to both sides:

$$2y \leq 6 - 3x$$

and then we multiply both sides by 1/2. Since 1/2 is positive, this preserves the inequality:

$$y \leq 3 - \frac{3}{2}x$$

If we consider, instead of the inequality $3x + 2y \leq 6$, the corresponding linear equation

$$3x + 2y = 6$$

(whose graph is shown in Figure II.4.1), and solve for y in terms of x,

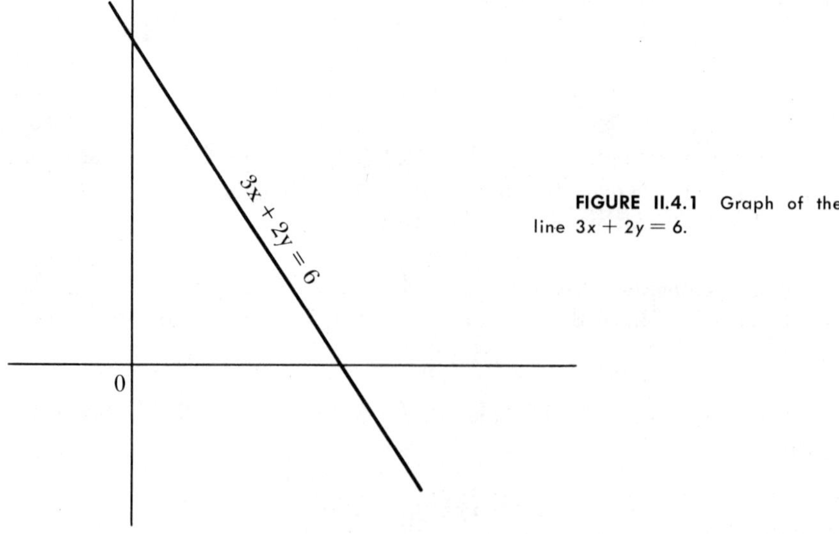

FIGURE II.4.1 Graph of the line $3x + 2y = 6$.

we obtain

$$y = 3 - \frac{3}{2}x$$

Suppose, now, that the point P, with coordinates (x_1,y_1), lies on this line, i.e., we have

$$y_1 = 3 - \frac{3}{2}x_1$$

Let Q be another point with coordinates (x_1,y_2), such that $y_2 < y_1$. This can best be expressed by saying that Q lies *below* P (since it has the same abscissa but a smaller ordinate). We see then that

$$y_2 < 3 - \frac{3}{2}x_1$$

so that the coordinates of Q satisfy the linear inequality. But P was an arbitrary point on the line, and Q was any point below P. We find, thus, that the linear inequality $3x + 2y \leq 6$ is satisfied by all the points *on* or *below* the line $3x + 2y = 6$. The converse is also true: if a point satisfies the given linear inequality, it must lie on or below the line (Figure II.4.2). The points below the line satisfy the *strict* inequality $3x + 2y < 6$, while those on the line, of course, satisfy the equation.

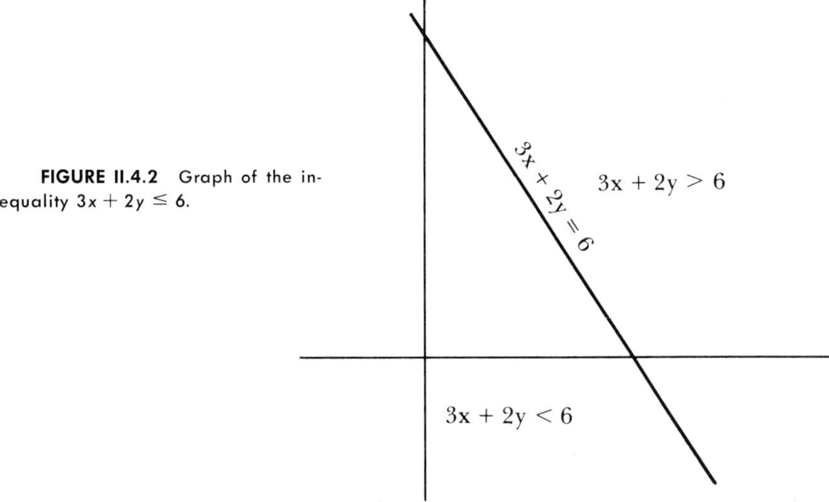

FIGURE II.4.2 Graph of the inequality $3x + 2y \leq 6$.

In a similar manner, we can consider the reverse inequality

$$3x + 2y \geq 6$$

If we do so, we find that this inequality is satisfied by all the points on or *above* the line $3x + 2y = 6$; the *strict* inequality $3x + 2y > 6$ is satisfied by the points above the line only.

In general, linear inequalities behave more or less in this manner. If we consider the general linear equation (2.3.5)

$$ax + by = c$$

we know that its graph is a straight line. This line divides the remainder of the Cartesian plane into two parts, or half-planes. In one of these half-planes, the inequality

2.4.14 $$ax + by > c$$

holds; in the other half-plane, it is the opposite inequality

2.4.15 $$ax + by < c$$

that holds. If we consider the *closed* half-planes, i.e., the half-planes along with the line that is their common boundary, we see that one closed half-plane is the graph of

2.4.16 $$ax + by \geq c$$

while the other closed half-plane is the graph of

2.4.17 $$ax + by \leq c$$

It is, naturally, important to determine which half-plane corresponds to each of the two inequalities (2.4.16) and (2.4.17). Generally speaking, this can best be done by solving for y. Given the inequality (2.4.16), we reduce it to the form

$$by \geq c - ax$$

Now, if b is positive, we can divide b (or equivalently, multiply by its reciprocal $1/b$) and obtain

$$y \geq \frac{c}{b} - \frac{a}{b}x$$

while, if b is negative, we obtain

$$y \leq \frac{c}{b} - \frac{a}{b}x$$

since in this case multiplication by $1/b$ reverses the inequality. For positive b, the inequality (2.4.16) corresponds to the half-plane *above* the line (2.3.5), while (2.4.17) corresponds to the half-plane below the line. For negative b, however, we find that (2.4.16) corresponds to the half-plane *below* the line, and (2.4.17) to that above the line.

In the event that $b=0$, of course, this analysis falls through. In fact, (2.3.5) reduces now to the form

$$ax = c$$

and this line is vertical (parallel to the y-axis), so that two half-planes cannot be classified as being above or below the line. Here we can only talk about the half-plane to the right or left of the line. It is not difficult to see that, if a is positive, then $ax \leq c$ corresponds to the left half-plane, while $ax \geq c$ corresponds to the right half-plane. If a is negative, the roles are reversed, and this time $ax \geq c$ corresponds to the left half-plane.

II.4.2 Example. Give the graph of the inequality

$$2x - 3y \geq 12$$

In this case, solution for y gives us

$$y \leq -4 + \frac{2}{3}x$$

which means that we want the half-plane below the line. The line itself is most easily plotted if we take its intercepts, which are $x_0=6$ and $y_0=-4$ (Figure II.4.3). The graph of the inequality is, then, the area on and below this line.

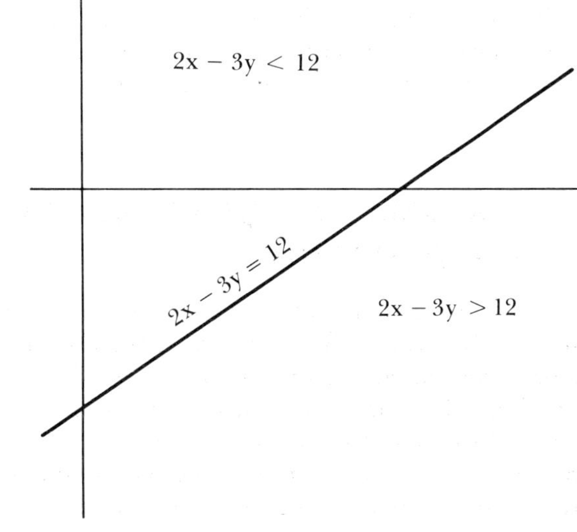

FIGURE II.4.3 Example II.4.2.

II.4.3 Example. Find the graph of the inequality

$$-3x \leq 15$$

This inequality reduces to

$$x \geq -5$$

so we want the half-plane to the right of the vertical line $x = -5$.

PROBLEMS ON INEQUALITIES

1. On graph paper, show the points which satisfy the following inequalities:

(a) $3x + 4y \leq 12$

(b) $4x - 2y \geq 10$

(c) $x + 2y \leq 14$

(d) $-x + 3y \geq 11$

(e) $x + 4y < 10$

(f) $x - 2y > 8$

(g) $x \geq 6$

(h) $3x - y \geq 3$

(i) $x - 3y < 12$

(j) $4x + 2y > 8$

2. A manufacturer has fixed costs of \$1000, and variable costs of \$30 per unit. The market price for his product is \$35 per unit. How many units must he sell to break even?

3. Suppose that the miner of problem 8, page 71 can sell the ore at \$55 per ton. What is his break-even point?

4. A manufacturer has variable costs of \$55 per unit in his production. He is offered some new machinery which costs \$25,000 and will reduce his variable costs to \$53 per unit. Assuming that the machinery has an active lifetime of 5 years, what yearly volume will justify the purchase of this machinery?

5. SYSTEMS OF EQUATIONS AND INEQUALITIES

We have, so far, studied linear equations and linear inequalities separately. Quite often, however, we are given a *system* of two or more such equations and inequalities. The idea, then, is to look for numbers that satisfy all these equations or inequalities simultaneously. Equivalently, we look for points that lie on the graphs of all these equations (or inequalities) simultaneously. Alternatively, we may consider that the graph of an equation is the solution-set of that equation: the set of points whose coordinates satisfy the equation. In solving a system of equations, then, we are looking for the intersection of their several solution-sets.

Let us consider the simplest such system, one consisting of two equations:

2.5.1 $$a_1x + b_1y = c_1$$

2.5.2 $$a_2x + b_2y = c_2$$

Each of these equations has, as its graph, a straight line. We know that two lines can be parallel or intersecting. If parallel they have no points in common; if they intersect they have exactly one point in common. It is also possible that the two equations represent the same straight line (we say that the two lines are *coincident*) and, in this case, all the points on the line lie on both graphs.

We see, therefore, that there are three possibilities for the system of linear equations (2.5.1) to (2.5.2), corresponding to the three possibilities for the pair of lines. If the lines are parallel, the system does not have a solution: there is no pair (x,y) that satisfies both these equations, and we say that the system is *inconsistent.* If the lines intersect, there is a unique pair (x,y) that satisfies both equations; this pair is called, quite naturally, the *solution* of the system. If the lines are coincident, the system will have an infinity of solutions.

Let us consider now the actual method of solution of systems of equations. Suppose we are given two equations such as (2.5.1) and (2.5.2). Let us assume that these two lines have a point (x_0,y_0) in common (it may, of course, happen that the lines have no point in common; in such a case, we shall see that this assumption leads to a contradiction). This means that the two equations, (2.5.1) and (2.5.2), will both hold if we substitute the values (x_0,y_0). Now, we can multiply the first equation by any number r, and the second equation by any number, s, obtaining

$$ra_1x + rb_1y = rc_1$$
$$sa_2x + sb_2y = sc_2$$

and we can add these two equations, obtaining a third equation

2.5.3 $$(ra_1 + sa_2)x + (rb_1 + sb_2)y = rc_1 + sc_2$$

which will hold whenever (2.5.1) and (2.5.2) both hold. That is, it will hold if we substitute the values (x_0,y_0) into the equation. Thus (2.5.3), if the left side does not vanish, is the equation of a line passing through the point of intersection (x_0,y_0) of the lines with equations (2.5.1) and (2.5.2).

The equation (2.5.3), which is obtained by multiplying (2.5.1) and (2.5.2) each by a constant and adding them, is said to be a *linear combination* of (2.5.1) and (2.5.2).

We now see that a linear combination of (2.5.1) and (2.5.2) is the equation of another line that will pass through the point of intersection of these two lines. If, moreover, (2.5.1) and (2.5.2) have different slopes, so that they have a unique point in common, it is not too difficult to see that the expression

2.5.4 $$m = -\frac{ra_1 + sa_2}{rb_1 + sb_2}$$

which is the slope of (2.5.3), can be given any value desired by a judicious choice of the two constants, r and s. Thus any one of the lines passing through the point of intersection of (2.5.1) and (2.5.2) may be obtained as a linear combination of their equations, i.e., in the form (2.5.3). In particular, the two lines passing through this point, parallel to the x-axis and y-axis respectively, may be obtained in this manner. The importance of this, of course, is that, in the equation of a line parallel to the x-axis, the variable x does not appear; in the equation of a line parallel to the y-axis, the variable y does not appear. Such an equation can then be solved for the single variable that appears in it; this is the process of *elimination*, which the reader doubtless encountered in high school algebra. Once the value of one of the unknowns is found, this value is substituted in one of the original equations; the other variable is then obtained directly.

II.5.1 Example. Find the solution of the system

$$\begin{aligned} 3x + 2y &= 12 \\ 2x - y &= 1 \end{aligned}$$

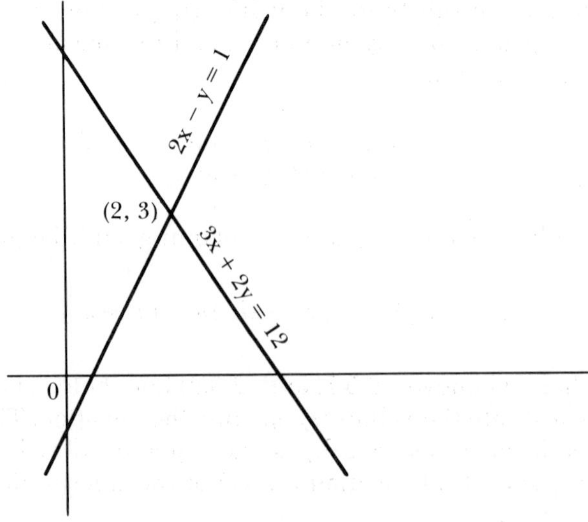

FIGURE II.5.1 Example II.5.1.

We look for a linear combination of these equations that represents a horizontal line (i.e., parallel to the x-axis). This has slope 0, and so is obtained by choosing r and s so that the numerator in the right side of (2.5.4) is equal to 0. This is most easily done by setting $r = a_2 = 2$ and $s = -a_1 = -3$; thus we have the two equations

$$\begin{aligned} 6x + 4y &= 24 \\ -6x + 3y &= -3 \end{aligned}$$

Adding these two equations together we obtain

$$7y = 21$$

an equation from which x has been eliminated. This of course gives us $y = 3$; substitution of this value into the first of the original equations gives us

$$3x + 2 \cdot 3 = 12$$

or

$$3x = 6$$

which has the solution $x = 2$. Thus (2,3) is the solution of the system.

II.5.2 Example. Find the solution of the system

$$\begin{aligned} x + 2y &= 4 \\ 2x - y &= 3 \end{aligned}$$

Again, we eliminate x by setting $r = 2$, $s = -1$. This gives us

$$\begin{aligned} 2x + 4y &= 8 \\ -2x + y &= -3 \end{aligned}$$

which upon addition reduces to

$$5y = 5$$

or $y = 1$. Substituting in the first equation, we have

$$x + 2 \cdot 1 = 4$$

or $x = 2$. Thus (2,1) is the solution of the system.

This method of solving systems of equations will of course break down if the two equations do not have a unique solution. In fact, the two lines (2.5.1) and (2.5.2) will fail to have a unique point in common if they have the same slope (in which case they are either parallel or coincident). Suppose, then, that these two lines have the common slope k; this means $a_1 = -kb_1$ and $a_2 = -kb_2$. But then, from (2.5.4)

$$m = \frac{rkb_1 + skb_2}{rb_1 + sb_2} = k$$

and so any linear combination must also have slope k. But the method of elimination consists precisely in finding a linear combination with slope 0 (or with infinite slope). It follows that the method of elimination cannot work for such systems; if we let the numerator in (2.5.4) vanish, we find that the denominator vanishes simultaneously, and (2.5.3) reduces either to an absurdity of the form $0 = 1$ (when the lines are parallel) or to an identity $0 = 0$ (when the lines are coincident).

II.5.3 Example. Find the solution of the system

$$\begin{aligned} 3x + 2y &= 5 \\ 9x + 6y &= 10 \end{aligned}$$

In this case we can eliminate x by letting $r = 3$, $s = -1$. We obtain

$$\begin{aligned} 9x + 6y &= 15 \\ -9x - 6y &= -10 \end{aligned}$$

Addition of these two equations, however, shows us that in eliminating x we have also eliminated y, giving rise to the absurdity $0 = 5$. We can only conclude that the system is infeasible: the two lines are parallel.

II.5.4 Example. Find the solution of the system

$$\begin{aligned} 6x - 4y &= 8 \\ -9x + 6y &= -12 \end{aligned}$$

In this case, we eliminate x by setting $r = 3$ and $s = 2$, obtaining

$$\begin{aligned} 18x - 12y &= 24 \\ -18x + 12y &= -24 \end{aligned}$$

Addition of these two equations leads to the identity $0 = 0$. This means that the two equations are equivalent. The two lines are coincident and there is no *unique* solution; there is, rather, an infinity (all the points on a line). We may, in such cases, leave the system in the form of *one* of the original equations; we may also solve for one of the variables in terms of the other. Thus, we could solve for y:

$$y = \frac{3}{2}x - 2$$

and state that this is the general solution in the sense that, for any (arbitrary) value of x, the value of y given here will satisfy the system. Similarly, we could solve for x:

$$x = \frac{4}{3} + \frac{2}{3}y$$

and treat this also as the general solution.

II.5.5 Example. A druggist has two solutions, one containing 50 per cent alcohol, and the other containing 80 per cent alcohol. He wishes to obtain 24 fluid ounces of a solution with 60 per cent alcohol. How much of each should he use?

Letting x and y be the amounts, respectively, of the 50 per cent and of the 80 per cent solutions, we have

$$\begin{aligned} x + \quad\;\; y &= 24 \\ 0.50x + 0.80y &= 14.4 \end{aligned}$$

since $.60(24) = 14.4$. Solving, we obtain

$$x = 16, \; y = 8$$

as the desired quantities (in fluid ounces).

II.5.6 Example. Supply and Demand Equilibrium. Costs for the producers of a certain commodity are such that, if it can be sold at a price of $\$p$ per lb., production will be $5000p + 1000$ pounds. On the other hand, consumers will buy x pounds if the price is $4 - 0.001x$ dollars per pound. Find the equilibrium price, i.e., the price at which supply will exactly equal demand.

The supply equation is

$$x = 5000p + 1000$$

whereas the demand equation is

$$p = 4 - 0.001x.$$

Solving this system, we obtain

$$x = 3500, \; p = 0.5$$

and so the equilibrium price is 50¢ per lb.; demand and supply will then be 3500 lb.

For systems of more than two equations, the procedure is similar, though with certain differences. If we consider the geometry of the situation, we see that, in general, three lines in the plane form a triangle (Figure II.5.2); it is only in special cases that they have a point

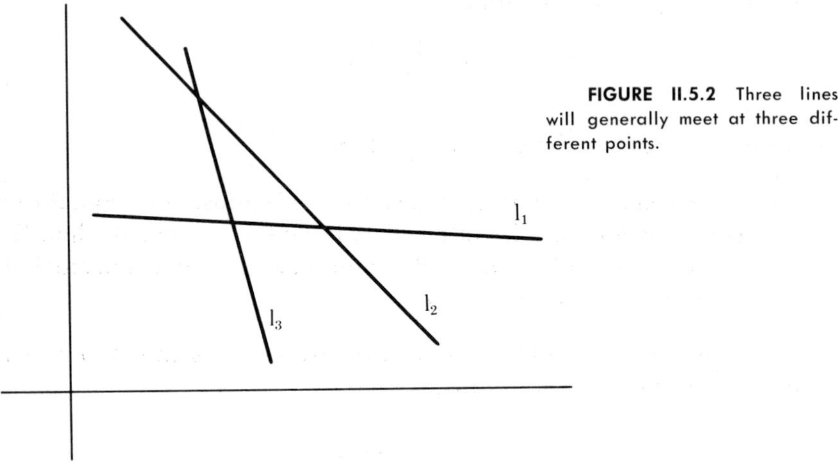

FIGURE II.5.2 Three lines will generally meet at three different points.

in common (Figure II.5.3). Sometimes it may even happen that all three lines are coincident, so that the system may have an infinity of solutions. In general, however, a system of three or more equations in the two unknowns, x and y, will have no solution.

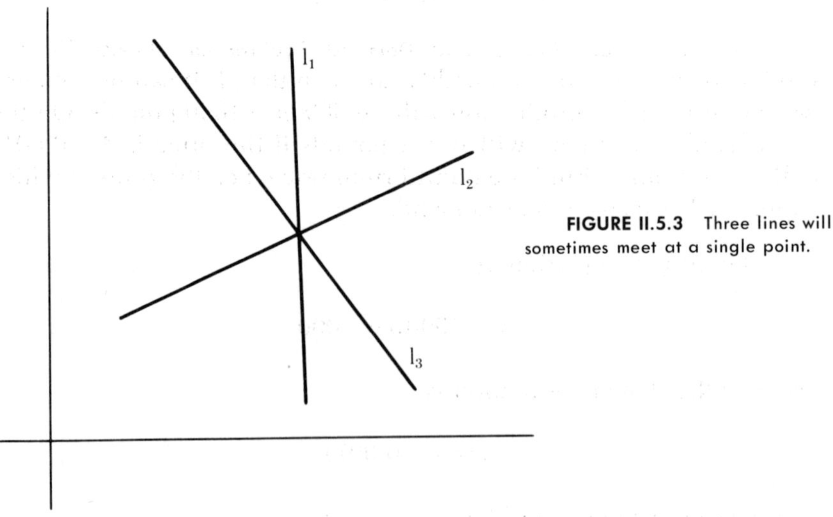

FIGURE II.5.3 Three lines will sometimes meet at a single point.

The method used for solving such systems is, generally, to find the point of intersection of two of the lines. Once this point has been found, it is simply a matter of checking whether the remaining lines of the system all pass through this point. If they do, well and good;

if not, the system has no solution. A systematic procedure would run as follows:

Solve the first two equations in the system as has just been explained. There are three possibilities: the two equations may have a unique solution, or no solution at all, or an infinity of solutions.

If the two equations have a unique solution, substitute this solution in the remaining equations of the system to check whether it satisfies them all. If it does, it is the unique solution of the system. If not, the system has no solution.

If the two equations have no common solution, the system has no solution.

If the two equations have an infinity of solutions, they represent coincident lines and are equivalent. This means one of the two equations may be discarded, giving rise to a smaller system. We then repeat the procedure with the smaller system.

II.5.7 Example. Solve the system

$$\begin{aligned} 3x + y &= 5 \\ x - 2y &= -3 \\ 2x + 2y &= 6 \\ x + 2y &= 5 \end{aligned}$$

In this system, we take the first two equations:

$$\begin{aligned} 3x + y &= 5 \\ x - 2y &= -3 \end{aligned}$$

and solve by the elimination procedure: choosing $r = 1$, $s = -3$ to give

$$\begin{aligned} 3x + y &= 5 \\ -3x + 6y &= 9 \end{aligned}$$

and adding these,

$$7y = 14$$

gives us $y = 2$; substitution in the first equation gives

$$3x + 2 = 5$$

or $x = 1$. We then substitute (1,2), the unique solution of the first two equations, in the remaining equations:

$$\begin{aligned} 2 \cdot 1 + 2 \cdot 2 &= 6 \\ 1 + 2 \cdot 2 &= 5 \end{aligned}$$

Since both of these are correct, it follows that (1,2) is the solution of the system.

II.5.8 Example. Solve the system

$$\begin{aligned} 3x + y &= 5 \\ x - 2y &= -3 \\ 2x + y &= 6 \end{aligned}$$

In this case, we once again consider the first two equations, which have the unique solution (1,2). We substitute this into the third equation, obtaining

$$2 \cdot 1 + 2 = 6$$

This is false, and it follows that the system has no solution.

II.5.9 Example. Solve the system

$$\begin{aligned} x - 2y &= 3 \\ -2x + 4y &= -6 \\ x + y &= 6 \end{aligned}$$

In this case, the first two equations

$$\begin{aligned} x - 2y &= 3 \\ -2x + 4y &= -6 \end{aligned}$$

are seen to be equivalent. Thus, we may discard one of them, obtaining the reduced system

$$\begin{aligned} x - 2y &= 3 \\ x + y &= 6 \end{aligned}$$

which can be seen (subtracting one equation from the other) to have the solution $x = 5$, $y = 1$. Thus, (5,1) is the unique solution to the system.

II.5.10 Example. Solve the system

$$\begin{aligned} x + 2y &= 6 \\ 2x + 4y &= 10 \\ x - 3y &= 7 \\ 3x + y &= 5 \end{aligned}$$

In this case, the first two equations

$$\begin{aligned} x + 2y &= 6 \\ 2x + 4y &= 10 \end{aligned}$$

are seen to represent parallel lines. Since they have no common solution, it follows that the system has no solutions at all.

PROBLEMS ON SYSTEMS OF EQUATIONS

1. Solve the following systems of simultaneous equations. Give all solutions:

(a) $$\begin{aligned} x + 2y &= 5 \\ 2x + y &= 4 \end{aligned}$$

(b) $$\begin{aligned} x - 3y &= 7 \\ x + y &= 11 \end{aligned}$$

(c) $$\begin{aligned} 2x + 3y &= 7 \\ x - 2y &= -1 \end{aligned}$$

(d) $$\begin{aligned} 3x - 2y &= 5 \\ x + y &= 5 \end{aligned}$$

(e) $$\begin{aligned} 2x + 4y &= 8 \\ x + 2y &= 4 \end{aligned}$$

(f) $$\begin{aligned} x + 3y &= 6 \\ -x + 2y &= -1 \end{aligned}$$

(g) $$\begin{aligned} x - 4y &= 3 \\ 2x + y &= -3 \end{aligned}$$

(h) $$\begin{aligned} 4x + 6y &= 8 \\ 6x + 9y &= 10 \end{aligned}$$

(i) $$\begin{aligned} -2x + 3y &= 5 \\ x - 2y &= -4 \end{aligned}$$

(j) $$\begin{aligned} 3x - 4y &= 6 \\ x + y &= 9 \end{aligned}$$

(k) $$\begin{aligned} 3x + 2y &= 5 \\ 6x + 4y &= 10 \end{aligned}$$

(l) $$\begin{aligned} 2x + 4y &- 5 \\ 3x + 6y &= 7 \end{aligned}$$

(m) $$\begin{aligned} x + 2y &= 5 \\ 3x + y &= 5 \\ -x + 4y &= 7 \\ x - 2y &= -3 \end{aligned}$$

(n) $$\begin{aligned} -2x + 4y &= 6 \\ -3x + 6y &= 9 \\ x + 2y &= 1 \\ -x + 3y &= 4 \end{aligned}$$

(o) $$\begin{aligned} x - y &= 3 \\ 2x + y &= 6 \\ -x - y &= 2 \\ x + 2y &= 3 \end{aligned}$$

(p) $$\begin{aligned} x + 3y &= 4 \\ 3x + y &= 2 \\ x - 4y &= 3 \\ x + 3y &= 2 \end{aligned}$$

(q) $$\begin{aligned} 3x + 4y &= 10 \\ -x + 2y &= 0 \\ 2x + y &= 5 \\ x - 3y &= -1 \end{aligned}$$

(r) $$\begin{aligned} -x + 2y &= 6 \\ x + 3y &= 9 \\ 2x + y &= 3 \\ x - 2y &= -6 \end{aligned}$$

2. Show that (a) the medians and (b) the altitudes of the triangle with vertices (5,1), (2,4), and (1,0) intersect at one point.

3. A pharmacist wishes to obtain 20 fluid ounces of a solution containing 75 per cent alcohol. He has two solutions, containing 40 per cent and 90 per cent alcohol respectively. How much of each solution should he use?

4. A food processing company has a large quantity of peanuts costing 10¢ per lb., and cashews costing 60¢ per lb. It wishes to obtain 25 pounds of a mixture costing 25¢ per lb. How much of this mixture should be peanuts?

5. Foods A and B have respectively 500 and 1000 calories per pound. It is desired to form a mixture weighing 7 lbs. and containing 4000 calories. How much of each food should be used?

6. At a price of $20 per ton, the supply of a certain commodity is 4000 T., while demand is 4800 T. If the price is raised to $30 per T., supply and demand will be 4500 T. and 4700 T. respectively. Assuming that both supply and demand are linearly related to price, find the equilibrium price level for this commodity.

The procedure with systems of linear inequalities is generally different. In fact, each linear inequality determines, not a line, but the entire half-space on one side of a line. Thus the solution to a system of linear inequalities will generally be, not a unique point, but rather, a figure bounded by straight lines, i.e., a polygon or some similar figure.

In the case of two inequalities, each is determined by a line. Generally, they are two intersecting lines; these divide the plane into four regions, labeled A, B, C, and D in Figure II.5.4. The general

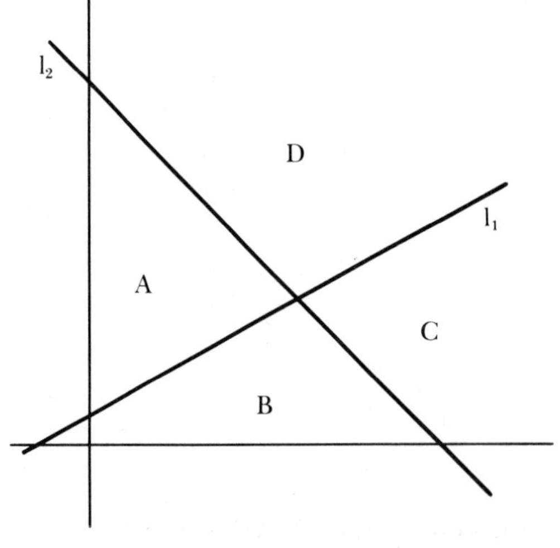

FIGURE II.5.4 The lines l_1 and l_2 divide the plane into four regions.

solution consists of all the points in one of these regions. For instance, the system of inequalities

$$\begin{aligned} x - \ \ y &\geq 0 \\ x - 2y &\leq 0 \end{aligned}$$

may be seen to be equivalent to the relation (2.2.2). Its solution is the shaded area in Figure II.2.2, which we repeat as Figure II.5.5.

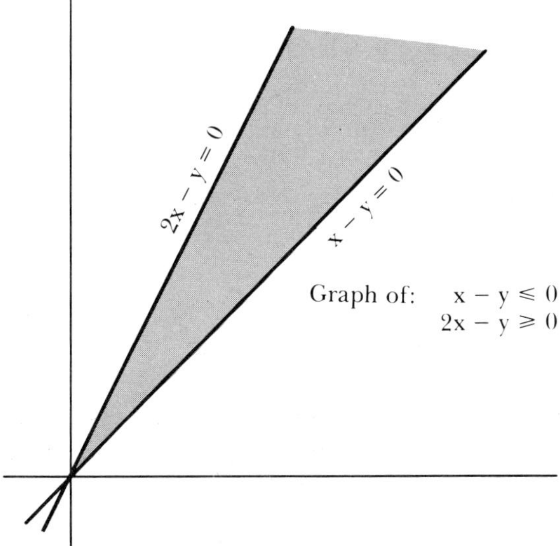

FIGURE II.5.5

Special cases arise, of course. If the lines are parallel, the solution may be the entire, infinite strip between two lines (Figure II.5.6).

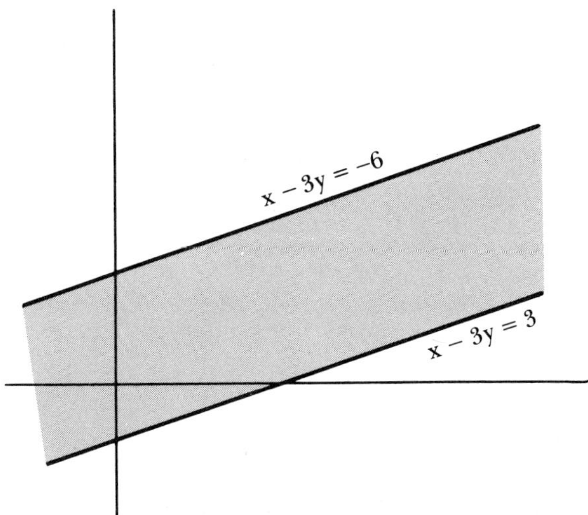

FIGURE II.5.6 The shaded area is the solution of the system

$$x - 3y \leq 3$$
$$x - 3y \geq -6$$

It may also happen that the two inequalities are contradictory as in the system

$$x + 2y \geq 5$$
$$x + 2y \leq 3$$

In this case, we want the points above the first line and below the second line. But the first line is parallel to and above the second, so there can be no solution.

It may even happen that the two inequalities give a single line as solution, thus:

$$x + 2y \geq 5$$
$$x + 2y \leq 5$$

These two lines are coincident. We want those points that are on or above the line, and also those on or below the line. But in this case all the points must lie on the line $x + 2y = 5$.

For larger systems of inequalities, we proceed similarly. Consider the system

$$x + y \leq 4$$
$$x + 2y \leq 5$$
$$x \geq 0$$
$$y \geq 0$$

Let us take the first of these inequalities,

$$x + y \leq 4$$

and multiply the fourth by -1, which gives us

$$-y \leq 0$$

These two inequalities, being of the same type, can be added to give

$$x \leq 4$$

Since we already have $x \geq 0$, we know that x is free to vary in the interval from 0 to 4, i.e.,

$$0 \leq x \leq 4$$

For any value x in this range, y can vary within a certain range that depends on x: we must have, from the inequalities, $y \geq 0$, and, moreover,

$$y \leq 4 - x$$

$$y \leq \frac{5}{2} - \frac{1}{2}x$$

These are two inequalities that must be satisfied by y. Depending on the value of x, one of them will be stronger than the other: thus, if $x = 1$, the inequalities are

$$y \leq 3$$
$$y \leq 2$$

and it is clear that the second is stronger than the first. On the other hand, for $x = 4$, the inequalities become

$$y \leq 0$$

$$y \leq \frac{1}{2}$$

and in this case it is clear that the first is stronger than the second. The transition between the two cases occurs when the two inequalities are the same, i.e., when

$$4 - x = \frac{5}{2} - \frac{1}{2}x$$

which is solved to give us $x = 3$. For x greater than 3, it is the first inequality that dominates, while for x smaller than 3, the second inequality dominates. Thus the general solution to the system of inequalities might be:

$$0 \leq x \leq 4$$

$$\text{If } 0 \leq x \leq 3, \text{ then } 0 \leq y \leq \frac{5}{2} - \frac{1}{2}x$$

$$\text{If } 3 \leq x \leq 4, \text{ then } 0 \leq y \leq 4 - x$$

This is a general solution of the system; it remains to be asked, however, whether we have gained much by expressing it in this form. There is the advantage of being able to obtain a point in the set directly: we can, say, choose $x = 2$ (this satisfies the first line of the solution); according to the second line, we must then have

$$0 \leq y \leq \frac{3}{2}$$

so that (2,1) is a solution—as may be checked directly.

Against this advantage, we have lost much of the conciseness of the original system of inequalities, and this after much work. Thus we shall not, in general, attempt to give the solutions to systems of inequalities in such form.

It is much more interesting, from a practical point of view, to know what the constraint set "looks like." We can say that it is a quadrilateral bounded by the four lines that determine the inequalities; to avoid ambiguities, we may give the four vertices or *extreme points* of the quadrilateral.

Each of the four vertices of the quadrilateral is at the intersection of two of the bounding lines of the quadrilateral: in other words, each of the vertices can be found by solving a system of two equations in the unknowns x and y: the equations are simply the inequalities that determine the constraint set, with the inequality sign replaced by an equals sign. In this example the four vertices will be the solutions of the systems

$$\begin{aligned} x + y &= 4 \\ x + 2y &= 5 \end{aligned}$$

which gives us the point (3,1);

$$\begin{aligned} x + y &= 4 \\ y &= 0 \end{aligned}$$

which gives (4,0);

$$\begin{aligned} x + 2y &= 5 \\ x &= 0 \end{aligned}$$

which gives (0,5/2); and

$$\begin{aligned} x &= 0 \\ y &= 0 \end{aligned}$$

which, of course, gives (0,0). The graph of the system of four inequalities is best characterized in this form: it is the quadrilateral with the vertices at (0,0), (0,5/2), (4,0), and (3,1) as shown in Figure II.5.7.

It may be noted that we did not consider all the points of intersection of pairs of the lines bounding the quadrilateral. We could, for

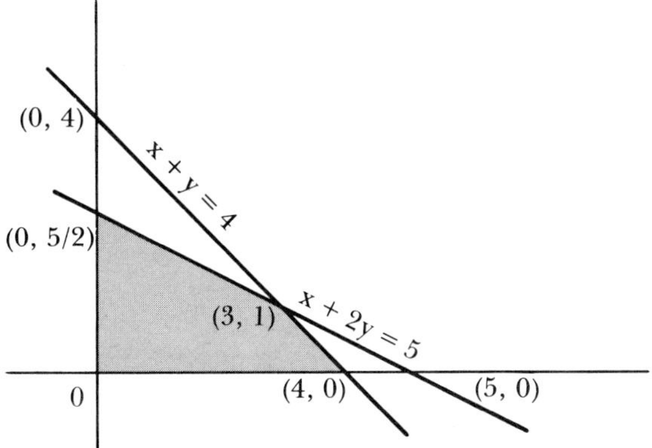

FIGURE II.5.7 The shaded area is the solution of the system.

example, have taken the two lines

$$\begin{aligned} x + 2y &= 5 \\ y &= 0 \end{aligned}$$

and solved to obtain (5,0). This is not, however, a vertex of the graph, for the simple reason that it does not satisfy the inequality

$$x + y \leq 4$$

which must be satisfied by all points of the figure (including, of course, the vertices). Similarly, the system

$$\begin{aligned} x + y &= 4 \\ x &= 0 \end{aligned}$$

gives the point (0,4), which does not satisfy the inequality

$$x + 2y \leq 5$$

and is not, therefore, in the figure.

Points obtained by treating two of the inequalities as equations and solving the resulting system need not be vertices of the graph, as we have no quarantee that they will satisfy the other inequalities. If they do satisfy them, however, they will be vertices. The rule is: in a system of loose linear inequalities (in two variables), the vertices of the graph can be found by treating two of the inequalities as equations and solving. If this system of two equations has a unique solution, this is a vertex of the graph if, and only if, it satisfies the remaining inequalities.

II5.11 Example. Give the graph of the system

$$\begin{aligned} x + y &\leq 5 \\ x &\geq 0 \\ x &\leq 3 \\ y &\geq 0. \end{aligned}$$

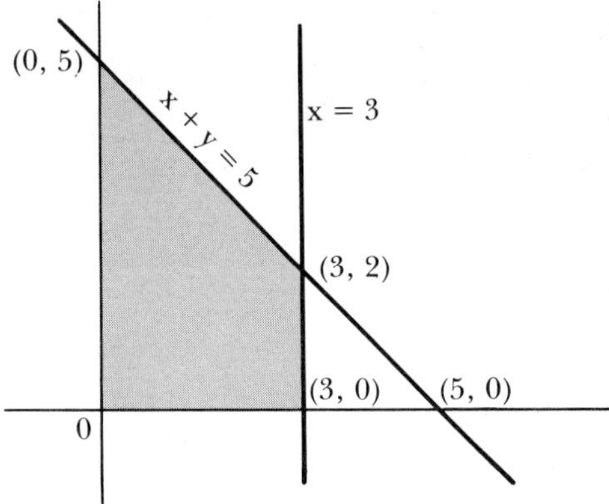

FIGURE II.5.8 The shaded area is the solution of Example II.5.11.

From Figure II.5.8 we see that the set is a quadrilateral. To find its vertices, we take the inequalities, two at a time, and treat them as equations. The first and second inequalities give us

$$\begin{aligned} x + y &= 5 \\ x &= 0 \end{aligned}$$

which has solution (0,5). This satisfies the other two inequalities.

The first and third inequalities give

$$\begin{aligned} x + y &= 5 \\ x &= 3 \end{aligned}$$

with solution (3,2). Again, this satisfies the remaining inequalities.

The first and fourth inequalities give

$$\begin{aligned} x + y &= 5 \\ y &= 0 \end{aligned}$$

with solution (5,0). This does not, however, satisfy the inequality

$$x \leq 3$$

and is therefore not a vertex of the graph.

The second and third inequalities give

$$x = 0$$
$$x = 3$$

which is clearly not feasible.

The second and fourth give the point (0,0). This satisfies the other inequalities.

Finally, the third and fourth inequalities give the point (3,0). It may be checked that it satisfies the other inequalities.

Of the six pairs of inequalities, then, we see that four give the points (0,5), (3,2), (0,0), and (3,0), which satisfy the remaining inequalities and are therefore vertices of the graph. Of the two other pairs, one gives a point that fails to satisfy the inequalities, while the other pair fails to have a solution.

This, then, characterizes the graph: it is, simply, a quadrilateral with vertices at (0,5), (3,2), (0,0), and (3,0).

II.5.12 Example. We give an example now to show how practical problems give rise to systems of linear inequalities (and, possibly, equations).

A mixture must be made with foods A and B. Each unit of food A weighs 5 gm. and contains 1 gm. of protein. Each unit of food B weighs 3 gm. and contains 0.5 gm. protein. The mixture must weigh 60 gm. or less and contain at least 8 gm. protein.

We can summarize the data of the problem, if we wish, in the form of a table (Table I.5.1).

TABLE II.5.1

Food	Units	Unit Weight	Protein/Unit
A	x	5	1
B	y	3	0.5
Total		≤ 60	≥ 8

Letting x and y be the number of units of food A and B to be put into the mixture, we find that the weight of food A is $5x$ grams, while the weight of food B is $3y$ grams. Thus we obtain the relation

$$5x + 3y \leq 60$$

In a similar way, the need for at least 8 grams of protein can be restated as

$$x + 0.5y \geq 8$$

Two additional inequalities arise due to the fact that the amounts used must be positive or zero. These are the so-called *non-negativity constraints.* The problem can be stated in the form

$$\begin{aligned} 5x + 3y &\leq 60 \\ x + 0.5y &\geq 8 \\ x &\geq 0 \\ y &\geq 0 \end{aligned}$$

We proceed then to solve the system as we did with Example II.5.11. The first two inequalities give the system

$$\begin{aligned} 5x + 3y &= 60 \\ x + 0.5y &= 8 \end{aligned}$$

with solution $(-12,40)$, which does not satisfy the constraint $x \geq 0$. The first and third inequalities give

$$\begin{aligned} 5x + 3y &= 60 \\ x &= 0 \end{aligned}$$

with solution (0,20), which does satisfy the remaining inequalities. The first and fourth inequalities give the point (12,0); the second and third inequalities, (0,16); and the second and fourth, (8,0). All these points satisfy the inequalities and are therefore vertices of the set. On the other hand, the third and fourth inequalities give (0,0), which does not satisfy the inequality $x + 0.5y \geq 8$ and thus is not a vertex. This set can, therefore, be characterized as the quadrilateral with vertices (0,20), (12,0), (0,16), and (18,0) as shown in Figure II.5.8.

PROBLEMS ON SYSTEMS OF LINEAR INEQUALITIES

1. On graph paper, show the sets of points corresponding to the following systems of linear inequalities. Find the extreme points of these sets.

(a)
$$\begin{aligned} x + 2y &\leq 12 \\ 3x + y &\leq 9 \\ x &\geq 0 \\ y &\geq 0 \end{aligned}$$

(b)
$$\begin{aligned} 2x + y &\leq 20 \\ 2x + 3y &\leq 24 \\ x - y &\geq 5 \\ x &\geq 0 \\ y &\geq 0 \end{aligned}$$

(c)
$$\begin{aligned} 3x + y &\geq 16 \\ x + 2y &\leq 20 \\ x &\geq 0 \\ y &\geq 0 \end{aligned}$$

(d)
$$\begin{aligned} 2x + y &\geq 20 \\ 2x + 3y &\geq 24 \\ x - y &\geq 5 \\ x &\geq 0 \\ y &\geq 0 \end{aligned}$$

(e)
$$\begin{aligned} 3x + 4y &\leq 16 \\ 2x + y &\leq 8 \\ x - y &\leq 2 \\ x &\geq 0 \\ y &\geq 0 \end{aligned}$$

2. A nut company has 6000 lb. of mixed nuts and 4000 lb. of peanuts. The company sells three products: an expensive mixture containing 5 lb. of mixed nuts for each pound of peanuts, a cheaper mixture containing equal parts of mixed nuts and peanuts, and peanuts without any other nuts. Give a system of linear inequalities expressing the amounts which the company can make of each of these products.

3. A dietitian is given two foods, A and B. Food A contains 200 grams of protein and 1000 calories per pound. Food B has 50 grams of protein and 2800 calories per pound. The dietitian wishes to make a mixture weighing at least 6 pounds, which will contain at least 800 grams of protein, and not more than 8000 calories. Give a system of linear inequalities for the permissible mixtures, and draw a diagram on graph paper.

4. Each unit of item A requires 20 cubic feet storage space and costs $50. Each unit of item B requires 35 cu. ft. of space and costs $40. A firm has $10,000 available to purchase these two items, which must be stored in a warehouse with a total capacity of 8000 cu. ft. Give a system of linear inequalities for the permissible purchases, and draw a diagram.

6. HIGHER DIMENSIONS

We have seen that the set of real numbers can be put into a natural correspondence with the set of points on a line; the set of pairs of real numbers with the set of points in a plane. We related one-dimensional geometry with the analysis of one variable, two-dimensional geometry with the analysis of two variables. Similarly, we can relate solid (three-dimensional) geometry with the analysis of three real variables. It is then only a short step to talking about space of four, five, six, or any number of dimensions, even though there is no physical interpretation of this idea. The justification is, simply, that it allows us to speak of the relations among four or more variables in the highly descriptive language of geometry, though it be only by generalization of the concepts that we have studied in solid geometry. Let us look, first, at the three-dimensional case. Once again a point, O, is chosen; through this point (called, as usual, the origin) three mutually perpendicular lines are then taken and called respectively, the x-axis, the y-axis, and the z-axis (collectively, the coordinate axes).

Three points, I, I', and I'', are then taken, equidistant from O, on the three axes. These serve to give us, first, the unit of distance (the common length of the three segments $\overline{OI}$, $\overline{OI'}$, and $\overline{OI''}$) and the "positive" direction on each of the three axes.

In Figure II.6.1, the x-axis and the y-axis are assumed to lie in the plane of the page. The z-axis, however, is supposed to come *out* of the page (toward the reader). Such a coordinate system is known as a *right-handed system.* This nomenclature is derived from the fact that a right-handed screw, rotated from the direction of positive x toward the direction of positive y (counterclockwise in this case), will move in the direction of positive z. We could just as easily have chosen the point I'' in such a way that the positive z-direction was *into* the book (away from the reader), a *left-handed system.* The principal reason for choosing a right-handed system is convenience: most authors assume right-handed systems, and a left-handed system, while perfectly consistent, clashes with established use. What this means is immediately apparent to anyone who has tried to screw a left-handed bolt into a right-handed nut.

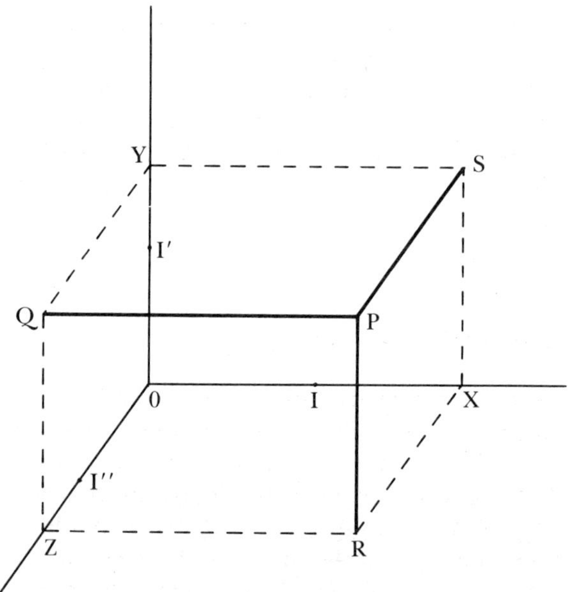

FIGURE II.6.1 Location of a point in three-dimensional space.

The three coordinate axes, taken two at a time, determine three planes called the coordinate planes. The x- and y-axes determine a plane called the xy-plane, the x- and z-axes determine the xz-plane, and the y- and z-axes, the yz-plane.

To each point in space, we can assign three numbers, the directed (perpendicular) distances to that point from the three coordinate planes. The distance to the yz-plane is called the x-coordinate, that to the xz-plane the y-coordinate, and that to the xy-plane the

z-coordinate. Thus, in Figure II.6.1, to the point P are assigned, as coordinates, the distances $\overline{QP}$, $\overline{RP}$, and $\overline{SP}$, respectively. Here Q is the point of intersection of a line through P parallel to the x-axis, with the yz-plane. The points R and S are obtained analogously.

Conversely, any triple of numbers (x,y,z) corresponds to a point in space. The point (a,b,c) can be obtained by moving a units in the x-direction, then b units in the y-direction, and finally, c units in the z-direction. (The order in which these three motions are carried out is unimportant. We could just as well move, first, b units in the y-direction, followed by a units in the x-direction, and c units in the z-direction. The same point would be obtained, thanks to the geometric theorem which states that opposite sides of a rectangle are equal.)

The general first-degree equation

2.6.1 $$ax + by + cz = d$$

represents, not a line, but a plane in space. (This can best be seen if we consider that the equation $x = 0$ represents the yz-plane.) Nevertheless, it is still called a linear equation.

The general loose linear inequality

2.6.2 $$ax + by + cz \geq d$$

will represent one of the two *half-spaces* bounded by the plane of equation 2.6.1). That is, it will represent all the space that lies on one side of the plane 2.6.1), plus the plane itself. The corresponding strict inequality

2.6.3 $$ax + by + cz > d$$

will represent the half-space, minus the bounding plane.

The solution of systems of linear equations can best be understood if we look at certain geometric considerations. Normally, two planes intersect along a line, while three planes intersect at a point. Four planes normally form a tetrahedron and therefore have no points in common. Analytically, this means that two equations (in three variables) generally have an infinity of solutions, while three equations have a unique solution, and four or more equations none at all. This is, of course, only the general case; it may happen that a system of only two equations has no solutions (parallel planes), just as it may happen that a larger system has a solution (several planes passing through a common point) or even an infinity of solutions (if the planes are coaxial, i.e., have a line in common). It may happen, for that matter, that all the planes are coincident.

Such systems are solved in essentially the same way as systems of equations in two unknowns, i.e., by elimination and substitution. Naturally, it is a considerably longer process, since now there are more unknowns to eliminate and substitute.

II.6.1 Example. Solve the system

$$\begin{aligned} x + 2y + 3z &= 9 \\ 2x + y + 2z &= 7 \\ x + y + 3z &= 8 \end{aligned}$$

We start by eliminating x. This can be done by multiplying the first equation by 2, the second by -1, and adding:

$$\begin{aligned} 2x + 4y + 6z &= 18 \\ -2x - y - 2z &= -7 \end{aligned}$$

to obtain

$$3y + 4z = 11$$

Similarly, we can multiply the second equation by -1, the third equation by 2, and add

$$\begin{aligned} -2x - y - 2z &= -7 \\ 2x + 2y + 6z &= 16 \end{aligned}$$

obtaining

$$y + 4z = 9$$

We have now a system of two equations

$$\begin{aligned} 3y + 4z &= 11 \\ y + 4z &= 9 \end{aligned}$$

in the two unknowns, y and z. We then proceed to solve this system by subtracting the second equation from the first, to obtain

$$2y = 2$$

or $y = 1$. This value is then substituted to give us

$$3 + 4z = 11$$

which reduces to $z = 2$. Finally, the values $y = 1$, $z = 2$ are substituted in one of the original equations, giving,

$$x + 2 + 6 = 9$$

or, equivalently, $x = 1$. Thus (1,1,2) is the (unique) solution to the system.

II.6.2 Example. Find the solution of the system

$$\begin{aligned} 3x + 4y + z &= 11 \\ -x - 2y + 2z &= 3 \\ -x + 2y + z &= 8 \end{aligned}$$

In this case, we can eliminate x from the first and second equations if we multiply the second by 3 and add:

$$\begin{aligned} 3x + 4y + z &= 11 \\ -3x - 6y + 6z &= 9 \end{aligned}$$

to obtain

$$-2y + 7z = 20$$

Similarly, adding the first equation to three times the third

$$\begin{aligned} 3x + 4y + z &= 11 \\ -3x + 6y + 3z &= 24 \end{aligned}$$

we obtain

$$10y + 4z = 35$$

We now have a system

$$\begin{aligned} -2y + 7z &= 20 \\ 10y + 4z &= 35 \end{aligned}$$

from which we can eliminate y by multiplying the first equation by 5:

$$\begin{aligned} -10y + 35z &= 100 \\ 10y + 4z &= 35 \end{aligned}$$

to get

$$39z = 135$$

or $z = 45/13$. This is then substituted in the previous equations, obtaining $y = 55/26$ and $x = -10/13$. Thus $(-10/13, 55/26, 45/13)$ is the solution of the system.

II.6.3 Example. Find the solution of the system

$$\begin{aligned} x + 2y - z &= 6 \\ 2x - y + 2z &= 4 \\ 3x - 4y + 5z &= 3 \end{aligned}$$

In this case, we can eliminate x by subtracting the second equation from two times the first, obtaining

$$5y - 4z = 8$$

and subtracting the third equation from three times the first, which gives us

$$10y - 8z = 15$$

Because the resulting system of two equations is, however, infeasible, the original system is also infeasible.

II.6.4 Example. Solve the system

$$\begin{aligned} x + 2y - z &= 6 &\quad (e_1) \\ 2x - y + 2z &= 4 &\quad (e_2) \\ 3x - 4y + 5z &= 2 &\quad (e_3) \end{aligned}$$

This system is the same as that of Example II.6.3 except for the right-hand side of the third equation. Proceeding as before, we obtain the two equations

$$\begin{aligned} 5y - 4z &= 8 &\quad (e_4) \\ 10y - 8z &= 16 &\quad (e_5) \end{aligned}$$

which are equivalent, i.e., represent the same plane in space. We must see, now, exactly what this means. The equation (e_4) was obtained as a linear combination of (e_1) and (e_2). This means (as in the case of systems of equations of two unknowns), that every solution of (e_1) and (e_2) must also solve (e_4). Thus the plane of (e_4) must contain the line of intersection of the planes (e_1) and (e_2). Similarly, the plane (e_5) will contain the line of intersection of (e_1) and (e_3). But because (e_4) and (e_5) are the same plane, the two lines at which (e_1) intersects (e_2) and (e_3) will both lie in the plane (e_4).

Although two distinct lines can lie *at most* in one common plane, these two lines are seen to lie in both (e_1) and (e_4). The implication is that either the two lines are the same, or the two planes (e_1) and (e_4) are the same. But, since it is easy to see that (e_1) and (e_4) are different planes, e.g., (2,2,0) satisfies (e_1) but not (e_4), it follows that the two lines are the same.

Thus (e_1) intersects (e_2) and (e_3) along the same line. This line is common to all three planes, and therefore every point on the line satisfies all three equations; there is an infinity of solutions. We obtain these by letting one of the variables, say, z, be arbitrary, and then solving (e_4) for y (in terms of z):

$$y = \frac{8}{5} + \frac{4}{5}z \tag{e_6}$$

This value is then substituted in (e_1) to give us

$$x = \frac{14}{5} - \frac{3}{5}z \tag{e_7}$$

These two expressions, (e_6) and (e_7), together with the statement that z is arbitrary, form the general solution of the system. Of course, this is not the only form in which the solution can be given; we might just as easily have made y arbitrary and obtained

$$z = -2 + \frac{5}{4}y$$

$$x = 4 - \frac{3}{4}y$$

as the solution. We could also solve for y and z in terms of an arbitrary x. In any of these three cases, the resulting expressions would be called the general solution of the original system of equations.

II.6.5 Example. Solve the system

$$\begin{aligned} x + 2y + 3z &= 9 \\ 2x + y + 2z &= 7 \\ x + y + 3z &= 8 \\ x - y + 2z &= 4 \end{aligned}$$

In a system such as this, we solve the sub-system consisting of the first three equations, and, if this system has a unique solution, we check to see whether this satisfies the remaining equations. (The procedure is much the same as when handling systems of three or more equations in two unknowns.)

From Example II.6.1, the first three equations have solution (1,1,2). Since this satisfies the fourth equation as well, viz., $1 - 1 + 2 \cdot 2 = 4$, we conclude that (1,1,2) is the solution of this system.

II.6.6 Example. Solve the system

$$\begin{aligned} x + 2y - z &= 6 \\ 2x - y + 2z &= 4 \\ 3x - 4y + 5z &= 2 \\ x + y - z &= 2 \end{aligned}$$

Once again, we solve the first three equations; as in Example II.6.4, we find that they do not have a unique solution, but rather, a general solution consisting of an entire line. We may actually do two things in this case. One possibility is to discard one of the first three equations, reducing to a system of three equations that we solve in the usual manner. Another possibility is to take the general solution of Example II.6.4 and introduce it into the fourth equation here; we do this since it turns out to be considerably shorter (after all, most of the work has already been done). We obtain the single equation

$$\left(4 - \frac{3}{4}y\right) + y - \left(-2 + \frac{5}{4}y\right) = 2$$

which reduces to

$$6 - y = 2$$

or $y = 4$. Introducing this into the solution obtained for Example II.6.4 (with y arbitrary) we obtain $x = 1$, $z = 3$. Thus (1,4,3) is the solution of this system.

II.6.7 Example. Solve the system

$$\begin{aligned} x + 2y + 3z &= 9 \\ 2x + y + 2z &= 7 \\ x + y + 3z &= 8 \\ x - y + 2z &= 6 \end{aligned}$$

In this case, the first three equations give the solution (1,1,2). Introducing these values into the fourth equation, we obtain,

$$1 - 1 + 2 \cdot 2 = 6$$

which is patently false. We conclude that this system has no solution.

II.6.8 Example. A mixture must be made of foods A, B, and C.

The weight, protein content, and carbohydrate content (in grams) of each unit of these foods is given in the table below:

Food	Weight	Protein	Carbohydrate
A	9	2	4
B	3	1	1
C	4	1	2

The mixture must contain 9 grams of protein and 17 grams of carbohydrate. Is it possible to make such a mixture, with a weight of 38 grams?

Letting x, y and z be the number of units of each food, we obtain the system

$$\begin{aligned} 9x + 3y + 4z &= 38 \\ 2x + y + z &= 9 \\ 4x + y + 2z &= 17. \end{aligned}$$

This system has solution

$$x = 3, \quad y = 1, \quad z = 2$$

so the mixture should contain 3, 1, and 2 units respectively of A, B, and C.

II.6.9 Example. Using the same data as Example II.6.8, is it possible to make a mixture weighing 32 grams, containing 7 grams of protein and 15 grams of carbohydrate?

In this case we obtain the system

$$\begin{aligned} 9x + 3y + 4z &= 32 \\ 2x + y + z &= 7 \\ 4x + y + 2z &= 15 \end{aligned}$$

which has the solution

$$x = 3, \quad y = -1, \quad z = 2$$

and it follows that the problem cannot be solved (it requires negative values).

For systems of linear inequalities, the procedure could be more or less as in the two-variable case. Again, we could try to give the solution set explicitly. It may be more interesting, however, to look for the extreme points (vertices) of the set, which the inequalities determine. This is considerably complicated by the fact that we cannot draw figures to help us determine the way in which the planes intersect.

II.6.10 Example. A factory produces widgets, gadgets, and flibbers using two machines called a bender and a twister. Each widget requires two hours in the bender and one hour in the twister. A gadget requires three hours in the bender and two hours in the twister, while a flibber requires one hour in the bender and two in the twister. The bender can be used at most 50 hours per week, while the twister can be used at most 40 hours per week. What combinations of the products can be made?

If we denote the number of widgets, gadgets, and flibbers made in a week by x, y, and z, respectively, the data of the problem can be written in the form of a table (Table II.6.1).

TABLE II.6.1

Product	Number	Hr. in Bender	Twister
Widgets	x	2	1
Gadgets	y	3	2
Flibbers	z	1	2
Total		≤ 50	≤ 40

This table can be used to give the constraints of the problem: there are two constraints corresponding to the availability of the machines, and three non-negativity constraints due to the fact that negative amounts of a product are not possible. Thus we obtain the system

$$\begin{aligned} 2x + 3y + z &\leq 50 && (i_1)\\ x + 2y + 2z &\leq 40 && (i_2)\\ x &\geq 0 && (i_3)\\ y &\geq 0 && (i_4)\\ z &\geq 0 && (i_5) \end{aligned}$$

The set of possible combinations forms a polyhedron in space; it will be adequately described if we give its extreme points (vertices). In a manner analogous to that used in the two-variable case, we obtain the extreme points by taking three of the inequalities, solving them as equations, and then checking that the point obtained satisfies the remaining inequalities. The inequalities (i_1), (i_2), and (i_3) give us the system of equations

$$\begin{aligned} 2x + 3y + z &= 50\\ x + 2y + 2z &= 40\\ x &= 0 \end{aligned}$$

The x is immediately eliminated, the resulting system of two equations has solution $y = 15$, $z = 5$, and these three planes intersect at (0,15,5). It is easily checked that this point satisfies the other inequalities.

There are 10 combinations of three inequalities from among these five. Taking one combination at a time, we put our results in table form (Table II.6.2).

TABLE II.6.2

Inequalities	Point	Check
1,2,3	(0,15,5)	Yes
1,2,4	(20,0,10)	Yes
1,2,5	(−20,30,0)	No
1,3,4	(0,0,50)	No
1,3,5	(0,50/3,0)	Yes
1,4,5	(25,0,0)	Yes
2,3,4	(0,0,20)	Yes
2,3,5	(0,20,0)	No
2,4,5	(40,0,0)	No
3,4,5	(0,0,0)	Yes

Here we find that the set of possible combinations is a five-faced figure having the six points (0,15,5), (20,0,10), (0,50/3,0), (25,0,0), (0,0,20), and (0,0,0) as its vertices.

Unfortunately, not every such problem will have a solution as easily characterized as this. The trouble is that there is no guarantee that the desired set will be bounded. If not bounded, it is not a polyhedron and cannot be characterized in terms of its vertices. An example of this danger follows.

II.6.11 *Example.* Find the solution of the system

$$\begin{aligned} 2x + 3y + z &\geq 50 \\ x + 2y + 2z &\geq 40 \\ x &\geq 0 \\ y &\geq 0 \\ z &\geq 0 \end{aligned}$$

This system is the same as that in Example I.6.8, except that the first two inequalities have been reversed. If we proceed as before, although we will obtain the same 10 points of intersection for the bounding planes, the process of checking whether they satisfy the remaining inequalities will give different results. Indeed, we will find that the vertices in this case are the points (0,15,5), (20,0,10), (0,0,50), (0,20,0), and (40,0,0).

We are tempted to describe the solution set of this problem as the polyhedron with these vertices. This is not, however, the case. We have here a five-faced figure with only five vertices. Such a polyhedron must be a pyramid with a quadrilateral base, but in such a pyramid, four of the faces pass through the same point. This is clearly not true here, since each combination of three faces gives a different point. It follows that this figure cannot be a polyhedron. It is, in fact, unbounded and, hence, not characterized by its vertices. Our method fails us.

The reader may wonder: since our method fails here, how do we know that it actually worked in Example II.6.10? In other words, is it not possible that the same difficulties might have arisen there? The answer is that, in general, it is very difficult to be certain of such things. For the present, the reader will simply have to take our word; in a subsequent chapter, we shall see that such difficulties are not insuperable, as we are not generally interested in obtaining every solution to such a system: one will generally do.

For the case of four or more unknowns, we will briefly describe the behavior of such systems here. A method of solution for these is given in Chapters VII and VIII.

In general, when dealing with many unknowns, we label them $x_1, x_2, \ldots, x_n$. The general first-degree equation in n unknowns has the form

2.6.4 $$a_1x_1 + a_2x_2 + \ldots + a_nx_n = b$$

and represents something that we will call a *hyperplane*, i.e., a flat $(n-1)$-dimensional subset which is the generalization (in n-space) of a plane in 3-dimensional space.

A system of m such equations, known as an *m by n system*, has the general form

2.6.5
$$\begin{array}{l} a_{11}x_1 + a_{12}x_2 + \ldots + a_{1n}x_n = b_1 \\ a_{21}x_1 + a_{22}x_2 + \ldots + a_{2n}x_n = b_2 \\ \cdots\cdots\cdots\cdots\cdots\cdots\cdots\cdots \\ a_{m1}x_1 + a_{m2}x_2 + \ldots + a_{mn}x_n = b_m \end{array}$$

in which the coefficients are now given double subscripts to denote the equation and variable, respectively.

That is, a_{ij} is the coefficient of the variable x_j in the ith equation, while b_i is the constant term in the ith equation. The 2×3 system

$$\begin{array}{l} 3x_1 + 2x_2 - x_3 = 5 \\ 4x_1 - 5x_2 + 2x_3 = 6 \end{array}$$

is an example of the more general system

$$\begin{array}{l} a_{11}x_1 + a_{12}x_2 + a_{13}x_3 = b_1 \\ a_{21}x_1 + a_{22}x_2 + a_{23}x_3 = b_2 \end{array}$$

in which

$$\begin{array}{llll} a_{11}=3 & a_{12}=2 & a_{13}=-1 & b_1=5 \\ a_{21}=4 & a_{22}=-5 & a_{23}=2 & b_2=6 \end{array}$$

A system such as (2.6.5) usually has solutions for $m \leq n$, but not for $m > n$. More exactly, the usual case is that a system has a solution set of dimension $n-m$, whenever $m < n$. If $m = n$, the system normally has a unique solution, whereas for $m > n$, there usually is no solution.

This is, however, only the usual case. It may be that a system of two equations in many unknowns has no solutions (parallel hyperplanes), just as it may happen that a system with more equations than unknowns has one or more solutions. We have seen all this in dealing with two- and three-variable systems.

The system of linear inequalities

2.6.6

$$\begin{aligned} a_{11}x_1 + a_{12}x_2 + \ldots + a_{1n}x_n &\leq b_1 \\ a_{21}x_1 + a_{22}x_2 + \ldots + a_{2n}x_n &\leq b_2 \\ \ldots\ldots\ldots\ldots\ldots\ldots\ldots\ldots\ldots \\ a_{m1}x_1 + a_{m2}x_2 + \ldots + a_{mn}x_n &\leq b_n \end{aligned}$$

generally represents a solid figure in n-dimensional space, bounded only by hyperplanes. Often it will be a hyperpolyhedron, i.e., the generalization of a polyhedron to n-dimensional space. Sometimes, however, this figure is not bounded. The considerations are quite similar to those in the three-variable case.

The general method of solution of a system such as (2.6.5) is much the same as in the two- or three-variable case, by elimination and substitution. Errors are, however, likely to occur (i.e., a particular equation might be disregarded) unless a very systematic procedure is used. Such a procedure is studied in Chapter VII.

If we attempt to solve a system such as (2.6.6) as we did in the three-variable case, the same difficulties will arise, complicated by the fact that if m and n are large the number of combinations of n inequalities from the full set of m can be enormous. Generally, we shall concern ourselves with finding one particular vertex (extreme point) of the solution set—usually, the one that maximizes or minimizes some objective such as profits or costs. A procedure for this is discussed fully in Chapter VIII.

PROBLEMS ON HIGHER DIMENSIONS

1. Solve the following systems of simultaneous equations.

(a)
$$\begin{aligned} 3x + 4y - 2z &= 9 \\ x - 2y + z &= -2 \\ 2x + 3y + z &= 4 \end{aligned}$$

(b)
$$\begin{aligned} 5x + 2y - 3z &= -3 \\ x - 3y - z &= 2 \\ 3x + y + z &= 4 \end{aligned}$$

(c)
$$\begin{aligned} x - 2y + 2z &= 5 \\ 2x + y - 2z &= 6 \\ -x - 3y + 4z &= 2 \end{aligned}$$

(d)
$$\begin{aligned} x - 2y + 2z &= 5 \\ 2x + y - 2z &= 6 \\ -x - 3y + 4z &= -1 \end{aligned}$$

(e)
$$\begin{aligned} -2x + 4y + z &= -1 \\ x - y + z &= 3 \\ 2y + z &= 3 \end{aligned}$$

(f)
$$\begin{aligned} 3x + y - 2z &= -3 \\ x + 2y + 3z &= -4 \\ 2x + y + 3z &= 1 \\ x - y - 2z &= 3 \end{aligned}$$

2. Each unit of food A weighs 6 grams and contains 2 grams of protein. Each unit of food B weighs 8 grams, and contains 1 gram of protein, and each unit of food C weighs 12 grams and contains 2 grams of protein. A mixture is required weighing 50 grams, which will contain exactly 10 grams of protein. Find the general solution to this problem.

3. Referring to the foods in problem 2, above, A contains no fat, while units of foods B and C contain 1 gram and 5 grams, respectively, of fat. Is it possible to make a mixture weighing 50 grams, which will contain exactly 10 grams of protein and 12 grams of fat?

4. Three items, A, B, and C, must be processed on three machines, I, II, and III. Processing time for each item on the machines is given in the table:

	I	II	III
A	3	1	4
B	1	5	9
C	2	6	5

Assuming that machines I and II can run 40 hours per week, and machine III can run 60 hours per week, what combinations of items A, B, and C can be made?

CHAPTER III

NON-LINEAR CURVES

1. THE CONIC SECTIONS

We have, until now, considered only linear (i.e., first-degree) equations. Our reasons for this consisted mainly in the fact that these are the simplest to treat, and that, moreover, many practical cases can be at least approximated by linear relations. Quite often, however, relations exist which cannot be treated linearly. We therefore need some knowledge of more complicated equations and their graphs.

We shall, in these sections, study some of the properties of quadratic, or second-degree, equations. The most general second-degree equation (in two variables) has the form

3.1.1 $$Ax^2 + Bxy + Cy^2 + Dx + Ey + F = 0.$$

Note that (3.1.1) differs from the general linear equation (2.3.5) by the first three terms, i.e., those in x^2, xy, and y^2. These are considered second-degree, or *quadratic*, terms, because the variables x and y appear there multiplied by themselves, or by each other. We shall see below that these three terms tell us a great deal about the equation and its graph.

It may be shown that, if equation (3.1.1) has any solutions, then its graph will be a *conic section*, that is to say, the curve of intersection of a plane with a circular cone, such as is shown in Figure III.1.1. Note that the cone here is actually a double cone, going to infinity in either direction.

The conic sections were first studied by the Greek geometers. Apollonius (3rd Century B.C.), in particular, wrote a detailed treatise on them. In his treatise, Apollonius recognized four important cases, depending on the way that the plane cuts the cone.

The simplest case of all occurs when the plane is perpendicular

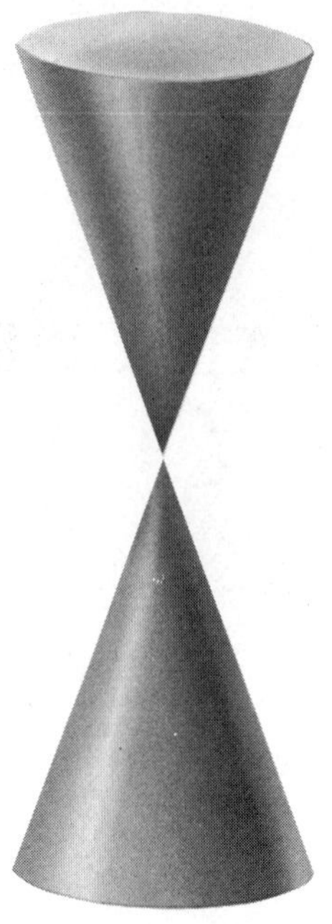

FIGURE III.1.1 A (double) cone.

to the axis of the cone (see Fig. III.1.2). In this case it is easy to see that the section obtained is a circle.

Suppose now that the plane is turned slightly, so that it is no longer perpendicular to the cone (see Fig. III.1.3). The section will be an elongated, closed curve called an *ellipse*.

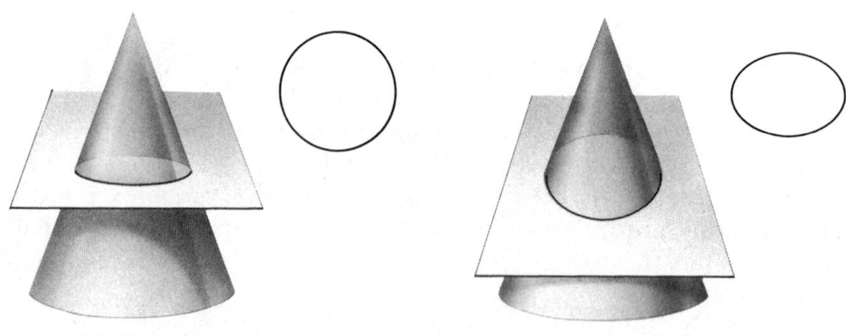

FIG. III.1.2

FIG. III.1.3

FIGURE III.1.2 The plane cuts the cone in a circle.

FIGURE III.1.3 The plane cuts the cone in an ellipse.

As we continue turning the plane, the ellipse becomes more and more elongated (Apollonius would say that the *eccentricity* increases). Eventually, however, we will reach a position in which the plane is parallel to the edge of the cone (see Fig. III.1.4). In this case the curve is not closed; it goes to infinity, and is known as a *parabola*.

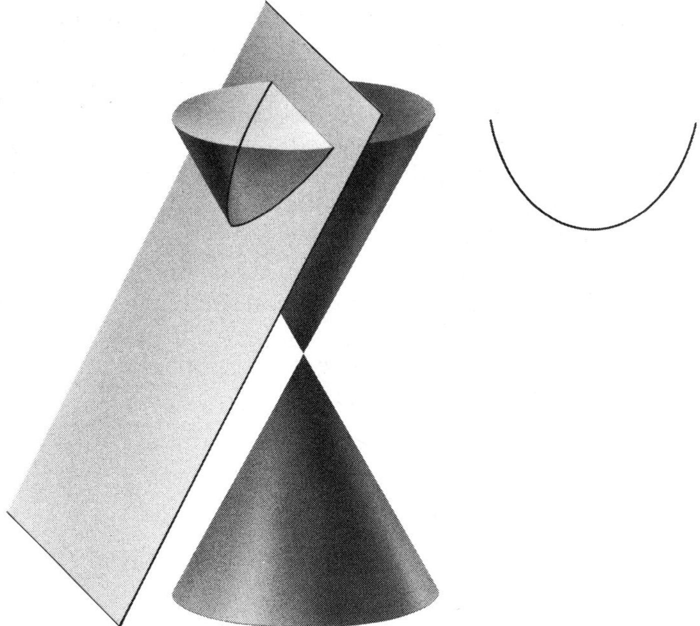

FIGURE III.1.4 Parabola.

Finally, as we continue to turn the plane, we find that it actually intersects the two parts of the cone along disjoint curves, each looking somewhat like a parabola (Fig. III.1.5). This double curve is known as a *hyperbola.*

FIGURE III.1.5 Hyperbola.

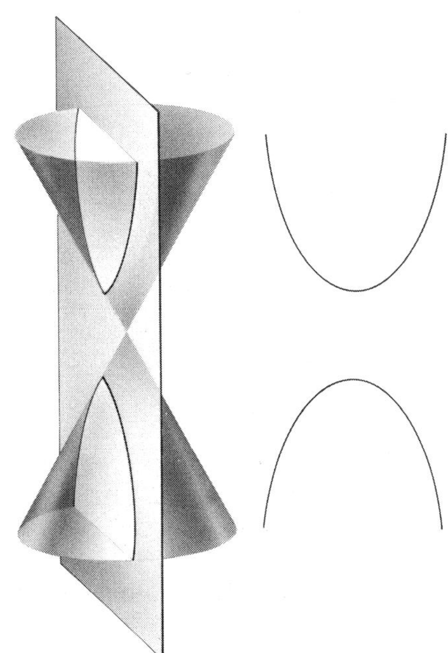

This was Apollonius' analysis. While agreeing with him, we feel that two small modifications should be made. One—admittedly a minor quibble—is to point out that the circle is merely a special case of the more general ellipse, and should be treated as such.

A more substantial addition to the theory consists in pointing out that the preceding analysis did not allow for the possibility that the plane might pass through the vertex of the cone. Consideration of this event shows us that there are three new cases, depending once again on the inclination of the plane to the cone. According to this inclination, the conic section may reduce to a single point (Fig. III.1.6), a straight line (Fig. III.1.7), or a pair of straight lines (Fig. III.1.8). These can be thought of as the degenerate analogues of the ellipse, parabola, and hyperbola respectively.

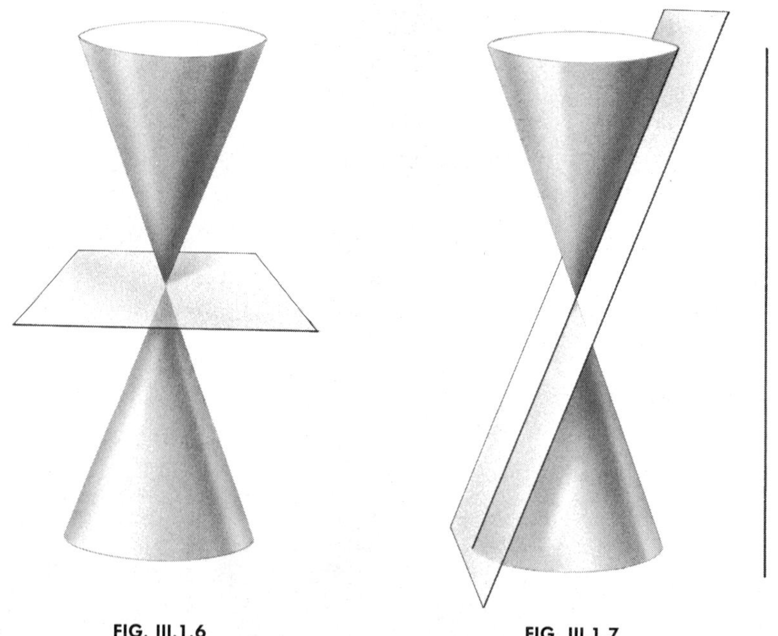

FIG. III.1.6 FIG. III.1.7

FIGURE III.1.6 The plane meets the cone at a single point.

FIGURE III.1.7 The plane intersects the cone along a straight line.

For the sake of completeness, we add that the graph of equation (3.1.1) may also consist of two parallel lines, or have no points at all. We will not attempt to give a geometric interpretation of these cases.

III.1.1 Example. Give the graphs of the following quadratic equations.

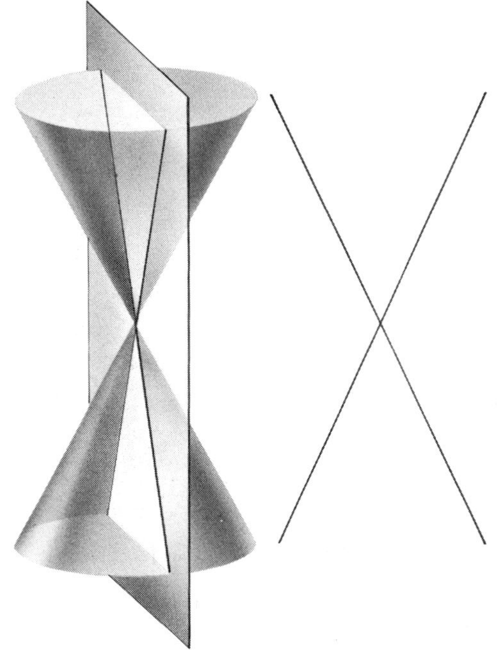

FIGURE III.1.8 The plane intersects the cone along two lines.

(a) $$x^2 + y^2 - 25 = 0$$

(b) $$3x^2 + y^2 - 2y - 8 = 0$$

(c) $$x^2 + 2xy + y^2 + 3x - 4 = 0$$

(d) $$2xy + y^2 + 5x - 10 = 0$$

(e) $$x^2 - xy - 2y^2 + x + 10y - 12 = 0$$

(f) $$x^2 + 2xy + 2y^2 + 5 = 0.$$

We will consider these equations, one at a time. In case (a), we find that the graph is a circle—in fact, the circle with center at the origin, and with a radius of 5 units (see Fig. III.1.9).

In case (b), the graph is seen to be an ellipse (Fig. III.1.10). Case (c) gives a parabola (Fig. III.1.11), while case (d) gives a double curve—a hyperbola (Fig. III.1.12).

Case (e) is seen to be one of the degenerate cases—it consists of the two lines $x + y = 3$, and $x - 2y = -4$ (Fig. III.1.13). Finally, the equation (f) is a very special case—there are no points satisfying the equation.

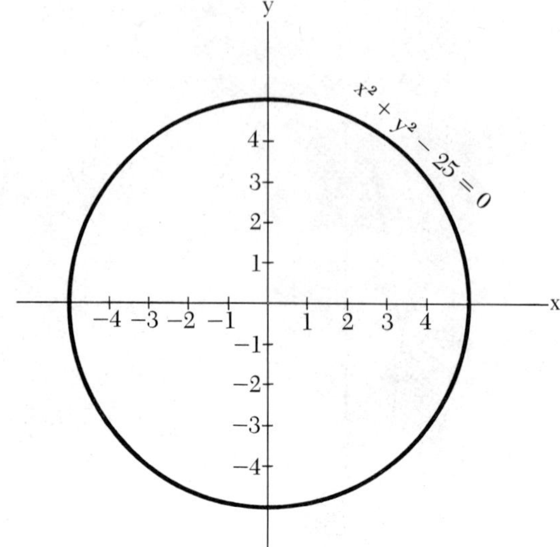

FIGURE III.1.9 *A circle.*

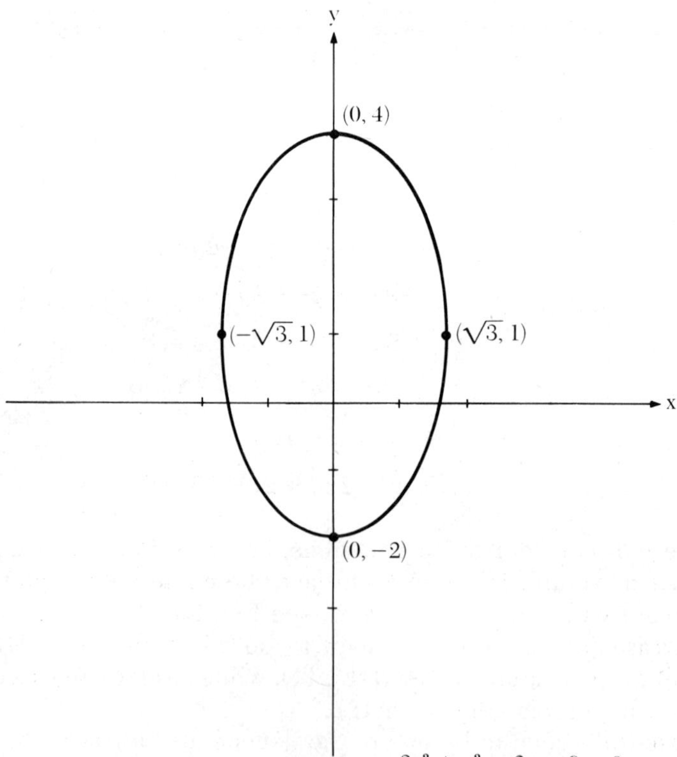

Center of symmetry at (0, 1)

$$3x^2 + y^2 - 2y - 8 = 0$$

$$\text{or } \frac{x^2}{3} + \frac{(y-1)^2}{9} = 1$$

FIGURE III.1.10 An ellipse.

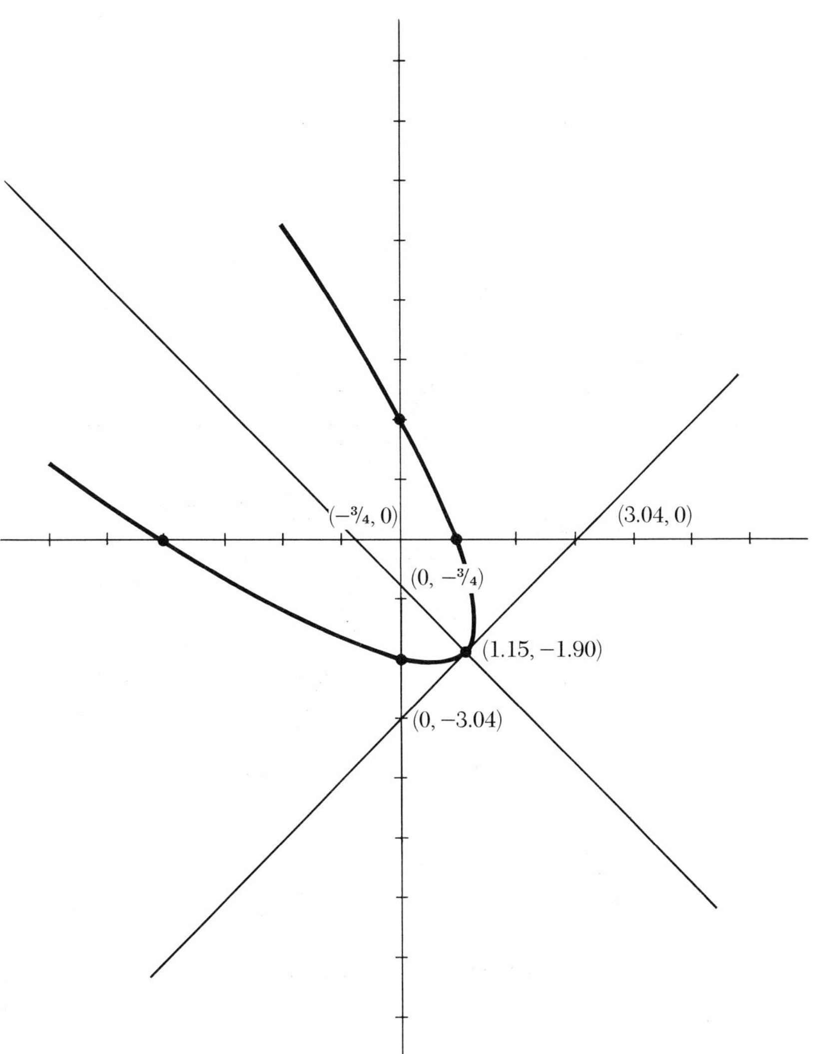

FIGURE III.1.11 A parabola.

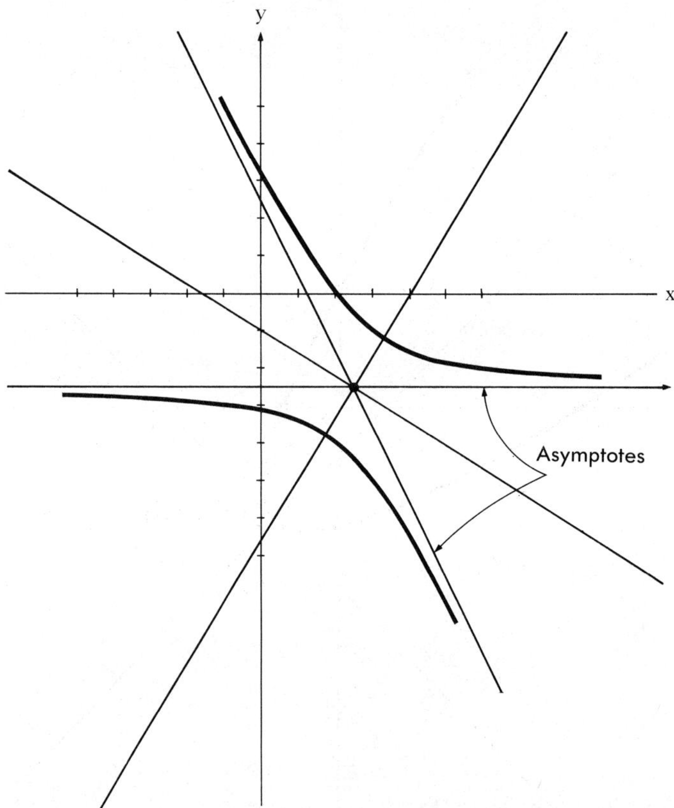

FIGURE III.1.12 A hyperbola.

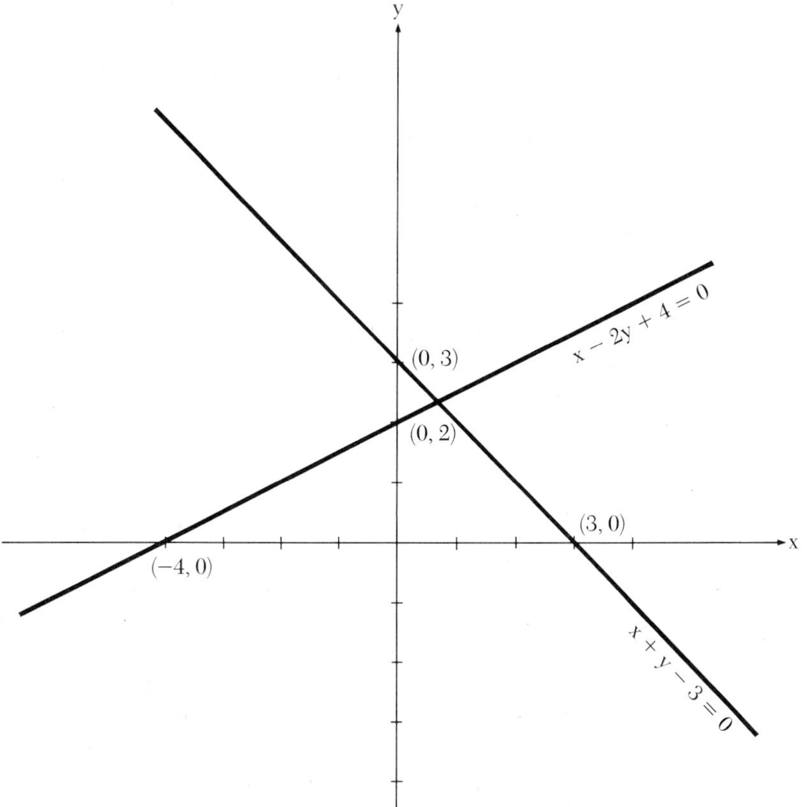

FIGURE III.1.13 Two intersecting lines.

2. CLASSIFICATION OF CONICS

Let us see now how to distinguish among the several cases. We shall assume that the given conic is non-degenerate, i.e., a hyperbola, parabola, or ellipse (possibly a circle). Given the equation (3.1.1), we wish to know which of these it represents.

Considering the quadratic equation

$$Ax^2 + Bxy + Cy^2 + Dx + Ey + F = 0$$

we can solve for y (in terms of x), obtaining (unless $C = 0$)

3.2.1 $$y = \frac{-Bx - E \pm \sqrt{\Delta}}{2C}$$

where the discriminant, Δ, is given by

$$\Delta = (Bx + E)^2 - 4(Ax^2 + Dx + F)C$$

or

$$\text{3.2.2} \qquad \Delta = (B^2 - 4AC)x^2 + (2BE - 4CD)x + E^2 - 4CF.$$

We pointed out, above, that the geometric distinction among these cases lay in the boundedness of the curves: an ellipse is bounded, a parabola is bounded in one direction, while a hyperbola is unbounded in both directions. Let us assume, then, that x is large, and study the behavior of (3.2.1) for large x (either positive or negative).

If $B^2 < 4AC$, then, for large x, the quadratic term will dominate (3.2.2), making the discriminant negative. Thus there will be no solution to (3.2.1). We conclude that the curve is bounded: (3.1.1) represents either an ellipse or a circle.

If $B^2 = 4AC$, then the leading term in (3.2.2) vanishes. Then the discriminant, Δ, will be a linear function of x. The coefficient of x is $2BE - 4CD$. If this is positive, then Δ will be positive for large positive x and negative for large negative x. Thus (3.2.1) has solutions for large positive x, but not for large negative x. The curve is therefore bounded in the direction of negative x. In the contrary case, which $2BE - 4CD$ is negative, the curve will be bounded in the direction of positive x. In either case, we see that the curve is a parabola. (We disregard the case $2BE = 4CD$, which leads to a degenerate curve.)

Finally, suppose that $B^2 > 4AC$. Then, for large x, the leading term in (3.2.2) will make Δ positive, and we conclude that the curve is unbounded in both directions; i.e., it is a hyperbola.

We see, then, that the character of the non-degenerate conic (3.1.1) is directly dependent on the sign of the expression $B^2 - 4AC$. We have, in effect, the cases:

$B^2 < 4AC$	circle or ellipse
$B^2 = 4AC$	parabola
$B^2 > 4AC$	hyperbola

III.2.1 Example. As a check on the validity of this analysis, let us consider the equations of Example III.1.1. Thus, for equation (a),

$$x^2 + y^2 - 25 = 0,$$

we find $A = C = 1$, $B = 0$. Thus $B^2 < 4AC$. This is a circle, as we saw above.

In example (b),

$$3x^2 + y^2 - 2y - 8 = 0,$$

we have $A = 3$, $B = 0$, $C = 1$. Again $B^2 < 4AC$, and this is an ellipse.

For equation (c),

$$x^2 + 2xy + y^2 + 3x - 4 = 0,$$

$A = C = 1$, and $B = 2$. Thus $B^2 = 4AC$, and the curve is a parabola.

For equation (d),

$$2xy + y^2 + 5x - 10y = 0$$

we have $A = 0$, $B = 2$, and $C = 1$, and the curve is a hyperbola.

For (e),

$$x^2 - xy - 2y^2 + x + 10y - 12 = 0,$$

we have $A = 1$, $B = -1$, $C = -2$. Thus $B^2 > 4AC$, and we conclude that this is a (degenerate) hyperbola.

Finally, for (f),

$$x^2 + 2xy + 2y^2 + 5 = 0$$

we have $A = 1$, $B = C = 2$. Thus $B^2 < 4AC$, and if we wish to give this a name, we can call it an imaginary ellipse.

PROBLEMS ON CONIC SECTIONS

1. Draw the graphs of the following quadratic equations:

(a) $3x^2 + 7y^2 - 6x - 12 = 0$.

(b) $y = 3x^2 - 5x + 11$.

(c) $3x^2 + xy - 2y^2 + 4x - 6y - 36 = 0$.

(d) $x^2 + 2xy + y^2 = 0$.

(e) $x^2 - y^2 = 1$.

(f) $x^2 + y^2 = 0$.

(g) $3xy - 4x + y + 15 = 0$.

(h) $x^2 + 4xy + 4y^2 - 2x = 0$.

(i) $3x^2 + 4xy - 6x + 12 = 0$.

(j) $y = x - 3x^2 + 1$

2. Classify the curves of example 1, above, as ellipses (including circles), parabolas, or hyperbolas. Which are degenerate?

3. THE CIRCLE

We have seen above that the equation of a circle will necessarily have the form (3.1.1), with $B^2 < 4AC$. We know, however, that such an equation will normally give us an ellipse; a circle is merely a special case.

Geometrically, we know that a circle is the set of all points lying at a fixed distance, r, from a fixed point, C. The point C is of course the *center* of the circle, while r is the *radius*. Let us suppose that the center, C, has coordinates (a,b). Then we know, from (2.2.4), that the distance from the point (x,y) to (a,b) is the square root of the expression $(x-a)^2+(y-b)^2$. Thus, the condition for (x,y) to be r units from (a,b) is

3.3.1 $$(x-a)^2+(y-b)^2=r^2$$

and we see that (3.3.1) is the general equation of a circle.

III.3.1 Example. Find the equation of the circle with center (3,−5) and radius 2.

Using equation (3.3.1), we have

$$(x-3)^2+(y+5)^2=4$$

or

$$x^2+y^2-6x+10y+30=0.$$

If we are given an equation in the form (3.3.1), there is no problem: we recognize it immediately as the circle with the given center (a,b) and radius r. Unfortunately, however, it is more often given in a form such as (3.1.1). We must learn how to recognize a circle, and how to find its radius and center, in this case.

Starting from equation (3.3.1), and expanding the squares, we obtain the equivalent form

$$x^2+y^2-2ax-2by+(a^2+b^2-r^2)=0.$$

This can then be multiplied by any non-zero constant, k, to obtain

3.3.2 $$kx^2+ky^2-2akx-2bky+k(a^2+b^2-r^2)=0.$$

We conclude that (3.1.1) will be a circle if it has the form (3.3.2). Effectively, this means that we must have

3.3.3 $$A=C\neq 0$$

3.3.4 $$B=0.$$

Conversely, if (3.3.3) and (3.3.4) hold, then we can always find a, b, and r^2, so that the graph of the equation will be a circle. We point out, however, that the circle might degenerate to a single point (corresponding to $r = 0$), or even be imaginary (corresponding to $r^2 < 0$).

In effect, let us suppose that we are given a quadratic equation of the form (3.1.1), in which (3.3.3) and (3.3.4) hold. We may assume that $A = C = 1$, since, if this is not so, we can divide by the common value of A and C. Thus, equation (3.1.1) will have the form

$$x^2 + y^2 + Dx + Ey + F = 0.$$

Writing $a = -D/2$ and $b = -E/2$, this gives us

$$x^2 + y^2 - 2ax - 2by + F = 0,$$

or, by the usual process of completing the square,

$$(x - a)^2 + (y - b)^2 = a^2 + b^2 - F.$$

Then the circle has center (a,b) and radius $r = \sqrt{a^2 + b^2 - F}$.

III.3.2 Example. Find the center and radius of the circle

$$x^2 + y^2 + 4x - 6y + 12 = 0.$$

Completing the square, we see that this may be rewritten in the form

$$(x + 2)^2 + (y - 3)^2 = 1$$

and we conclude that this circle has center $(-2, 3)$ and radius 1.

III.3.3 Example. Find the center and radius of the circle with equation

$$2x^2 + 2y^2 + 12x + 8y + 12 = 0.$$

Here, we divide first by the leading coefficient, to obtain

$$x^2 + y^2 + 6x + 4y + 6 = 0.$$

We then proceed as before, completing the square:

$$(x + 3)^2 + (y + 2)^2 = 7.$$

Thus, the circle has center $(-3, -2)$ and radius $\sqrt{7}$.

III.3.4 Example. Find the center and radius of the circle with the equation

$$x^2 + y^2 - 8x + 6y + 25 = 0.$$

In this case, the equation reduces to

$$(x - 4)^2 + (y + 3)^2 = 0$$

so that the circle has its center at (4,−3). The radius, however, is zero, and we conclude that the graph of this equation is the single point (4,−3), thought of as a "null circle."

Product Transformation Curves. In many cases, it is found that the same process gives two different products—as an example, a chicken coop can produce either eggs or chickens for sale. A slight adjustment in the process will vary the amounts of each product obtained. Generally, the two products must compete for the use of the resources present, so that increasing the amount of one product will cause the other product to decrease. The amounts which can be produced will satisfy some relation, usually given in the form of an equation. The graph of this equation is known as a *product transformation curve.*

III.3.5 Example. A company produces two different grades of brass by variations in the production process. The possible amounts x and y (in hundreds of tons) are related by

$$x^2 + y^2 + 8x + 2y = 32.$$

Draw the product transformation curve. What is the maximum amount which can be produced of the first grade of brass?

It is easy to see that the equation reduces to

$$(x + 4)^2 + (y + 1)^2 = 49$$

so that the product transformation curve is a circle, with center at (−4,−1) and radius 7. More precisely, it is that portion of the circle which lies in the positive quadrant (both x and y non-negative). Clearly, the amount x will be maximized by letting $y = 0$, which gives $x = 2.93$. Thus, at most 293 T of the first grade of brass can be produced. The maximum amount of the second grade, obtained by setting $x = 0$, is 474 T.

PROBLEMS ON THE CIRCLE

1. Each of the following equations represents a circle. In each case, find the center and radius.

(a) $x^2 + y^2 - 4x + 2y + 4 = 0$.

(b) $x^2 + y^2 - 8x + 12y + 10 = 0$.

(c) $x^2 + y^2 + 2x - 8 = 0$.

(d) $x^2 + y^2 - 12x + 4y + 16 = 0$.

(e) $x^2 + y^2 + 16y + 48 = 0$.

2. In each of the following, give the equation of the circle with the given center and radius.

(a) Center (3, 5), radius 7.

(b) Center (−1, 2), radius 4.

(c) Center (−1, 4), radius 6.

(d) Center (0, −3), radius 10.

(e) Center (4, 2), radius 1.

(f) Center (3, −1), radius 3.

(g) Center (10, 5), radius 5.

(h) Center (−1, −1), radius 7.

3. A grape grower can produce either table grapes or wine; the possible amounts x of table grapes (in pounds) and y of wine (in gallons) satisfy

$$x^2 + 10x + y^2 + 200y = 12{,}475.$$

Plot the product transformation curve. What is the maximum possible amount of table grapes? of wine?

4. At a price of x dollars per car, the demand for a certain model is y cars, where

$$x^2 + y^2 + 3000x + 2000y = 17{,}000{,}000.$$

Plot the demand curve for this car. What is the highest price at which sales are still possible?

128 4. THE ELLIPSE

As we saw above, the equation (3.1.1), with $B^2 < 4AC$, represents either an ellipse or a circle. We have given above the conditions for a circle (3.3.3 and 3.3.4), and conclude that (3.1.1) represents an ellipse if $B^2 < 4AC$, *but* (3.3.3) and (3.3.4) do not both hold. Thus, each of the following equations represents an ellipse, as may be verified by plotting their graphs:

$$x^2 - xy + 2y^2 - 4y - 10 = 0$$
$$x^2 + 4y^2 - 16 = 0$$
$$3x^2 + 3xy + y^2 - 6x + 4y - 8 = 0.$$

Let us consider the equation

$$Ax^2 + Cy^2 + F = 0$$

where A and C are positive, $A \neq C$, and F is negative. This is clearly the equation of an ellipse. Dividing by $-F$, and rearranging, we find that this equation may be written in the form

3.4.1 $$\frac{x^2}{a^2} + \frac{y^2}{b^2} = 1.$$

Let us look at the graph of equation (3.4.1), which is shown as Figure III.4.1. As may be seen, the ellipse is inscribed in the rectangle formed by the lines $x = \pm a$, $y = \pm b$. (Geometrically, it is as if

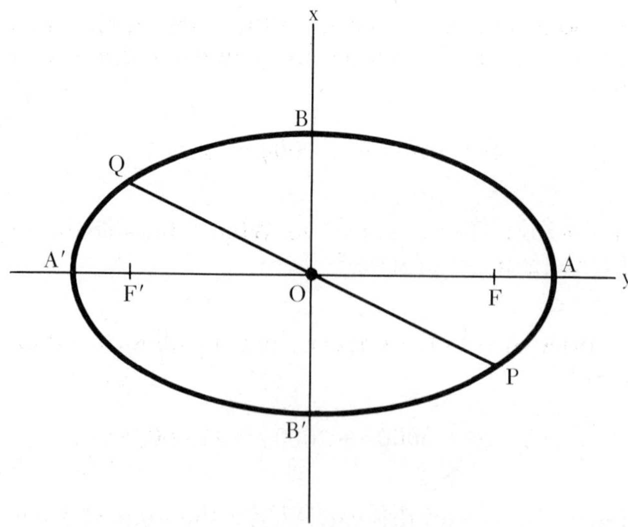

FIGURE III.4.1 An ellipse. AA′ and BB′ are the major and minor axes, while F and F′ are the foci.

the circle $x^2 + y^2 = a^2$ had been compressed in the y-direction.) The origin is the *center* of the ellipse, so called because it has the property of being the mid-point of any chord passing through it, as for example, PQ in the figure.

Chords passing through the center of the ellipse are known as *diameters.* In the figure, the segments PQ, AA', and BB' are all diameters. Now in an ellipse (as opposed to a circle) the diameters vary in length. The longest and shortest of these are always perpendicular to each other, and are called, respectively, the *major axis* and the *minor axis.* In the figure, AA' is the major axis, while BB' is the minor axis. (This is, of course, due to the fact that we have, quite arbitrarily, assumed $a > b$; if, instead, we had assumed $b > a$, then BB' would be the major axis, and AA' the minor axis.) Thus the major axis is $2a$ units, and the minor axis $2b$ units, in length. The numbers a and b are called the *semi-major axis* and *semi-minor axis,* respectively, and correspond, in the ellipse, to the radius of a circle. (If, in effect, $a=b$, the ellipse becomes a circle with radius a.)

Taking the point B (an end-point of the minor axis) as center, we may draw a circle of radius a. This circle intersects the major axis at two points, F and F', with coordinates $(0, f)$ and $(0, -f)$ respectively. It is easily verified that

3.4.2 $$f = \sqrt{a^2 - b^2}.$$

The points F and F' are known as the *foci* of the ellipse, and $2f$, the distance between them, is the *focal distance* of the ellipse.

The ellipse has many interesting properties; it can be shown, for example, that the sum of the distances from a point on the ellipse to the two foci is always equal to the length of the major axis. (This and other properties are left as exercises to the reader.) This gives us a mechanical method for drawing an ellipse of given major axis and focal length: put two thumbtacks a distance $2f$ from each other, then loop a string of length $2a + 2f$ around them, and move a pencil inside the string, keeping the string taut at all times (see Fig. III.4.2).

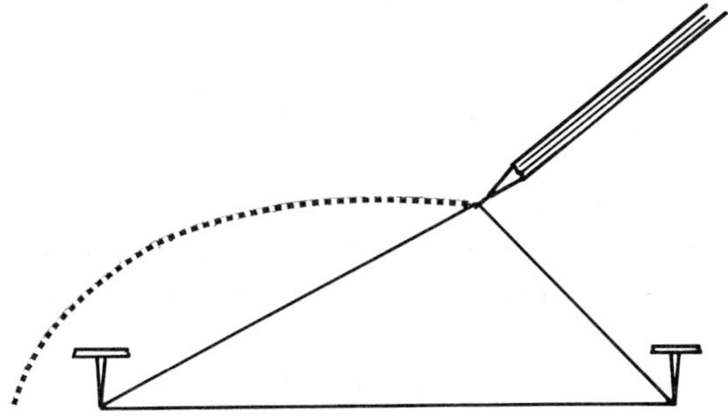

FIGURE III.4.2 Drawing an ellipse. Thumbtacks are placed at the foci, and a string looped around them.

 Another, and most important property, is the fact that the orbit of a planet is always an ellipse, with the sun at one of the foci.

Sometimes, of course, we will run into an ellipse whose center is not at the origin, but at some other point, say (h,k). In the equation

3.4.3 $$\frac{(x-h)^2}{a^2}+\frac{(y-k)^2}{b^2}=1$$

we find that the graph is an ellipse with center (h,k), and with axes of length $2a$ and $2b$, parallel to the x- and y-axes, respectively (Fig. III.4.3). The ellipse is inscribed in the rectangle bounded by the lines $x=h\pm a$, $y=k\pm b$. The foci (again assuming that $a>b$) will be at $y=k$, $x=h\pm f$, with f as in (3.4.2).

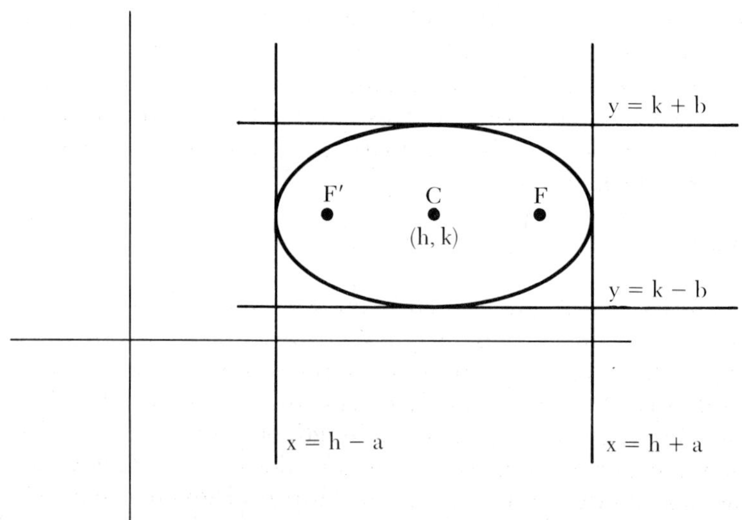

FIGURE III.4.3 The ellipse has center C and foci F, F'.

Given any equation of the form (3.1.1), with A and C both positive, and $B=0$, it is possible to reduce it to the form

$$\frac{(x-h)^2}{a^2}+\frac{(y-k)^2}{b^2}=\delta$$

where δ is 0, 1, or -1. When $\delta=1$, we have an ellipse. For $\delta=0$, we have a "null-ellipse," consisting only of the point (h,k), while, for $\delta=-1$, the ellipse is imaginary (no points satisfy the equation).

III.4.1 ***Example.*** Find the graph of the curve

$$x^2+2y^2-4x+12y+4=0.$$

Using the method of completing the square, we find

$$(x-2)^2+2(y+3)^2=18.$$

This equation can be divided by 18, to give us

$$\frac{(x-2)^2}{18}+\frac{(y+3)^2}{9}=1.$$

Thus the graph of the equation is an ellipse with center at $(2, -3)$, with semi-major axis $\sqrt{18}$, and semi-minor axis 3. The major axis is parallel to the x-axis. The ellipse is inscribed in the rectangle with sides $x=2\pm\sqrt{18}$ and $y=3\pm 3$. If we wish to draw the ellipse by the mechanical method discussed above, we note that $f=3$, and so the foci are at $(-1, -3)$ and $(5, -3)$.

III.4.2 Example. The demand y for eggs in a certain community (in dozens per month) is approximately related to the price x (in cents per dozen) by the equation

$$16x^2+y^2+640x-153{,}600=0.$$

Plot the corresponding demand curve.

We can rewrite the equation in the form

$$\frac{(x+20)^2}{10{,}000}+\frac{y^2}{160{,}000}=1$$

so that the curve is an ellipse with center at $(-20, 0)$. The major axis, along the line $x=-20$, is 800 units, while the minor axis, along the x-axis, is 200 units. Again, we point out that it is only that part of the curve which lies in the positive quadrant that is of importance. The maximal demand, obtained by setting $x=0$ (i.e., giving the eggs away) will be about 392 dozen per month; on the other hand, if the price is increased to 80¢ per doz., demand will drop to 0.

PROBLEMS ON THE ELLIPSE

1. In each of the following, give the center, major axis, minor axis, and focal length of the given ellipse.

(a) $3x^2+12y^2-18x+72y+6=0.$

(b) $9x^2+4y^2+18x-16y+16=0.$

(c) $x^2+16y^2-6x-7=0.$

(d) $25x^2+4y^2-50x+8y-71=0.$

(e) $49x^2+16y^2-294x+64y-2631=0.$

2. Define the *eccentricity* of an ellipse by the equation $f = ae$, where $2f$ is the focal length, and $2a$ is the major axis of the ellipse. (Note that the definition must be slightly modified when $b > a$, so that the major axis is $2b$.) Find the eccentricity of each of the ellipses in problem 1, above.

3. Consider the ellipse given by equation (3.4.1), with $a > b$. Let F be the focus $(f, 0)$, where f is given by (3.4.2). Consider the line L, $x = a^2/f$. Show that, if P is an arbitrary point on the ellipse, the ratio of its distance from the focus F, to its distance from the line L, is constant. What is, indeed, this ratio? The line L is called a *directrix* of the ellipse. Does the ellipse have any other directrix?

4. Show that, if P is an arbitrary point on the ellipse (3.4.1), the sum of the distances of P from the two foci, F and F', is always equal to the major axis of the ellipse.

5. A shoe company can produce different amounts, x and y, of men's and women's shoes by varying its process. Possible amounts are given by

$$9x^2 + 16y^2 + 180x + 160y = 118{,}700.$$

Plot the product transformation curve for this company.

6. At a price of x cents a gallon, the supply of apple cider is y gallons, where x and y satisfy

$$64x^2 + 9y^2 - 19{,}200x + 3600y - 3{,}960{,}000 = 0$$

for $x \leqslant 150$, and $y = 600$ for $x \geqslant 150$.

Plot the supply curve for cider.

5. THE PARABOLA

Let us consider now the parabola. We saw that this is represented by (3.1.1) with $B^2 = 4AC$. (As with the circle or ellipse, the parabola may become degenerate, or even imaginary, but we shall not worry about that.) We shall consider here the special case in which $B = 0$. The equation $B^2 = 4AC$ means then that either A or C (but not both) will vanish.

We take, first of all, the equation

$$y^2 = 2px \tag{3.5.1}$$

where p is a parameter (i.e., a constant which can have any value desired). The graph of (3.5.1) is shown in Figure III.5.1 and is clearly a

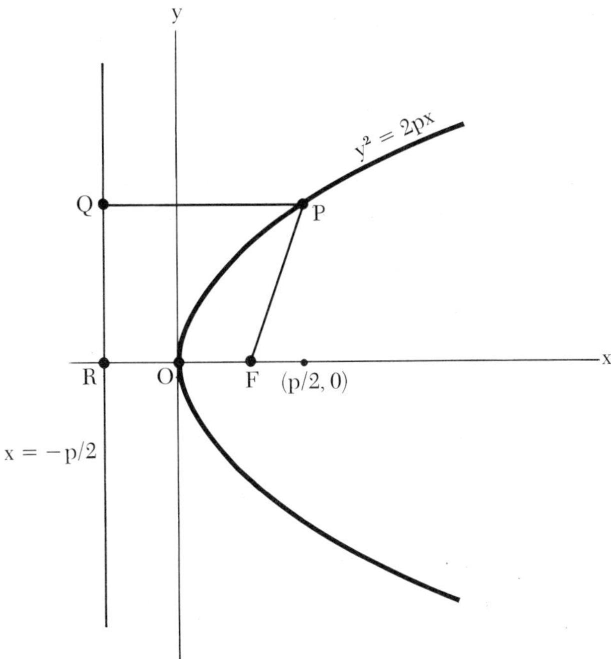

FIGURE III.5.1 A parabola. O is the vertex, F the focus, and OF the axis, while QR is the directrix. Note that PF = PQ.

parabola. We note that, since $(-y)^2 = y^2$, the x-axis is an axis of symmetry for the parabola. It is, indeed, called the *axis* of the parabola, and (if we think of the parabola as an extremely elongated ellipse) it corresponds to the major axis of an ellipse.

The point of intersection of the parabola with its axis (in this case, the origin) is called the *vertex* of the parabola. The point $(p/2, 0)$ is the *focus*, while the line $x = -p/2$ is the *directrix* of the parabola. These two have the important property that the distance of an arbitrary point on the parabola from the focus is equal to its perpendicular distance from the directrix. In effect, if $P = (x,y)$ is a point, its distance from the focus is $PF = \sqrt{(x - p/2)^2 + y^2}$, while its perpendicular distance from the directrix is $PQ = x + p/2$ (see Fig. III.5.1). It is easy to prove, then, that the equation

$$x + \frac{p}{2} = \sqrt{\left(x - \frac{p}{2}\right)^2 + y^2}$$

is equivalent to (3.5.1).

In general, of course, the parabola will not appear in the simple form of equation (3.5.1). If we have $A = B = 0$ in equation (3.1.1), we obtain something of the form

$$Cy^2 + Dx + Ey + F = 0.$$

We can then divide by C, and completing the square, reach an equation

3.5.2 $$(y-k)^2 = 2p(x-h).$$

It is easy now to plot the graph of this equation: we see that it will be a parabola with vertex at (h,k); its focus is the point $(h+p/2, k)$, and its directrix has the equation $x = h - p/2$.

III.5.1 *Example.* Find the graph of the equation

$$3y^2 + 15x - 6y + 12 = 0.$$

In this case we divide by 3, to obtain

$$y^2 + 5x - 2y + 4 = 0.$$

Completing the square, this is

$$(y-1)^2 = -5(x+3/5).$$

We conclude that the parabola has its vertex at $(-3/5, 1)$, and its axis is the line $y = 1$. We have $p = -5/2$, and so the focus is at $(-37/20, 1)$ while the directrix is the line $x = 13/20$.

III.5.2 *Example.* Find the equation of the parabola with focus (5, 2) and directrix $x = 1$.

We remember that a point on the parabola must be equidistant from the focus and the directrix. Thus

$$(x-5)^2 + (y-2)^2 = (x-1)^2$$

and, after expansion and collection of terms, this reduces to

$$y^2 - 8x - 4y + 28 = 0.$$

An alternative method would be to note that the vertex must lie half-way between the focus and the directrix, i.e., at (3, 2). Then the value of the parameter p is equal to the distance from the focus to the directrix, or 4. Thus (3.5.2) becomes

$$(y-2)^2 = 8(x-3)$$

which reduces to the same as above.

Quite frequently the parabola appears in the form of an equation

3.5.3 $$y = ax^2 + bx + c.$$

In this case, we find that we have (3.1.1), with $B = C = 0$. The main difference between this parabola and that in (3.5.2) is that, for (3.5.3), the directrix is horizontal (parallel to the x-axis). In effect, the curve has been rotated through a right angle. It is possible, again by completing the square, to express (3.5.3) in a form similar to (3.5.2), with the sole difference that the roles of x and y are interchanged.

III.5.3 Example. Find the focus, directrix, and vertex of the parabola with equation

$$y = 3x^2 + 5x - 1.$$

Dividing by 3, and completing the square, this equation reduces to

$$\left(x + \frac{5}{6}\right)^2 = \frac{1}{3}\left(y + \frac{37}{12}\right)$$

Thus the vertex is the point $(-5/6, -37/12)$. We have here $p = 1/6$, and so the focus is $(-5/6, -3)$. Finally, the directrix is $y = -19/6$.

III.5.4 Example. Find the equation of the parabola with focus (2, 5) and directrix $y = 1$.

Comparing this with Example III.5.2, we see that it differs only in that the roles of x and y have been interchanged. We conclude that the equation must be

$$x^2 - 8y - 4x + 28 = 0$$

or, solving for y,

$$y = \frac{1}{8}x^2 - \frac{1}{2}x + \frac{7}{2}.$$

III.5.5 Example. A poultry farmer finds that, at a price of 40¢ per dozen, he can sell 8000 dozen eggs per month, whereas, at 30¢ per dozen, he can sell 10,000 dozen per month. He has fixed costs of \$800 per month, plus variable costs of 20¢ per doz. Assuming that the relation between demand and price, and that between costs and production, are both linear, find the farmer's profits as a function of price.

We use here the equation: profits equals revenue minus costs. In turn, revenue equals price times amount sold. Using the two-point formula, we find that the relation between demand d (in dozens per month) and price x (in cents per dozen) is

$$d = 16{,}000 - 200x$$

and so revenue, in cents per month, is

$$r = dx = 16{,}000x - 200x^2.$$

Let us assume that the farmer produces just enough eggs to sell at the price x. The cost of these d dozen, in cents, is

$$c = 80{,}000 + 20d$$

or, substituting the value of d,

$$c = 400{,}000 - 4000x.$$

The farmer's profit, then, in dollars per month, is given by $y = (r - c)/100$, or

$$y = -2x^2 + 200x - 4000$$

where x is still in cents per dozen.

This equation can be rewritten

$$y - 1000 = -2(x - 50)^2$$

so that the curve is a parabola with vertex at (50, 1000), and vertical axis. The parabola opens downward, so that the farmer's maximum profit is \$1000 per month, obtained by setting $x = 50$¢ per doz. We note, however, that the maximum is quite flat: even at 40¢ or 60¢ per doz., the farmer would still make a profit of \$800 per month.

PROBLEMS ON THE PARABOLA

1. Find the focus, vertex, and directrix of each of the following parabolas.

(a) $y^2 + 6x + 8y - 12 = 0.$

(b) $y = 5x^2 + 4x - 12$

(c) $y^2 - 12x + 2y - 4 = 0.$

(d) $y = 14x^2 - 3x + 15.$

(e) $y^2 + 2x - 4y - 8 = 0.$

2. An apartment house owner finds that, at \$100 per month, he can rent all 30 of his apartments, but, at \$200 per month, he can only rent 10 of them. Assuming demand to be a linear function of price, find revenue as a function of price. What rent should be charged, so as to maximize rental revenues?

3. On a certain farm, the cost of producing x bushels of wheat is y dollars, where

$$y = .001\,x^2 + .3\,x + 20.$$

Assuming that the wheat can be sold for \$1 a bushel, find the farm's break-even point. What is the maximum amount of wheat that can be produced at a profit?

6. THE HYPERBOLA

We now study the hyperbola. As we know, this corresponds to equation (3.5.1) with $B^2 > 4AC$. Once again, we shall limit our treatment to some special cases; the first of these occurs when $B = 0$. This means that the product AC must be negative, i.e., A and C must be of opposite sign.

Consider the equation

3.6.1 $$\frac{x^2}{a^2} - \frac{y^2}{b^2} = 1$$

whose graph is shown in Figure III.6.1. The equation is quite similar to (3.4.1), but the difference in sign gives us a very different curve.

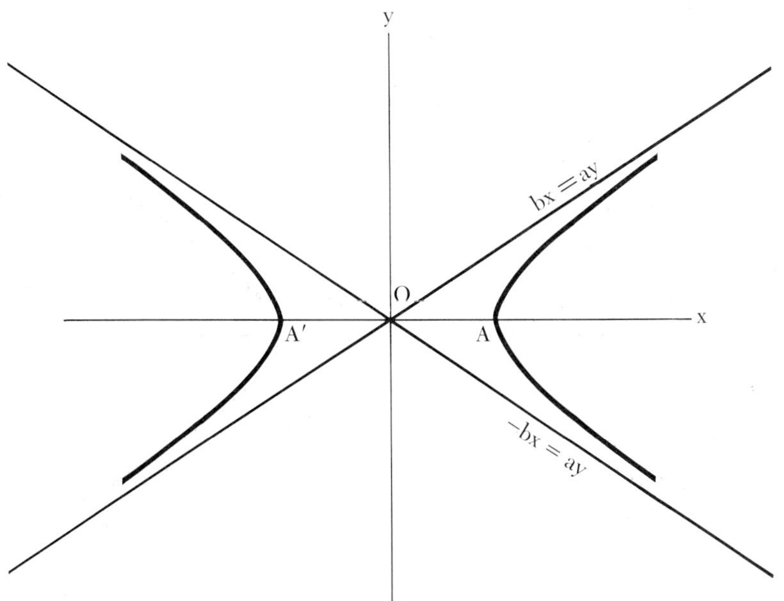

FIGURE III.6.1 A hyperbola, with asymptotes.

As in the case of the ellipse (3.4.1), we shall say that the origin is the center of the hyperbola (3.6.1). This corresponds to the symmetry properties of the curve, and also to the fact that the origin bisects every chord through it. We note that the curve is symmetric with respect to both coordinate axes.

The hyperbola intersects the x-axis at the two points $A = (a,0)$ and $A' = (-a,0)$. The segment AA' is the *transverse axis* of the hyperbola; its length is, of course, $2a$.

Consider, next, the two lines

3.6.2
$$\frac{x}{a} = \pm\frac{y}{b}$$

which pass through the center of the hyperbola. It may be seen that the curve (3.6.1) will never meet these lines; however, it seems to approach them as the coordinates increase. The lines (3.6.2) are known as the *asymptotes* of the hyperbola.

Look, finally, at the equation

3.6.3
$$\frac{y^2}{b^2} - \frac{x^2}{a^2} = 1.$$

This is also a hyperbola, closely related to (3.6.1); it is known as the *conjugate* of (3.6.1). We note that it has the same center and asymptotes. Its transverse axis is the segment BB' (Fig. III.6.2), where $B = (0,b)$ and $B' = (0,-b)$. This is known as the *conjugate* axis of

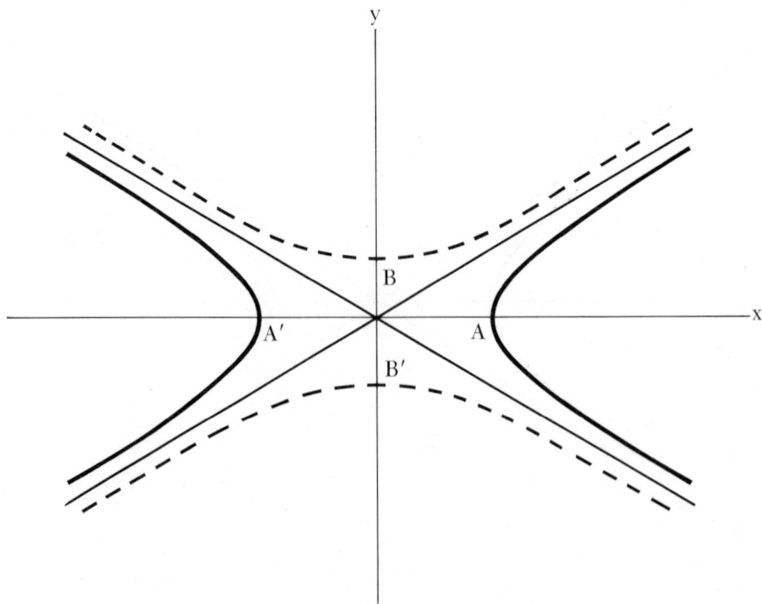

FIGURE III.6.2 Two conjugate hyperbolas. The asymptotes are the same.

(3.6.1). (Similarly, the transverse axis of (3.6.1) is the conjugate axis of (3.6.3).)

III.6.1 Example. Find the transverse axis, conjugate axis, and asymptotes of the hyperbola

$$5x^2 - 7y^2 + 35 = 0.$$

Dividing by 35, we find that

$$-\frac{x^2}{7} + \frac{y^2}{5} = 1$$

This equation is of the form (3.6.3), and thus the transverse axis is along the y-axis. Since $b^2 = 5$, the transverse axis goes from $(0, \sqrt{5})$ to $(0, -\sqrt{5})$. The conjugate axis goes from $(\sqrt{7}, 0)$ to $(-\sqrt{7}, 0)$. Finally, the asymptotes are $x/\sqrt{7} = \pm y/\sqrt{5}$. The graph is shown in Figure III.6.3.

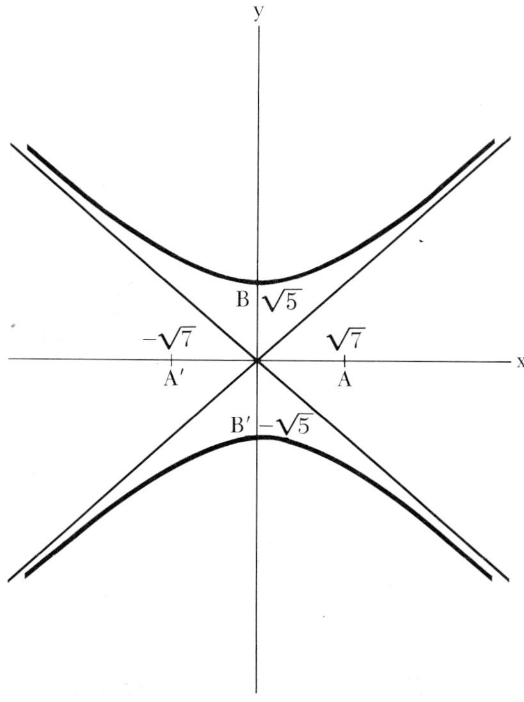

FIGURE III.6.3 The hyperbola $5x^2 - 7y^2 + 35 = 0$, with asymptotes. AA' and BB' are the conjugate and transverse axes, respectively.

In the more general case, we are given the equation

$$Ax^2 + Cy^2 + Dx + Ey + F = 0$$

where A and C have opposite signs. This equation can be reduced to the form

3.6.4
$$\frac{(x-h)^2}{a^2} - \frac{(y-k)^2}{b^2} = \delta$$

where δ is 1, -1, or 0. The hyperbola (3.6.4) will have its center at (h, k), and asymptotes

$$\frac{x-h}{a} = \pm\frac{y-k}{b}.$$

The transverse axis will be parallel to the x-axis if $\delta = 1$ (Fig. III.6.4), or to the y-axis if $\delta = -1$. The conjugate axis will of course be perpendicular to the transverse axis. In either case, both axes will pass through the center. In case $\delta = 0$, we find that the "hyperbola" reduces to its asymptotes; this is a degenerate case.

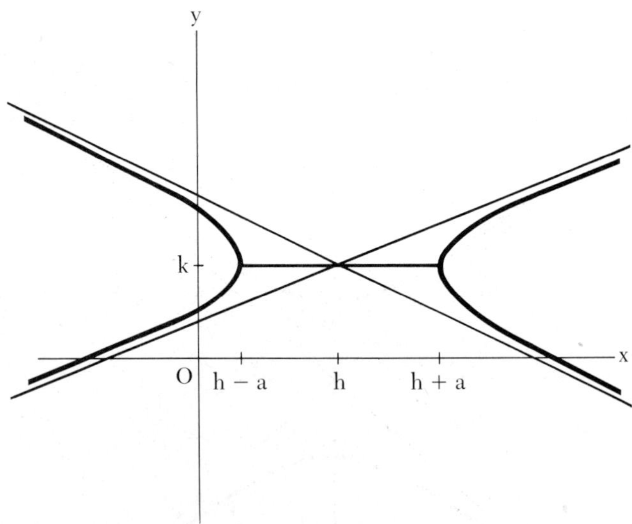

FIGURE III.6.4 The hyperbola, $\frac{(x-h)^2}{a^2} - \frac{(y-k)^2}{b^2} = 1$, with asymptotes and transverse axis.

III.6.2 Example. Find the asymptotes, transverse axis, and conjugate axis of the curve

$$x^2 - 4y^2 - 2x - 24y - 19 = 0.$$

Completing the squares, we have

$$(x-1)^2 - 4(y+3)^2 = -16.$$

We divide this by 16, and obtain

$$\frac{(x-1)^2}{16} - \frac{(y+3)^2}{4} = -1$$

so that the hyperbola's center is at $(1, -3)$. The asymptotes will be the two lines

$$\frac{x-1}{4} = \pm\frac{y+3}{2}$$

The transverse axis will be parallel to the y-axis, and is therefore the segment joining the points $(1, -1)$ and $(1, -5)$, while the conjugate axis lies between the points $(5, -3)$ and $(-3, -3)$.

III.6.3 Example. A dairy processing company finds that, if it pays farmers a price of x cents per gallon of milk, the production y, in gallons of milk, will be given by

$$625x^2 - y^2 + 12{,}500x - 937{,}500 = 0.$$

Plot the milk supply curve for this region.

It is easy to see that this equation reduces to

$$\frac{(x+10)^2}{1600} - \frac{y^2}{1{,}000{,}000} = 1.$$

so that the curve is a hyperbola with center $(-10, 0)$ and asymptotes $y = \pm\, 25(x-10)$. (Of course, only part of the curve is relevant, and only the asymptote $y = 25(x-10)$ is relevant.) The vertex is at $(30, 0)$, and we see that the company must pay at least 30¢ per gal. if there is to be any production.

Another case of the hyperbola occurs when, in equation (3.1.1), $A = C = O$. Once again we start with the simplest example, the equation

3.6.5 $$xy = c$$

where c is an arbitrary constant. This is shown in Figure III.6.5. We note that, in this case, the origin is still the center of the hyperbola; this time, however, the asymptotes are the two coordinates axes.

More generally, the equation

$$Bxy + Dx + Ey + F = 0$$

can be reduced to

3.6.6 $$(x-h)(y-k) = c$$

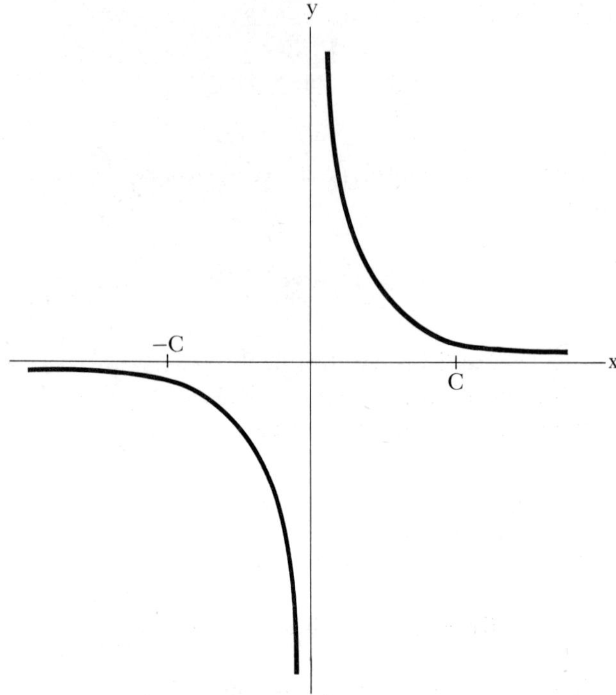

FIGURE III.6.5 The hyperbola $xy = c$.

which is shown in Figure III.6.6. This is almost the same as the graph of equation (3.6.5), except that the center is at (h,k). The asymptotes are the lines $x = h$ and $y = k$.

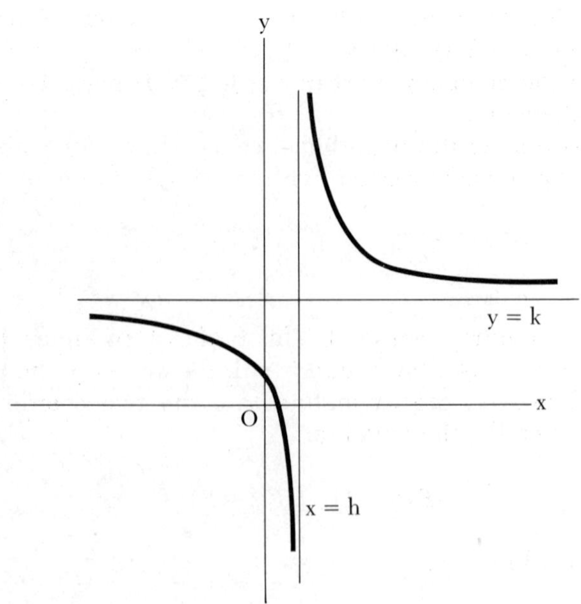

FIGURE III.6.6 The hyperbola $(x - h)(y - k) = c$. Note the asymptotes.

The asymptotes divide the plane into four quadrants. In (3.6.5), we note that the curve lies entirely in the first (i.e., upper right) and third (lower left) quadrants. However, if the right side of (3.6.5) is negative, the curve will lie entirely in the second (upper left) and fourth (lower right) quadrants. An obvious example is the equation

$$xy = -1$$

shown in Figure III.6.7.

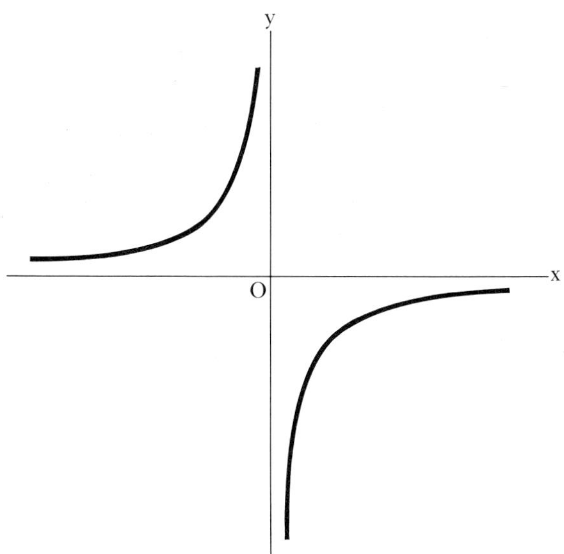

FIGURE III.6.7 The hyperbola $xy = -1$.

III.6.4 Example. Give the graph of the equation

$$xy + 5x - 3y + 12 = 0.$$

It is easy to see that this equation can be written in the form

$$(x - 3)(y + 5) = 3.$$

Thus the curve has center (3, –5); the asymptotes are $x = 3$ and $y = -5$. As 3 is positive, the curve lies in the first and third quadrants determined by its asymptotes (see Fig. III.6.8).

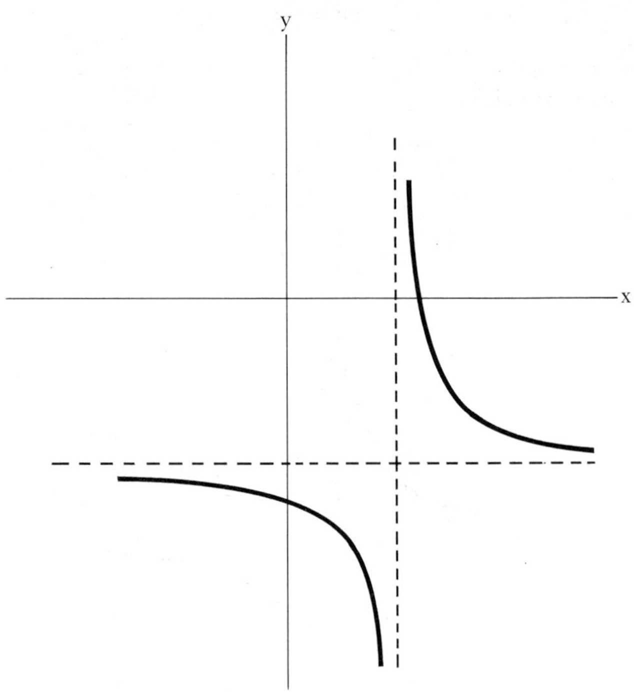

FIGURE III.6.8 The hyperbola $xy + 5x - 3y + 12 = 0$.

III.6.5 Example. Give the graph of the equation

$$xy + 7y + 4 = 0.$$

In this case, the equation reduces to

$$(x + 7)y = -4.$$

Thus the center is at (–7, 0). The asymptotes are $x = -7$, $y = 0$, and the curve lies in the second and fourth quadrants (Fig. III.6.9).

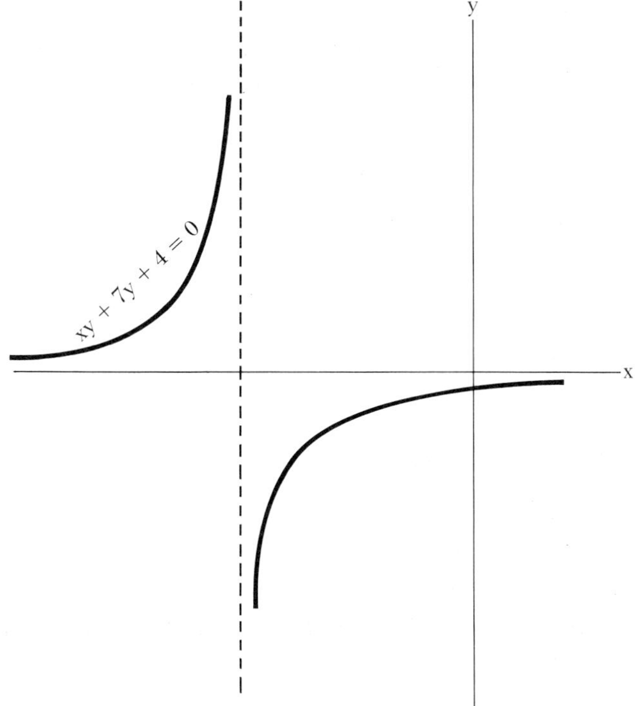

FIGURE III.6.9 The hyperbola $xy + 7y + 4 = 0$.

III.6.5 Example. Find the graph of the equation

$$xy + 7x - 3y - 21 = 0.$$

In this case, the equation becomes

$$(x - 3)(y + 7) = 0$$

and we find that the hyperbola degenerates: the equation is satisfied by every point on either of the two lines $x = 3$ and $y = -7$. Thus the hyperbola coincides with its asymptotes.

III.6.7 Example. The relation between the price x (in cents per lb.) of a certain product, and the amount y that can be sold (in tons) is given by

$$xy + 100x + 20y = 8000.$$

Plot the demand curve for this product.

It may be seen that the equation reduces to

$$(x + 20)(y + 100) = 10{,}000$$

so that the curve is a hyperbola with asymptotes $x = -20$ and $y = -100$. It may be seen that the maximum demand for this product (at $x = 0$) is 400 T; on the other hand, no sales are possible if the price exceeds 80¢ per lb.

PROBLEMS ON THE HYPERBOLA

1. Find the center and asymptotes of the following hyperbolas. Plot the curves on graph paper.

(a) $4x^2 - y^2 + 16x + 2y - 10 = 0$.

(b) $9x^2 - 16y^2 + 12x - 64y + 25 = 0$.

(c) $x^2 - 4y^2 - 8x + 24y + 64 = 0$.

(d) $25x^2 - 4y^2 - 50x + 8y - 71 = 0$.

(e) $25x^2 - 4y^2 - 50x + 8y - 129 = 0$.

2. Give the conjugate hyperbola of each of the hyperbolas given in Problem 1, above.

3. Give the center and asymptotes of each of the following hyperbolas.

(a) $3xy - 6x + 12y + 12 = 0$.

(b) $xy + 4x - 2y + 1 = 0$.

(c) $xy - 2x + 8y = 0$.

(d) $xy + 3x - 4y + 4 = 0$.

(e) $xy - 6x + 8 = 0$.

4. Given the hyperbola (3.6.1), define the number f by

$$f = a^2 + b^2$$

The points $F = (f, 0)$ and $F' = (-f, 0)$ are the foci of the hyperbola. Prove that, if P is an arbitrary point on the hyperbola, the difference between the distances from P to the foci F and F' is equal to the length of the transverse axis.

5. Given the hyperbola (3.6.1), define the eccentricity e by the equation $f = ae$, where f is as in Problem 4, above. Show that the line $x = a/e$ is a *directrix* for the hyperbola, having the property that, if P is an arbitrary point on the hyperbola, then the ratio of the distances of P from F and from the directrix is constant.

6. At a price of x cents per gallon, demand for gasoline is y gallons, where x and y satisfy

$$900x^2 - y^2 - 144{,}000x - 200y + 5{,}760{,}000 = 0$$

if $x \leq 80$, and $y = 0$ if $x \geq 80$. Plot the demand curve for gasoline in this region.

7. A housewife has \$50 to spend on food. She will, first of all, buy 10 lb. of hamburger. Moreover, of whatever money she has left (after this 10 lb. has been paid for), she will spend exactly one third on hamburger. If the price of hamburger is x cents per lb., how many pounds will she buy? Plot the graph of this relation.

7. SYSTEMS OF EQUATIONS

We shall study here systems of two equations, one linear and one quadratic, in the two variables x and y. We point out that, as with single quadratic equations in one variable, there may be two, one, or no solutions. The only exception to this rule occurs when the quadratic equation degenerates; it is then possible that the line corresponding to the linear equation coincides with one of the two lines of the quadratic. There will then be an infinity of solutions.

Given the system

3.7.1 $$Ax^2 + Bxy + Cy^2 + Dx + Ey + F = 0$$

3.7.2 $$ax + by = c$$

we would like to find pairs (x, y) which satisfy both equations simultaneously.

The method of elimination, which was so useful in dealing with systems of linear equations, turns out to be rather impractical here. A better method is to solve (3.7.2) for one of the variables in terms of the other, and then substitute this in (3.7.1). This gives a quadratic equation in the remaining variable, which can be solved as in Appendix I. The solution obtained is then substituted in (3.7.2) to find the other variable.

III.7.1 Example. Solve the system of equations

$$x^2 + 2xy + y^2 + 3x + 5 = 0$$

$$3x + y = 7.$$

From the linear equation, we obtain

$$y = 7 - 3x$$

which is substituted in the first equation, giving

$$x^2 + 2x(7 - 3x) + (7 - 3x)^2 + 3x + 5 = 0$$

or, simplifying,

$$4x^2 - 25x + 54 = 0.$$

By the usual quadratic formula, this gives us

$$x = \frac{25 \pm \sqrt{-239}}{8}$$

and we conclude that the system has no solutions: the line does not touch the curve, which is in this case a parabola (see Fig. III.7.1).

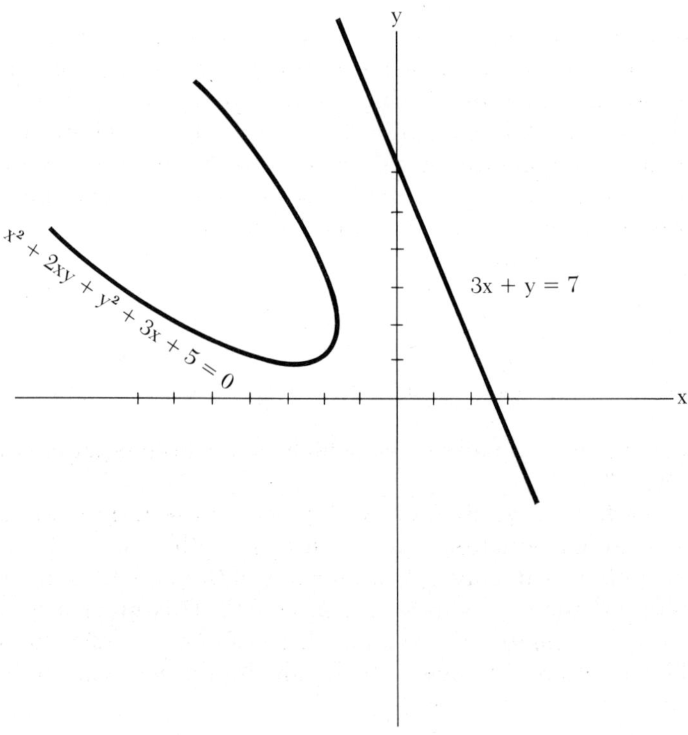

FIGURE III.7.1 The solution of Example I.13.1.

III.7.2 Example. Solve the system

$$x^2 - 3y^2 + 7x - 4y - 10 = 0$$

$$2x - 3y = 0.$$

As before, we solve the linear equation for y, obtaining

$$y = \frac{2x}{3}$$

and substitute this in the first equation, to get

$$x^2 - 3\frac{(2x)^2}{9} + 7x - 4\frac{2x}{3} - 10 = 0$$

or, equivalently,

$$x^2 - 13x + 30 = 0.$$

Using the quadratic formula, this gives us

$$x = \frac{13 \pm \sqrt{169 - 120}}{2} = \frac{13 \pm 7}{2}$$

or

$$x = 3 \text{ or } 10.$$

We now use these values of x to obtain y (we have already solved for y in terms of x). For $x = 3$, we find $y = 2$, and, for $x = 10$, we find $y = 20/3$. Thus, the system has the two solutions: (3, 2) and (10, 20/3). This is shown in Figure III.7.2.

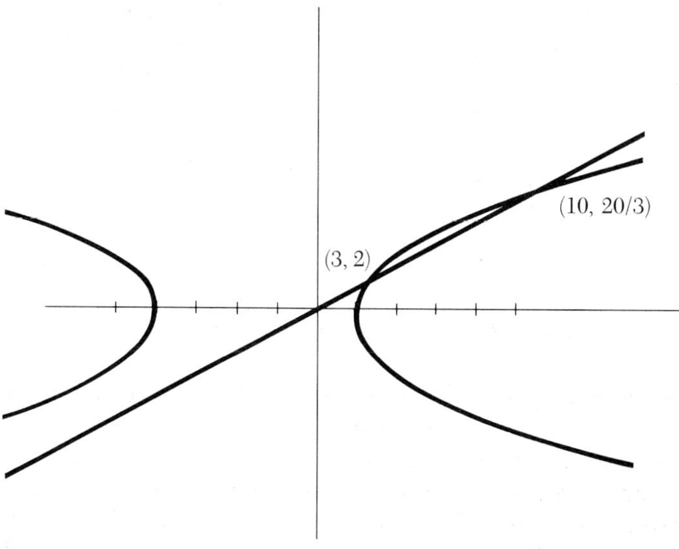

FIGURE III.7.2 Solution of Example I.13.2.

III.7.3 Example. An oil company finds that the cost y (in dollars) of producing x gallons of gasoline is given by the equation

$$x^2 - 400y^2 + 100{,}000{,}000 = 0.$$

Assuming that it can sell the gasoline at 6¢ per gal., what is the company's break-even point?

It may be seen that the cost curve is a hyperbola with center at the origin; the asymptotes are $x = \pm 20y$, and the vertices are at $x = 0$, $y = \pm 500$. The profit curve, $y = 0.06x$, will start below the cost curve, but will cross it and lie above it for large x. We are looking for the point of intersection of the two curves. Substituting $y = .06x$ in the first equation, we have

$$x^2 - 1.44x^2 = -100{,}000{,}000$$

or (approximately) $x = \pm 15{,}100$. Clearly the negative root is extraneous, and we find that the company must sell at least 15,100 gallons to break even.

III.7.4 Example. At a price of x dollars per pound, demand y (in pounds) for a certain chemical compound is given by the equation

$$60x^2 + y^2 + 2500x + 200y = 625{,}000.$$

At that same price, supply z (also in pounds) is given by

$$14x - z = 200.$$

Find the equilibrium price for this compound.

Equilibrium occurs when supply equals demand. Setting $z = y$ in the above equations, we find that $x = 50$ or $-3125/64$. The negative solution is clearly extraneous, and so the equilibrium price is $50 per lb. At this price, supply and demand are both 500 lb.

More generally, we may consider the problem of solving a system of two quadratic equations:

$$A_1x^2 + B_1xy + C_1y^2 + D_1x + E_1y + F_1 = 0$$

$$A_2x^2 + B_2xy + C_2y^2 + D_2x + E_2y + F_2 = 0.$$

The obvious method would be to solve one of these equations for x in terms of y, and then substitute in the other. Unfortunately, this does not give rise to a quadratic, but rather, to a biquadratic (or fourth-degree) equation. In effect, two conics can meet at as many as four different points (Fig. III.7.3). While it is possible to solve such a system, this is considerably more complicated, and we shall not at-

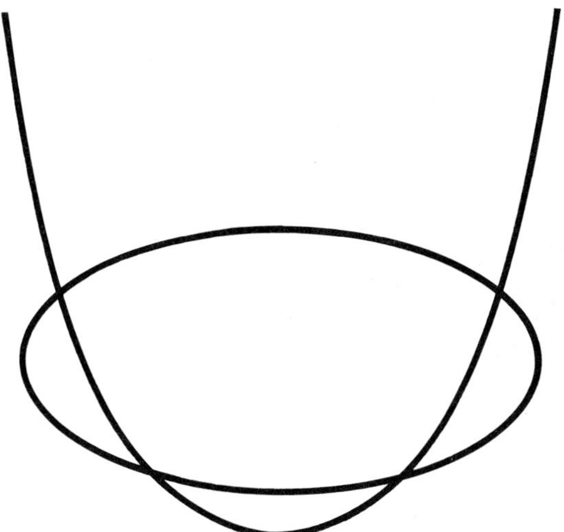

FIGURE III 7.3 Two conics can meet at as many as four points.

tempt it here. (See Appendix I for some details on the solution of biquadratic equations).

A special case occurs, however, when the two conics are both circles. For we know, from elementary geometry, that a circle is determined by three points. It follows that two distinct circles can meet, at most, at two points (Fig. III.7.4). It seems, therefore, reasonable to believe that such a system will be easier to handle.

Consider, then, the system

3.7.3 $$x^2 + y^2 + D_1x + E_1y + F_1 = 0$$

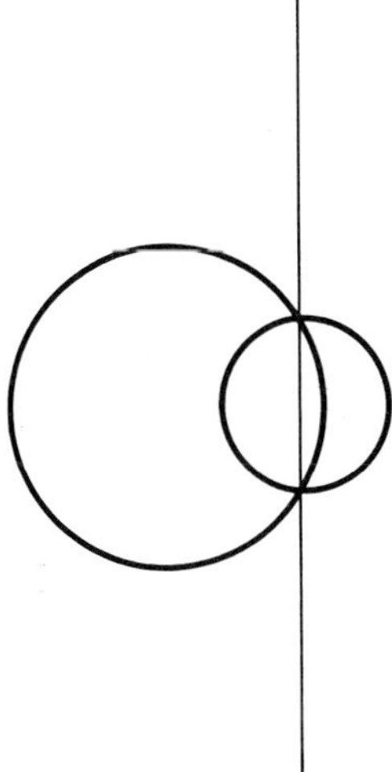

FIGURE III.7.4 The common chord of two circles.

3.7.4 $$x^2 + y^2 + D_2x + E_2y + F_2 = 0.$$

Subtracting (3.7.4) from (3.7.3), we obtain the linear equation

3.7.5 $$(D_1 - D_2)x + (E_1 - E_2)y = F_2 - F_1.$$

Now, what does (3.7.5) represent? We know that (unless $E_1 = E_2$ and $D_1 = D_2$) this equation represents a straight line. Moreover, (3.7.5) is satisfied by any pair (x, y) which satisfies both (3.7.3) and (3.7.4). We conclude that, if (3.7.3) and (3.7.4) represents intersecting circles, then (3.7.5) must be their common chord: the line determined by their two points of intersection. Through a slightly more subtle argument we see that if (3.7.3) and (3.7.4) are tangent circles (i.e., if they have only one point in common) then (3.7.5) must be their common tangent. It must pass through their point of tangency, but cannot meet either of the circles at any other point. It is not very clear what (3.7.5) represents when the two circles (3.7.3) and (3.7.4) fail to intersect.

Once equation (3.7.5) has been obtained, it is possible to solve the system consisting of (3.7.3) and (3.7.5) by the method of substitution. This will give us the points of intersection of one of the circles with the common chord. But these are precisely the points of intersection of the two circles.

III.7.5 Example. Solve the system

$$x^2 + y^2 - 6x + 3y - 19 = 0$$

$$x^2 + y^2 - 4x + 6y - 37 = 0.$$

Subtracting the first equation from the second, we have

$$2x + 3y - 18 = 0$$

as the equation of the common chord. This is then solved for x:

$$x = 9 - \frac{3}{2}y$$

and this is substituted in the first equation, obtaining

$$\left(9 - \frac{3}{2}y\right)^2 + y^2 - 6\left(9 - \frac{3}{2}y\right) + 3y - 19 = 0$$

or

$$13y^2 - 60y + 32 = 0.$$

This is solved by the usual method

$$y = \frac{60 \pm \sqrt{3600 - 1664}}{26} = \frac{60 \pm 44}{26}$$

$$y = 4 \text{ or } \frac{8}{13}.$$

We then obtain x from these two values of y; thus

$$x = 9 - \frac{3}{2}(4) = 3$$

or

$$x = 9 - \frac{3}{2} \cdot \frac{8}{13} = \frac{105}{13}.$$

Thus the two points of intersection are (3, 4) and (105/13, 8/13).

III.7.6 Example. Solve the system

$$x^2 + y^2 - 2x + 4y - 4 = 0$$

$$x^2 + y^2 - 8x - 4y + 16 = 0.$$

As before, we subtract, to obtain

$$6x + 8y - 20 = 0,$$

or

$$x = \frac{10}{3} - \frac{4}{3}y.$$

We substitute this in the first equation:

$$\left(\frac{10}{3} - \frac{4}{3}y\right)^2 + y^2 - 2\left(\frac{10}{3} - \frac{4}{3}y\right) + 4y - 4 = 0$$

or

$$25y^2 - 20y + 4 = 0,$$

which has the solution

$$y = \frac{20 \pm \sqrt{400 - 400}}{50} = \frac{2}{5}.$$

This is substituted above, to obtain $x = 14/5$. Thus the two circles are tangent, meeting at the point (14/5, 2/5). The line obtained was the common tangent to the two circles.

III.7.7 Example. Solve the system

$$x^2 + y^2 - 3x + 8y + 10 = 0$$

$$x^2 + y^2 - 3x - 6y + 8 = 0.$$

Subtracting, we have

$$14y + 2 = 0$$

or

$$y = -\frac{1}{7}.$$

Substituting in the first equation, we have

$$x^2 + \left(\frac{1}{7}\right)^2 - 3x - 8\left(\frac{1}{7}\right) + 10 = 0$$

or

$$49x^2 - 147x + 435 = 0.$$

It is easy to see that this equation has no solutions. Thus the two circles fail to intersect. The reader may interpret the geometrical meaning of the "common chord," in this case, as he may please.

III.7.8 Example. Solve the system

$$x^2 + y^2 + 6x - 2y + 2 = 0$$

$$x^2 + y^2 + 6x - 2y + 4 = 0.$$

In this case, subtraction gives the absurdity $2 = 0$. We conclude that the common chord fails to exist: the circles do not intersect. (In fact, they are concentric.)

III.7.9 Example. At a price of x cents a pound, the demand for wool yarn in a certain region is y tons, where

$$x^2 + y^2 + 20x + 10y = 8000.$$

At the same price, the supply of wool yarn in this region will be z tons, where

$$x^2 + z^2 - 200x + 40z = 1300$$

for prices under \$1 per lb., and 88.2 tons for prices above \$1 per lb. Find the equilibrium price.

We set $y = z$, and obtain a system of two equations, both of which represent circles. Solution of the system will give us $x = 40$, $y = z = 70$ (there is also a solution with negative x, which is clearly extraneous). Thus there is an equilibrium price of 40¢ per lb.; demand and supply will then both equal 70 T.

PROBLEMS ON SIMULTANEOUS EQUATIONS

1. Find all solutions to the following systems of simultaneous equations:

(a) $3x^2 + 7y^2 - 6x - 12 = 0.$
$x + 2y = 7.$

(b) $3x^2 + xy - 2y^2 + 4x - 6y - 36 = 0.$
$3x + 2y = 6.$

(c) $x^2 - y^2 = 9.$
$3x - 2y = 7.$

(c) $3x^2 + 4xy - 6x - 8 = 0.$
$4x + y = 9.$

(d) $y = x - 3x^2 + 1.$
$x - y = 11.$

(e) $3xy - 4x + y - 6 = 0$
$3x - 2y = 2.$

(f) $x^2 + y^2 - 4x + 2y - 11 = 0.$
$3x + y = 9.$

(g) $x^2 + y^2 - 4x + 2y - 11 = 0.$
$x^2 + y^2 - 30x - 6y + 65 = 0.$

(h) $x^2 + y^2 + 2x - 2y - 2 = 0.$
$x^2 + y^2 + 6x - 8 = 0.$

(i) $x^2 + y^2 - x + 2y - 41 = 0.$
$x^2 + y^2 + 6x - 10y - 2 = 0.$

(j) $x^2 + y^2 + 2x + 2y - 2 = 0.$
$x^2 + y^2 + 4x + 2y - 4 = 0.$

2. Let the circle C have equation

$$x^2 + y^2 + Dx + Ey + F = 0.$$

Given a point P with coordinates (p, q), define the *power* of P with respect to the circle C as w, where

$$w = p^2 + q^2 + Dp + Eq + F$$

Let L be any line through P, and suppose that L intersects the circle C at the points Q and R. Prove that the product of the two distances, PQ and PR, is equal to w. (This product should be considered negative if P lies between Q and R, positive otherwise.)

3. At a price x, the demand y for a certain commodity is given by the equation

$$xy + 20x + 12y = 160$$

while supply z is given by

$$3x - 2z = 20.$$

Find the equilibrium price for this commodity. What are supply and demand then?

4. The cost of making x dozens of a certain product is y dollars, where

$$16x^2 - y^2 + 144 = 0.$$

If the product can be sold for \$4.50 per dozen, what is the break-even point?

5. An electrical goods company finds that it can produce x miles of copper wire at an average cost of y dollars per mile, where y satisfies

$$x^2 + y^2 - 100x - 300y + 12{,}000 = 0,$$

$y \leq 150$. On the other hand, these can be sold at a price of z dollars per mile, where

$$x^2 + z^2 = 2000.$$

Find the company's break-even point. What is the maximum amount that can be produced at a profit?

8. THE EXPONENTIAL AND LOGARITHMIC FUNCTIONS

We have, so far in this chapter, studied quadratic functions and their graphs, the conic curves; these are among the most important of non-linear curves. The reason for this is, partly, that they continuously appear in physical applications, and partly, that many processes can be approximated by quadratics, thus giving rise to equations which can be solved algebraically. Nevertheless, there are many other functions which are of importance in analysis. We shall in this section consider two types: the exponential and the logarithmic functions.

We shall assume here that the reader is familiar with the use of exponents and logarithms. (If this is not the case, we recommend a brief reading of their properties in Appendix 3.) Our interest here is in their graphs and their more common applications.

Consider first the exponential function

3.8.1 $$y = a^x$$

where a is an arbitrary positive constant.

The graph of equation (3.8.1) can have two general forms, depending on whether $a < 1$ or $a > 1$. For $0 < a < 1$ (Fig. III.8.1), we see that, for large negative x, y is very large and positive. As x increases, however, we see that y decreases so as to approach zero, and indeed the curve has the x-axis as asymptote in the positive direction.

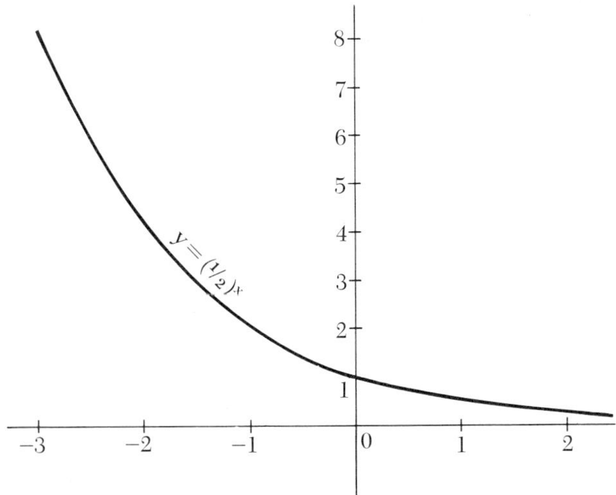

FIGURE III.8.1 Graph of the equation $y = (1/2)^x$.

For $a > 1$ (Fig. III.8.2), the situation is quite the opposite. For large, negative x, y will be small (close to zero) and positive. As x

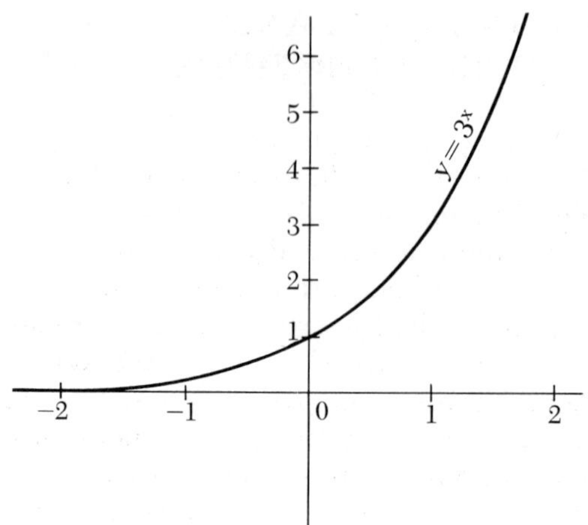

FIGURE III.8.2 Graph of the equation $y = 3^x$.

increases, y also increases, and this increase is very rapid for positive x. Again the x-axis is an asymptote, but this time in the negative direction.

(For $a = 1$, we obtain the trivial case of a horizontal line $y = 1$ as graph.)

Slightly more complicated is the function

3.8.2 $$y = ka^x$$

where k is also an arbitrary constant. For positive k (Fig. III.8.3), the curves obtained are very similar to those for (3.8.1); the main difference is that the y-intercept is at k, instead of 1.

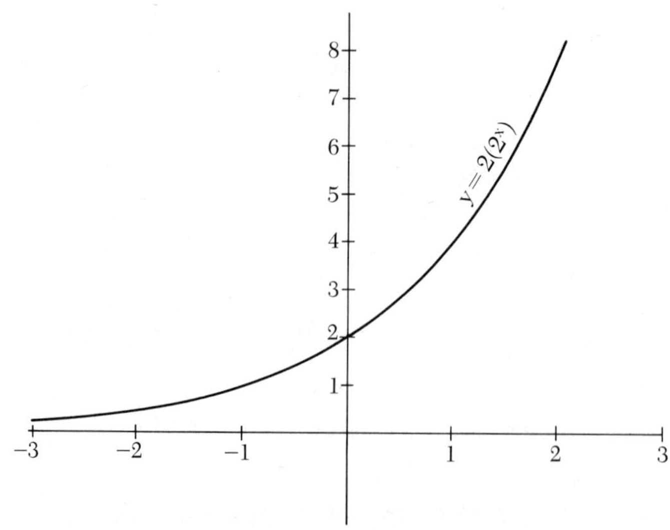

FIGURE III.8.3 Graph of $y = 2(2^x)$.

For negative k (Fig. III.8.4) the behavior is similar, but the curves obtained are "mirror images" of the others, lying in the lower half of the plane.

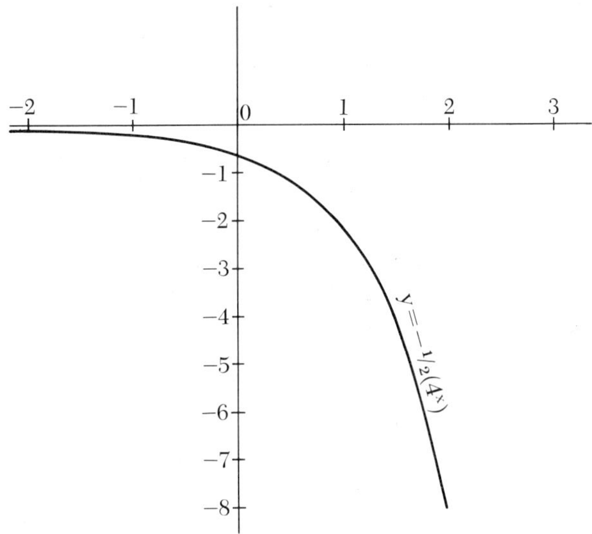

FIGURE III.8.4 Graph of $y = -\frac{1}{2}(4^x)$.

The most important property of the function (3.8.2) can be seen if we use the functional notation $y = f(x)$. Then

$$f(x+1) = ka^{x+1} = aka^x = af(x)$$

and so

3.8.3 $$f(x+1) - f(x) = (a-1)\,f(x).$$

We see that the increase in y, as x increases one unit, is proportional to the value of y. (Of course, if $a < 1$, this is actually a decrease rather than an increase.) This is most useful, as there are many cases in which the growth (or decay) of a quantity is directly proportional to the value of that quantity.

Compound Interest. Let us suppose that a sum P of money is placed in a savings account, where it receives interest at a rate of i (typically, $i = 0.04$ or 0.05) per annum, compounded annually. Then, if $y(x)$ is the value of the account after x years, we see that we must have

$$y(x+1) = y(x) + iy(x)$$

which is equivalent to equation (3.8.3), with $a = i + 1$. We conclude that we must have

$$y = k(1 + i)^x.$$

It still remains to determine the constant k. This is done by noticing that at $x = 0$ (i.e., at the time the money is deposited), the value of the account is precisely P. But we saw that the y-intercept of (3.8.2) is k. Thus $k = P$, and we have

$$y = P(1 + i)^x \tag{3.8.4}$$

as the formula for a compound-interest account.

Parenthetically, we point out that, except for integer values of x, the formula (3.8.4) is not quite valid. In effect, the interest is credited to the account once a year, whereas (3.8.4) shows continuous growth. This is a valid objection, and we should agree that, when x is not an integer, it must be reduced to the largest integer which is smaller than x.

III.8.1 Example. A sum of $1000 is placed at interest of 4 per cent, compounded annually. What will the value of the account (principal plus interest) be after 2 years? After 4 years? After 5½ years?

The value of the account is, of course,

$$y = 1000\ (1.04)^x.$$

For $x = 2$, we find $y = 1081.60$; for $x = 4$, it is $y = 1169.86$. For $x = 5\frac{1}{2}$, we substitute $x = 5$, since no interest is paid during the middle of the year. Then, for $x = 5$, we find $y = 1216.65$.

More generally, we find that interest is paid, not yearly, but more frequently, say quarterly or monthly. (In some cases, banks have even taking to compounding every second, or, for that matter, continuously. But interest is only paid quarterly or monthly.) The treatment then is essentially the same, with the modification that we are to count, not years, but interest periods. Thus (3.8.4) continues to apply, but x is now the number of interest periods, while i is the interest per period (*not* per year).

III.8.2 Example. The sum of $500 is placed at interest of 4 per cent per annum, compounded quarterly. How much interest will this account earn in one year?

The interest rate is 1 per cent per quarter. Thus $i = 0.01$, and $x = 4$, as one year is equal to 4 quarters. Then

$$y = 500(1.01)^4 = 520.30$$

is the value of the account. The interest paid is, therefore, $20.30.

It is clear that the equation (3.8.3) is characteristic, not only of compound interest, but also of certain other growth models. The following example shows this well.

III.8.3 Example. A farm is started with 20 hamsters. Assuming that these double in number every 160 days, how many hamsters will there be on the farm after one year?

It is clear that we must have an equation of the form (3.8.2). The y-intercept is 20, and so

$$y = 20a^x.$$

To find the base a, we see that, if we measure x in 160-day periods, then y doubles whenever x increases by one unit. Thus $a = 2$. Then

$$y = 20(2^x).$$

After one year, we have $x = 365/160 = 73/32$. Thus,

$$y = 20(2^{73/32}).$$

Using logarithms, this is

$$\log y = \log 20 + \frac{73}{32} \log 2$$

and this gives us $y = 97.2$. Thus, after one year, the farm will have approximately 97 hamsters. (This is of course not exact, as reproduction of hamsters is not a mechanistic process.)

When $0 < a < 1$, the equation (3.8.2) describes, not growth, but decay:

III.8.4 Example. Suppose that, because of inflation, the U.S. dollar decreases in purchasing power by 5 per cent each year. What will its purchasing power be, 5 years from now (relative to present power)?

We have here

$$y = (0.95)^x.$$

For $x = 5$, this gives us $y = (0.95)^5$, or approximately 0.7736. Thus, in 5 years, the dollar will lose approximately 22.7 per cent of its purchasing power.

A different type of growth model occurs when a certain variable has a natural upper limit. For example, a manufacturing company might make a new kitchen appliance. The appliance has a natural market equal to the number of kitchens in the country (or, perhaps, in the world), but cannot exceed this number under the twin assumptions (a) that there is no use for more than one in a kitchen, and (b) that the appliance has a reasonably long life-time. The growth encountered here is usually represented by a function of the form

$$y = L - ka^x \tag{3.8.5}$$

where $k > 0$ and $0 < a < 1$. The graph of this equation is shown in Figure III.8.5. Note that, as x increases, y increases and tends to L; the line $y = L$ is an asymptote. Since y cannot be negative, only that part of the curve to the right of the x-intercept can apply. The x-intercept, then, is the time at which the growth process begins (from zero).

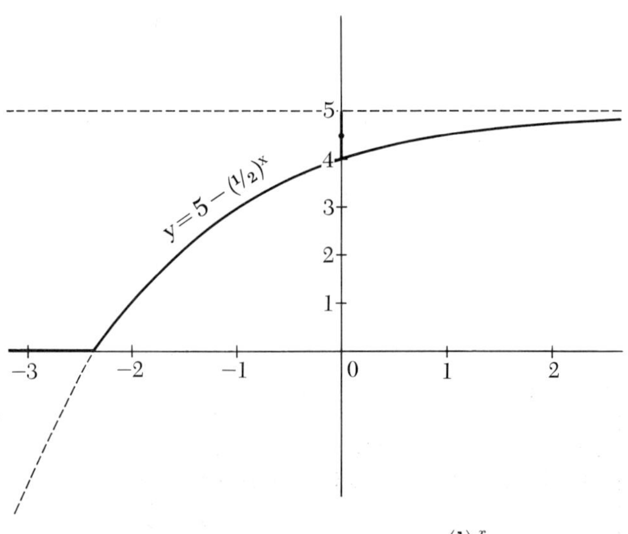

FIGURE III 8.5 Graph of $y = 5 - \left(\frac{1}{2}\right)^x$.

III.8.5 Example. Starting a campaign, a political candidate finds that only 3500 of the 28,500 registered voters in his district have heard of him. His campaign is such that every week he will become known to 20 per cent of the people who had not previously known him. How many voters will he have reached after a seven-week campaign?

The potential here is, of course, the 28,500 voters in the district. If y is the number of voters who have heard of the candidate, we see that $28{,}500 - y$ (the number who have not heard of him) satisfies the decay equation

$$28{,}500 - y = k(0.8)^x$$

where x is the number of weeks elapsed. Now, at the beginning of the campaign ($x = 0$), there are 25,000 who have not heard of the candidate. Thus $k = 25{,}000$, and

$$y = 28{,}500 - 25{,}000\,(0.8)^x.$$

For $x = 7$, this gives us $y = 23{,}258$. Thus the great majority of the voters will have heard of him after 7 weeks.

We have thus far studied two types of growth processes: an unbounded growth, which is proportional to the size of the (dependent) variable itself, and a limited growth, which is proportional to the difference between the variable and some upper limit, L. In some cases, we find that the growth of a variable must be (approximately) proportional to both of these. This happens, say, in the case of a population which increases by reproduction, but is bounded by a natural limit (e.g., the number of rats in a controlled habitat with a limited amount of food). Such a growth is generally represented by an equation of the form

3.8.6
$$y = \frac{L}{1 + ka^x}$$

where $k > 0$ and $0 < a < 1$.

The graph of (3.8.6) is shown in Figure III.8.6. For large negative

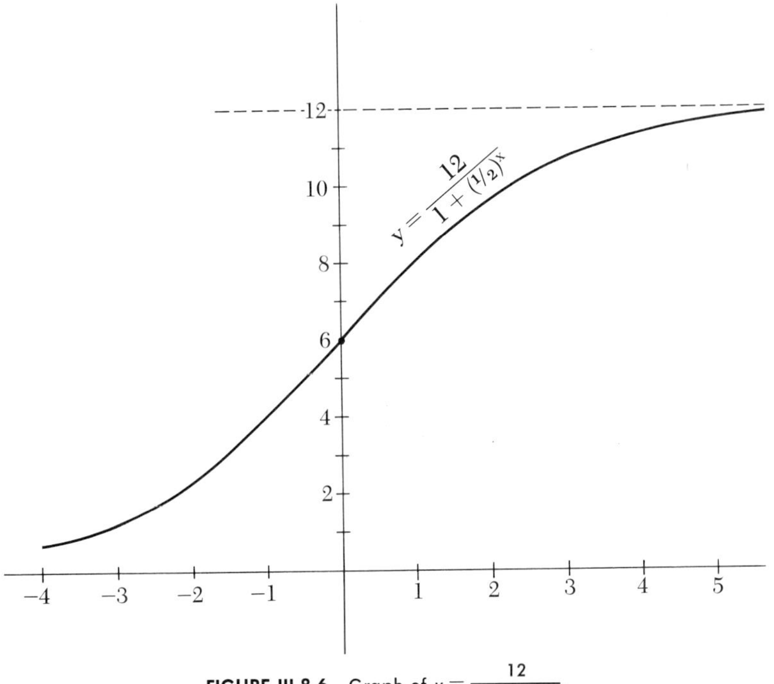

FIGURE III 8.6 Graph of $y = \dfrac{12}{1 + \left(\frac{1}{2}\right)^x}$.

x, the denominator on the right side of (3.8.6) increases without bound, and so y is small (close to zero). For large positive x, the denominator tends to 1, and so y tends to L. Thus y increases as x increases, with the two lines $y = 0$ and $y = L$ as asymptotes. We see that, when y is very small, it increases rather slowly (there are only a few rats, and they simply cannot reproduce any faster). Again, when y is close to L, the increase is small (the food begins to run out, and the rats stop reproducing). In between, the rate of increase can be quite rapid.

III.8.7 Example. A company makes a new type of sports car; it estimates that there are 10,000 potential buyers for this model. Instead of advertising, the company decides to give 100 models away, as it feels that word-of-mouth advertising from satisfied owners will be all the publicity that it needs. (Hopefully, all owners will be satisfied.) Thus, sales will be low at the beginning, when few people have heard of the car, and low as y approaches 10,000, as only a few customers will then be left, but rapid in between. If 25 cars are sold during the first week, how many cars can the company expect to sell in 12 weeks?

We assume here that y, the number of cars in operation after x weeks, will satisfy the equation (3.8.6), with k and a to be determined. We have $y = 100$ for $x = 0$, and $y = 125$ for $x = 1$, which, substituted in (3.8.6), give us

$$\begin{aligned} 1 + k &= 100 \\ 1 + ka &= 80. \end{aligned}$$

We will have, then, $k = 99$, and $a = 0.8$ (approximately). Therefore,

$$y = \frac{10{,}000}{1 + 99(0.8)^x}$$

Letting $x = 12$, we obtain $y = 1280$. This includes the 100 cars given away, and so sales in 12 weeks will be 1180 cars.

Closely related to the exponential function is the logarithmic function

3.8.7 $$y = \log_a x,$$

read, "y is the logarithm of x to the base a." Here, a is a positive number, different from 1, and (3.8.7) is equivalent to

$$x = a^y.$$

The importance of the logarithmic function for numerical computations is well known. From our point of view, its importance derives from the well-known equation

3.8.8 $$\log_a uv = \log_a u + \log_a v.$$

Figure III.8.7 shows the graph of the equation (3.8.7). As can be seen, the logarithm, y, increases as x increases (if $a > 1$), for positive values of x. The question is as to the rate of increase of y. We see that

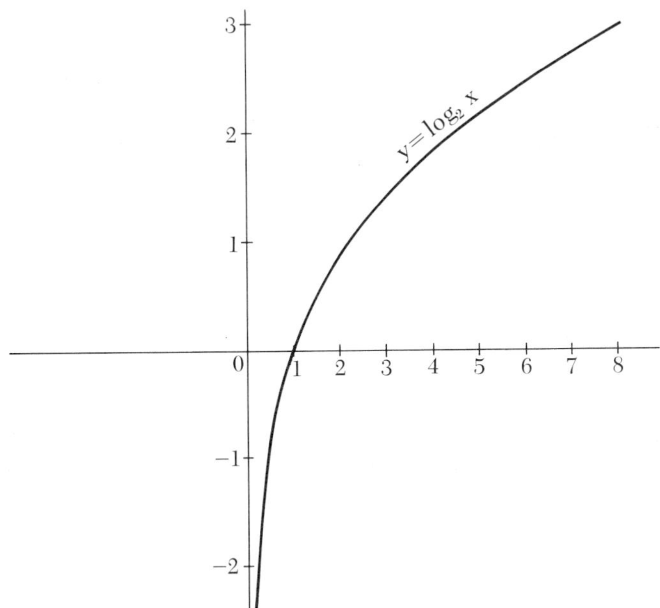

FIGURE III.8.7 Graph of $y = \log_2 x$. Notice the similarity to Fig. III.8.1.

if x increases by an amount h, then y increases by an amount

$$\log_a (x + h) - \log_a x = \log_a (1 + h/x).$$

Now, it can be shown that, for small values of h/x, this $\log (1 + h/x)$ is approximately proportional to h/x. We conclude that, if x is increased by a small amount h, its logarithm, y, increases by an amount which is directly proportional to h, but inversely proportional to x.

The Weber-Fechner Law. Psychologists have found that, if a certain stimulus (such as a light affecting an eye, or a noise affecting an ear) is increased by a small amount, there is an increase in response which is directly proportional to the increase in stimulus, but inversely proportional to the size of the original stimulus. (In less technical language, a small increase in noise will be noticed if the original noise level was low, but will not be noticed if the original level was high.) It follows that the response y will satisfy an equation of the form (3.8.7), or, more generally,

$$y = k + \log_a x, \tag{3.8.9}$$

where x is the stimulus. This is known as the *Weber-Fechner law*, after the physiologists E. H. Weber (1795–1878) and G. T. Fechner (1801–1887).

As an example, we find that a star is assigned magnitude 6 if it is barely visible to the unaided eye, and magnitude 1 if it is 100 times as bright as a magnitude 6 star. Then, if a star is x times as bright as a magnitude 6 star, its magnitude is given by (3.8.9), with $k=6$ and $a = \sqrt[5]{0.01} = 0.398$. In particular, we see that if a star is as bright as 35 stars of magnitude 6, then it will have magnitude

$$y = 6 + \log_{0.398} 35 = 2.14.$$

Consumers' Preferences. It is often assumed that the Weber-Fechner law also applies to economic situations. In this case, the stimulus might be an amount of a certain commodity, and the response would be the amount that someone is willing to pay for that amount. Thus, the amount of money, y, that someone is willing to spend for an amount x of some commodity should satisfy an equation of the form

$$y = \log_a (k + bx).$$

It is clear that $y = 0$ when $x = 0$, and so $k = 1$. Thus

$$y = \log_a (1 + bx) \tag{3.8.10}$$

where $a > 1$, and $b > 0$.

III.8.9 ***Example.*** A housewife is willing to pay 50¢ for 1 lb. of hamburger, but only \$1 for 3 lb. Assuming a law of the form (3.8.10), how much would she be willing to pay for 5 lb. of hamburger?

We have to find both a and b in (3.8.10). We know that $y = 1/2$ for $x = 1$, and $y = 1$ for $x = 3$. Therefore,

$$\log_a (1 + b) = 1/2$$
$$\log_a (1 + 3b) = 1.$$

These equations can be rewritten as

$$1 + b = a^{1/2}$$
$$1 + 3b = a.$$

Thus,

$$1 + 3b = (1 + b)^2$$

which has the solutions $b = 0, 1$. For $b = 0$, we obtain $a = 1$, which is clearly not valid (we had ruled out such values of a and b). For $b = 1$, we have $a = 4$, and so

$$y = \log_4 (1 + x).$$

For 5 lb., then, the housewife would be willing to pay

$$\log_4 6 = \frac{\log_{10} 6}{\log_{10} 4} = \frac{0.7782}{0.6021}$$

or about $1.29.

PROBLEMS ON THE EXPONENTIAL AND LOGARITHMIC FUNCTIONS

1. Plot the graphs of the following equations:

(a) $y = 1 + 4^x$

(b) $y = 2 + (1/2)^x$.

(c) $y = 7 + 5(2^x)$.

(d) $y = \log_{10} (3 + x)$.

(e) $y = \log_2 x + 3 \log_3 x$.

2. The sum of $10,000 is placed at interest of 5 per cent per annum, compounded semiannually. Find the value of this account (principal plus interest) after two years.

3. The number of sheep on a ranch doubles every three years. Assuming that the ranch starts with 100 sheep, how many will there be after five years?

4. An assembly line produces 300 radios per hour. Due to the workers' errors, there are always some radios which must be rejected as defective, but, as the work is repeated, these errors become fewer in number. As a result, the number of defective radios produced (per hour) decreases by one half every 20 days. If an average of five defective radios are being produced per hour today, how many will be produced 10 days from today?

5. A new type of kitchen appliance is being promoted by word of mouth only; there is a potential market of 12,000 buyers for this appli-

ance. On January 1, 1971, 6000 had been sold, and, on July 1, 1971, 7200 had been sold. Assuming a growth curve of the form (3.8.6), how many will have been sold by January 1, 1972?

6. A housewife is willing to pay \$1 for 2 lb. of cheese, but only \$2 for 5 lb. of the same cheese. How much will she pay for 10 lb. of cheese? (Assume a law of the form (3.8.10).)

THE THEORY OF PROBABILITY

1. PROBABILITIES

The theory of probability arose from the desire of men to understand the behavior of systems that cannot entirely be controlled. More exactly, it arose from the desire of a French gambler, the Chevalier de Méré, to understand why his dice had suddenly turned against him. The Chevalier's concern was a very practical one: he was losing money.

For some time, de Méré had bet (at even odds) that, in four tosses of a die, he would obtain a six at least once. This was a profitable bet, as was attested to by his systematic winnings. Unfortunately, he had been too successful and, as a result, found himself unable to obtain any takers for his bets. This led him to change the bet: he now bet that, in 24 tosses of a pair of dice, he would obtain a double-six at least once. The Chevalier reasoned that this bet would be as profitable as the previous one (since, indeed, $4/6 = 24/36$); the change in the bet, however, would enable him to find new takers. He did find new takers for the bet, but, to his surprise, he began to lose systematically.

We shall consider this example later on, and see what de Méré's mistake consisted in. For the present we will simply note that, not being a mathematician, he proceeded to consult one—Blaise Pascal (1623-1662), one of the greatest of all time. Pascal discussed the problem with Pierre de Fermat (1601-1665), and, from the analysis made by these two men, a new branch of mathematics was born.

The fundamental problem in the theory of probability is, then, as follows. A certain action, called an *experiment*, is carried out. The experiment may have any one of several possible *outcomes*; the idea is to predict, more or less, the frequency with which these outcomes (or certain sets of them, called *events*) will appear—assuming that the experiment can be repeated many times. The experiment can be of many types: we give examples.

(a) A die is tossed; the *outcome* is the face that lands on the *bottom*.
(b) A dart is thrown at a bull's-eye; the *outcome* is the section of the bull's-eye in which the dart lands.
(c) A coin is tossed three times; the *outcome* is the run of heads and tails obtained.
(d) A telephone book is opened; the *outcome* is the page to which it is opened.

We give these examples to illustrate several things. Example (a) shows that the outcomes may be labeled in many ways (since it is usual to think of the *top* face of the die as the outcome of this experiment). Example (b) shows that many outcomes may be classed together as a single outcome (we could consider each point on the board as a different outcome). Example (c) shows that an experiment may consist of several other, simpler experiments. Example (d) shows that the experimenter may be able to influence, to some degree, the outcome, since he can choose, to some extent, the section of the book to which he will open (though possibly not the exact page). It is often necessary to explain the manner in which the experiment is carried out.

Now, how are probabilities to be assigned? This is an extremely complicated question—not from a mathematical, but from a practical point of view. Mathematically, we can give a system of axioms that are to be satisfied by any assignment of probabilities. The difficulty lies in our practical desire that the probability assigned to an outcome correspond to the relative frequency with which it occurs. This can be ascertained only through experimentation. But such experiments, precisely because of their nature, are always subject to random fluctuations, which tend to, but need not, cancel out over the long run. Moreover, the "long run" might be so long as to be quite impractical.

An example may be of use in explaining this difficulty. A coin is said to be *fair* if, when it is tossed, each of the two outcomes (heads and tails) has a probability 1/2 of occurring. This means (for reasons to be seen later) that, in tossing a fair coin 20 times, the probability of obtaining heads every time is $(1/2)^{20}$, or approximately 0.000001. Thus, if a coin is tossed 20 times, and turns up heads every time, we would be justified in claiming that the coin is skew (i.e., not fair).

Suppose, however, that each adult person in the United States (say, some 100,000,000 people) were to toss a (fair) coin 20 consecutive times. In this case, it is almost certain that approximately 100 of them would have the singular experience of seeing a string of 20 consecutive heads; another 100 would obtain a string of 20 consecutive tails. We might imagine each of these 200 claiming that the newly minted coins were entirely skew; in fact, however, we would claim that it was a cause for surprise if no such strings of 20 heads occurred. The idea is that, no matter how unlikely an outcome, it is liable to happen, and so we have to be careful before assigning probabilities on the basis of observed results, no matter how many times the experiment may be repeated.

A second conceptual difficulty is the fact that, in many cases, the experiment is such that it cannot or will not be repeated. The Sox meet the Cats tomorrow to determine the Baseball Championship; what is the probability that the Sox will win? A building contractor makes a bid on a projected building; what is the probability that he will be the low bidder? Mr. Jones will go hunting tomorrow for the first time in his life; what is the probability that he will shoot a fox? In each of these cases, it is clear that the "experiment," with all its implications, cannot be repeated at all (the baseball players, for example, could play a seven game championship set, but then the individual games would no longer be as important as the single championship game). Thus a probability can be assigned to these outcomes only by defining the concept of probability in a different manner.

In such cases, it is common to talk about *subjective* probability. The subjective probability of an event is, more or less, the subjective belief in its likelihood; the event has, say, a subjective probability of 1/2 if a person feels that a bet at even odds is a "fair" bet. This is not, to say the least, a very scientific idea; it is further complicated by the fact that many people may have different beliefs about the event's likelihood, and so the subjective probability is not well defined. (Practical men have, of course, found a way to solve this problem; the bookmaker, seeing that he cannot reconcile the subjective probabilities held by different people, looks for an equilibrium value that will guarantee that he wins whether, in fact, the Sox or the Cats win the championship.)

The problem of assigning probabilities to the outcomes of an experiment remains. In the absence of anything better, the rule of Pierre Simon, Marquis de Laplace (1749-1827), has been used: when there is no indication to the contrary, assume that each of the possible outcomes has the same probability. When tossing a coin, one assumes (unless there is some indication to the contrary) that each of the two sides has the same probability. If an urn has 50 balls, all identical except for color, then each of the balls has the same probability (1/50) of being drawn out by a blindfolded man. Laplace's rule does not hold when it seems reasonable that one of the outcomes is more likely than another. For instance, if a die has one face much larger than the others, it seems reasonable to assume that the die will land on that face more often than on the others.

In the rest of this chapter, we shall assume that probabilities can be assigned to the outcomes: we shall then attempt to draw logical conclusions about the outcomes of other, more complicated experiments. This is the domain of probability theory. The problem of assigning these original probabilities belongs to the field of statistical analysis and is treated elsewhere.

2. PROBABILITY SPACES

We shall begin our study of probability by considering only experiments with a finite number of possible outcomes.

IV.2.1 Definition. A *finite probability space* is a finite set of outcomes

$$X = \{O_1, O_2, \ldots, O_n\}$$

together with function, P, which assigns a non-negative and real number to each of the outcomes O_i, in such a way that

4.2.1
$$\sum_{i=1}^{n} P(O_i) = 1$$

The number $P(O_i)$ is the probability of the outcome O_i; it is often replaced by p_i. The reason for condition (4.2.1) should be clear: since each trial of the experiment will have one (and only one) of the outcomes O_i, it follows that the sum of their relative frequencies must be equal to 1.

IV.2.2 Definition. An *event* is a subset of the set X.

An event is defined as a set of outcomes. The probability function P, which is defined for outcomes, can be extended to events by saying that an event happens whenever any one of the outcomes belonging to this event occurs. Thus, the probability of the event will be equal to the sum of the probabilities of the several outcomes:

4.2.2
$$P(A) = \sum_{O_i \in A} P(O_i)$$

In particular, X is itself an event called the *certain event,* since it contains all the possible outcomes. The empty set, $\varnothing$, is also an event, called the *impossible* event. It is easy to see that

4.2.3
$$P(X) = 1$$

4.2.4
$$P(\varnothing) = 0$$

A certain event must have probability 1; an impossible event, probability 0. The converse of this is not quite true; an event may be possible and still have probability 0, or it may have probability 1 and still not be certain. For finite probability spaces, however, this will not happen. 173

We give now examples of finite probability spaces.

IV.2.3 Example. Consider the experiment of tossing a die and looking at the top face. The experiment has six possible outcomes:

$$X = \{1, 2, 3, 4, 5, 6\}$$

If the die is fair, then each of these outcomes will have the same probability; since the sum of these must be 1, it follows that

$$p_i = \frac{1}{6} \text{ for } i = 1, 2, \ldots, 6$$

Consider the event "even number":

$$E = \{2,4,6\}$$

We can calculate that

$$P(E) = p_2 + p_4 + p_6 = \frac{1}{2}$$

Similarly, if $S = \{1,2\}$, we have

$$P(S) = p_1 + p_2 = \frac{1}{3}$$

IV.2.4 Example. Let the experiment consist of three tosses of a coin; the outcome will be the run of heads and tails obtained. There are 8 possible outcomes:

$$\begin{array}{ccc} H & H & H \\ H & H & T \\ H & T & H \\ H & T & T \end{array} \qquad \begin{array}{ccc} T & H & H \\ T & H & T \\ T & T & H \\ T & T & T \end{array}$$

If the coin is fair, each of these outcomes will have the same probability, namely, 1/8. If we consider the event "two heads," we have

$$E = \{H\ H\ T,\quad H\ T\ H,\quad T\ H\ H\}$$

and it follows that

$$P(E) = \frac{3}{8}$$

IV.2.5 Example. Let the experiment consist of drawing a card from a deck. There are 52 possible outcomes; the event "hearts" will contain 13 of these outcomes. If we assume that each card has the same probability, 1/52, of being drawn, then

$$P\text{ (hearts)} = \frac{13}{52} = \frac{1}{4}$$

For the same experiment, the event "ace" will contain four of the possible outcomes. Thus

$$P\text{ (ace)} = \frac{4}{52} = \frac{1}{13}$$

It is not too difficult to see that, if A and B are disjoint events, i.e., if $A \cap B = \varnothing$, then

4.2.5 $$P(A \cup B) = P(A) + P(B)$$

In fact, $P(A \cup B)$ is the sum of the probabilities of the outcomes that belong to $A \cup B$. This sum can be split into two sums, corresponding to $P(A)$ and $P(B)$. Note that this would not be possible if $A \cap B$ were non-empty, since then any outcome that belonged to $A \cap B$ would be counted twice.

If we have several disjoint events, i.e., events $A_1, A_2, \ldots, A_m$ such that $A_j \cap A_k = \varnothing$ whenever $j \neq k$, then (4.2.5) generalizes directly, to give

4.2.6 $$P(A_1 \cup A_2 \cup \ldots \cup A_m) = P(A_1) + P(A_2) + \ldots + P(A_m)$$

Again, we point out that (4.2.6) will not hold unless the events A_k are disjoint.

In case two events are not disjoint, we have the more general relation

4.2.7 $$P(A \cup B) = P(A) + P(B) - P(A \cap B)$$

The reader will note the similarity of this to equation (1.4.3). In fact, the proof is entirely analogous, and may be left as an exercise to the reader.

IV.2.6 ***Example.*** Given the same experiment as in Example IV.2.5, find the probability of obtaining *either* a heart *or* an ace.

For this problem, we can write

$$H = \text{“hearts”}$$
$$A = \text{“ace”}$$

and we are looking for $P(H \cup A)$. We already know $P(H)$ and $P(A)$;

to find $P(H \cap A)$, note that $H \cap A$ consists of one single outcome (the ace of hearts) and so

$$P(H \cap A) = \frac{1}{52}$$

Applying (4.2.7), now, we will have

$$P(H \cup A) = P(H) + P(A) - P(H \cap A) = \frac{1}{4} + \frac{1}{13} - \frac{1}{52} = \frac{16}{52}$$

and so

$$P(H \cup A) = \frac{4}{13}$$

This could, of course, have been obtained directly by noticing that there are exactly 16 cards that are either hearts or aces: 13 hearts, and 3 other aces.

IV.2.7 Example. Let an experiment consist of a toss of two dice. There are, all told, 36 possible outcomes (assuming that the two dice are distinguishable):

(1,1), (1,2), (1,3), (1,4), (1,5), (1,6)
(2,1), (2,2), (2,3), (2,4), (2,5), (2,6)
(3,1), (3,2), (3,3), (3,4), (3,5), (3,6)
(4,1), (4,2), (4,3), (4,4), (4,5), (4,6)
(5,1), (5,2), (5,3), (5,4), (5,5), (5,6)
(6,1), (6,2), (6,3), (6,4), (6,5), (6,6)

Consider the event A: "the sum of the numbers on the two dice is equal to 10." We have

$$A = \{(6,4), (5,5), (4,6)\}$$

and so, assuming that each outcome has probability 1/36, we find

$$P(A) = \frac{3}{36} = \frac{1}{12}$$

In a similar way, the event B, "doubles are thrown," will contain six outcomes:

$$B = \{(1,1), (2,2), (3,3), (4,4), (5,5), (6,6)\}$$

and so

$$P(B) = \frac{6}{36} = \frac{1}{6}$$

We can see also that $A \cap B = (5,5)$; hence

$$P(A \cap B) = \frac{1}{36}$$

and so

$$P(A \cup B) = \frac{1}{12} + \frac{1}{6} - \frac{1}{36} = \frac{2}{9}$$

IV.2.8 Example. There are 39 families in a small town; of these, 23 are Republican and 16 are Democrat. There are 7 families headed by a woman, and of these, 5 are Democrat.

A poll sampler chooses one family at random. Find the probability that: (a) the family is Democrat, (b) it is headed by a woman, (c) it is *either* Democrat *or* headed by a woman.

Let D be the event "the family is Democrat," and let W be the event "it is headed by a woman." Assuming that each family has a probability 1/39 of being chosen, we have:

$$P(D) = \frac{16}{39}$$

$$P(W) = \frac{7}{39}$$

while for the "joint event," "Democrat *and* headed by a woman,"

$$P(D \cap W) = \frac{5}{39}\,.$$

We thus have

$$P(D \cup W) = \frac{16}{39} + \frac{7}{39} - \frac{5}{39} = \frac{18}{39}$$

so that the probability that the family is *either* Democrat *or* headed by a woman is 18/39, or 6/13.

PROBLEMS ON DISCRETE PROBABILITY SPACES

1. A fair die is tossed. Let E be the event $\{2,4,6\}$ and let S be the event $\{1,2\}$. Find the probabilities of the events E, S, $E \cap S$, $E - S$, $E \cup S$, and $S - E$.

2. A fair coin is tossed three times. List the eight possible outcomes. Find the probabilities of the events:

(a) The first toss is H.

(b) Either the first toss is H, or the third toss is T.

(c) Both the second and third tosses are T.

(d) No tosses are H.

(e) At least one of the tosses is T.

(f) The first toss is H, and the third toss is T.

(g) There are at least two T's.

3. A card is drawn at random from a deck (of 52 cards). What is the probability that:

(a) The card is an ace?

(b) It is the ace of hearts?

(c) It is *either* an ace *or* a heart?

4. A pair of fair dice is tossed. What is the probability that:

(a) The sum of the spots showing on the top of the dice is 8?

(b) It is 8 or more?

(c) It is 4 or less?

3. CONDITIONAL PROBABILITY. BAYES' FORMULA

As was mentioned before, we are interested in finding relationships among the probabilities of various events in order to analyze more complicated experiments. Let us consider the idea of *conditional* probability: given that an event A happens, what is the probability of a second event, B?

From a practical point of view, probability theory is applicable when the analyst lacks information about all the relevant variables. A poker player must rely on the theory of probability because he does not know what the card on top of the deck is; if he could peek, he would know whether to bet or fold. Suppose, however, that he notices the top card has a frayed corner. If he knows this deck of cards, the knowledge gives him additional information, which will make his decision easier for him. He no longer has to consider all the cards in the deck, but only those that have frayed corners.

To come back to our original question, the probability of event B, given that event A happens, is the relative frequency of B, considering only those cases in which A happens. This leads to the following definition.

IV.3.1 Definition. The *conditional probability* of the event B, given the event A, is the probability $P(B \mid A)$, given by

4.3.1 $$P(B \mid A) = \frac{P(A \cap B)}{P(A)}$$

IV.3.2 Example. In the experiment of Example IV.2.7, find the conditional probability of obtaining doubles, given that the sum of the numbers on the two dice is 10.

In this case, we are looking for $P(B \mid A)$. We have, from IV.2.7,

$$P(A \cap B) = \frac{1}{36}$$

$$P(A) = \frac{1}{12}$$

Therefore

$$P(B \mid A) = \frac{1/36}{1/12} = \frac{1}{3}$$

Note that $P(B \mid A)$ is twice $P(B)$. Whether A occurs will alter the probability of B.

IV.3.3 Example. Using the data of Example IV.2.8, find the probability that the family sampled is Democrat, given that it is headed by a woman. Find also the probability that it is headed by a woman, given that it is Democrat. (Note that these are not the same thing.)

We are looking here for $P(D \mid W)$ and $P(W \mid D)$. Now

$$P(D \mid W) = \frac{P(D \cap W)}{P(W)} = \frac{5/39}{7/39} = \frac{5}{7}$$

while

$$P(W \mid D) = \frac{P(D \cap W)}{P(D)} = \frac{5/39}{16/39} = \frac{5}{16}$$

Two events are thought of as independent if neither has any influence on the other. Probabilistically, we can best express this by saying that the conditional probability is the same as the absolute (i.e., not conditional) probability.

IV.3.4 Definition. Two events, A and B, are said to be independent if

$$P(A \cap B) = P(A)P(B) \tag{4.3.2}$$

For independent events, the *joint probability* (i.e., the probability $P(A \cap B)$ that they will *both* happen) is equal to the product of their individual absolute probabilities.

IV.3.5 Example. Consider the experiment of Example IV.2.4. Let A be the event "the first toss is a head," and let B be the event "the third toss is tails." We find then that

$$A = \{H\,H\,H, H\,H\,T, H\,T\,H, H\,T\,T\}$$

$$B = \{H\,H\,T, H\,T\,T, T\,H\,T, T\,T\,T\}$$

and

$$A \cap B = \{H\,H\,T, H\,T\,T\}$$

Since each outcome has probability 1/8, we find that

$$P(A) = P(B) = \frac{1}{2}$$

$$P(A \cap B) = \frac{1}{4} = \frac{1}{2} \cdot \frac{1}{2}$$

The joint probability is the product of the absolute probabilities, and we conclude that the two events are independent. This corresponds to the intuitive idea that the result of the first toss should have no influence on the third toss, or vice-versa. On the other hand, note that for Example IV.2.8, the events D and W are *not* independent.

For more than two events, independence is defined in a slightly different manner:

IV.3.6 Definition. Three events A, B, and C are independent if any two of them are independent and, moreover,

$$P(A \cap B \cap C) = P(A)P(B)P(C) \tag{4.3.3}$$

In general, n events $A_1, A_2, \ldots, A_n$ are independent if any $n-1$ are independent and also

$$P(A_1 \cap A_2 \cap \ldots \cap A_n) = P(A_1)P(A_2)\ldots P(A_n) \tag{4.3.4}$$

Definition IV.3.6 is an inductive definition: it defines independence for n events in terms of independence for $n-1$ events, plus an additional condition. An alternative definition would be to say that a collection of events is independent if, for any sub-collection, the joint probability is the product of the absolute probabilities.

Note that it is possible for three events not to be independent, although any two of them are independent. (Conversely, three events may satisfy (4.3.3) and still not be independent because two of them are not.)

IV.3.7 Example. In the experiment of Example IV.2.7, let A be the event "the first die shows a 6"; let B be the event "the second die shows a 3," and let C be the event "the sum of the numbers on the two dice is 7." We have, then,

$$A = \{(6,1), (6,2), (6,3), (6,4), (6,5), (6,6)\}$$

$$B = \{(1,3), (2,3), (3,3), (4,3), (5,3), (6,3)\}$$

$$C = \{(1,6), (2,5), (3,4), (4,3), (5,2), (6,1)\}$$

and so $P(A) = P(B) = P(C) = 1/6$. To check for independence, we see that

$$A \cap B = \{(6,3)\}$$

$$A \cap C = \{(6,1)\}$$

$$B \cap C = \{(4,3)\}$$

and so

$$P(A \cap B) = \frac{1}{36} = P(A)P(B)$$

$$P(A \cap C) = \frac{1}{36} = P(A)P(C)$$

$$P(B \cap C) = \frac{1}{36} = P(B)P(C)$$

We see that any two of these events are independent. On the other hand,

$$A \cap B \cap C = \varnothing$$

so that $P(A \cap B \cap C) = 0$, and it follows that the three events are not independent.

4. COMPOUND EXPERIMENTS

In the last few examples, we have assumed that we are given the joint probabilities of the events directly, i.e., that we have been given the probabilities of each of the outcomes. In general, however, this is not so: the problem usually involves computing the probabilities of the outcomes.

Let us now, for instance, consider the problem of the Chevalier de Méré. The Chevalier's original bet was that he would obtain at least one six in four tosses of a die. We can find the probability of this event by considering the complementary event, "no sixes are obtained." Assuming the die to be fair, the probability of *not* obtaining a six on a given toss is 5/6; since the four tosses are independent, we must multiply four such factors, i.e.,

$$P(\text{no sixes are obtained}) = \left(\frac{5}{6}\right)^4 = \frac{625}{1296}$$

And so the Chevalier had a probability of $1 - 625/1296 = 671/1296$, or slightly more than 1/2, of winning. He had an advantage: in the long run, he would win more often than not.

Consider now the Chevalier's second bet: the probability of *not* obtaining a double six on any given toss of the dice, assuming the dice to be fair, is 35/36. For 24 independent tosses, we have

$$P(\text{no double sixes are obtained}) = \left(\frac{35}{36}\right)^{24}$$

This probability, computed by using logarithms, is, approximately 0.512. We conclude that de Méré had only a probability 0.488 of winning. In the long run, he should lose somewhat more often than he won—a prediction verified by his experience. The two bets were by no means equivalent. Some mathematical analysis would have saved the Chevalier a great sum of money—but, as Pascal pointed out, de Méré was no mathematician.

IV.4.1 Example. A test consists of five multiple-choice questions, each one giving a choice of four possible answers. A student who has not read the subject matter guesses at each answer. What is the probability that he will answer all five questions correctly?

In a sense, this problem is not entirely well defined. The difficulty is that, even though a student has not read the subject matter, the correct answer may yet be somewhat more logical than any of the others, so that an "educated guess" gives a greater probability of success than a wild guess. It may also be that the five questions are logically related, so that the student's success on one of them is not independent of success on the others. Notwithstanding these possibilities, we shall assume that the student has a 1/4 probability of success on each question (since there are four answers, of which only one is correct), and that the several questions are independent.

Letting S_i be the event "success on the ith question," we have

$$P(S_i) = \frac{1}{4} \qquad \text{for } i = 1, 2, 3, 4, 5$$

We are looking for the probability of success on all the questions, i.e., for $P(S_1 \cap S_2 \cap S_3 \cap S_4 \cap S_5)$. Since the questions are assumed independent, we have, by (5.3.4),

$$P(S_1 \cap S_2 \cap S_3 \cap S_4 \cap S_5) = \left(\frac{1}{4}\right)^5 = \frac{1}{1024}$$

When events are not independent, we are quite often given their conditional probabilities. In such cases, we find joint probabilities by rewriting (4.3.1) in the form

4.4.1 $$P(A \cap B) = P(A)P(B \mid A)$$

IV.4.2 Example. Three identical urns are labeled A, B, C. Urn A contains 5 red and 7 black balls. Urn B contains 1 red and 5 black balls. Urn C contains 12 red and 3 black balls. An urn is chosen at random, and then a ball is taken at random from the urn. What is the probability that the ball will be red?

Two assumptions are made here: the first is that each urn has the same probability, 1/3, of being chosen. The second is that, from a given urn, each of the balls has the same probability of being chosen. Thus, the conditional probability of a red ball, given that urn A is chosen, will be 5/12 (since 5 of the 12 balls in the urn are red); it is similar for the other urns. Letting R be the event "a red ball is chosen," we find that

$$P(A) = P(B) = P(C) = \frac{1}{3}$$

$$P(R \mid A) = \frac{5}{12}$$

$$P(R \mid B) = \frac{1}{6}$$

$$P(R \mid C) = \frac{4}{5}$$

The event R can be written as the union of disjoint events

$$R = (R \cap A) \cup (R \cap B) \cup (R \cap C)$$

and so, by (4.2.6),

$$P(R) = P(R \cap A) + P(R \cap B) + P(R \cap C)$$

Now, by (4.4.1),

$$P(R \cap A) = P(A)P(R \mid A) = \frac{1}{3} \cdot \frac{5}{12} = \frac{5}{36}$$

$$P(R \cap B) = P(B)P(R \mid B) = \frac{1}{3} \cdot \frac{1}{6} = \frac{1}{18}$$

$$P(R \cap C) = P(C)P(R \mid C) = \frac{1}{3} \cdot \frac{4}{5} = \frac{4}{15}$$

so that

$$P(R) = \frac{5}{36} + \frac{1}{18} + \frac{4}{15} = \frac{83}{180}$$

IV.4.3 Example. A lot of 20 flashbulbs contains 17 good and 3 defective items. A sample of size 3 is taken. What is the probability that all 3 will be good?

We assume that each item in the lot has the same probability of being chosen. Let I be the event "the first item chosen is good," II be the event "the second item is good," and III be the event "the third item is good." We will apply (4.4.1) twice:

$$P(I \cap II \cap III) = P(I \cap II)P(III \mid I \cap II) = P(I)P(II \mid I)P(III \mid I \cap II)$$

Since 3 of the 20 bulbs are defective, we have, clearly,

$$P(I) = \frac{17}{20}$$

Suppose that the first item is good. In this case, there will be only 19 bulbs left, including 16 good ones, and

$$P(II \mid I) = \frac{16}{19}$$

Similarly, if the first two bulbs chosen are good, there will be 15 good bulbs among the 18 left, and so

$$P(III \mid I \cap II) = \frac{15}{18}$$

Thus,

$$P(I \cap II \cap III) = \frac{17}{20} \cdot \frac{16}{19} \cdot \frac{15}{18} = \frac{34}{57}$$

IV.4.4 Example. Find the probability that, from a group of 24 people, no 2 have the same birthday.

For this problem, we shall assume that each day is as likely to be a birthday as any other. Let A_i be the event "the ith person does not have the same birthday as any of the first $i-1$." Then we are looking for the probability of the event $A_1 \cap A_2 \ldots \cap A_n$.

Suppose, now, that all the events $A_1, A_2, \ldots, A_{i-1}$ occur. This means that no two of the first $i-1$ persons have a common birthday; they occupy $i-1$ different days. For event A_i to occur, the ith person must then be born in one of the remaining $365-(i-1)$ days. Thus,

$$P(A_i \mid A_1 \cap \ldots \cap A_{i-1}) = \frac{366-i}{365}$$

Applying rule (4.3.4), now,

$$P(A_1 \cap \ldots \cap A_n) = \frac{365}{365} \cdot \frac{364}{365} \cdot \ldots \cdot \frac{366-n}{365}$$

For $n = 24$, we have

$$P(A_1 \cap \ldots \cap A_{24}) = \frac{365}{365} \cdot \frac{364}{365} \cdots \frac{342}{365} = 0.46$$

We see, thus, that the probability that no 2 have the same birthday is *less than 1/2*! This means that, from 24 people, 2 will have a common birthday more often than not—a strangely unintuitive result, since, indeed, 24 is much smaller than 365.

PROBLEMS ON CONDITIONAL PROBABILITIES

1. A pair of fair dice is tossed. What is the probability that one of the dice show three spots, given that the sum of the spots is 5?

2. A fair coin is tossed three times. What is the probability that the first toss will be H, given that there are at least two H's thrown?

3. A test consists of five multiple choice questions; each question has three possible answers. A student guesses at all the answers; what is the probability that he will obtain a grade of 20 per cent or better? of 100 per cent?

4. Three otherwise identical urns are labeled A, B, and C. Urn A contains 5 red and 6 black balls. Urn B contains 3 red and 1 black ball. Urn C contains 10 red and 20 black balls. An urn is chosen at random (probability 1/3 each) and then a ball is taken from the urn. What is the probability that it will be red?

5. A chewing gum company has 30 different baseball cards, and

gives 1 with each pack of gum. A boy buys 10 packs of gum. What is the probability that he will obtain 10 different cards?

6. A machine makes bolts according to specifications. When it is working well, the bolts meet the specifications with probability 0.98. However, there is a probability 0.35 that the machine will be working poorly, in which case the bolts will meet specifications with probability 0.40. What is the probability that a given bolt will meet the specifications?

7. The game of *craps* is played in the following manner: a player tosses a pair of dice. If the number obtained is 2, 3, or 12, he loses immediately; if it is 7 or 11, he wins immediately. If any other number is obtained on the first toss, then that number becomes the player's "point," and he must keep on tossing the dice until either he "makes his point" (i.e., obtains the first number again), in which case he wins; or he obtains 7, in which case he loses. Find the probability of winning.

8. A machine makes pieces to fit specifications; the probability that a given piece will be suitable is 0.95. A sample of 10 pieces is taken from the machine's production. What is the probability that all 10 pieces will be good?

9. A traveling salesman assigns 0.3 as the probability of making a sale at each call. How many calls must he make in order that the probability of making at least one sale be 0.9 or better?

10. Three urns are labeled A, B, and C. Urn A has four red and two black balls. Urn B has five red and seven black balls; urn C has two red and six black balls. An urn is chosen at random and a ball is taken from that urn. What is the probability that the ball will be black?

A very common application of conditional probabilities deals with what is known as *a priori* and *a posteriori* probabilities.

As an illustration of this type of problem, let us consider the experiment of Example IV.4.2. We know that each of the three urns has the same probability, 1/3, of being chosen. Suppose that the three urns are indistinguishable; we could, nevertheless, find out which is which by the simple expedient of looking inside the urn (the number of balls of each color will determine the urn, i.e., if we find that it has 12 red and 3 black balls then we know that it must be urn C). Unfortunately, it may not be possible to look at all the balls in the urn; the question is whether seeing just one of them will give us any information. In fact, this generally does happen. An extreme example would be the case in which urns A and B had no red balls; a red ball would then, necessarily, be drawn from urn C. In our example, the situation is not so extreme, but still the contents of the three urns are sufficiently different so that some new information is gained. Since urn C has such a high proportion of red balls (12 out of 15), a red ball would most likely be drawn from urn C. In fact, we have

$$P(R \cap C) = P(C)P(R \mid C) = \frac{4}{15}$$

and

$$P(R) = \frac{83}{180}$$

Therefore

$$P(C \mid R) = \frac{4/15}{83/180} = \frac{48}{83}$$

so that the *conditional* probability $P(C \mid R)$ is considerably higher than the absolute probability $P(C)$.

In general, it is assumed that the outcomes of an experiment can be classified in two ways (as in our example, by urn and by color). This gives us two partitions of the outcome space:

$$X = A_1 \cup A_2 \cup \ldots \cup A_m$$
$$X = B_1 \cup B_2 \cup \ldots B_n$$

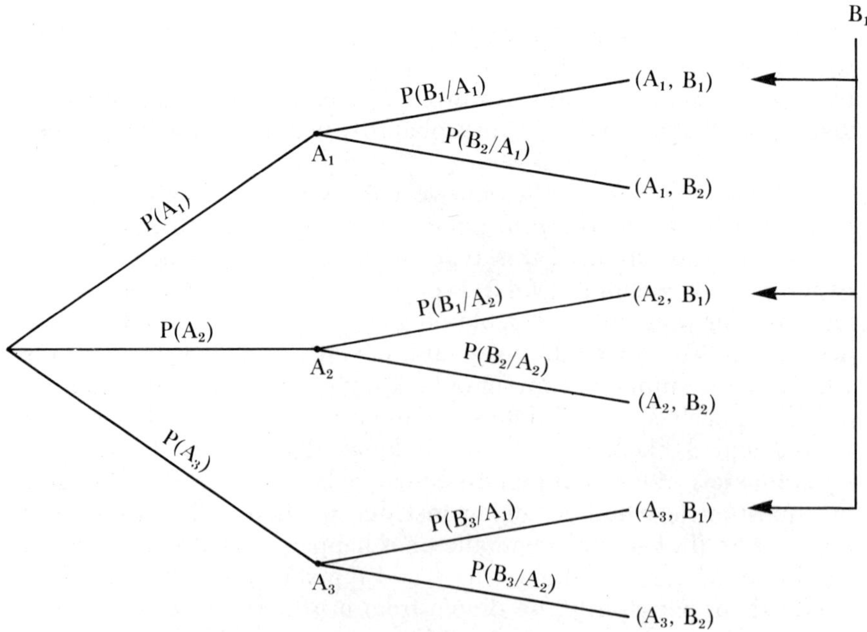

FIGURE IV.4.1 Scheme of two-classification experiment.

It is assumed that, for the first classification, the *absolute* probabilities $P(A_i)$ are known. For the second partition, it is the *conditional* probabilities $P(B_j \mid A_i)$ that are known. Now the nature of the experiment is such that it is easy to observe the value of the B_j, whereas the value of the A_i is either impossible or extremely difficult to ascertain. The idea is then to use the observed value B_j to obtain a new (conditional) probability $P(A_i \mid B_j)$. This new distribution of the A_i is known as the *a posteriori* distribution, while the previous (absolute) distribution is known as the *a priori* distribution. In our example, each of the three urns had an *a priori* probability 1/3 of being chosen. After the event, i.e., assuming that a red ball has been drawn, we conclude that urn C has a considerably larger *a posteriori* probability, 48/83.

In general, we wish to compute $P(A_i \mid B_j)$. Now, by Definition IV.3.1,

$$P(A_i \mid B_j) = \frac{P(A_i \cap B_j)}{P(B_j)}$$

However, we are given neither $P(A_i \cap B_j)$, nor $P(B_j)$; these must be computed. The numerator presents no trouble; it is given by

$$P(A_i \cap B_j) = P(A_i)P(B_j \mid A_i)$$

The denominator, on the other hand, requires some work. To simplify things somewhat, let us assume that $m = 2$, i.e., the A classification can give only two values, A_1 and A_2. We can write

$$B_j = (A_1 \cap B_j) \cup (A_2 \cap B_j)$$

Since A_1 and A_2 are disjoint, it follows that $A_1 \cap B_j$ and $A_2 \cap B_j$ are also disjoint. Then

$$P(B_j) = P(A_1 \cap B_j) + P(A_2 \cap B_j)$$

and, by (4.4.1),

$$P(B_j) = P(A_1)P(B_j \mid A_1) + P(A_2)P(B_j \mid A_2)$$

For the special case in which $m = 2$, then, we have the formula

$$P(A_i \mid B_j) = \frac{P(A_i)P(B_j \mid A_i)}{P(A_1)P(B_j \mid A_1) + P(A_2)P(B_j \mid A_2)}$$

In the more general case, the formula for $P(A_i \mid B_j)$ is derived analogously; it is

$$P(A_i \mid B_j) = \frac{P(A_i)P(B_j \mid A_i)}{\sum_{k=1}^{m} P(A_k)P(B_j \mid A_k)} \tag{4.4.2}$$

Equation (4.4.2) is known as *Bayes' formula.*

IV.4.5 Example. Of all applicants for a job, it is felt that 75 per cent are able to do the job, and 25 per cent are not. To aid in the selection process, an aptitude test is designed such that a capable applicant has probability 0.8 of passing the test, while an incapable applicant has probability 0.4 of passing the test. An applicant passes the test. What is the probability that he will be able to do the job?

This is a straightforward application of Bayes' formula. Let A_1 be the event "the applicant is able to do the job," and A_2 the complementary event. Let B_1 be the event "the applicant passes the test," and B_2 the complement. Then

$$P(A_1) = 0.75 \qquad P(A_2) = 0.25$$
$$P(B_1 \mid A_1) = 0.8 \qquad P(B_1 \mid A_2) = 0.4$$

By Bayes' formula,

$$P(A_1 \mid B_1) = \frac{(0.75)(0.8)}{(0.75)(0.8) + (0.25)(0.4)} = \frac{0.6}{0.6 + 0.1} = \frac{6}{7}$$

The test is a reasonably effective screening device in this case; only 1/7 of the persons passing the test will be incapable.

We may also check for the probability $P(A_1 \mid B_2)$, i.e., the probability that an applicant who fails the test would still be able to do the job. We have

$$P(B_2 \mid A_1) = 0.2 \qquad P(B_2 \mid A_2) = 0.6$$

and so

$$P(A_1 \mid B_2) = \frac{(0.75)(0.2)}{(0.75)(0.2) + (0.25)(0.6)} = \frac{0.15}{0.15 + 0.15} = \frac{1}{2}$$

Thus, half the rejected applicants are able to do the work. Note that the test is not infallible: incapable applicants may pass; capable applicants may fail.

IV.4.6 Example. A machine produces bolts to meet certain specifications. When the machine is in good working order, 90 per cent of the bolts meet the specifications. Occasionally, however, the machine goes out of order, and only 50 per cent of the bolts produced then meet the specifications. It is calculated that the machine is in good order 80 per cent of the time. A sample of two bolts is taken from the machine's current production, and both are found to meet the specifications. What is the probability that the machine is out of order?

Once again, we must use Bayes' formula. Let A_1 be the event "the machine is in good order," and A_2 the complementary event. Let B_1 be the event "both bolts are good." We know that

$$P(A_1) = 0.8 \qquad P(A_2) = 0.2$$

Next, we must compute $P(B_1 \mid A_i)$. If the machine is working well, the probability that a bolt will meet the specifications is 0.9. The probability that both bolts meet the specifications is, then, $(0.9)^2 = 0.81$:

$$P(B_1 \mid A_1) = 0.81$$

and similarly, $(0.5)^2 = 0.25$, so

$$P(B_1 \mid A_2) = 0.25$$

Thus,

$$P(A_1 \mid B_1) = \frac{(0.8)(0.81)}{(0.8)(0.81) + (0.2)(0.25)} = \frac{0.648}{0.648 + 0.05} = 0.93$$

IV.4.7 Example. Under the same conditions as in Example IV.4.6, a sample of two bolts is taken and both are found defective. What is the probability that the machine is out of order?

Let B_2 represent the event "both bolts are defective." As in the previous example, we compute

$$P(B_2 \mid A_1) = (0.1)^2 = 0.01$$
$$P(B_2 \mid A_2) = (0.5)^2 = 0.25$$

Applying Bayes' law, we have

$$P(A_2 \mid B_2) = \frac{(0.2)(0.25)}{(0.8)(0.01) + (0.2)(0.25)} = \frac{0.05}{0.008 + 0.05} = 0.86$$

IV.4.8 Example. A person is either suffering from tuberculosis (with probability 0.01) or not. To test for this disease, chest x-rays are taken. If the person is ill, the test will be positive with probability 0.999; if he is well, the test will be positive with probability 0.2. A person takes the x-ray test, which gives a positive result. What is the probability that the person is well?

Again, let A_1 be the event "tuberculosis," A_2 the complementary event. Let B_1 be the event "positive." We have

$$P(A_1) = 0.01 \qquad P(A_2) = 0.99$$

$$P(B_1 \mid A_1) = 0.999 \qquad P(B_1 \mid A_2) = 0.2$$

then

$$P(A_2 \mid B_1) = \frac{(0.2)(0.99)}{(0.01)(0.999) + (0.2)(0.99)} = \frac{0.198}{0.00999 + 0.198} = 0.95$$

Note that even with a positive test result, the person tested is still more likely to be well than ill. The reason for this is simply that it is considered better for a healthy patient to undergo further examinations than for a sick man to go undetected. The x-ray test is made purposely difficult to pass, i.e., the smallest dark spot is considered a positive result.

PROBLEMS ON COMPOUND EXPERIMENTS

1. A company has a position open; the probability that an applicant be capable of doing the required work is 0.65. The firm's personnel department devises an aptitude test that a capable applicant will pass with probability 0.8, whereas an incapable applicant will pass with probability 0.4. An applicant passes the test; what is the probability that he will be able to do the work?

2. Three identical urns are labeled A, B, and C. Urn A contains 5 red and 3 black balls. Urn B contains 6 red and 10 black balls; urn C contains 12 red and 4 black balls. An urn is chosen at random, and then a ball is taken from the urn, also at random. If the ball is red, what is the probability that the urn was urn B?

3. A person is either well (probability 0.9) or ill (probability 0.1). If he is well, his temperature will be normal with probability 0.95, whereas if he is ill, his temperature will be normal with probability 0.35.

   (a) If a person's temperature is normal, what is the probability that he is ill?

   (b) If the temperature is not normal, what is the probability that he is well?

4. A machine makes bolts to fit specifications. When the machine is working well (probability 0.8) the bolts fit specifications with probability 0.95; when it is not working well, the bolts fit specifications with probability 0.5. Two bolts taken from the machine's production are both defective. What is the probability that the machine is working well?

5. A contractor must build a road through a piece of ground. The soil is either clay (probability 0.68) or rock. If it is rock, a geological test will give a positive result with probability 0.75; if it is clay, the same test will give a positive result with probability 0.25. Given that the test shows a positive result, what is the probability that the soil is rock?

5. REPETITION OF SIMPLE EXPERIMENTS: THE BINOMIAL DISTRIBUTION

Let us consider a very elementary type of experiment: one with only two outcomes. These outcomes are generally called *success* (S) and *failure* (F). Such an experiment is called a *simple experiment.* It may be described entirely in terms of the single number p, where

4.5.1 $$p = P(S)$$

Since the probability of success is p, it follows that the probability of failure is $1 - p$; we write

4.5.2 $$q = 1 - p = P(F)$$

For a single trial of this experiment, there is very little else to say. Suppose, however, that the experiment is repeated many times. We will obtain a *run* of successes and failures, which may look like this:

$$SFFSSFFSSFSFFSSS$$

We are interested, generally, not so much in the exact run of successes and failures, but rather, in the number of successes obtained.

Consider the run just given. The probability of each success is p; that of each failure is q. If we assume (as we shall) that the trials are independent, then the probability of this run is

$$pqqppqqppqpqqppp$$

obtained by multiplying the probabilities for each trial. This product may, of course, be rewritten as $p^9 q^7$. Note that any other run of nine successes and seven failures would have the same probability. In general, we find that, in n independent trials of a simple experiment, any run that contains k successes and $n - k$ failures has probability $p^k q^{n-k}$. To find the probability of obtaining exactly k successes in n trials, we must find the number of runs having k successes. This number is obtained by noticing that the k successes can appear in any k of the n trials; thus, the number of such runs must be equal to the number of combinations of k elements that can be taken from a set of n elements, i.e., $\binom{n}{k}$. Multiplying this number by the probability of each of the runs, we obtain the *binomial coefficient*

4.5.3 $$B(n,k;p) = \binom{n}{k} p^k q^{n-k}$$

This expression is also known as a *Bernoulli* coefficient, after Jacob Bernoulli (1654-1705), who analyzed the problem. A sequence of independent trials is also known as a sequence of Bernoulli trials.

IV.5.1 Example. A fair coin is tossed 10 times. What is the probability of obtaining exactly seven heads?

The coin is fair, so $p = 1/2$. According to formula (4.5.3), the probability of this event is

$$B\left(10,7;\frac{1}{2}\right) = \frac{10!}{3!\,7!}\left(\frac{1}{2}\right)^7\left(\frac{1}{2}\right)^3 = \frac{120}{1024}$$

IV.5.2 Example. A fair die is tossed five times. What is the probability of obtaining three aces?

In this case, we have $p = 1/6$. Hence

$$B\left(5,3;\frac{1}{6}\right) = \frac{5!}{3!\,2!}\left(\frac{1}{6}\right)^3\left(\frac{5}{6}\right)^2 = \frac{250}{7776}$$

IV.5.3 Example. A machine produces bolts to meet certain specifications; 90 per cent of the bolts produced meet those specifications. A sample of five bolts is taken from the machine's production. What is the probability that two or more of these fail to meet the specifications?

In this problem we are looking for the event "at least two fail to meet specifications," or more concisely, "at least 2 failures." The formula that we have developed, however, will only give the probability of exactly k successes (or failures). Thus we must divide the event into the disjoint events "2 failures," "3 failures," "4 failures," and "5 failures." We have $n = 5$, and $p = 0.9$, and so the relevant probabilities are

$$P(2 \text{ failures}) = B(5,3;0.9) = \frac{5!}{3!\,2!}(0.9)^3(0.1)^2 = 0.0729$$

$$P(3 \text{ failures}) = B(5,2;0.9) = \frac{5!}{2!\,3!}(0.9)^2(0.1)^3 = 0.0081$$

$$P(\text{4 failures}) = B(5,1;0.9) = \frac{5!}{1!\,4!}\,(0.9)\,(0.1)^4 = 0.00045$$

$$P(\text{5 failures}) = B(5,0;0.9) = \frac{5!}{0!\,5!}\,(0.1)^5 = 0.00001$$

and therefore

$$P(\text{at least 2 failures}) = 0.0729 + 0.0081 + 0.00045 + 0.00001 = 0.08146$$

This same problem could be solved by considering the complementary event, "at most 1 failure." This can be written as the union of the two events "1 failure" and "no failures." We have

$$P(\text{1 failure}) = B(5,4;0.9) = \frac{5!}{4!\,1!}\,(0.9)^4(0.1) = 0.32805$$

$$P(\text{no failures}) = B(5,5;0.9) = \frac{5!}{5!\,0!}\,(0.9)^5 = 0.59049$$

so that

$$P(\text{at most 1 failure}) = 0.32805 + 0.59049 = 0.91854$$

Since the events "at most 1 failure" and "at least 2 failures" are complementary, we have

$$P(\text{at least 2 failures}) = 1 - 0.91854 = 0.08146$$

Note that, as expected, we obtain the same answer both ways.

IV.5.4 Example. In a large city, 60 per cent of the heads of household are Democrats. A poll taker visits eight houses, and asks the party affiliation of the head of household. What is the probability that at least six are Democrats?

Because the city is large, it may reasonably be assumed that the probabilities remain equal at each trial, i.e., the trials are independent. The probability of *at least six Democrats* is then

$$B(8,6;0.6) + B(8,7;0.6) + B(8,8;0.6)$$

Now

$$B(8,6;0.6) = \frac{8!}{6!\,2!}\,(0.6)^6(0.4)^2 = 0.209$$

$$B(8,7;0.6) = \frac{8!}{7!\,1!}\,(0.6)^7(0.4) = 0.089$$

$$B(8,8;0.6) = \frac{8!}{8!\,0!}\,(0.6)^8 = 0.017$$

so that

$$P(\text{at least 6 Democrats}) = 0.315$$

6. DRAWINGS WITH AND WITHOUT REPLACEMENT

Let us now consider the following problem:

An urn contains 10 red and 4 black balls. Three balls are drawn from the urn. What is the probability that exactly two of the balls drawn will be red?

In a problem such as this, in which elements are taken from a finite set (drawings from a finite population), there are two possible and very distinct methods of carrying out the experiment. The experimenter may, if he wishes, take a ball from the urn, record its color, and *replace it* before drawing the next ball. This method is known as *drawing with replacement.* Alternatively, the experimenter may draw the three balls, simultaneously or one at a time, without replacing them. This method is known as *drawing without replacement.* It is clear that the two methods will give different results; for instance, if the urn contains only two balls, it will be impossible even to make three drawings without replacement, whereas drawings with replacement can always be made.

Suppose that the drawings are made *with* replacement. In this case, it is not too difficult to see that, at each of the drawings, the urn contains the original number of red and black balls. The probability of obtaining a given colored ball will be the same at each drawing, regardless of what might have happened in previous drawings, i.e., the drawings are independent. It follows that such drawings are, in effect, Bernoulli trials. Thus, for drawings *with replacement,* the solution to the problem is as follows. Since 10 of the 14 balls are red, we have $p = 5/7$. Then the probability of 2 red balls is

$$B\left(3,2;\frac{5}{7}\right) = \frac{3!}{2!\,1!}\left(\frac{5}{7}\right)^2\left(\frac{2}{7}\right) = \frac{150}{343}$$

Let us consider, now, the case of drawings *without* replacement. For this method, it is clear that the drawings are no longer independent, and so they cannot be treated as Bernoulli trials. A new analysis of the problem is therefore necessary.

We analyze the problem of *drawings without replacement* by assuming that all the possible samples of three balls that can be drawn from the urn have the same probability. It is then merely a matter of seeing how many such samples there are, and how many of those belong to the desired event (i.e., satisfy the requirement of containing two red and one black ball). The desired probability is then the quotient of these two numbers.

In fact, the number of possible samples is equal to the number of combinations of 3 elements from a set of 14 elements, i.e., the binomial coefficient $\binom{14}{3}$. In turn, the number of samples containing two red balls and one black ball is obtained by multiplying the number of ways in which two red balls can be taken by the number of ways in which one black ball can be taken. These are, respectively, the number of combinations of 2 elements from a set of 10 elements, and the number of combinations of 1 element from a set of 4 elements, i.e., $\binom{10}{2}$ and $\binom{4}{1}$. There are, then, $\binom{14}{3}$ possible samples, of which $\binom{10}{2}\binom{4}{1}$ belong to our event. Thus

$$P(\text{exactly 2 red balls}) = \frac{\binom{10}{2}\binom{4}{1}}{\binom{14}{3}} = \frac{\dfrac{10!\,4!}{2!\,8!\,1!\,3!}}{\dfrac{14!}{3!\,11!}}$$

$$= \frac{144}{364} = \frac{36}{91}$$

Note that the probability in this case is slightly less than in the case of drawings with replacement.

In general, we may postulate an urn containing m balls, of which m_1 are red and $m_2 = m - m_1$ are black. A sample of n balls is taken (without replacement) from the urn; we wish to know the probability that exactly k of them are red.

There are, in all, $\binom{m}{n}$ combinations of n elements from a set of m elements. This is the number of possible samples. Of these, we consider those that have exactly k red and $n - k$ black balls. There are $\binom{m_1}{k}$ combinations of k red balls, and $\binom{m_2}{n-k}$ combinations of $n - k$ black balls. The product of these two numbers will be the number of samples in the event. Thus,

$$P(k \text{ red balls}) = \frac{\binom{m_1}{k}\binom{m_2}{n-k}}{\binom{m}{n}} \qquad \textbf{4.6.1}$$

IV.6.1 Example. A poker hand consists of 5 cards, taken at random from a deck of 52 (which is divided into 4 suits of 13 cards each). What is the probability that all the cards in a hand will be of the same suit (a hand known as a flush)?

We can best solve this problem by considering the probability that all five cards in the hand are, say, spades. The probability of a flush will be four times this, since the flush can be in any one of the four suits.

It is easy to see that this is a problem of sampling without replacement. We will have $m = 52$, $m_1 = 13$ (the number of spades), and $m_2 = 39$. Also, $n = k = 5$.

$$P(\text{flush in spades}) = \frac{\binom{13}{5}\binom{39}{0}}{\binom{52}{5}} = \frac{\dfrac{13!\,39!}{5!\,8!\,0!\,39!}}{\dfrac{52!}{5!\,47!}} = \frac{33}{66640}$$

and, therefore

$$P(\text{flush}) = \frac{4 \cdot 33}{66640} = \frac{33}{16660}$$

The probability of a flush is slightly less than 1 in 500. (This probability is not exactly the same as that given in poker primers because we include straight flushes among the flushes.)

IV.6.2 Example. A company tests a lot of 50 flash-bulbs by taking a sample of 10. If the sample contains 2 or more duds, the lot is rejected. Given that a lot contains 13 duds, what is the probability that it will be rejected?

As in Example IV.6.1, we shall do this by considering the complementary event, "at most one dud." This is, in turn, divided into the two events, "no duds," and "exactly one dud." Now

$$P(\text{no duds}) = \frac{\binom{37}{10}\binom{13}{0}}{\binom{50}{10}} = 0.033$$

$$P(1\text{ dud}) = \frac{\binom{37}{9}\binom{13}{1}}{\binom{50}{10}} = 0.153$$

and so the probability that the lot will be rejected is

$$1 - 0.033 - 0.153 = 0.814$$

IV.6.3 Example. A small town has 20 houses. In 12 of these, the head of household is a Democrat. A poll taker visits 8 houses. What is the probability that in at least 6, the head of household be a Democrat? (Compare Example IV.5.4.)

This is clearly an example of sampling without replacement. We have

$$P(6\text{ Democrats}) = \frac{\binom{12}{6}\binom{8}{2}}{\binom{20}{8}} = 0.205$$

$$P(7\text{ Democrats}) = \frac{\binom{12}{7}\binom{8}{1}}{\binom{20}{8}} = 0.050$$

$$P(8\text{ Democrats}) = \frac{\binom{12}{8}\binom{8}{0}}{\binom{20}{8}} = 0.004$$

so that

$$P(\text{at least 6 Democrats}) = 0.259$$

PROBLEMS ON DRAWINGS WITH AND WITHOUT REPLACEMENT

1. A fair coin is tossed 12 times. What is the probability of obtaining:

   (a) Exactly six heads?

   (b) At least eight heads?

   (c) Between five and seven heads?

2. A fair die is tossed six times. What is the probability of obtaining:

   (a) Exactly two aces?

   (b) At least three aces?

3. A multiple-choice test consists of eight questions, each one having three possible answers. What should the passing mark be to ensure that someone who guesses at the answers will have a probability smaller than 0.05 of passing?

4. A machine produces pieces to meet specifications; the probability that a given piece be suitable is 0.95. A sample of 10 pieces is taken. What is the probability that 2 or more be defective?

5. Two baseball teams meet in the world series; team A, the stronger team, has probability 0.6 of winning in each game. What is the probability that team A will win the series (i.e., win at least four of seven games)?

6. A package of flash bulbs contains 16 good and 4 defective bulbs. A sample of 5 bulbs is taken. What is the probability that at least three of the bulbs will be good?

7. An urn contains 15 red and 10 black balls. Three balls are taken from the urn; what is the probability that 2 of these be red?

8. A lot of 16 tires contains 13 good and 3 defective tires. A sample of 6 is taken. What is the probability that it contains at least 4 good tires?

9. Lots of 20 elements are tested in the following manner: a sample of size 4 is taken, and the number of defective elements in this sample is called x. If $x = 0$, the lot is accepted, if $x \geq 3$, it is rejected. If $x = 1$ or 2, a further sample of size 3 is taken and the number of defectives in this lot is called y. Then the lot is accepted if $x + y \leq 3$, and rejected if $x + y \geq 4$. What is the probability that a lot will be accepted if:

(a) It contains 15 good and 5 defective elements?

(b) It contains 12 good and 8 defective?

(c) It contains 17 good and 3 defective?

10. A simple experiment, with probability of success p, is repeated until m successes are obtained. Show that the probability of obtaining the mth success on the kth trial is given by the *negative binomial distribution*

$$P_m(k) = \binom{k-1}{m-1} p^m (1-p)^{k-m}$$

7. RANDOM VARIABLES

Very often there is a numerical quantity associated with each outcome of an experiment. It may represent money, as in the case of a gamble, or it may represent some other measurable quantity—the strength of a rope, the number of games won by a baseball team, or the life of a light bulb. Such a numerical quantity is known as a *random variable.* We give a precise definition.

IV.7.1 Definition. A *random variable* is a function whose domain is the set of outcomes of an experiment and whose range is a subset of the real numbers.

We shall generally denote a random variable by a capital letter: X, Y, and so on. The values of the random variable will be denoted by lower-case letters: x, y, and so on.

Since, by its very nature, a random variable takes its values according to a randomization scheme, it becomes natural to look for the probability of each of these values. These probabilities make up what is known as the *distribution* of the random variable.

Associated with the random variable X, there are two functions that describe its distribution. The first is the *probability function,* $f(x)$, defined by

4.7.1 $$f(x) = P(X = x)$$

The second is the *cumulative distribution function,* $F(x)$, defined by

4.7.2 $$F(x) = P(X \leq x)$$

Thus f gives the probabilities of each of the several possible values of X, while F gives the probability that X will be smaller than or equal to a given number. We shall, almost invariably, deal with the probability function, f, although the cumulative distribution function will be of use on some occasions.

IV.7.2 Example. Let X be the number of spots appearing on one toss of a fair die. Then X has the probability function

$$f(x) = \begin{cases} \frac{1}{6} \text{ for } x = 1,2, \ldots, 6 \\ 0 \text{ otherwise} \end{cases}$$

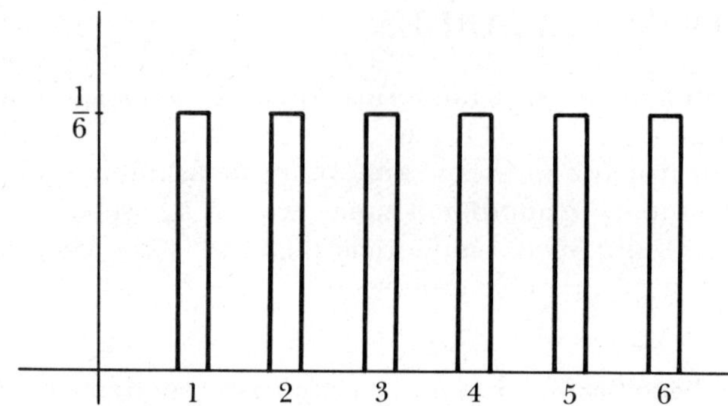

FIGURE IV.7.1 Probabilities on the toss of one die.

and cumulative distribution function

$$F(x) = \begin{cases} 0 \text{ for } x < 1 \\ \frac{1}{6} \text{ for } 1 \leq x < 2 \\ \frac{1}{3} \text{ for } 2 \leq x < 3 \\ \frac{1}{2} \text{ for } 3 \leq x < 4 \\ \frac{2}{3} \text{ for } 4 \leq x < 5 \\ \frac{5}{6} \text{ for } 5 \leq x < 6 \\ 1 \text{ for } \quad x \geq 6 \end{cases}$$

See Figures IV.7.1 and IV.7.2 for these two functions.

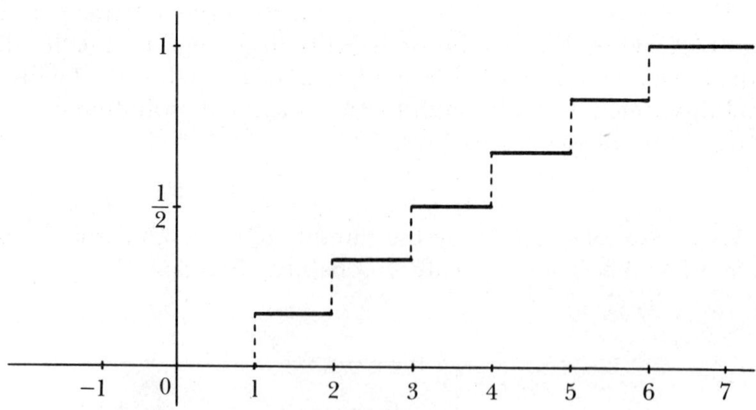

FIGURE IV.7.2 Cumulative distribution for the toss of one die.

IV.7.3 Example. Let Y be the number of spots appearing on two independent tosses of a fair die. The distribution of Y may be found by studying the table in Example IV.2.7; it is

$$f(2) = \frac{1}{36}$$

$$f(3) = \frac{2}{36} = \frac{1}{18}$$

$$f(4) = \frac{3}{36} = \frac{1}{12}$$

$$f(5) = \frac{4}{36} = \frac{1}{9}$$

$$f(6) = \frac{5}{36}$$

$$f(7) = \frac{6}{36} = \frac{1}{6}$$

$$f(8) = \frac{5}{36}$$

$$f(9) = \frac{4}{36} = \frac{1}{9}$$

$$f(10) = \frac{3}{36} = \frac{1}{12}$$

$$f(11) = \frac{2}{36} = \frac{1}{18}$$

$$f(12) = \frac{1}{36}$$

and $f(x) = 0$ for all other x (see Figure IV.7.3).

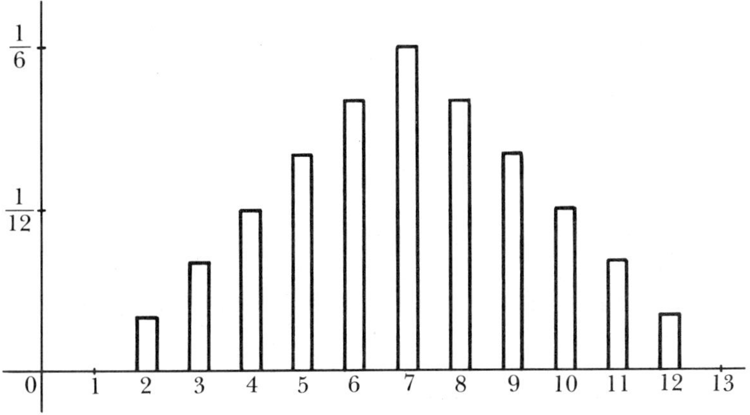

FIGURE IV.7.3 Probabilities for toss of two dice.

IV.7.4 Example. Let X be the number of heads obtained in 10 independent tosses of a fair coin. Its distribution is obtained by using formula (4.5.3) with $n=10$, $p=1/2$.

$$
\begin{aligned}
f(0) &= \binom{10}{0}\left(\frac{1}{2}\right)^{10} &&= \frac{1}{1024}\\
f(1) &= \binom{10}{1}\left(\frac{1}{2}\right)\left(\frac{1}{2}\right)^{9} &&= \frac{10}{1024}\\
f(2) &= \binom{10}{2}\left(\frac{1}{2}\right)^{2}\left(\frac{1}{2}\right)^{8} &&= \frac{45}{1024}\\
f(3) &= \binom{10}{3}\left(\frac{1}{2}\right)^{3}\left(\frac{1}{2}\right)^{7} &&= \frac{120}{1024}\\
f(4) &= \binom{10}{4}\left(\frac{1}{2}\right)^{4}\left(\frac{1}{2}\right)^{6} &&= \frac{210}{1024}\\
f(5) &= \binom{10}{5}\left(\frac{1}{2}\right)^{5}\left(\frac{1}{2}\right)^{5} &&= \frac{252}{1024}\\
f(6) &= \binom{10}{6}\left(\frac{1}{2}\right)^{6}\left(\frac{1}{2}\right)^{4} &&= \frac{210}{1024}\\
f(7) &= \binom{10}{7}\left(\frac{1}{2}\right)^{7}\left(\frac{1}{2}\right)^{3} &&= \frac{120}{1024}\\
f(8) &= \binom{10}{8}\left(\frac{1}{2}\right)^{8}\left(\frac{1}{2}\right)^{2} &&= \frac{45}{1024}\\
f(9) &= \binom{10}{9}\left(\frac{1}{2}\right)^{9}\left(\frac{1}{2}\right) &&= \frac{10}{1024}\\
f(10) &= \binom{10}{10}\left(\frac{1}{2}\right)^{10} &&= \frac{1}{1024}
\end{aligned}
$$

In Figure IV.7.4, note that while 5 is the most probable value of the random variable, its probability is still less than 1/4. On the other hand, there is approximately a 0.66 probability that the variable will lie between 4 and 6.

If X and Y are random variables that depend on the outcome of the same experiment, we consider their *joint distribution*, given by the function f:

4.7.3 $$f(x,y) = P(X=x,\ Y=y)$$

Thus in the experiment of tossing a fair die twice, we may let X and Y be the number of spots showing on the first and second tosses of the die, respectively. Then the joint probability of X and Y is given by

$$
f(x,y) = \begin{cases} \dfrac{1}{36} & \text{for } x,y = 1, 2, \ldots, 6 \\ 0 & \text{otherwise} \end{cases}
$$

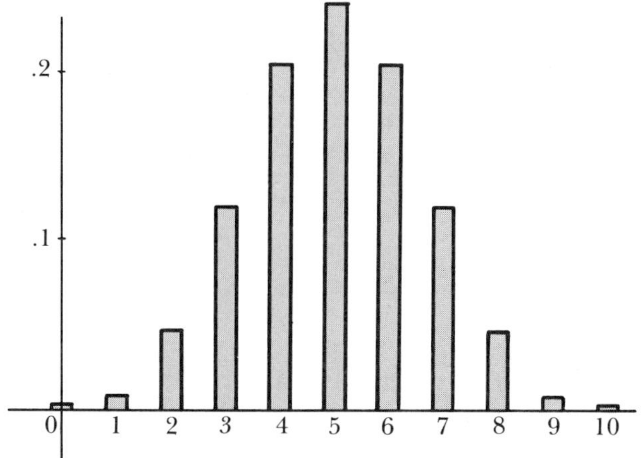

FIGURE IV.7.4 Binomial distribution with $n = 10$, $p = 1/2$.

The *absolute* distribution of X may be obtained from the joint distribution of X and Y; it will be given by

4.7.4 $$g(x) = \sum_j f(x, y_j)$$

where the summation is taken over all the possible values, y_j, of the variable Y. Similarly, the distribution of Y is given by

4.7.5 $$h(y) = \sum_i f(x_i, y)$$

where the summation is taken over all the possible values, x_i, of X.

Closely connected with these are the *conditional* distributions of X and Y. For a possible value, y_j, of Y, we define the conditional distribution of X by

4.7.6 $$g(x \mid y_j) = \frac{f(x, y_j)}{h(y_j)}$$

Similarly, for one of the possible values, x_i, of X, the conditional distribution of Y is given by

4.7.7 $$h(y \mid x_i) = \frac{f(x_i, y)}{g(x_i)}$$

Note the similarity of these definitions to those of Section IV.3.

IV.7.5 Example. Let X be the number of heads obtained in three tosses of a fair coin, and let Y be either 1 or 0, depending on whether the first toss is heads or tails. Then the joint distribution of X and Y is given by Table IV.7.1.

TABLE IV.7.1

Y	X				$h(y)$
	0	1	2	3	
0	1/8	1/4	1/8	0	1/2
1	0	1/8	1/4	1/8	1/2
$g(x)$	1/8	3/8	3/8	1/8	

The marginal row and marginal column give the absolute distributions of X and Y, respectively. The conditional distributions are easily obtained; we might give them in the form of two tables. The first, Table IV.7.2, gives the conditional distribution, $g(x \mid y_j)$; the second, Table IV.7.3, gives $h(y \mid x_i)$:

TABLE IV.7.2

Y	X				
	0	1	2	3	
0	1/4	1/2	1/4	0	$g(x \mid y_j)$
1	0	1/4	1/2	1/4	

TABLE IV.7.3

Y	X				
	0	1	2	3	
0	1	2/3	1/3	0	$h(y \mid x_i)$
1	0	1/3	2/3	1	

As may be seen in Example IV.7.5, the conditional distributions of a random variable may be quite different from its absolute distribution. In some cases, however, these are all equal: the random variables are then independent. We give a more formal definition as follows.

IV.7.6 Definition. Two random variables, X and Y, are *independent* if, for every pair (x,y) of values of the two variables,

$$f(x,y) = g(x)\,h(y)$$

An equivalent definition of independence would be to say that two variables are independent if their joint probability function can be factored into two functions, each of which depends on only one of the variables.

For instance, the random variables X,Y, of Example IV.7.4, are independent. On the other hand, the variables of Example IV.7.5 are not independent.

IV.7.7 Example. Consider the data of Example IV.2.7. We can let X be 0 or 1 according to whether the family is Republican or Democrat, and Y be 0 if the family is not headed by a woman, 1 if it is. Then the joint distribution of (X,Y) is shown in Table IV.7.4.

TABLE IV.7.4

Y	X		$h(y)$
	0	1	
0	21/39	11/39	32/39
1	2/39	5/39	7/39
$g(x)$	23/39	16/39	

The two marginal distributions, $g(x)$ and $h(y)$, are given at the bottom and to the right, respectively, of the joint distribution table.

The conditional distributions are shown in Tables IV.7.5 and IV.7.6.

TABLE IV.7.5

Y	X		
	0	1	
0	21/32	11/32	$g(x \mid y_j)$
1	2/7	5/7	

TABLE IV.7.6

Y	X		
	0	1	
0	21/23	11/16	$h(y \mid x_i)$
1	2/23	5/16	

Note that X and Y are *not* independent.

8. EXPECTED VALUES. MEANS AND VARIANCES

In general, if a random variable has any practical interpretation (if, for instance, it represents money or some measurable service or commodity) it becomes natural to look, in some fashion, for an *average* value of the variable. An example will perhaps explain this best.

Consider the game of roulette, as played in European casinos. The roulette wheel has 37 numbers ranging from 0 to 36. There are many types of bet possible, but the simplest is a bet on a single number. If the ball lands on the chosen number, the player gets back his bet, plus 35 times his bet. If it lands on any other number, he loses his bet.

Let us assume that the player bets \$1 on a single number. Let X be the amount won by the player; then X can have the values -1 (if the player loses) and $+35$ (if he wins.) The probabilities of these two values are obtained by assuming that the roulette wheel is perfectly symmetric, so that the ball has the same probability of landing on each of the 37 numbers. Thus,

$$f(+35) = \frac{1}{37}$$

$$f(-1) = \frac{36}{37}$$

Let us suppose, now, that the player makes this bet, not once, but very many times. If the total number of bets made is n, then, by the frequency definition of probability, he should win approximately $n/37$ times, and lose approximately $36n/37$ times. Since each win is \$35, while each loss is \$1

$$\text{Total winnings} = \frac{n}{37} \cdot 35 = \frac{35n}{37}$$

$$\text{Total losses} \quad = \frac{36n}{37} \cdot 1 = \frac{36n}{37}$$

$$\text{Net winnings} \quad = -\frac{n}{37}$$

It follows that, in a large number, n, of bets, the player will probably lose approximately $n/37$ dollars. (It may be, of course, that he will have a lucky streak and win, or a bad streak and lose much more, but, for large n, these will be rare occurrences.) On the average, he will lose approximately 1/37 dollars per bet. Since a loss is interpreted as a negative win, we obtain a value of $-1/37$ as the average value of the random variable X. We call this the *expected value* of X.

In the general case, let us assume that the random variable X has possible values $x_1, x_2, \ldots, x_m$, with probabilities $f(x_1)$, $f(x_2), \ldots, f(x_m)$ respectively. If the experiment is repeated a large number, n, of times, then X should take on the value x_i approximately $nf(x_i)$ times. If X represents money, or has some such practical interpretation (so that there is a point to adding the values of X obtained), we find that the sum of the values of X, for the (approximately) $nf(x_i)$ times that it has the value x_i, is $nx_if(x_i)$. We now take the sum of these expressions for all possible values of x_i, and obtain (approximately),

$$\sum_{i=1}^{m} nx_if(x_i) = n \sum_{i=1}^{m} x_if(x_i)$$

Since we are generally interested, rather, in the average value (per trial of the experiment) we divide this expression by n, obtaining the somewhat simpler form

4.8.1 $$E[X] = \sum_{i=1}^{m} x_if(x_i)$$

This expression is called the *expected value,* or *expectation,* of the random variable X.

IV.8.1 Example. Let X be the number of spots showing after a toss of a fair die. Its expectation can be calculated from Table IV.8.1.

TABLE IV.8.1

x_i	$F(x_i)$	$x_if(x_i)$
1	1/6	1/6
2	1/6	2/6
3	1/6	3/6
4	1/6	4/6
5	1/6	5/6
6	1/6	6/6

Adding the entries in the last column of this table, we obtain the expectation

$$E[X] = \frac{7}{2}$$

IV.8.2 Example. Let X be the number of heads obtained in 10 tosses of a fair coin. We form Table IV.8.2.

TABLE IV.8.2

x_i	$f(x_i)$	$x_i f(x_i)$
0	1/1024	0
1	10/1024	10/1024
2	45/1024	90/1024
3	120/1024	360/1024
4	210/1024	840/1024
5	252/1024	1260/1024
6	210/1024	1260/1024
7	120/1024	840/1024
8	45/1024	360/1024
9	10/1024	90/1024
10	1/1024	10/1024

Once again, we add the entries in the last column, to obtain

$$E[X]=5$$

If X is a random variable, and U is a function of X, say

$$U=\varphi(X)$$

then U is also a random variable. We may obtain the expected value of U by computing its distribution (which can be done if we know the distribution of X). This is not generally necessary, however, as it is not difficult to see that

4.8.2 $$E[\varphi(X)]=\sum_{i=1}^{m}\varphi(x_i)f(x_i)$$

IV.8.3 Example. Let X be as in Example IV.8.1, and let $U=X^2$. We compute its expectation from Table IV.8.3.

TABLE IV.8.3

x_i	x_i^2	$f(x_i)$	$x_i^2 f(x_i)$
1	1	1/6	1/6
2	4	1/6	4/6
3	9	1/6	9/6
4	16	1/6	16/6
5	25	1/6	25/6
6	36	1/6	36/6

Adding the entries in the last column, we obtain

$$E[U]=\frac{91}{6}$$

Two of the most important descriptive properties of a random variable are the *mean* and the *variance*. These are defined in terms of expectations.

IV.8.4 Definition. Let X be a random variable. Then the *mean*, μ_X, and the *variance*, σ_X^2, of X are defined by

4.8.3 $$\mu_X = E[X]$$

4.8.4 $$\sigma_X^2 = E[(X - \mu_X)^2]$$

The mean is the same as the expected value of X. The *variance*, which is always non-negative, is a measure of the *dispersion* of X: if σ_X^2 is small, then X will usually be close to μ; whereas if σ_X^2 is large, then X will often differ considerably from its mean. The square root of the variance, σ_X, is known as the *standard deviation.*

9. RULES FOR COMPUTING THE MEAN AND VARIANCE

Computation of the mean and variance of a random variable is often simplified by using the following theorem:

IV.9.1 Theorem. Let X and Y be random variables, and α a constant. Then

4.9.1 $$E[\alpha] = \alpha$$

4.9.2 $$E[\alpha X] = \alpha E[X]$$

4.9.3 $$E[X+Y] = E[X] + E[Y]$$

Moreover, if X and Y are independent, then

4.9.4 $$E[XY] = E[X]\,E[Y]$$

Proof. If α is constant (i.e., not a variable), then it will have the value α with probability 1, and no other values. This proves (4.9.1).

To prove (4.9.2), note that

$$E[\alpha X] = \sum \alpha x_i f(x_i) = \alpha \sum x_i f(x_i) = \alpha E[X]$$

To prove (4.9.3), let X and Y have the joint distribution f; let g and h be the absolute distributions of X and Y respectively. Then

$$\begin{aligned}E[X+Y] &= \sum_{i,j} (x_i + y_j)f(x_i,y_j) \\ &= \sum_{i,j} x_i f(x_i,y_j) + \sum_{i,j} y_j f(x_i,y_j) \\ &= \sum_i \left\{x_i \sum_j f(x_i,y_j)\right\} + \sum_j \left\{y_j \sum_i f(x_i,y_j)\right\} \\ &= \sum_i x_i g(x_i) + \sum_j y_j h(y_j) \\ &= E[X] + E[Y]\end{aligned}$$

Finally, suppose X and Y are independent. This means $f(x,y) = g(x)h(y)$. Then

$$\begin{aligned}E[XY] &= \sum_{i,j} x_i y_j f(x_i,y_j) \\ &= \sum_{i,j} (x_i g(x_i))(y_j h(y_j)) \\ &= \sum_i \left\{x_i g(x_i) \sum_j y_j h(y_j)\right\} \\ &= \left\{\sum_i x_i g(x_i)\right\} \left\{\sum_j y_j h(y_j)\right\} \\ &= E[X]\, E[Y]\end{aligned}$$

IV.9.2 Corollary. If X has mean μ and variance σ^2, then

4.9.5 $$\sigma^2 = E[X^2] - \mu^2$$

Proof. We have

$$\begin{aligned}\sigma^2 &= E[(X-\mu)^2] = E[X^2 - 2\mu X + \mu^2] \\ &= E[X^2] - 2\mu E[X] + \mu^2 \\ &= E[X^2] - 2\mu^2 + \mu^2 \\ &= E[X^2] - \mu^2\end{aligned}$$

IV.9.3 Corollary. If X and Y are independent random variables with variances σ_X^2 and σ_Y^2 respectively, then the variance of $X+Y$ is

4.9.6 $$\sigma_{X+Y}^2 = \sigma_X^2 + \sigma_Y^2$$

Proof. We have

$$\begin{aligned}E[(X+Y)^2] &= E[X^2 + 2XY + Y^2] \\ &= E[X^2] + 2E[XY] + E[Y^2]\end{aligned}$$

and

$$\mu^2_{X+Y} = (\mu_X + \mu_Y)^2 = \mu^2_X + 2\mu_X \mu_Y + \mu^2_Y$$

Now, by independence, $E[XY] = \mu_X \mu_Y$, and so

$$\begin{aligned}\sigma^2_{X+Y} &= E[(X+Y)^2] - \mu^2_{X+Y}\\ &= E[X^2] - \mu^2_X + E[Y^2] - \mu^2_Y\\ &= \sigma^2_X + \sigma^2_Y\end{aligned}$$

IV.9.4 Corollary. Let X be a random variable with mean μ_X and variance σ^2_X, and let $Y = \alpha X + \beta$. Then Y has mean

4.9.7 $$\mu_Y = \alpha\mu_X + \beta$$

and variance

4.9.8 $$\sigma^2_Y = \alpha^2 \sigma^2_X$$

We will omit the proof of Corollary IV.9.4.

IV.9.5 Example. Find the mean and standard deviation of the number of successes, X, in n trials of a simple experiment with probability of success p.

Let us assume first that $n = 1$. In this case, X can have only the values 0 and 1, with probabilities q and p, respectively.

$$E[X] = 0 \cdot q + 1 \cdot p = p$$

and

$$E[X^2] = 0^2 \cdot q + 1^2 \cdot p = p$$

It follows that X has mean $\mu = p$, and variance

$$\sigma^2 = E[X^2] - \mu^2 = p - p^2 = pq$$

Thus, for a single trial, $\mu = p$ and $\sigma = \sqrt{pq}$.

For larger values of n, we define the random variable Y_i, where

$$Y_i = \begin{cases} 0 & \text{if } i\text{th trial is a failure} \\ 1 & \text{if } i\text{th trial is a success} \end{cases}$$

It is not difficult to see that

$$X = Y_1 + Y_2 + \ldots + Y_n$$

Now, the Y_i are independent random variables, each having mean p and variance pq. By Theorem V.9.7 and its corollaries, X must have mean np and variance npq. Thus

4.9.9 $$\mu = np; \ \sigma = \sqrt{npq}$$

Note that the standard deviation increases only as the *square root* of the number of trials.

IV.9.6 Example. Let X be the number of spots showing on n independent tosses of a die. Find the mean and variance of X.

Once again, we start by letting $n = 1$. In this case, we know (from previous problems) that $E[X]=7/2$ and $E[X^2]=91/6$. Thus

$$\mu = E[X] = \frac{7}{2}$$

$$\sigma^2 = E[X^2] - \mu^2 = \frac{91}{6} - \frac{49}{4} = \frac{35}{12}$$

For other values of n, we use the fact that the tosses are independent, obtaining

$$\mu = \frac{7n}{2}$$

$$\sigma^2 = \frac{35n}{12}$$

IV.9.7 Example. A gambler bets \$1 on a simple experiment that has probability p of success. If the outcome is a success, he wins \$1 and stops playing. If the outcome is a failure, he loses \$1 and bets \$2 on the next trial of the experiment. He continues in this manner, doubling his bet after each failure, until he obtains a success. He stops betting if he either obtains a success or loses n times in a row (an event that he considers extremely improbable). What is the expected value of his winnings?

The gambler's bet on the kth trial (assuming he is still playing) will be 2^{k-1} dollars. If he loses on each of the first k trials, he will have lost

$$1 + 2 + 2^2 + \ldots + 2^{k-1} = 2^k - 1$$

dollars before betting on the $(k+1)$th trial. The bet then will be 2^k dollars so that, if he wins, he will win back all his losses plus an additional dollar. If, however, he loses n consecutive times, he will wind up losing $2^n - 1$ dollars. Letting X be his winnings, we see that X can have the two values 1 and $1-2^n$. The probability of losing n consecutive times is q^n, and so

$$\begin{aligned} E[X] &= 1(1-q^n) + (1-2^n)q^n \\ &= 1 - 2^n q^n \\ &= 1 - (2q)^n \end{aligned}$$

Now, if $q > 1/2$, then $2q > 1$, and so $(2q)^n$ increases (without limit) as n increases. For large values of n, therefore, the variable X will have an expected value that is negative and very large. This is true even though large values of n mean that the probability q^n of losing will be very small. We conclude that this system of betting (known as a *martingale*) is not a very good system.

10. TWO IMPORTANT THEOREMS

The importance of the mean and standard deviation of a random variable is given by the following two theorems.

IV.10.1 Theorem (Chebyshev's Inequality). Let X have mean μ and standard deviation σ. Then, for any $t > 0$,

4.10.1
$$P(|X-\mu| \geq t\sigma) \leq \frac{1}{t^2}$$

Proof. Assume that the event

$$|X - \mu| \geq t\sigma$$

or, equivalently,

4.10.2
$$(X-\mu)^2 \geq t^2\sigma^2$$

has probability p. We may partition the possible values of X into two sets; those for which (4.10.2) holds and those for which it does not. We may write

$$\begin{aligned} \sigma^2 &= \sum_i (x_i - \mu)^2 f(x_i) \\ &= \sum_i{}' (x_i - \mu)^2 f(x_i) + \sum_i{}'' (x_i - \mu)^2 f(x_i) \end{aligned}$$

where the first sum is over those x_i for which (4.10.2) holds, and the second over the other values of X. Now, the second sum is non-negative, and so

$$\sigma^2 \geq \sum_i{}' (x_i - \mu)^2 f(x_i)$$

and, as (4.10.2) holds for this sum,

$$\sigma^2 \geq \sum_i{}' t^2 \sigma^2 f(x_i)$$

$$= t^2 \sigma^2 \sum_i{}' f(x_i)$$

The sum here is precisely the probability of (4.10.2); thus

$$\sigma^2 \geq t^2 \sigma^2 p$$

and so

$$p \leq \frac{1}{t^2}$$

Chebyshev's inequality states that the probability of a large deviation from the mean can never be too probable; this is given in terms of the standard deviation.

IV.10.2 Theorem (The Law of Large Numbers). Let X be a random variable with mean μ. Let $X_1, X_2, X_3, \ldots$ be independent repetitions of the variable X, and let

$$\overline{X}_n = \frac{1}{n} \sum_{i=1}^{n} X_i$$

Then, for any $\epsilon > 0$, and any $\delta > 0$, there exists n such that

4.10.3 $$P(|\overline{X}_n - \mu| \geq \epsilon) < \delta$$

Proof. Assume X has variance σ^2. Then

$$Y_n = \sum_{i=1}^{n} X_i$$

will have variance $n\sigma^2$ and so $\overline{X}_n = \frac{1}{n} Y_n$ will have variance σ^2/n, and standard deviation $\sigma/\sqrt{n}$. Applying Chebyshev's inequality, we see that

$$P(|\overline{X}_n - \mu| > \epsilon) \leq \frac{\sigma^2}{n\epsilon^2}$$

and it is easy to see that, as n increases, the right side of this inequality approaches 0.

The Law of Large Numbers states that the idea of an expected value is really a valid idea: if the experiment is repeated often enough, it is almost certain that $\overline{X}_n$, the *average observed value* of X, will be quite close to μ. Note that this does *not* say that any particular value of $\overline{X}_n$ will be very probable, only that those values that are close to μ will be much more probable than those that are not. For example, in 100 tosses of a fair coin, the most probable number of heads is 50 (giving a value $\overline{X}_n = 0.5$). Yet the probability of having *exactly* 50 heads is only 0.08. This is not large, but, on the other hand, the probability of exactly 40 heads is only about 0.01, while that of exactly 35 heads is only 0.001. The probability of having between 40 and 60 heads (i.e., of $0.4 \leq \overline{X}_n \leq 0.6$), however, will be about 0.96.

IV.10.3 Example. In a certain large city, 55 per cent of the population are Democrats. A pollster takes a sample of n people, asking their party affiliation. How large should n be, in order that there be a probability of at least 0.95 that between 54 per cent and 56 per cent of those sampled be Democrats?

Since the city is large, the trials may be considered independent. In a sample of size n, the number of Democrats, X, will be a binomial random variable, with mean

$$\mu = 0.55\, n$$

and variance

$$\sigma^2 = (0.55)(0.45)\, n = 0.2475\, n$$

The event "between 54 per cent and 56 per cent are Democrats" can be expressed as

$$\left|\frac{X - \mu}{n}\right| \leq 0.01$$

or, equivalently,

$$|X - \mu| \leq 0.01\, n$$

We apply Chebyshev's inequality, with $t\sigma = 0.01\, n$, or

$$t^2 = \frac{0.0001\, n^2}{0.2475\, n} = \frac{n}{2475}$$

The probability of the desired event is at least $1 - 1/t^2$; we therefore want $1/t^2 \leq 0.05$. It is sufficient to take

$$\frac{2475}{n} \leq 0.05$$

or, equivalently, $n \geq 49{,}500$. A sample of 49,500 will be adequate for our purpose.

In reality, this value of n is unnecessarily large; it may be shown that $n = 9604$ would be sufficient. This is because Chebyshev's inequality uses only the mean and variance of the distribution; it does not take special features of the distribution (i.e., that it is a binomial distribution) into account.

Theorem IV.10.2 (The Law of Large Numbers) nearly guarantees that, in the long run, the average value of a random variable will approach its expectation. It becomes natural, then, *if an experiment can be repeated many times,* to take the course of action that will maximize the decision-maker's *expected* profit. This is of importance when the profit (or cost) depends on both the decision-maker's action and the outcome of the experiment.

IV.10.4 Example. An automobile company produces a DeLuxe Model that costs \$4000 to produce and sells for \$8500. The demand for this model is uncertain; the company estimates (from past experience) that as many as four cars might be sold; the probability that k cars will be sold is given by Table IV.10.1.

TABLE IV.10.1

k	$f(k)$
0	0.1
1	0.4
2	0.3
3	0.15
4	0.05

How many cars should the company produce to maximize its profits?

Let us assume that the company produces n cars, and the demand is for k cars. Assuming a profit of \$4500 for each car sold, and a loss of \$4000 for each car produced but not sold, we obtain the table of profits (Table IV.10.2).

TABLE IV.10.2

k	n					$f(k)$
	0	1	2	3	4	
0	0	−4000	−8000	−12000	−16000	0.1
1	0	4500	500	−3500	−7500	0.4
2	0	4500	9000	5000	1000	0.3
3	0	4500	9000	13500	9500	0.15
4	0	4500	9000	13500	18000	0.05
$E[X]$	0	3650	3900	1600	−2075	

The right-hand column gives, once again, the probabilities of each value. The bottom row gives the expected profits for each choice of n. We see that this expectation gives a maximum for $n = 2$. We conclude that the company should produce two cars every year.

In dealing with expectations, it is assumed, at least tacitly, that the experiment can be repeated many times. Otherwise the Law of Large Numbers has no chance to work. It follows that, when the experiment cannot be repeated, the criterion of maximizing the expected profit cannot be applied indiscriminately. Consider, for example, a person who is given the possibility of investing $5000 in a new company. The company has, say, a 0.5 probability of succeeding, in which case the investor will receive $20,000 in profits. On the other hand, the company might also fail with probability 0.5, in which case the investor will lose his $5000. It is easy to see that the expected profit from this investment is

$$\frac{1}{2}(20{,}000) + \frac{1}{2}(-5000) = 7500$$

so that, to maximize the expected profit, the investor should choose this gamble rather than some other course of action that might give him a certain profit of, say, $6000. On the other hand, if the investor is not too rich, he might be unwilling to risk $5000; he could, quite reasonably, feel that it is more money than he can afford to lose and that, moreover, he will have no chance to recoup his losses if the company fails (he will then be broke). For a very rich investor, however, the gamble might be worth-while: he can afford a $5000 loss. It is clear that it is not sufficient to look at money only; economists have suggested that, instead of maximizing the expected monetary profit, the decision maker should maximize the expected *utility* of the outcome. Why this should be, and what is meant by utility, is beyond the scope of this book.

PROBLEMS ON RANDOM VARIABLES

1. A fair die is tossed 100 times; let X be the number of aces obtained. Find the mean and variance of X.

2. A player tosses a fair coin 10 times. If the number X of heads is even, then the player wins X dollars; if X is odd, then the player loses X dollars. What are the player's expected winnings?

3. A merchant stocks units of a perishable item. Each unit costs him \$2 and sells for \$5. The merchant estimates that the demand k for this item has distribution $f(k)$ given by

$$\begin{aligned} f(0) &= 0.1 \\ f(1) &= 0.15 \\ f(2) &= 0.25 \\ f(3) &= 0.3 \\ f(4) &= 0.15 \\ f(5) &= 0.05 \end{aligned}$$

How many units should the merchant stock to maximize his expected profits?

4. A fair coin is tossed 10,000 times. Using Chebyshev's inequality, find an upper bound for the probability that the number of heads obtained be smaller than 4500 or larger than 5500.

5. If X and Y are random variables with mean μ_X and μ_Y respectively, and variance σ_X^2 and σ_Y^2 respectively we define the *covariance*

$$\sigma_{XY} = E[(X - \mu_X)(Y - \mu_Y)]$$

Prove that

$$\sigma_{XY} = E[XY] - E[X]E[Y]$$

and that, if $U = aX + b$, and $V = cY + d$ (where a, b, c, and d are constants) then

$$\sigma_{UV} = ab\,\sigma_{XY}$$

6. If X and Y are random variables with variance σ_X^2 and σ_Y^2 respectively, and covariance σ_{XY}, we define the *correlation coefficient*

$$\rho_{XY} = \frac{\sigma_{XY}}{\sigma_X \sigma_Y}$$

Prove that, if $U = aX + b$, and $V = cY + d$, with $ac > 0$, then

$$\rho_{UV} = \rho_{XY}$$

7. Let X, Y, and Z be mutually independent random variables with variance σ_X^2, σ_Y^2, and σ_Z^2, respectively. Let

$$U = X + Z$$
$$V = Y + Z$$

Find σ_U^2, σ_V^2, σ_{UV} and ρ_{UV} in terms of σ_X^2, σ_Y^2, and σ_Z^2.

8. A coin is tossed 10 times. Let X be the number of heads obtained, and let $Y = X^2$. Find the correlation coefficient ρ_{XY}.

9. An urn contains 100 balls each painted with two colors. One of the colors is either red or black; the other color is either green or yellow. The distribution of colors is given by the following table:

	Green	Yellow
Red	38	15
Black	15	32

A ball is drawn from the urn; let $X = 0$ if the first color is red, and 1 if it is black. Let $Y = 0$ if the second color is green, and 1 if it is yellow. Find the correlation coefficient ρ_{XY}.

10. A machine is either working well (probability 0.8) or it is not. If it is working well, its product will be good with probability 0.75; if not working well, its product will be good with probability 0.4. Find the correlation coefficient ρ_{XY}, where $X = 0$ or 1 depending on whether the machine is working well or not, and $Y = 0$ or 1 depending on whether its product is good or defective.

CHAPTER V

DIFFERENTIAL CALCULUS

1. DERIVATIVES

A manufacturer has analyzed his production and marketing costs and arrived at a formula

$$C = f(x)$$

giving the total cost C of making and selling a quantity x of his product. He has also studied the effect on price of putting a quantity x on the market and come up with another formula

$$R = g(x)$$

giving the total revenue R from marketing x units of his product in, say, a month.

He has production and sales fixed at x units per month, and raises the question of whether or not he should make any change in this. Increased production will undoubtedly increase costs, but hopefully it will also increase revenue. The obvious question is, "Which will go up faster?"

To study a little more carefully what we mean by the word "faster" here, let us turn to a graphical representation of the situation. Suppose the curve in Figure V.1.1 gives the cost, C, in terms of the quantity produced, x. The increased cost caused by an increase in production from x to $x + h$ is $f(x + h) - f(x)$; so the fraction

5.1.1 $$\frac{f(x+h) - f(x)}{h}$$

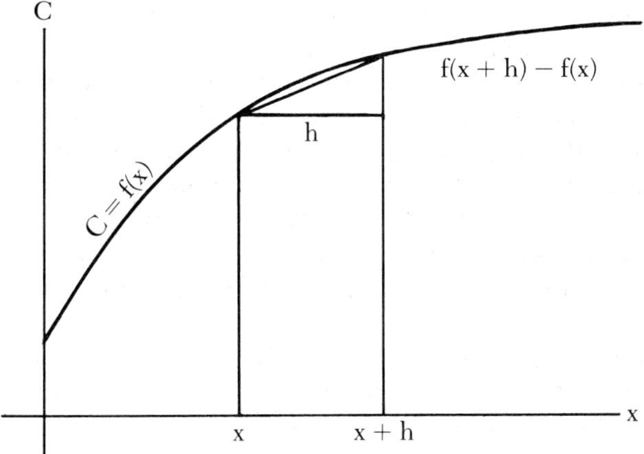

FIGURE V.1.1 Cost curve for production of x units per month.

gives the increase in cost per unit increase in production. However, it appears from Figure V.1.1 that the fraction (5.1.1) is also the slope of the chord to the cost curve drawn from the point $(x, f(x))$ to the point $(x + h, f(x + h))$. So, a study of slopes seems pertinent to the discussion of changes in cost (or revenue, as the case may be).

Unless the cost curve is a straight line, the fraction (5.1.1) obviously changes as h does. What we are interested in here is the "immediate" effect at x of production changes. Figure V.1.2 suggests a geometric approach to an explanation of what we mean by "immediate." Recall that we are interested in increase in cost per unit increase in production, and that for a definite increase h in production this is given by the slope of a chord to the cost curve. Now, as suggested by Figure V.1.2, as h gets smaller the chords swing around toward the position of the tangent line. For this reason, we will define the immediate effect at x of production change on cost as *the slope of the tangent line to the cost curve at x.*

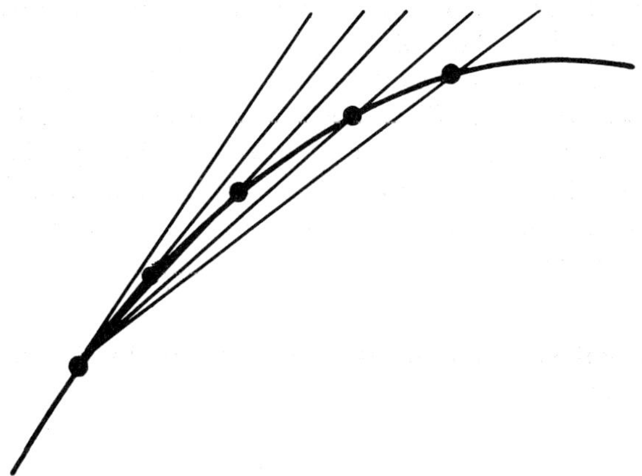

FIGURE V.1.2 Behavior of the chord to a curve as h decreases.

The economist calls this quantity the *marginal cost* at x. Similarly, the slope of a tangent line to the revenue curve is called *marginal revenue.* Given the situation depicted at the beginning of this section, in which cost and revenue are assumed to depend simply on volume of production, the following procedures would seem to be indicated. If marginal cost is less than marginal revenue, increase production. If marginal cost is greater than marginal revenue, decrease production. If marginal cost equals marginal revenue, leave a good thing alone. The significance of this last dictum will appear more forcibly in Section 7.

The natural question now is, "Given cost in terms of quantity produced, how do you find marginal cost?" Let us attack the more general question, "Given the equation of a curve, how do you find the slope of a tangent line?"

Recall that (5.1.1) gives the slope of a chord and that (Fig. V.1.2) the shorter the chord the closer its slope seems to be to that of the tangent line. We now resort to the idea that the mathematician calls a *limit,* and write for the slope of the tangent line at x,

5.1.2
$$\lim_{h\to 0}\frac{f(x+h)-f(x)}{h}$$

We shall make no attempt to present a comprehensive discussion of the limit concept here. An informal description is that a limit is the end result of a trend. Specifically, the symbol

$$\lim_{h\to 0}$$

in (5.1.2) is read, "the limit as h tends to zero." Note that to substitute $h=0$ in (5.1.1) yields 0/0, which is meaningless. Furthermore, for $h\neq 0$ we never actually get the slope of the tangent line out of (5.1.1). Nevertheless, as we shall see shortly, when certain specific functions are used in place of the general f in (5.1.2), a little strategically chosen algebraic manipulation reveals a definite trend for small h, and limits of the form (5.1.2) can be evaluated. Finally, we must emphasize (and this, too, we shall illustrate shortly) that there are functions f for which the fraction (5.1.1) exhibits no trend for small h (or exhibits more than one), and for such functions we say the limit (5.1.2) does not exist.

V.1.1 Definition. The *derivative* of a function f is a function f' defined for each x by

$$f'(x)=\lim_{h\to 0}\frac{f(x+h)-f(x)}{h}$$

provided this limit exists. If the limit defining $f'(x)$ exists, f is said to be *differentiable* at x.

V.1.2 Definition. If x and y are variables related by an equation $y=f(x)$, then the *derivative of y with respect to x* is denoted by D_xy and defined by

$$D_xy=f'(x).$$

Note that we have looked at three interpretations of the idea of derivative. If x is volume of production, C is cost, and R is revenue, then D_xC is marginal cost and D_xR is marginal revenue. If x and y are abscissas and ordinates, respectively, on a curve, then D_xy is the slope of the tangent line to the curve at x.

V.1.3 Example. Given that $f(x)=x^2$, find $f'(x)$. We have $f(x+h) = (x+h)^2$; so

$$f'(x)=\lim_{h\to 0}\frac{f(x+h)-f(x)}{h}=\lim_{h\to 0}\frac{(x+h)^2-x^2}{h}$$

Expanding the binomial square in the last numerator, we have

$$(x+h)^2-x^2=x^2-2xh+h^2-x^2=2xh-h^2.$$

The last expression has h as a factor, and so the h in the denominator can be divided out to give

$$\frac{f(x+h)-f(x)}{h}=\frac{2xh-h^2}{h}=2x-h.$$

Note that there is no division by zero involved here. A limit as $h\to 0$ is the end result of a trend involving values of h close to but different from zero. Thus, we have

$$f'(x)=\lim_{h\to 0}(2x-h)=2x.$$

We take it that this last step is reasonably obvious. If h is close to 0, then $2x-h$ is close to $2x$, and the end result of this trend on the part of $2x-h$ is $2x$ itself.

V.1.4 Example. If cost is given in terms of volume of production by $C=\sqrt{x}$, find an expression for marginal cost in terms of x. Here we have

$$D_xC=\lim_{h\to 0}\frac{\sqrt{x+h}-\sqrt{x}}{h}$$

224 The secret to unscrambling this one is to rationalize the numerator by multiplying numerator and denominator by $\sqrt{x+h}+\sqrt{h}$:

$$\frac{\sqrt{x+h}-\sqrt{x}}{h}\cdot\frac{\sqrt{x+h}+\sqrt{x}}{\sqrt{x+h}+\sqrt{x}}=\frac{x+h-x}{h(\sqrt{x+h}+\sqrt{x})}$$

$$=\frac{h}{h(\sqrt{x+h}+\sqrt{x})}$$

$$=\frac{1}{\sqrt{x+h}+\sqrt{x}}$$

Thus,

$$D_xC=\lim_{h\to 0}\frac{1}{\sqrt{x+h}+\sqrt{x}}=\frac{1}{2\sqrt{x}}.$$

V.1.5 *Example.* Find an expression for the slope of the tangent line at (x, y) to the curve whose equation is $y = 1/x$. The slope of the tangent is given by

$$D_xy=\lim_{h\to 0}\frac{\dfrac{1}{x+h}-\dfrac{1}{x}}{h}=\lim_{h\to 0}\frac{\dfrac{x-(x+h)}{(x+h)\,x}}{h}$$

$$=\lim_{h\to 0}\frac{-h}{h(x+h)x}=\lim_{h\to 0}\frac{-1}{(x+h)x}=\frac{-1}{x^2}.$$

V.1.6 *Example.* Find the equation of the tangent line to the curve $y = 1/x$ at the point $(2, 1/2)$. By V.1.5, the slope of this tangent line is

$$\frac{-1}{x^2}=\frac{-1}{4};$$

therefore, using the point-slope form of the equation of a line we have as the required equation

$$y-\frac{1}{2}=-\frac{1}{4}(x-2),$$

$$4y-2=-x+2,$$

$$x+4y-4=0.$$

V.1.7 *Example.* Let f be defined by

$$f(x)=|x|=\begin{cases} x \text{ for } x\geqslant 0,\\ -x \text{ for } x<0.\end{cases}$$

(Note: $|x|$ is called the *absolute value* of x.) Show that $f'(0)$ does not exist. If $f'(0)$ were defined, it would be the limit of

$$\frac{|0+h|-|0|}{h}=\frac{|h|}{h}=\begin{cases} h/h=\ \ 1 \text{ for } h>0, \\ -h/h=-1 \text{ for } h<0. \end{cases}$$

Therefore, some values of the pertinent fraction are 1 and others are −1, and there is no single trend, hence no limit. Figure V.1.3 explains this geometrically; the curve $y=|x|$ has a sharp corner at the origin, hence no unique tangent here.

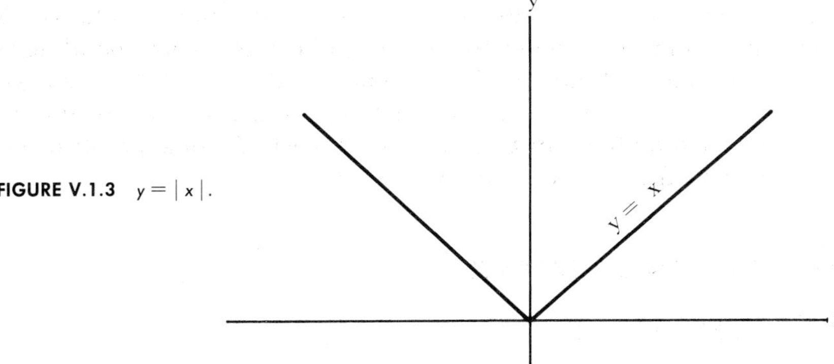

FIGURE V.1.3 $y=|x|$.

V.1.8 Example. Using the data of Example III.5.5, find the poultry farmer's revenues as a function of quantity produced, and find the marginal profit per dozen eggs.

We are told that the farmer can sell x dozen eggs at a price p, where

$$x=16{,}000-200p$$

or

$$p=80-\frac{x}{200}$$

and that his revenue is then

$$r=px=80x-\frac{x^2}{200}.$$

The marginal profit is of course the derivative of r. Then, if $r=f(x)$,

$$f(x+h)=80x+80h-\frac{x^2+2hx+h^2}{200}$$

and so

$$\frac{f(x+h)-f(x)}{h}=\frac{80x+80h-80x}{h}-\frac{x^2+2hx+h^2-x^2}{200h}$$

$$=80-\frac{x}{100}-\frac{h}{200}.$$

Therefore,

$$D_x r=\lim_{h\to 0}\left(80-\frac{x}{100}-\frac{h}{200}\right)=80-\frac{x}{100}$$

is the marginal revenue in cents per dozen. We saw in Example III.5.5 that maximum profits were obtained at a price of 50¢, corresponding to a production of 6000 doz., and we see here that, for $x=6000$, we have $D_x r=20$. We were told that marginal cost is always 20¢ per dozen, and we conclude that, for this example at least, maximal profits occur when marginal cost equals marginal revenue.

PROBLEMS ON DERIVATIVES

Find $f'(x)$:

1. $f(x)=x$
2. $f(x)=x^3$
3. $f(x)=1/x^2$
4. $f(x)=x^{1/3}$ [Hint: $a^3-b^3=(a-b)(a^2+ab+b^2)$]
5. $f(x)=x^2-2x$
6. $f(x)=2x^3+4x^2$
7. $f(x)=x^2+3$
8. $f(x)=(x+3)^2$

Find the equation of the tangent line to each of the following curves at the indicated point.

9. $y=x^2$ $(2, 4)$
10. $y=x^2$ $(-3, 9)$
11. $y=1/x$ $(-1, -1)$
12. $y=\sqrt{x}$ $(4, 2)$
13. $y=2-x^2$ $(1, 1)$
14. $y=3-(1/x)$ $(1, 2)$

15. A landlord owns a large piece of farm land. If the field is tended by x men, the monthly revenue from it is y dollars, where

$$y = 100\sqrt{x} - 0.01\,x^2.$$

Find the marginal revenue per man employed.

16. A steel mill can produce x tons of steel at an average price of y dollars per ton, where

$$y = 1.01\,x^2 - 2x + 150.$$

Find the marginal cost of producing a ton of steel.

2. POWER FUNCTIONS.

The examples in the previous section showed us how to compute the derivatives of certain functions, by the use of Definition V.1.1. Unfortunately, this is a very lengthy procedure, as can be seen from those same examples. Moreover, such a procedure often calls for considerable ingenuity. It therefore becomes important to look for rules which will allow us to differentiate the most common functions —at least—with a minimum of work. We look now for such rules.

From Examples V.1.3 to V.1.5 a pattern begins to emerge. These results may be summarized as follows.

V.1.3: $D_x x^2 = 2x.$

V.1.4: $D_x x^{1/2} = \frac{1}{2}\,x^{-1/2}.$

V.1.5: $D_x x^{-1} = -x^{-2}.$

The general formula suggested here is

5.2.1 $$D_x x^n = nx^{n-1}.$$

In the examples of Section 1 this was illustrated for $n = 2$, 1/2 and -1. Now, formula (5.2.1) is indeed correct for all values of n, positive or negative, integral or fractional. Using the binomial theorem, we can derive it for the case of positive integer n.

$$D_x x^n = \lim_{h\to 0} \frac{(x+h)^n - x^n}{h}$$

$$= \lim_{h\to 0} \frac{x^n + nx^{n-1}h + \dfrac{n(n-1)}{2}\,x^{n-2}h^2 + \cdots + h^n - x^n}{h}$$

$$= \lim_{h\to 0} \frac{nx^{n-1}h + \frac{n(n-1)}{2} x^{n-2}h^2 + \cdots + h^n}{h}$$

$$= \lim_{n\to 0} \left(nx^{n-1} + \frac{n(n-1)}{2} x^{n-2}h + \cdots + h^{n-1}\right)$$

$$= nx^{n-1}.$$

We shall not derive the other cases of (5.2.1) at this time. However, it is a formula that should be learned, and we shall proceed henceforth to use it for all values of the exponent n.

V.2.1 Theorem. If

$$\lim_{h\to 0} F(h) = a, \lim_{h\to 0} G(h) = b$$

and if c is a constant, then

$$\lim_{h\to 0} [F(h) + G(h)] = a + b \text{ and } \lim_{h\to 0} c\,F(h) = ca.$$

In the light of the highly informal approach we have taken to the notion of limit, we can obviously give no proof of this theorem. We ask the student to examine it for plausibility, and we will proceed to use it.

V.2.2 Theorem. If $y = f(x) + g(x)$, then $D_x y = f'(x) + g'(x)$.
Proof.

$$D_x y = \lim_{h\to 0} \frac{f(x+h) + g(x+h) - [f(x) + g(x)]}{h}$$

$$= \lim_{h\to 0} \left[\frac{f(x+h) - f(x)}{h} + \frac{g(x+h) - g(x)}{h}\right]$$

$$= f'(x) + g'(x)$$

by V.2.1.

V.2.3 Theorem. If c is a constant and $y = cf(x)$, then $D_x y = cf'(x)$.
Proof.

$$D_x y = \lim_{h\to 0} cf(x+h) - cf(x)$$

$$= c \lim_{h\to 0} \frac{f(x+h) - f(x)}{h}$$

$$= c\,f'(x)$$

again by V.2.1.

Informally, the derivative of a sum is the sum of the derivatives, and a constant factor carries over to the derivative. There might be some question why we give a "proof" of these facts based on a theorem on limits for which we give no semblance of proof. The situation is this. Limits combine with algebra very nicely. Indeed, given $\lim F(h) = a$ and $\lim G(h) = b$, it follows that $\lim F(h)G(h) = ab$. However, it is *not* true that if $y = f(x)g(x)$, then $D_x y = f'(x)g'(x)$. This stems, not from the theory of limits, but from the algebra of fractions. Therefore, we have given our "proofs" above to show that the algebra works right to give simple formulas for derivatives of sums and constant multiples. The correct formula for the derivative of a product is more complicated than this, and shall be given in a later section.

V.2.4 Theorem. If $f(x) \equiv c$ (c a constant), then $f'(x) \equiv 0$.

Proof.

$$f'(x) = \lim_{h\to 0} \frac{c - c}{h} = \lim_{h\to 0} 0 = 0.$$

Note that this fits formula (5.2.1) with $n = 0$. We single this result out for special mention for two reasons. First, the student may not have enough confidence in (5.2.1) to accept it for $n = 0$ without a special demonstration. Second, we want to call attention to the converse of V.2.4.

V.2.5 Theorem. If $f'(x) \equiv 0$, then for some constant c, $f(x) \equiv c$.

This is a very deep theorem and we make no attempt to give a proof; however, we shall refer to it on several occasions. Why, though, is the proof of V.2.4 trivial while that of V.2.5 is profound? The proof of V.2.4 is based on the fact that if a function, $(c - c)/h$, is everywhere zero, then its limit is. On the other hand, a function may have the limit zero without ever being zero; so V.2.5 is far from obvious. (For example, given the function $f(x) = x^2$, we see that $\dfrac{f(0 + h) - f(0)}{h}$ has limit 0 as h goes to 0, but is never actually zero.) What it says is that if, for every x, the limit of the fractions defining the derivative is zero, then for every x and every h the fractions themselves are zero. This follows from the special structure of these fractions and not from any simple theorem on limits.

Formula (5.2.1) and theorems V.2.2 to V.2.4 give us some practical methods for computing derivatives. The process of computing a derivative is called *differentiation*, and a large portion of an introductory course in calculus is concerned with differentiation by formula. As the following examples will show, the formulas of this section enable us to write down a large number of derivatives by inspection.

V.2.6 *Example.* Find $D_X y$ if $y = 5x^4 - 2x^3 + 6x - 7$.

Solution:

$$\begin{aligned} D_x y &= D_x(5x^4) + D_x(-2x^3) + D_x(6x) + D_x(7) && \text{by V.2.2} \\ &= 5D_x x^4 - 2D_x x^3 + 6D_x x + D_x(7) && \text{by V.2.3} \\ &= 20x^3 - 6x^2 + 6 + D_x(7) && \text{by (5.2.1)} \\ &= 20x^3 - 6x^2 + 6 && \text{by V.2.4} \end{aligned}$$

For this first example we have written out a step-by-step justification. In practice, the answer should be written down by inspection.

V.2.7 *Example.*

$$\begin{aligned} D_x\left(6x^5 - \frac{3}{x^2} + 2\sqrt[3]{x}\right) &= D_x(6x^5 - 3x^{-2} + 2x^{1/3}) \\ &= 30x^4 + 6x^{-3} + \left(\frac{2}{3}\right)x^{-2/3}. \end{aligned}$$

V.2.8 *Example.* Let x be quantity of production, C cost, and R revenue. Given that

$$C = 10 + \frac{x^2}{4000}, \quad R = \sqrt{x},$$

find x so that marginal cost and marginal revenue are equal. Using our formulas we have

$$D_x C = \frac{x}{2000}, \quad D_x R = \frac{1}{2\sqrt{x}};$$

we set these equal and solve for x.

$$\begin{aligned} \frac{x}{2000} &= \frac{1}{2\sqrt{x}} \\ x^{3/2} &= 1000 \\ x &= 1000^{2/3} = (\sqrt[3]{1000})^2 = 10^2 = 100. \end{aligned}$$

PROBLEMS ON POWER FUNCTIONS

Find $D_x y$.

1. $y = 5x^3 + 6x^2 - 7$
2. $y = 3x^4 - 5x^2 + 6x$
3. $y = 6x^7 - 3x^5 + 5$

4. $\sqrt{2x} + 2\sqrt{x}$

5. $y = \dfrac{x^2}{2} + \dfrac{2}{x^2}$

6. $y = \sqrt[3]{x^2} + \sqrt{x^3}$

7. $y = x^{1/3} + x^{-1/3}$

8. $y = \dfrac{x^2+1}{x} = x + \dfrac{1}{x}$

9. $y = \dfrac{x+2}{\sqrt{x}}$

10. $y = \dfrac{x^2 + x + \sqrt{x}}{x^{3/2}}$

11. $y = (x+1)^2 = x^2 + 2x + 1$

12. $y = (x+3)(x-2)$

13. $y = x(x^2 - 3x)$

14. $y = x^2(1 - \sqrt{x})$

15. $y = x\sqrt{x} - 3$

Find the equation of the tangent line to each of the following curves at the indicated point.

16. $y = 2x^2 + 1\,(3, 19)$

17. $y = x^3 - 5\,(2, 3)$

18. $y = x^2 - 2x + 1\,(1, 0)$

19. $y = x^3 - x^2\,(3, 18)$

20. $y = x^2 + 4x + 1\,(-3, -2)$

Find the value of x for which $D_xC = D_xR$.

21. $C = x^2 + 1,\ R = x^3$

22. $C = 1 + 8\sqrt{x},\ R = 3x^{2/3}$

23. A miner finds that, at $10 per ton, he can sell 500 T of ore, but at $12 per ton, he can sell only 300 T. Assuming demand to be a linear function of price, obtain revenue as a function of amount produced. What is the marginal value of the ore, per ton?

24. The cost of producing x yards of copper wire is y dollars, where

$$y = 300 + 25x + 0.02x^2 + 0.001x^{7/2}.$$

Find the marginal cost of this wire, per yard.

3. DIFFERENTIALS

We have introduced the notation $D_x y$ for the derivative of y with respect to x. Probably the most widely used notation for this quantity is that introduced by the German philosopher and mathematician, Gottfried W. Leibnitz (1646–1716):

$$\frac{dy}{dx}.$$

It is probably fair to say that the Leibnitz notation for a derivative has been both the most useful and the most controversial item in the history of calculus. The crux of the matter is that dy/dx looks like a fraction, but is it? The confusing thing is that given $y = f(x)$ and using Leibnitz notation, we have

$$\frac{dy}{dx} = \lim_{h \to 0} \frac{f(x+h) - f(x)}{h}$$

Thus, there is a temptation to say that $dy = \lim \, [f(x+h) - f(x)]$ and $dx = \lim h$ However, this is sheer nonsense because these individual limits are both zero. So, for over 200 years there was a lot of double talk about how the pieces of the Leibnitz fraction are not "really" zero; they are "infinitesimals," whatever that means.

Happily, in recent years the matter has been satisfactorily cleared up. According to modern theory if x and y are abscissa and ordinate variables on a curve, then dx and dy are the increments in x and y, not along the curve, but, rather, along the tangent line to the curve. This is pictured in Figure V.3.1. Note that dx and dy are not particularly "small" and that at each point of the tangent line $dy \div dx$ does, indeed, give the slope of the tangent.

Actually, the modern theory is much more sophisticated than this, though it does yield this geometric interpretation of dx and dy when x

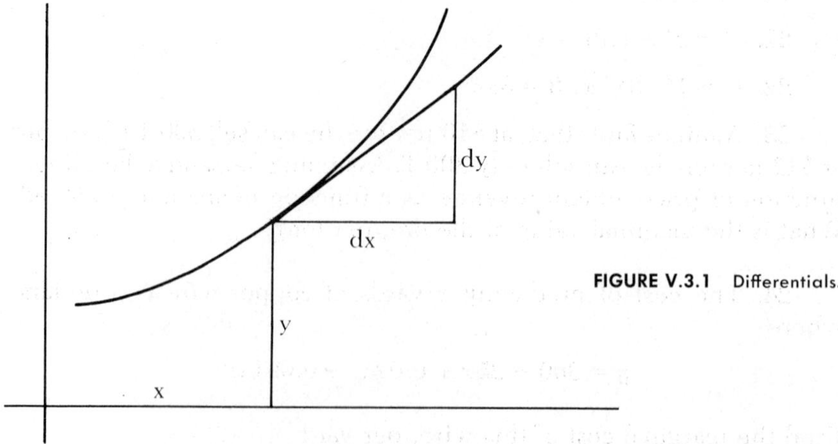

FIGURE V.3.1 Differentials.

and y are rectangular coordinates. What are needed to make Leibnitz notation really work are not just dx and dy but things like $d(x^2)$, $d(2y^3-5)$, and so on. There is available today a very ingenious construction by which each variable u on a smooth curve generates a variable du, called the *differential* of u, on each tangent line. Differentials are so defined that for any variables u and v and constants a and b we have

5.3.1 $$d(au+bv) = a\,du + b\,dv,$$

5.3.2 $$dv = D_u v\,du.$$

We shall not try to present any details of the modern definition of differentials here. A summary may be found in Munroe, *Calculus*, Saunders (1970), Chapter 3. We do want to use differentials here, and the facts that we need to know for their effective use are that they satisfy (5.3.1) and (5.3.2) and that for x and y the differentials have the geometric significance pictured in Figure V.3.1. Note that it follows at once from (5.3.2) (divide by du) that the Leibnitz fraction does, indeed, give appropriate derivatives.

V.3.1 Example. Find $d(4x^3)$ in terms of x and dx. If we set $v=4x^3$, then by the formulas of Section 2,

$$D_x v = 12x^2;$$

so by (5.3.2),

$$d(4x^3) = dv = 12x^2\,dx.$$

Normally, the result $d(4x^3) = 12x^2\,dx$ is written down by inspection. We go through the above discussion this one time to show how it follows from the basic properties of the differential listed above.

V.3.2 Example. Find in terms of x and y an expression for slopes of tangent lines to the ellipse

$$3x^2+4y^2=5.$$

Turning to the equation of the ellipse, we take differentials:

$$8y\,dy = -6x\,dx$$

$$\frac{dy}{dx} = \frac{-3x}{4y}$$

Since (Fig. V.3.1) dy/dx gives slopes of tangent lines, this solves the problem.

V.3.3 Example. Find the equation of the line tangent to the curve

$$4x^3 - 3x + 2y^3 - y^2 = 2$$

at the point (1, 1). We proceed to find the slope as in V.3.2.

$$12x^2\,dx - 3dx + 6y^2\,dy - 2y\,dy = 0$$

$$(6y^2 - 2y)\,dy = (3 - 12x^2)\,dx$$

$$\frac{dy}{dx} = \frac{3 - 12x^2}{6y^2 - 2y} = -\frac{9}{4}$$

at (1, 1). So, the equation of the tangent line is

$$y - 1 = -\frac{9}{4}(x - 1),$$

$$9x + 4y - 13 = 0.$$

PROBLEMS ON DIFFERENTIALS

Find $D_x y$ in terms of x and y.

1. $5x^4 + 3x^3 - 2y^5 = 7$
2. $4x^3 + 7x = y^4 - 2y^2$
3. $x^{3/2} + y^{3/2} = 1$
4. $x^3 + y^3 - 3x + y = 1$
5. $x^2 - y^3 = 2x + y - 3$
6. $\sqrt{x} - \sqrt{y} = 1$
7. $3x^2 - 4y^2 = 2$
8. $x = 2y^2 + 3y - 1$

Find the equation of the tangent line to each of the following curves at the indicated point.

9. $x^2 + y^2 = 25$ (3, 4)
10. $2x^2 - x + y^3 = 7$ (2, 1)
11. $\sqrt{x} + \sqrt{y} = 5$ (4, 9)
12. $x^2 + y^2 = 2y$ (0, 0)
13. Let $y = \sqrt{x}$ and compute $D_x y$; then solve the given equation to get $x = y^2$ and compute $D_y x$. Show that

$$D_y x = \frac{1}{D_x y}.$$

14. Use differentials to show that the result of Problem 13 holds in general, not just for that example.

15. Production costs y at a certain plant are related to the quantity x produced by the equation

$$y^2 - y = x^3 + \sqrt{x}$$

Find the marginal cost in terms of x and y.

16. The cost of producing x Snoopy dolls and y Charlie Brown dolls is

$$5x + 7y + 0.1x^2 + 0.05y^3.$$

Given that total production costs are fixed at \$15,000, find $D_x y$.

17. The women's fashion market is such that the average price x of boots and the average price y of shoes should be pegged so that

$$\sqrt{x} + \sqrt{y} = 10.$$

Find the marginal average price of boots with respect to the average price of shoes.

4. THE CHAIN RULE

In this section and the next we get a picture of the way derivative formulas are organized. The functions in common use are built up from a very few basic ones by means of various combining operations. Derivative formulas follow the same pattern. We learn a few specific derivative formulas (like $D_x x^n = nx^{n-1}$) and then we learn what might be called "combining formulas." If, then, a function is a combination of basic pieces and we know how to differentiate the pieces, the combining formulas tell us how to put these results together to get the derivative of the entire combination. Actually, we have already encountered this. The only specific derivative formula we have available now is that for x^n, but we have seen how to differentiate anything that consists of power functions combined by addition and multiplication by constants.

One of the most potent operations for combining functions is that of *composition* or taking a "function of a function." (See Definition I.3.9.) For example, suppose

5.4.1 $$y = \sqrt{1 + x^2}.$$

This can be regarded as a two-stage operation. That is, the equations

5.4.2 $$y = \sqrt{u}, \; u = 1 + x^2$$

describe the same relation between x and y as does the single equation (5.4.1). From (5.4.2) we can compute

5.4.3 $$D_u y = \frac{1}{2\sqrt{u}}, \; D_x u = 2x.$$

However, we want $D_x y$ as determined by (5.4.1). If we write derivatives as quotients of differentials, we see immediately how to get this from the information in (5.4.3). Specifically,

5.4.4 $$\frac{dy}{dx} = \frac{dy}{du}\frac{du}{dx}$$

because cancellation of the factor du reduces the right side to the fraction on the left. To apply this to the specific problem at hand, we substitute from (5.4.3) into (5.4.4) to get

$$\frac{dy}{dx} = \frac{1}{2\sqrt{u}}(2x) = \frac{x}{\sqrt{1 + x^2}}.$$

Formula (5.4.4) is called the *chain rule*. It is one of the most useful formulas in calculus. We want to look at it in several other forms as we discuss its uses, but first we might add a word about our proof. We said that you prove it by cancelling du. This is certainly the way to remember the rule, and if the modern theory of the differential is properly developed it becomes a perfectly valid proof. We mention this point because in Section 3 we referred to Chapter 3 of Munroe's *Calculus* for a discussion of differentials. In that chapter the chain rule is used in the development of the theory. This makes for a simplified discussion of differentials, but the theory can be developed (Munroe, *Calculus*, Section 5–9) without presupposing the chain rule.

Now, let us look at the general problem in various notations. Suppose we are given

5.4.5 $$y = f(u), \; u = g(x)$$

which is to say

5.4.6 $$y = f[g(x)].$$

We want dy/dx, and the answer is given by (5.4.4). However, we see it in another form if we note that given (5.4.5) we have

$$\frac{dy}{du}=f'(u)=f'[g(x)],\ \frac{du}{dx}=g'(x).$$

Therefore, given (5.4.6), we have

5.4.7 $$\frac{dy}{dx}=f'[g(x)]g'(x).$$

Note that in (5.4.6) we work "from the inside out." To compute y we operate on x by g and then operate on the result by f. By contrast, in the chain rule, (5.4.7), we go the other way. First, we differentiate the "outside" function f, plug in $g(x)$, and then differentiate g and multiply. To see this in practice let us return to the example we started with:

$$y=\sqrt{1+x^2}.$$

Here the "inside" operation is to square and add 1; the "outside" one is to take square roots. To compute the derivative we first differentiate the square root function. We suggest as a first step writing the result in the skeleton form

$$\frac{1}{2\sqrt{\quad\quad}}.$$

What we put in the blank space here is the "inside" function; this gives

$$\frac{1}{2\sqrt{1+x^2}}.$$

Finally, we must multiply by the derivative of whatever we put in the blank space, namely $1+x^2$ whose derivative is $2x$:

$$\frac{1}{2\sqrt{1+x^2}}\cdot 2x$$

A chain rule can have any number of links in it. Suppose

$$y=f\{g[\phi(x)]\}$$

or

$$y=f(u),\ u=g(v),\ v=\phi(x).$$

Cancellation of differentials gives us the rule

$$\frac{dy}{dx} = \frac{dy}{du}\frac{du}{dv}\frac{dv}{dx}$$

which could also be written

$$\frac{dy}{dx} = f'\{g[\phi(x)]\}\, g'[\phi(x)]\, \phi'(x).$$

V.4.1 *Example.* Find dy/dx, given that

$$y = (x + \sqrt{1 - x^3})^4.$$

The "outside" function is the fourth power; so we begin with the framework

$$4(\qquad\qquad\qquad)^3.$$

Putting in $x + \sqrt{1 - x^3}$, we have as the first factor

$$4(x + \sqrt{1 - x^3})^3.$$

The framework for the second factor is

$$1 + \frac{1}{2\sqrt{\qquad}}$$

Into this we put $1 - x^3$, and the second factor is

$$1 + \frac{1}{2\sqrt{1 - x^3}}$$

Finally, by inspection, the third factor is

$$-3x^2.$$

Therefore,

5.4.8 $$\frac{dy}{dx} = 4(x + \sqrt{1 - x^3})^3 \left(1 + \frac{1}{2\sqrt{1 - x^3}}\right)(-3x^2).$$

Algebraic simplification can be carried out if desired, but we suggest that the student practice until he can write answers in the form (5.4.8) without having to do any scratch work on the side.

V.4.2 *Example.* Using the data of Example III.6.3, solve for supply y as a function of price x. Find the marginal supply.

The supply equation can be rewritten

$$y^2 = 625(x^2 + 20x - 1500)$$

and so

$$y = 25\sqrt{x^2 + 20x - 1500}.$$

We write here

$$y = 25u^{1/2};\ u = x^2 + 20x - 1500$$

and it is easy to see that

$$D_x y = \frac{25}{2\sqrt{x^2 + 20x - 1500}} \cdot (2x + 20)$$

or

$$D_x y = \frac{25x + 250}{\sqrt{x^2 + 20x - 1500}}.$$

PROBLEMS ON THE CHAIN RULE

Compute each of the following derivatives.

1. $D_x\sqrt{1 + x^2}$
2. $D_x\sqrt{1 - x^2}$
3. $D_x(1/\sqrt{1 - x^2})$
4. $D_x(1/\sqrt{1 + x^2})$
5. $D_x(x^3 - 3x^2 + 5)^4$
6. $D_x(5x^4 - 2x^3)^7$
7. $D_x\sqrt{3x^2 - 2x + 1}$
8. $D_x(x^6 + 4x^3 - 5)^{1/3}$
9. $D_x[1/(1 - x^{1/3})]$
10. $D_x[1/(1 - x)^{1/3}]$
11. $D_x\sqrt{x - (1 - x)^3}$
12. $D_x(x - \sqrt{1 - x})^3$

13. $D_x[(1/\sqrt{1+x}) + (1/\sqrt{1-x})]$

14. $D_x[1/(\sqrt{1+x} + \sqrt{1-x})]$

15. $D_x 1/\sqrt{x + [1/(1-x)]}$

16. $D_x \sqrt{1-(1-x)^2}$

17. $D_x(x^{2/3} - x^{-2/3})^{1/3}$

18. $D_x \sqrt{(1/\sqrt{x}) + (1/x^2)}$

19. $D_x 1/[(1/x) - 1/(1-x)]$

20. $D_x \sqrt{x + \sqrt{1-x}}$

From each of the following sets of relations compute $D_t y$.

21. $y = x^2 - 2x + 1,\ x = 2t - \sqrt{t}$

22. $y = 2x^2 + 3x,\ x = t^3 + \sqrt[3]{t}$

23. $y = x^2 + 1,\ x = t^2 + 1$

24. $y = \sqrt{1+x^2},\ x = 3t^2 - 2$

25. $y = \sqrt{x,\ x = 4t^2} - 3t + 6$

26. $y = 1 - x,\ x = t - \sqrt{t}$

27. $y = \sqrt{1+x},\ x = 3t^2 + 2$

28. $y = 1/x,\ x = \sqrt{1-t^2}$

29. $y = -2/x^3,\ x = \sqrt{1+4t^2}$

30. $y = (1/x) - (1/\sqrt{x}),\ x = t + \sqrt{t}$

31. $y = 1/(1-x),\ x = t - t^2$

32. $y = x^2 - 4x + 3,\ x = \sqrt{t}$

33. $y = 1 - u,\ u = x^2 + 4,\ x = 2 - 3t$

34. $y = \sqrt{1+u^2},\ u = x - \sqrt{x},\ x = 1/(1-t)$

35. $y = \sqrt{u},\ u = 1 + 2x,\ x = t^2 - 3$

36. $y = 1/\sqrt{u},\ u = \sqrt{x},\ x = 1 - t^2$

37. $y = 3u^2 - 2,\ u = \sqrt{x^3 + 4},\ x = 1/t$

38. $y = u^2 - 1,\ u = x^3 - 1,\ x = t^4 - 1$

39. $y = 1 - v,\ v = 2u^2,\ u = 2 - 3x,\ x = \sqrt{t}$

40. $y = \sqrt{1 - v},\ u = u^3 - 3u^2,\ u = \sqrt{x^2} + 4,\ x = 1/(1 - t)$

41. At a price of x cents per lb., sales of cottage cheese in a certain town are y pounds, where

$$x^2 + 20x + y^2 + 50y = 5775.$$

Find the marginal volume of sales, $D_x y$, and the marginal revenue, $D_x(xy)$.

42. The supply of wheat, in bushels, is related to its price x, in dollars per bushel, by the equation

$$y = \sqrt{x^3 + 20x + 10{,}000}.$$

Find the marginal supply, $D_x y$.

43. A chemical company produces different amounts, x and y, of two compounds by the same process. The product transformation curve is given by the equation

$$5x^2 + 2y^2 = 9800$$

where x and y are in pounds. If the first chemical sells at \$7 per lb., and the second at \$5 per lb., solve for revenue, R, in terms of x. Find $D_x R$.

5. PRODUCTS AND QUOTIENTS

Suppose

$$y = f(x)g(x);$$

then by definition

$$\frac{dy}{dx} = \lim_{h \to 0} \frac{f(x+h)g(x+h) - f(x)g(x)}{h}$$

Now, we want to break this fraction up so that the only fractions left will be

$$\frac{f(x+h)-f(x)}{h} \quad \text{and} \quad \frac{g(x+h)-g(x)}{h}$$

because these last two fractions have limits $f'(x)$ and $g'(x)$, and the idea here is to get the derivative of the product in terms of the derivatives of the factors. The following algebraic trickery accomplishes our purpose.

$$\begin{aligned} f(x+h)g(x+h)-f(x)g(x) &= f(x+h)g(x+h)-f(x+h)g(x) \\ &\quad + f(x+h)g(x)-f(x)g(x) \\ &= f(x+h)[g(x+h)-g(x)] \\ &\quad + g(x)[f(x+h)-f(x)]. \end{aligned}$$

Thus,

$$\begin{aligned} \frac{dy}{dx} &= \lim_{h\to 0}\left[f(x+h)\frac{g(x+h)-g(x)}{h}+g(x)\frac{f(x+h)-f(x)}{h}\right] \\ &= f(x)g'(x)+g(x)f'(x) \end{aligned}$$

Let us set

$$u=f(x),\ v=g(x);$$

then the formula we have just derived can be written

5.5.1
$$\frac{d(uv)}{dx}=u\frac{dv}{dx}+v\frac{du}{dx}.$$

If we multiply this last equation through by dx, we have

5.5.2
$$d(uv)=u\,dv+v\,du.$$

Actually, the notion of differential is completely characterized by saying that it satisfies (5.3.1), (5.3.2), and (5.5.2). To put it another way, manipulation with differentials consists of routine algebra together with the use of these three formulas.

V.5.1 ***Example.*** Find dy/dx, given that

$$y=3x^2\sqrt{1-x^2}.$$

By (5.5.1) the answer is

$$3x^2\,D_x\sqrt{1-x^2}+\sqrt{1-x^2}\,D_x(3x^2).$$

To compute $D_x \sqrt{1-x^2}$ we must use the chain rule. A normal first step in solving this problem would be

$$D_x \sqrt{1-x^2} = \frac{-2x}{2\sqrt{1-x^2}}$$

which gives

$$\frac{dy}{dx} = 3x^2 \frac{1}{2\sqrt{1-x^2}}(-2x) + \sqrt{1-x^2}\,(6x)$$

From the product formula (5.5.2) we can easily obtain a formula for the derivative of a quotient. Let

$$y = u/v = u\,v^{-1};$$

then

$$dy = u\,d(v^{-1}) + v^{-1}\,du = u(-v^{-2}\,dv) + v^{-1}\,du = -\frac{u}{v^2}\,dv + \frac{du}{v}$$

$$= \frac{v\,du - u\,dv}{v^2}.$$

Thus, we have that if $y = u/v$, then

$$\frac{dy}{dx} = \frac{v\dfrac{du}{dx} - u\dfrac{dv}{dx}}{v^2}$$

V.5.2 *Example.*

$$D_x \frac{4x^2-3}{2x^4+x^2} = \frac{(2x^4+x^2)(12x^2)-(4x^3-3)(8x^3+2x)}{(2x^4+x^2)^2}$$

V.5.3 *Example.* Given that

$$y = \sqrt{\frac{2x+1}{x^2+2}}$$

there are at least three ways to organize the formulas for computing dy/dx. Taking the expression for y as we have written it, we have

$$\frac{dy}{dx} = \frac{1}{2}\sqrt{\frac{x^2+2}{2x+1}}\left[\frac{(x^2+2)(2)-(2x+1)(2x)}{(x^2+2)^2}\right]$$

If we rewrite y in the form

$$y = \frac{\sqrt{2x+1}}{\sqrt{x^2+2}}$$

we are led to

$$\frac{dy}{dx} = \frac{\sqrt{x^2+2}\left(\dfrac{1}{2\sqrt{2x+1}}\right)(2) - \sqrt{2x+1}\left(\dfrac{1}{2\sqrt{x^2+2}}\right)(2x)}{x^2+2}$$

Finally, we could write

$$y = (2x+1)^{1/2}(x^2+2)^{-1/2}$$

which leads to

$$\frac{dy}{dx} = (2x+1)^{1/2}\left(-\frac{1}{2}\right)(x^2+2)^{-3/2}(2x) + (x^2+2)^{1/2}\left(\frac{1}{2}\right)(2x+1)^{-1/2}(2).$$

These answers look different but actually they are not. They all reduce to

$$\frac{x^4+2x^2-x+4}{(2x+1)^{1/2}(x^2+2)^{3/2}}.$$

V.5.4 Example. At a price x, demand for a certain commodity is y units, where

$$xy + 20x + 100y = 4000.$$

Find the derivatives of demand and of revenue with respect to price.

The demand equation can be rewritten

$$(x+100)(y+20) = 2000$$

or

$$y = \frac{2000}{x+100} - 20.$$

Using the quotient formula, with $u = 2000$, $v = x + 100$, we obtain

$$D_x y = \frac{-2000}{(x+100)^2}$$

For revenue, we have $R = xy$, or

$$R = \frac{2000x}{x+100} - 20x.$$

Again we use the quotient formula, to obtain

$$D_x R = \frac{2000(x+100) - 2000x}{(x+100)^2} - 20$$

or

$$D_xR = \frac{200{,}000}{(x + 100)^2} - 20.$$

PROBLEMS ON PRODUCTS AND QUOTIENTS

Find each of the following derivatives.

1. $D_x[x\sqrt{1 + x^2}]$
2. $D_x[(2x - 3)(x^3 + 4x)]$
3. $D_y[(3y^3 - 2y)\sqrt{y^2 - 5}]$
4. $D_u[\sqrt{u - 1}\,(3u^2 + \sqrt{u})]$
5. $D_t[\sqrt{(t^3 - 3)(3t^4 + 5t)}]$
6. $D_v[\sqrt{v^3 - 3}\,\sqrt{3v^4 + 5v}\,]$
7. $D_u[2u^3(u^2 - 1)^5]$
8. $D_x[3x^5(x^3 + 2x^2 - 3)^4]$
9. $D_y[5y\sqrt[3]{y^2 - 3y}\,]$
10. $D_x[6x^4\sqrt[3]{x^3 - 5x^2 + 2}]$
11. $D_t[t^3(t - 1)(t - 2)]$
12. $D_u[u^2(u^3 - 2)(3u - u^2)]$
13. $D_x[\sqrt{(x - 1)(x - 2)\sqrt{2x}}$
14. $D_y\left(\frac{y}{y - 1}\right)$
15. $D_t\left(\frac{1 + t}{1 - t}\right)$
16. $D_v\left[\frac{\sqrt{v^2 - 1}}{\sqrt{v^2 + 1}}\right]$
17. $D_x\left[\sqrt{\frac{x^2 - 1}{x^2 + 1}}\right]$
18. $D_t\left[\frac{t}{\sqrt{1 - t}}\right]$
19. $D_u\left(\frac{u^3 - 3}{\sqrt{2}}\right)$

20. $D_u\left(\frac{\sqrt{3}}{u^2-1}\right)$

21. $D_x\left[\frac{\sqrt{2x}}{x^2+2}\right]$

22. $D_x\left[\frac{\sqrt[3]{3x-2}}{\sqrt{2x-3}}\right]$

23. $D_t\left[\frac{t-\sqrt{t}}{\sqrt[3]{t^3-t}}\right]$

24. $D_y\left[\frac{y^2(y-1)^3}{2-y^4}\right]$

25. $D_z\left[\frac{1-z}{z^3(2z-3)}\right]$

26. $D_x\left[\frac{(x-1)(x-2)}{(x-3)(x-4)}\right]$

27. A firm finds that it can produce x units at an average cost of y dollars per unit, where

$$x^2 + y^2 - 200x - 2000y + 1{,}000{,}000 = 0.$$

Let $C = xy$ be total cost, and find $D_x C$.

28. The firm of Problem 27 can sell x units at a price of z dollars per unit, where

$$200x + z^2 = 1{,}000{,}000.$$

Let $R = xz$ be revenue, and find $D_x R$.

29. For the same firm of Problems 27 and 28, let $r = R/C$ be the rate of return (dollars of revenue per dollar of cost). Find $D_x r$.

6. HIGHER ORDER DERIVATIVES.

If y is given in terms of x, then appropriate formulas yield an expression for dy/dx in terms of x. The idea occurs, "Why not use the formulas again and get the derivative of dy/dx?" This is quite feasible, and the result is called the *second derivative of y with respect to x.* Clearly, we could do it again and get a third derivative, and so on.

Various uses of higher order derivatives will appear later. We are concerned here with notation and computational techniques.

One notation for a derivative is

$$f'(x),$$

and in this notation we indicate a second derivative by adding another prime:

$$f''(x).$$

It is clumsy to use more than about three primes; so the symbol for n^{th} derivative is

$$f^{(n)}(x).$$

The superscript is put in parentheses to distinguish the nth derivative from the nth power.

Another notation for a first derivative is

$$D_x y.$$

We extend this to higher derivatives by regarding D_x as a *derivative operator.* When we write D_x with something behind it we mean, "Differentiate that something with respect to x." With D_x interpreted in this way,

$$D_x D_x y$$

means "differentiate $D_x y$," and this gives precisely the second derivative. This notation is always abbreviated to read

$$D_x^2 y.$$

Here the "square" of the operator means "apply it twice." So, in this notation an nth derivative is written

$$D_x^n y.$$

The Leibnitz notation is wonderful for first derivatives, but it breaks down for second derivatives. Unhappily, there is a differential–like notation for second derivatives, but it has very serious drawbacks. This comes from breaking dy/dx up as

$$\frac{d}{dx}\,y$$

and so regarding d/dx as a derivative operator. With this interpretation it is natural to write the second derivative as

$$\left(\frac{d}{dx}\right)^2 y = \frac{d^2y}{dx^2}$$

The beauty of dy/dx lies in the fact that it is a genuine fraction. However, a simple example shows that d^2y/dx^2 is not. If it were, we would have

5.6.1 $$d^2y = D_x^2y\ dx^2,$$

and this leads to trouble. Let

$$y = u^2,\ u = x^3$$

so that

$$y = x^6.$$

Now we have

$$dy = 2u\ du,\ du = 3x^2\ dx,\ dy = 6x^5\ dx,$$

and these two expressions for dy are consistent because

$$2u\ du = 2x^3(3x^2\ dx) = 6x^5\ dx.$$

However,

$$D_u^2y = 2 \quad \text{and} \quad D_x^2y = 30x^4;$$

so if (5.6.1) is supposed to hold we have

$$d^2y = 2du^2 \quad \text{and} \quad d^2y = 30x^4dx^2.$$

However,

$$2du^2 = 2(3x^2\ dx)^2 = 18x^4dx^2;$$

the two expressions for d^2y are unequal. The upshot of all this is that "second order differentials" *are not defined,* and the notation d^2y/dx^2 *cannot be interpreted as a fraction.* For this reason we shall not use this notation, though it does appear in the literature.

Differentials are still useful in computing second derivatives, but they have to be used in the following way:

$$D_x^2y = \frac{d(D_xy)}{dx}$$

V.6.1 *Example.* Compute $D_x^2 y$, given that

$$x^2 + y^2 = 1$$

We take differentials

$$2x\, dx + 2y\, dy = 0$$

and solve for dy/dx

$$D_x y = \frac{dy}{dx} = -\frac{x}{y}.$$

Now, we can compute

$$d(D_x y) = -\frac{y\, dx - x\, dy}{y^2};$$

so

$$D_x^2 y = \frac{d(D_x y)}{dx} = \frac{-y + x\dfrac{dy}{dx}}{y^2} = \frac{-y - \dfrac{x^2}{y}}{y^2}$$

$$= -\frac{y^2 + x^2}{y^3} = -\frac{1}{y^3}$$

V.6.2 *Example.* A cannonball is shot straight up in the air; after t seconds have passed, its distance above the ground is s feet, where

$$s = 96t - 16t^2.$$

What is the cannonball's velocity? What is its acceleration?

The velocity is, of course, the rate of change of position with respect to time. Thus

$$v = D_t s = 96 - 32t.$$

The acceleration is defined as the rate of change of velocity per unit time; this is

$$D_t v = D_t^2 s = -32.$$

We see that the acceleration is a constant, 32 feet per second per second. (This is the acceleration due to gravity, and is experienced by any body which is falling freely near the earth's surface.)

250 PROBLEMS ON HIGHER ORDER DERIVATIVES

Compute $D_x y$ and $D_x^2 y$.

1. $y = x^6 - 3x^5 + 2x^4 - 3x + 7$
2. $y = 4x^3 - 2x^2 + 1$
3. $y = 3x^4 - 2x^3 + 6x^2 - x + 3$
4. $y = (x^2 + 1)^7$
5. $y = (x^3 + 3x)^{15}$
6. $y = x^{5/2} + x^{3/2} + x^{1/2}$
7. $y = x^{1/3} + x^{-1/3}$
8. $y = x\sqrt{1 - x}$
9. $y = x^2\sqrt{1 - x^2}$
10. $y = \sqrt{x^3 - x}$
11. $y = (x^2 + 1)^{3/2}$
12. $y = \dfrac{x + 1}{x}$
13. $y = \dfrac{x}{x + 1}$
14. $y = \dfrac{x + 1}{x^2 + 1}$
15. $x^2 + xy + y^2 = 1$
16. $x^{1/2} + y^{1/2} = 1$
17. $x^2y + y^2 = 1$
18. $x^3 - 3y^3 = y$
19. $x^2 + xy = y$
20. $(x - 1)(y - 1) = 1$
21. In x days, ABC company has sales of y dollars, where

$$(y + 10)^3 = x + 1000.$$

What is the daily rate of sales of this company? What is the rate of increase of the company's daily sales?

22. An automobile moves along a straight road. In t seconds it covers s feet, where

$$s = 20t + 50t^2 - 2t^{7/2}.$$

What is the automobile's velocity? What is its acceleration?

7. MAXIMA AND MINIMA.

We have, so far in this chapter, studied the process of differentiation, and seen how a derivative can be interpreted. We will now consider what is probably the most important application of the differential calculus: that of finding maximal or minimal values of functions.

Actually, what we are discussing here is *local* maxima and minima. A point x_0 is a *local maximum* point for a function f if there is an interval I about x_0 so that

$$f(x) \leq f(x_0)$$

for all x in I. To define a local minimum we would reverse the inequality. Figure V.7.1 illustrates this idea. There x_1 is a local maximum and x_2 is a local minimum, though if we get far enough away f has values larger than $f(x_1)$ and others smaller than $f(x_2)$.

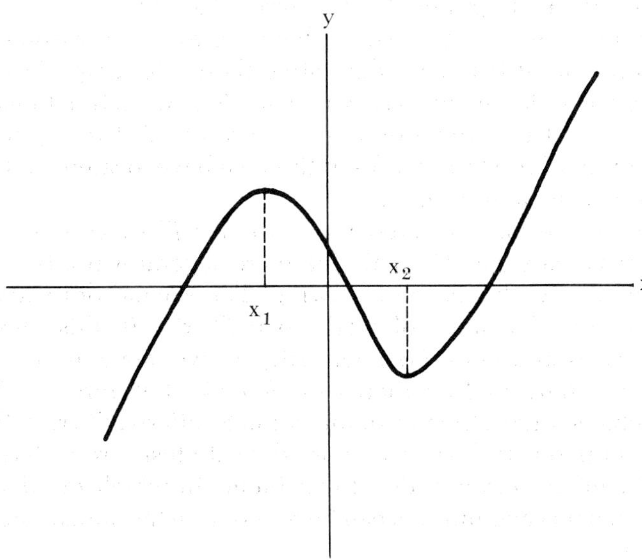

FIGURE V.7.1 Local maxima and minima.

FIGURE V.7.2 Tangents at maximum and minimum points.

Now, Figure V.7.2 suggests the idea for locating local maxima and minima. Informally, at a maximum or minimum point we have a horizontal tangent line and hence a zero derivative. More formally, we state this as follows.

V.7.1 Theorem. If x is a local maximum (minimum) for f, and if $f'(x)$ exists, then $f'(x) = 0$.

Proof. Suppose $f'(x) > 0$; then for h sufficiently small,

$$\frac{f(x+h) - f(x)}{h} > 0.$$

Thus, for all $x + h$ in some interval about x we have

$$f(x+h) \begin{cases} > f(x) & \text{for} \quad h > 0, \\ < f(x) & \text{for} \quad h < 0, \end{cases}$$

and it appears that x is neither a local maximum nor a local minimum. A similar argument applies if we assume $f'(x) < 0$.

Note that we have to assume that $f'(x)$ exists. For example, $|x|$ has a minimum at 0 but no derivative there; the graph has a sharp corner. By and large, we are going to deal with functions whose derivatives exist everywhere; so we interpret V.7.1 to say that if we find all points x for which $f'(x) = 0$ then we have located all the possible local maxima and minima.

Suppose we have located the zeros for $f'(x)$. How do we distinguish maxima from minima? There are several ways, but the simplest (when it works) is the following. The second derivative $f''(x)$ gives the rate of change of $f'(x)$; so if $f''(x) > 0$, this means that $f'(x)$ increases as x does. Geometrically, as we move to the right the slope of the tangent line increases. A look at Figure V.7.2 shows that this characterizes the minimum point. Similarly, $f''(x) < 0$ signals a maximum point. If $f''(x) = 0$, more delicate tests are called for, but we shall confine our attention to problems in which the distinction between maxima and minima can be found from the sign of the second derivative.

In summary, then, we shall work with the following characterizations of maxima and minima.

At a maximum, $f'(x) = 0, f''(x) < 0$;
At a minimum, $f'(x) = 0, f''(x) > 0$.

This is the first round of what can get to be a fairly complicated theory. We shall not investigate cases in which $f'(x)$ fails to exist or cases in which $f'(x) = f''(x) = 0$.

By these considerations we point out more forcibly than we have before why a businessman wants a volume of production for which marginal cost equals marginal revenue. After all, profit is revenue minus cost; so if

$$\frac{dC}{dx} = \frac{dR}{dx},$$

then

$$\frac{dP}{dx} = 0,$$

and this normally occurs at the point of maximum profit.

V.7.2 Example. A cylindrical tank with a bottom but no top is to have a volume of 10π cubic feet. Material for the bottom costs 15¢ a square foot, and that for the sides costs 12¢ a square foot. Find the dimensions for minimum cost. Let r be the radius of the base, h the height of the cylinder, and C the total cost. According to the information given, we have

$$C = 15\pi r^2 + 12 \cdot 2\pi r h.$$

Since the volume is to be 10π,

$$\pi r^2 h = 10\pi, \text{ whence } h = 10/r^2;$$

therefore,

$$C = 15\pi r^2 + \frac{240\pi}{r}$$

$$D_rC = 30\pi r - \frac{240\pi}{r^2}$$

$$D_r^2C = 30\pi + \frac{480\pi}{r^3}$$

Setting $D_rC = 0$, we have

$$30\pi r - \frac{240\pi}{r^2} = \frac{30\pi r^3 - 240\pi}{r^2} = 0;$$

so

$$30\,\pi\, r^3 - 240\,\pi = 0$$

$$r^3 = 8,$$

$$r = 2,$$

$$h = 10/r^2 = 5/2.$$

Since $D_r^2C > 0$ for $r = 2$, this is the desired minimum.

V.7.3 Example. A man has a total of \$1000 to invest in two enterprises. An investment of x dollars in the first enterprise will give a return of $\sqrt{x}$ dollars, while an investment of y dollars in the second enterprise will give a return of $\sqrt{2y}$ dollars. How should he invest for maximum return? Let x and y be the amounts invested and R the total return. We are given that

$$R = \sqrt{x} + \sqrt{2y}$$

and also that $x + y = 1000$; so

$$R = \sqrt{x} + \sqrt{2000 - 2x},$$

$$D_xR = \frac{1}{2\sqrt{x}} - \frac{1}{\sqrt{2000 - 2x}},$$

$$D_x^2R = \frac{-1}{4x^{3/2}} - \frac{1}{(2000 - 2x)^{3/2}}.$$

Setting $D_xR = 0$, we have

$$\frac{1}{2\sqrt{x}} = \frac{1}{\sqrt{2000 - 2x}},$$

$$2000 - 2x = 4x,$$

$$x = \frac{1000}{3},$$

$$y = 1000 - x = \frac{2000}{3}.$$

Since $D_x^2R < 0$ for $x = 1000/3$, this is the desired maximum. The enterprise yielding $\sqrt{2y}$ looks like the better deal; so we might be tempted to put all our money in that. Note that this procedure would yield $\sqrt{2000}$, while investing as the solution to the problem suggests would yield

$$\sqrt{\frac{1000}{3}}+\sqrt{\frac{4000}{3}}=3\sqrt{\frac{1000}{3}}=\sqrt{3000}.$$

V.7.4 Example. The cost in dollars of producing x tons of steel in a certain mill is given by

$$C=4000+25x+0.02x^2.$$

This steel can be sold for \$100 a ton. Find the production plan for maximum profit. The total revenue is the price times the amount sold; that is,

$$R=100x.$$

Profit is revenue minus cost:

$$\begin{aligned} P &= R-C \\ &= 100x-4000-25x-0.02x^2 \\ &= -4000+75x-0.02x^2, \\ D_xP &= 75-0.04x, \\ D_x^2P &= -0.04. \end{aligned}$$

Since D_x^2P is always negative, the zero for D_xP gives maximum profit:

$$\begin{aligned} 75-0.04x &= 0, \\ x &= 1875. \end{aligned}$$

V.7.5 Example. A manufacturer of ladies dresses figures that to produce a certain model his costs will consist of \$1000 fixed overhead plus \$10 per dress. The price he can sell them for depends on how "exclusive" they are, and his market analysts estimate that he can get in dollars $100-2\sqrt{x}$ if he puts x of them on the market. How many should he produce for maximum profit? Revenue is price times number sold:

$$\begin{aligned} R &= (100-2\sqrt{x})x \\ &= 100x-2x^{3/2}. \end{aligned}$$

Cost is the overhead plus 10 times the number produced:

$$C=1000+10x.$$

Profit is revenue minus cost:

$$\begin{aligned} P &= R-C \\ &= 100x-2x^{3/2}-1000-10x \\ &= 90x-2x^{3/2}-1000, \\ D_xP &= 90-3\sqrt{x}, \\ D_x^2P &= -3/(2\sqrt{x}). \end{aligned}$$

The second derivative is negative, so we get maximum profit for $D_xP = 0$:

$$3\sqrt{x} = 90,$$
$$x = 900.$$

So, he should put out 900 of them. If he does this the price will be

$$100 - 2\sqrt{900} = 40,$$

and his profit will be

$$40 \times 900 - 1000 - 10 \times 900 = 26{,}000.$$

Returning to V.7.4 (the steel mill), we find that marginal cost is

$$D_xC = 25 + 0.04x,$$

and for $x = 1875$ (the recommended production) $D_xC = 100$ which is the selling price. By contrast, in V.7.5 (ladies dresses)

$$D_xC = 10$$

while the optimal selling price is 40. It is often stated that marginal cost should (for maximum profit) equal selling price. We see here that this need not be so. The point is that, under certain conditions (notably that of a free market, in which price does not vary) marginal revenue will equal selling price. It is generally true that for maximum profit marginal cost should equal marginal revenue, but marginal revenue does not always equal the selling price. Indeed, let R be revenue, p price and x quantity produced; then

$$R = px.$$
$$D_xR = p + x\,D_xp.$$

If p is constant, then $D_xp = 0$ and $D_xR = D_xC = p$ for maximum profit. In many cases, however, we would expect D_xp to be negative (price decreases with quantity), and in this case $D_xR < p$.

From these examples there emerges a pattern. The method of attack on such problems may be outlined as follows.

(a) Introduce variables which measure quantities pertinent to the problem.

(b) Write an equation giving the variable to be maximized or minimized in terms of other variables.

(c) Use side conditions stated in the problem to eliminate all but one of the variables on the right-hand side of the equation in (b).

(d) Differentiate and locate maxima and minima by setting the first derivative equal to zero and checking the sign of the second derivative.

V.7.6 Example. A rubber company has a steady market for 100,000 rubber pads per year. These rubber pads are produced on machinery which can also be used for the production of other items, and therefore requires some preparation (setting-up) before the rubber pads can be made. Costs of production are \$5 for each pad, plus \$500 setting-up costs for each production run of pads. Additionally, the company must pay storage costs of 25¢ per year for each pad that has been produced but not sold. Assuming that demand is steady throughout the year, how often should production runs be scheduled so as to minimize total costs (production plus inventory)?

Let us assume that x pads are made at each production run; there will then be $100{,}000/x$ production runs every year. Production costs per year will then be

$$P = (100{,}000)5 + \frac{100{,}000}{x}(500)$$

or

$$P = 500{,}000 + \frac{50{,}000{,}000}{x}$$

where the second term represents setting-up costs.

To calculate inventory (storage) costs, we shall assume that production is so timed that each run of rubber pads is complete just as previous inventory runs out. Since demand is assumed to be steady, the inventory then decreases linearly from x to 0, and we see that average inventory will be $x/2$. Thus inventory costs will be

$$I = \frac{1}{4} \cdot \frac{x}{2} = \frac{x}{8}.$$

Total costs are therefore

$$C = \frac{x}{8} + 500{,}000 + \frac{50{,}000{,}000}{x}.$$

Since we wish to minimize C, we find its derivative:

$$D_x C = \frac{1}{8} - \frac{50{,}000{,}000}{x^2}$$

and, setting this equal to zero, we have

$$x^2 = 400{,}000{,}000$$

or

$$x = \pm 20{,}000.$$

The negative root is meaningless; this leaves just the root $x = 20{,}000$. To check whether this is a maximum or a minimum, we take the second derivative:

$$D_x^2 C = \frac{100{,}000{,}000}{x^3}$$

and this is clearly positive for positive x. We have therefore a minimum at $x = 20{,}000$. We conclude that each run should produce 20,000 pads; there will then be 5 runs per year.

PROBLEMS ON MAXIMA AND MINIMA

1. A farmer wishes to enclose a rectangular plot of ground, whose total area is to be 400 square yards, with chicken wire. What should the dimensions of the plot be in order to minimize the total amount of chicken wire needed?

2. A man wants to fence off a rectangular garden plot, using his neighbor's stone wall for one side. He has 120 ft. of fencing to use around the other three sides. Find the dimensions of the plot with maximum area.

3. A man wants to fence off a rectangular garden plot of 864 ft^2. He puts one side of it adjacent to his neighbor's property, and the neighbor agrees to pay half the cost of the fence down the property line. Find the dimensions of the plot for minimum cost to the owner.

4. A rectangular plot of perimeter 320 ft. is to contain a rectangular swimming pool surrounded by a walk 4 ft. wide along the sides and 6 ft. wide along the ends. Find the dimensions of the plot for a pool of maximum area.

5. A poster is to contain 96 in.^{2} of printed matter with margins of 3 in. each at top and bottom and 2 in. at each side. Find the dimensions of printed matter for minimum total area.

6. An open storage bin with square base and vertical sides is to be constructed from 300 ft.^{2} of material. Assuming that no material is wasted, find the dimensions for maximum volume.

7. A box with square base and open top is to hold 32 in.^{3} Find the dimensions which require the least amount of material.

8. Right circular cylindrical tin cans are to be manufactured to contain 8 in.^{3} each. There is no waste in cutting the sides, but each

end piece is cut from a square, and the remainder of the square is wasted. Find the dimensions of the cans to use the least material.

9. A box with square base is to be made from a square piece of cardboard 24 in. on a side by cutting out a square from each corner and turning up the sides. Find the dimensions if the box is to have maximum volume.

10. A box with a rectangular base is to be made from a rectangular piece of cardboard 24 by 12 in. by cutting out a square from each corner and turning up the sides. Find the dimensions if the box is to have maximum volume.

11. A box with a lid is to be made from a square piece of cardboard 24 in. on a side by cutting squares from two adjacent corners and then cutting from the other two corners rectangles so shaped that three of the resulting flaps will form sides and the fourth will form a side and the lid. Find the dimensions if the box is to have maximum volume.

12. A box with square base and top and vertical sides is to contain 625 ft.^{3}. Material for the base costs 35¢ a square foot, for the top 15¢ a square foot, and for the sides 20¢ a square foot. Find the dimensions for minimum cost.

13. A silo in the form of a cylinder surmounted by a hemisphere is to have a capacity of $1000\,\pi/3$ ft.^{3} Construction costs per unit surface area are twice as great for the hemisphere as for the cylinder, and it has a dirt floor. Find the dimensions for minimum cost.

14. A Norman window in the form of a rectangle surmounted by a semicircle is to have a perimeter of 24 ft. Find the dimensions for maximum area.

15. In the Norman window of Problem 14, the rectangle is to have clear glass, while the semicircle is to have colored glass which admits only half as much light per square foot as the clear glass does. Find the dimensions that will admit the most light.

16. A rectangular box with square base must be constructed to hold 500 cubic feet. What should its dimensions be, so as to minimize the surface area of the box, including the sides and bottom, but not including the top?

17. A cylindrical container must be constructed with a capacity of 1000 cubic feet. The top and bottom of the container must be made of material which costs 7¢ per square ft., while the sides of the container cost 5¢ per square ft. Find the dimensions which will minimize the total cost of the container.

18. A farmer has a crop of 100 bushels of potatoes, which he can sell for \$1 per bushel. He may, however, wait before picking the potatoes; if so, his crop will increase at the rate of 10 bushels per week, but the price will also decrease at the rate of 5¢ per bushel per week. How long should be wait before picking his crop?

19. An apartment building has 100 (identical) units. The landlord finds that at a price of \$70 per month, he can rent all the units; however, for each \$5 increase in rent, one of the units is vacated. What rent should he charge so as to maximize his income?

20. The profit that a mining company obtains from a production of x tons of mineral ore is given by

$$P = 9x^2 - x^3 - 15x$$

How much should be produced so as to maximize this profit?

21. An apartment house has 200 (identical) apartments; these can all be rented at a price of \$80 per month. For each rent increase of \$5 per month (per apartment) one apartment is vacated. What rent should be charged so as to maximize the total income from the building?

22. A steel mill can produce x tons of steel at a cost of C dollars, where

$$C = 500 + 75x + 0.04x^2$$

The steel can be sold at a price of \$91 per ton. How many tons should be produced, so as to maximize profits?

23. A man has \$500 to invest in two enterprises. The return from investing x dollars in the first enterprise is $2\sqrt{x}$ dollars, while that from investing y dollars in the second enterprise is $\sqrt{y}$ dollars. How much should he invest in each enterprise, so as to maximize his return?

24. A paving contractor estimates that if he hires x men it will take $1 + 1200x^{-1}$ days to complete a certain job. He pays each man \$10 a day, and he (the contractor) is to receive a bonus of \$120 for each day less than 14 required for the job. Find the number of men he should hire for maximum profit.

25. A manufacturer's total costs in producing x articles per week are $ax^2 + bx + c$. There is a demand law $p = A - Bx$ relating his selling price and the number that he can expect to sell. Find the number that should be produced for maximum profit. [Note: a, b, c, A, and B are positive constants.]

26. The government imposes a tax of t dollars per item on the manufacturer. Add the tax to his costs, and redetermine the output for maximum profit. [Note: The answer here will depend on t.]

27. Assuming that the manufacturer proceeds as indicated in Problem 26, find, in terms of the given constants, the value of t that will bring a maximum return to the government. Compute the price when this tax is imposed, and show that the increase in price is less than the tax.

8. EXPONENTIALS.

We want to study in this section a function f to be written

$$f(x) = e^x.$$

This raises at least two questions. The most obvious is, "What is the number e?" For the record, it is an irrational number whose five-point decimal approximation is 2.71828. Exactly why this number appears we shall show in Section 9. The second question is more profound: "What do you mean by exponentiation anyway?" Granted, e^2 means $e \cdot e$, and e^n means $e \cdot e^{n-1}$, but what do things like $e^{\sqrt{2}}$ and e^{π} mean?

There are various ways to answer this, but one of the most enlightening is to present the entire function in one fell swoop as the solution to an important type of problem. The problem we have in mind is the so-called *growth problem.* Let t be the time variable, and suppose we have a population which is growing (or shrinking) with time. Let $f(t)$ be the size of the population at time t, and suppose that the rate of growth at time t is proportional to $f(t)$. This situation is described by the *differential equation*

5.8.1 $$f'(t) = kf(t).$$

There are many problems that have this mathematical formulation. Population growth (people, bacteria, or what have you) generally follows this pattern. Radioactive decay (negative growth) is described by (5.8.1) with $k < 0$. Continuously compounded interest (investment growing at a rate proportional to total investment) is also described by (5.8.1). There are many other examples.

So, how do you "solve" (5.8.1) for $f(t)$ in terms of t? We shall consider the case $k = 1$; once this is properly handled the more general case falls into place easily. There is one other thing. It is easily seen that if $f(t)$ satisfies (5.8.1), then so does $Af(t)$ for any constant A. This is a question of the units used to measure population size; so we must specify this in posing our problem. Thus, we pose the following mathematical problem:

5.8.2 $$f'(t) = f(t),$$

5.8.3 $$f(0) = 1.$$

Physically, this says that rate of population growth equals population size and that we have a unit population at $t = 0$.

Now, it can be shown that there is an $f(t)$ defined for all t and satisfying (5.8.2). That proof is beyond the scope of this book, but it is significant to note that such a proof involves no specification as to just what $f(t)$ is. It is merely a proof that because of the form of the differential equation (5.8.2) there is a function that satisfies it.

Our experience in Chapter III, section 8, suggests that we should expect some type of exponential function, and indeed, we shall show that anything which satisfies (5.8.2) and (5.8.3) must have the algebraic properties that we would expect of an exponential function. Then, we shall show that there is only one function satisfying both (5.8.2) and (5.8.3). Thus, these two equations serve to *define* a function which we shall name exponential and write $f(t) = e^t$.

First, the key property: Let a be any constant; assume f satisfies (5.8.2), and compute

$$\begin{aligned} D_t[(f(t)f(a-t)] &= f'(t)f(a-t) + f(t)f'(a-t)(-1) \\ &= f(t)f(a-t) - f(t)f(a-t) \\ &= 0. \end{aligned}$$

Therefore, by V.2.5,

$$f(t)f(a-t) = \text{constant},$$

and setting $t = 0$ and using (5.8.3), we have

$$f(t)f(a-t) = f(a).$$

Setting $s = a - t$, we have

5.8.4 $$f(t)f(s) = f(t+s),$$

and this is the law of exponents, $e^t e^s = e^{t+s}$. Therefore, the solution to the growth problem does behave like an exponential.

Now, the first thing that follows from (5.8.4) is that $f(t)$ is never zero. If $f(t_0) = 0$, then by (5.8.4), $f(t_0 + s) = 0$ for all s, and setting $s = -t_0$ yields $f(0) = 0$, contradicting (5.8.3).

We are now ready to prove that there is only one solution to the problem. Suppose f and g were both solutions and set

5.8.5 $$\phi(t) = \frac{g(t)}{f(t)}.$$

This is possible because $f(t)$ is never zero. Then, $g(t) = \phi(t)f(t)$, and we have

$$\begin{aligned} g(t) = g'(t) &= \phi'(t)f(t) + \phi(t)f'(t) \\ &= \phi'(t)f(t) + \phi(t)f(t) \\ &= \phi'(t)f(t) + g(t); \end{aligned}$$

so $\phi'(t)f(t) = 0$, and since $f(t) \neq 0$,

$$\phi'(t) = 0.$$

By V.2.5, $\phi(t) =$ constant, and setting $t=0$ in (5.8.5), we have $\phi(t) = 1$; so $g(t) = f(t)$.

We must skip one more step now. We have shown that $f(t)$ is never zero and by (5.8.3), $f(0) = 1$; so it seems likely that $f(t)$ is always positive. Actually, as part of the existence proof (which we skipped) it can be shown that the graph of f has no breaks in it and therefore cannot go from positive to negative without going through zero.

Assuming this, we are now in a position to get a very good picture of this solution function for the growth problem. We have $f(t)$ everywhere positive; so $f'(t)$ is everywhere positive, since it equals $f(t)$. All tangent lines have positive slopes, and there are no maxima or minima. If we differentiate (5.8.2), we get $f''(t) = f'(t)$ and see that $f''(t)$ is everywhere positive; the curve is everywhere concave up. Noting that $f'(0) = f(0) = 1$, we see that the tangent line at 0 has the equation $y = 1 + t$. Since the curve is concave up, we thus have $f(t) \geq 1 + t$ for all t. By (5.8.4), $f(-t)f(t) = f(0) = 1$; so for large positive t,

$$f(-t) = \frac{1}{f(t)} \leq \frac{1}{t+1}$$

and we see that on the far left the curve goes to 0. Putting all this information together, we see that the graph must look like that in Figure V.8.1.

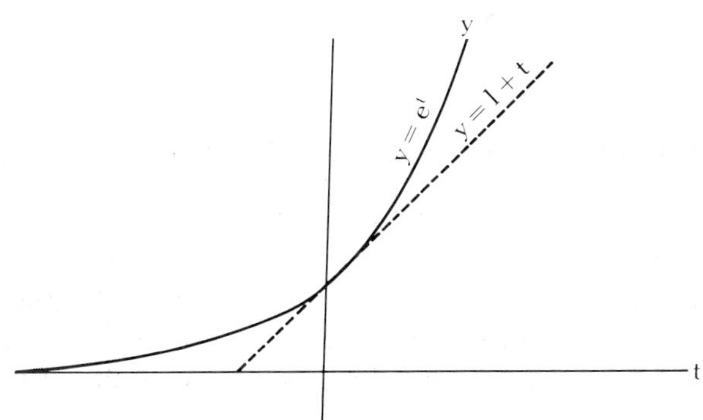

FIGURE V.8.1 Exponential function.

So, what does e^t mean? It is the unique solution to the growth problem; that is, it is completely determined by the conditions

5.8.6
$$D_t e^t = e^t,$$
$$e^0 = 1.$$

We have further seen that it follows from these defining equations that

$$e^s e^t = e^{s+t}$$

and that the graph of $y = e^t$ has the characteristics shown in Figure V.8.1.

Equation (5.8.6) is a new derivative formula that should be learned along with that for x^n. Let us look at a few mechanical problems that occur when we use (5.8.6) in conjunction with the chain rule and the product formula.

V.8.1 *Example.* Compute

$$D_x e^{x^3 - 3x}.$$

Answer: $e^{x^3-3x}(3x^2 - 3)$. The "outside" function here is the exponential and it is its own derivative; so the first step is just to copy the whole function. Now, the "inside" function appears in the exponent; so we multiply by the derivative of the exponent. The chain rule would be a purely mechanical rule to apply if only mathematicians wrote all composite functions in the same format. Unhappily, they do no such thing; so the student must learn that the "inside" function is in the parentheses in

$$(x^2 + 1)^3,$$

under the radical in

$$\sqrt{x^2 + 1},$$

and in the exponent in

$$e^{x^2+1}.$$

V.8.2 *Example.* Compute the derivative of

$$(x^3 + x)e^{-2x}.$$

Answer: $(x^3 + x)e^{-2x}(-2) + (3x^2 + 1)e^{-2x}$. The only difficulty here is psychological. A good way to learn the product formula is, "Copy one factor and differentiate the other; then reverse the process." With the exponential function you have to pay attention. When you write the exponential down again are you copying or differentiating?

V.8.3 Example. Show that the growth function

$$y = \frac{L}{1 + ke^{-ax}}$$

(which we considered in Section III.8) satisfies the equation

$$D_x y = \frac{a}{L}\, y(L - y).$$

When is the growth of y greatest?

It is easy to see, by differentiation, that

$$D_x y = \frac{akLe^{-ax}}{(1 + ke^{-ax})^2}$$

and it is easy to see that this can be rewritten in the form

$$D_x y = \frac{a}{L}\, y(L - y).$$

We see thus that this has, indeed, the property mentioned in Section III.8: growth is proportional both to the size of y, and to the difference between y and its upper bound, L. It is also easy to see that $D_x y$ is greatest when $y = L/2$.

PROBLEMS ON EXPONENTIALS

Find dy/dx.

1. $y = e^{x^2 - x}$
2. $y = e^{\sqrt{x}}$
3. $y = \sqrt{x - e^x}$
4. $y = e^x - e^{-x}$
5. $y = e^x + e^{-x}$
6. $y = \dfrac{e^x - e^{-x}}{e^x + e^{-x}}$
7. $y = x^2 e^{-x}$
8. $y = \sqrt{x}\, e^{\sqrt{x}}$

9. $y = e^{1/x}$

10. $y = (e^x)/x$

11. Show that $f(x) = a\,e^{bx}$ is the unique solution of

$$f'(x) = ab\,f(x),$$
$$f(0) = a.$$

12. Assuming that the world's population was 4 billion in 1970 and that it increases continuously at 3 per cent per year, when will it be 8 billion? 16 billion? Note: $e^{0.6931} = 2$ approximately.

13. A radioactive element decays at the rate of 1 per cent per year. Assume we have one unit of this element now, and find a formula in terms of t for the number of units that will be left in t years.

14. What is the half-life of the element in Problem 13? That is, after how many years will there be 1/2 unit left? From the note in Problem 12 you can find what power of e gives 1/2.

15. Let $y = e^{-x^2/2\sigma^2}$ where σ is a constant. Compute $D_x y$ and $D_x^2 y$ and show that the graph of this equation has a single maximum at $x = 0$, has no minima, and changes direction of concavity at $x = \pm\sigma$.

16. The total sales of a certain appliance, in x days, is given by

$$y = \frac{10{,}000}{1 + 3e^{-4x}}.$$

What is the daily rate of sales? When is this rate of sales greatest?

9. LOGARITHMS.

By definition the *logarithm* of x to the base a is the power of a that gives x. Here we are interested in logarithms to the base e where e is the number introduced in Section 8. These are called *natural logarithms.* We denote the natural logarithm of x by

$$\ln x.$$

Returning now to the definition, we see that

$$y = \ln x$$

means

$$x = e^y.$$

From this we can easily get the derivative of the logarithm function. Given $x = e^y$, we have

$$\frac{dx}{dy} = e^y;$$

so

$$\frac{dy}{dx} = \frac{1}{dx/dy} = \frac{1}{e^y} = \frac{1}{x}.$$

Thus, we have another derivative formula to learn:

$$D_x \ln x = 1/x.$$

Algebraic properties of the logarithm can be derived from those of the exponential. Let

$$\ln x = a, \ \ln y = b, \ \ln xy = c;$$

then

$$e^c = xy = e^a e^b = e^{a+b}.$$

That is, $c = a + b$, which is to say

$$\ln xy = \ln x + \ln y.$$

This is the usual law of logarithms. We should note also that since $e^0 = 1$ and $e^1 = e$, we have

$$\ln 1 = 0 \text{ and } \ln e = 1.$$

Another important property of logarithms may be derived as follows. Let k be any constant; then

$$D_x(k \ln x) = k/x$$

and

$$D_x(\ln x^k) = \frac{1}{x^k} k\, x^{k-1} = \frac{k}{x};$$

therefore, $k \ln x - \ln x^k =$ constant, and setting $x = e$ we see that this constant is 0. Thus,

$$\ln x^k = k \ln x.$$

We are now in a position to get a specific description of this mysterious number e. We note that

$$\ln e = 1 = \ln'(1) = \lim_{h\to 0} \frac{\ln(1+h) - \ln(1)}{h}$$
$$= \lim_{h\to 0} \frac{\ln(1+h)}{h}$$
$$= \lim_{h\to 0} \ln(1+h)^{1/h}.$$

Thus, it would appear (and can be proved) that

$$e = \lim_{h\to 0} (1+h)^{1/h}.$$

The following tabulation of specific values of $(1+h)^{1/h}$ is suggestive.

h	$(1+h)^{1/h}$
1	2
1/2	2.25
1/3	$2.370 \cdots$
1/10	$2.594 \cdots$
1/100	$2.705 \cdots$
1/1000	$2.717 \cdots$

As we have mentioned before,

$$e = 2.71828 \cdots.$$

If $y = f(x)$, then

$$\frac{d \ln y}{dx} = \frac{(1/y)\,dy}{dx} = \frac{dy/dx}{y} = \frac{f'(x)}{f(x)}$$

This so-called *logarithmic derivative* of y gives the rate of change of y divided by y. Multiplied by 100 this would be the percentage rate of change of y. That is, if

$$100 \frac{d \ln y}{dx} = a,$$

we would say that there is an a per cent change in y per unit change in x. The economist is often interested in the ratio of percentage change rate of one quantity to that of another. He calls this elasticity. Specifically, Ey/Ex is called the *elasticity of y with respect to x* and is defined by

$$\frac{Ey}{Ex} = \frac{d \ln y}{d \ln x} = \frac{dy/y}{dx/x} = \frac{x}{y}\frac{dy}{dx}.$$

V.9.1 Example. Compute the derivative of

$$\ln(1+x^2).$$

Answer:

$$\frac{1}{1+x^2}\cdot 2x = \frac{2x}{1+x^2}$$

Note that at the intermediate stage the "inside" function whose derivative must be multiplied in now appears in the denominator.

V.9.2 Example. Compute the derivative of

$$x^x.$$

We do not have a formula for this. We have one for x^n (constant exponent) and e^x (constant base) but none for variable exponent and base. We can do this by so-called *logarithmic differentiation* as follows.

$$y = x^x,$$
$$\ln y = x \ln x,$$

$$\frac{dy}{y} = d \ln y = dx \ln x + x\, d \ln x$$
$$= \ln x\, dx + x\frac{dx}{x};$$

so

$$\frac{dy}{dx} = y(\ln x + 1) = x^x(\ln x + 1).$$

V.9.3 Example. Compute the derivative of

$$\sqrt{(x^2+3)(x^3-2)}.$$

This can be done by methods studied earlier, but look at how it goes by logarithmic differentiation.

$$y = \sqrt{(x^2+3)(x^3-2)},$$

$$\ln y = \frac{1}{2}[\ln(x^2+3) + \ln(x^3-2)],$$

$$\frac{dy}{y} = \frac{1}{2}\left[\frac{2x\,dx}{x^2+3} + \frac{3x^2\,dx}{x^3-2}\right],$$

$$\frac{dy}{dx} = \frac{1}{2}\sqrt{(x^2+3)(x^3-2)}\left[\frac{2x}{x^2+3} + \frac{3x^2}{x^3-2}\right].$$

V.9.4 Example. The amount of money, y (in dollars), that a housewife is willing to pay for x lb. of hamburger is given by

$$y = \log_4(1 + 3x).$$

One hundred such housewives buy from a certain butcher. If the butcher can buy the hamburger at a price of 40¢ per lb., how many pounds should he buy so as to maximize his profit?

Let us suppose that the butcher sells x lb. to each housewife. Then his revenue is

$$R = 100 \log_4(1 + 3x) = 100 \frac{\ln(1 + 3x)}{\ln 4}$$

(this operation is called *change of base,* and is explained in Appendix 3), and his costs are $100(.40x) = 40x$. His profits are, therefore,

$$p = 100 \frac{\ln(1 + 3x)}{\ln 4} - 40x.$$

To find the maximum, we differentiate:

$$D_x p = \frac{300}{(\ln 4)(1 + 3x)} - 40.$$

Setting this equal to zero, we obtain

$$1 + 3x = \frac{15}{2 \ln 4} = 5.41$$

which gives (approximately) $x = 1.47$. To check whether this is a maximum, we find that

$$D_x^2 p = \frac{-900}{(\ln 4)(1 + 3x)^2}$$

which is clearly negative. Thus $x = 1.47$ is a maximum, and the butcher should buy 147 lb. of hamburger.

PROBLEMS ON LOGARITHMS

Find dy/dx.

1. $y = x^2 \ln x$

2. $y = \ln \frac{1}{x}$

3. $y = \frac{1}{\ln x}$

4. $y = \ln(\ln x)$

5. $y = (\ln x)^2$

6. $y = e^x \ln x$

7. $y = e^{-x} \ln x$

8. $y = e^{\ln x}$

9. $y = e^{x \ln b}$

10. $y = b^x$

11. $y = \ln(1 + e^x)$

12. $y = 1 + \ln x$

13. $y = x \ln x - x$

14. $y = \ln(x + \sqrt{x^2 - b^2})$

In each of the following find $D_x y$ both directly and by logarithmic differentiation. In each case show that the two results are equal. These expressions become progressively more complicated. In your opinion, at what stage does logarithmic differentiation become preferable to the direct method?

15. $y = x^{3/5}$

16. $y = (2x^3 - x)^{5/3}$

17. $y = \sqrt[3]{(x-3)(x+2)}$

18. $y = \sqrt[3]{\dfrac{(x-1)}{(x+1)}}$

19. $y = \sqrt[3]{(x^2 - 3x + 4)(x^3 - 4x^2)}$

20. $y = \sqrt[3]{\dfrac{(x^2 - 3x + 2)}{(x^3 + x^2 - x)}}$

21. $y = \sqrt[3]{(x^2 + 2x + 2)(x^2 + 4)(x^3 + 3)}$

22. Suppose the demand for a certain commodity is x units provided the price is p, and suppose x and p are related by $x = Ap^{-n}$ where A and n are constants. Show that the elasticity of demand with respect to price is $-n$. A 1 per cent increase has what effect on demand?

23. Let x be demand, p be price, and R be revenue. Recall that $R = xp$, and derive the following formula connecting marginal revenue, price, and elasticity of demand.

$$\frac{dR}{dx} = p\left(1 + \frac{1}{Ex/Ep}\right).$$

272 10. THE TRIGONOMETRIC FUNCTIONS.

We consider in this section the trigonometric functions: sine, cosine, tangent, and so on. The reader is doubtless familiar with these, as they are normally studied in high school. For our purposes, we shall modify the functions slightly: the only difference is in the unit which we use for angle measure. In high school, angles are invariably measured in degrees (°): one right angle is 90°, and a full circle is 360°. We introduce a new measure: radian measure.

The radian measure of an angle is the arc it subtends on a circle of radius 1. In calculus we always use radian measure of angles in discussing trigonometric functions. Thus, it is really more appropriate to talk of trigonometric functions of an arc rather than of an angle.

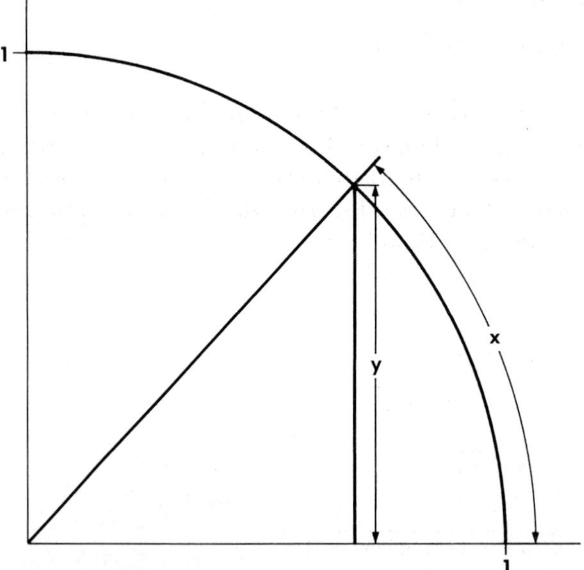

FIGURE V.10.1 On a unit circle, the arc length x is the radian measure of the angle. Note that y = sin x.

Specifically, the relation

$$y = \sin x$$

is pictured in Figure V.10.1. Now (see Figure V.10.2) if s is small, then

$$\frac{\text{chord}}{\text{arc}} = \frac{2y}{2s} = \frac{\sin s}{s}$$

seems to be close to 1. It is indeed true that

5.10.1 $$\lim_{s\to 0} \frac{\sin s}{s} = 1.$$

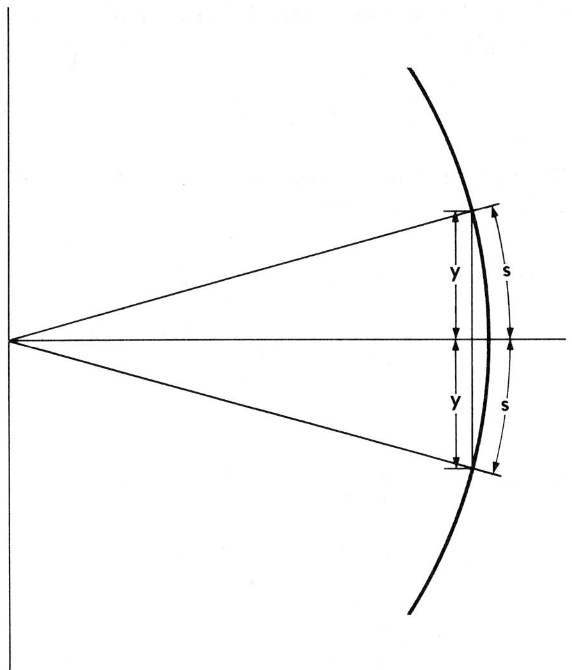

FIGURE V.10.2 For small s, $\frac{2y}{2s} \approx 1$.

We shall not try to prove this result but give only Figure V.10.2 as a plausibility argument.

If we accept (5.10.1) and recall the identity

$$\sin A - \sin B = 2 \cos \left(\frac{A+B}{2}\right) \sin \left(\frac{A-B}{2}\right),$$

we can obtain a derivative formula for the sine function as follows.

$$\begin{aligned} D_x \sin x &= \lim_{h \to 0} \frac{\sin(x+h) - \sin x}{h} \\ &= \lim_{h \to 0} \frac{2 \cos \left(x + \frac{h}{2}\right) \sin \left(\frac{h}{2}\right)}{h} \\ &= \lim_{h \to 0} \cos \left(x + \frac{h}{2}\right) \frac{\sin(h/2)}{h/2} \\ &= \cos x \end{aligned}$$

because the last factor has the limit 1 by (5.10.1).

In trigonometry, we encounter another variation on functional notation. Recall that $\sin^2 x$ means $(\sin x)^2$; so as we compute $D_x \sin^2 x$

 the "outside" function is the square function and the "inside" one is the sine. The result is

$$D_x \sin^2 x = 2 \sin x \cos x.$$

Derivatives of the other trigonometric functions may be obtained by various combining formulas.

$$\begin{aligned} D_x \cos x &= D_x \sqrt{1 - \sin^2 x} \\ &= \frac{1}{2\sqrt{1-\sin^2 x}}(-2 \sin x \cos x) \\ &= -\sin x. \end{aligned}$$

$$\begin{aligned} D_x \tan x &= D_x \frac{\sin x}{\cos x} \\ &= \frac{\cos x \cos x - \sin x(-\sin x)}{\cos^2 x} \\ &= \frac{\cos^2 x + \sin^2 x}{\cos^2 x} \\ &= \frac{1}{\cos^2 x} = \sec^2 x. \end{aligned}$$

The remaining formulas are

$$\begin{aligned} D_x \cot x &= -\csc^2 x, \\ D_x \sec x &= \sec x \tan x, \\ D_x \csc x &= -\csc x \cot x. \end{aligned}$$

Derivations are left to the reader.

Trigonometric functions often appear as solutions of differential equations. For example,

$$D_x^2 \sin x = D_x \cos x = -\sin x$$

and

$$D_x^2 \cos x = D_x(-\sin x) = -\cos x;$$

so $f(x) = \sin x$ and $f(x) = \cos x$ are both solutions of

5.10.2 $$f''(x) + f(x) = 0.$$

Also, it is readily verified that for any constants A and B,

$$f(x) = A \cos x + B \sin x$$

is a solution of (5.10.2).

Now, consider the differential equation (5.10.2) together with the conditions

$$f(0) = a,\ f'(0) = b.$$

We shall show that, like the growth problem (5.8.2 with 5.8.3), this problem has a unique solution. Suppose f and g are both solutions, and set $\phi(x) = f(x) - g(x)$; then

$$\begin{aligned}\phi''(x) + \phi(x) &= f''(x) - g''(x) + f(x) - g(x)\\ &= f''(x) + f(x) - [g''(x) + g(x)] = 0.\end{aligned}$$

Thus, multiplying by $2\phi'(x)$,

$$2\phi''(x)\phi'(x) + 2\phi'(x)\phi(x) = 0,$$

but the chain rule shows that

$$2\phi''(x)\phi'(x) = D_x[\phi'(x)]^2$$

and

$$2\phi'(x)\phi(x) = D_x[\phi(x)]^2;$$

so we must have

5.10.3 $$[\phi'(x)]^2 + [\phi(x)]^2 = \text{constant}.$$

However,

$$\begin{aligned}\phi'(0) &= f'(0) - g'(0) = b - b = 0,\\ \phi(0) &= f(0) - g(0) = a - a = 0;\end{aligned}$$

so the constant in (5.10.3) is zero, and it follows that $\phi(x) = 0$; i.e., f and g are identical.

V.10.1 *Example.* Given that

$$f''(x) + f(x) = 0 \text{ and } f(0) = f'(0) = 1,$$

find $f(x)$. We have just shown that the problem has a unique solution. Since $f(x) = A \cos x + B \sin x$ satisfies the differential equation for any A and B, we have the solution if we can find A and B so that $f(0) = f'(0) = 1$. Assuming

$$f(x) = A \cos x + B \sin x,$$

we have

$$f'(x) = -A \sin x + B \cos x.$$

Now, we set $x = 0$, and since $\sin 0 = 0$ and $\cos 0 = 1$, we have

$$f(0) = A = 1,$$
$$f'(0) = B = 1.$$

So $A = B = 1$ solves the problem; that is, the given conditions determine that $f(x) = \cos x + \sin x$.

V.10.2 Example. The company described in Example III.3.5 can sell the first type of brass for \$80 per ton, and the second for \$100 per ton. How much of each type should it produce to maximize revenues?

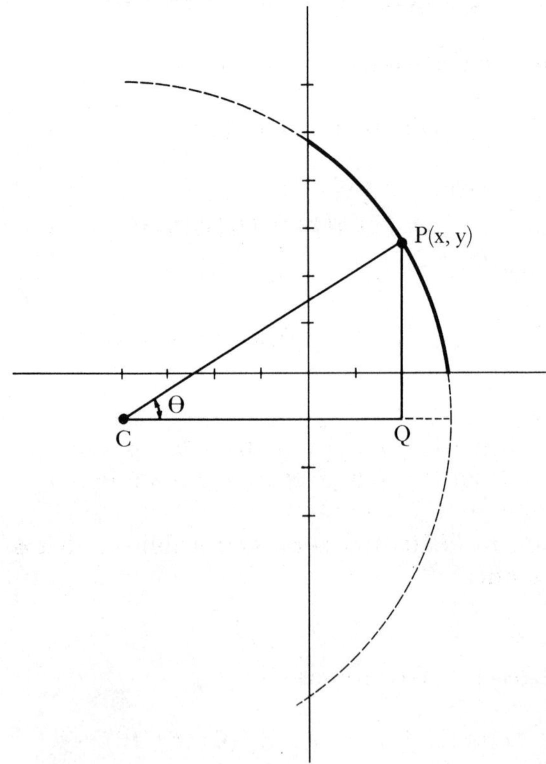

FIGURE V.10.3 Product transformation curve for Example V.10.2.

The product transformation curve (Fig. V.10.3) is a circle with radius 7 and center at $(-4, -1)$. We can see that, if P is a point on this curve, then

$$\frac{QP}{CP} = \sin \theta \qquad \frac{CQ}{CP} = \cos \theta.$$

Now clearly, $CP = 7$, $QP = y + 1$, and $CQ = x + 4$, where (x,y) are the coordinates of P. Thus,

$$x + 4 = 7 \cos \theta \qquad y + 1 = 7 \sin \theta$$

or

$$x = 7 \cos \theta - 4 \qquad y = 7 \sin \theta - 1.$$

If production is x thousand T of the first type of brass, and y thousand T of the second, then revenue in thousands of dollars is

$$R = 80x + 100y$$

or

$$R = 560 \cos \theta + 700 \sin \theta - 420.$$

We wish to maximize R. We therefore differentiate:

$$D_\theta R = -560 \sin \theta + 700 \cos \theta.$$

Setting this equal to zero, we obtain

$$\tan \theta = \frac{\sin \theta}{\cos \theta} = \frac{700}{560} = 1.25.$$

From a table of trigonometric functions, we find that

$$\sin \theta = 0.78, \qquad \cos \theta = 0.63$$

and so $x = 0.41$ and $y = 4.46$. To check whether this is a maximum or a minimum, we find

$$D_\theta^2 R = -560 \cos \theta - 700 \sin \theta$$

which is negative at the point in question. Thus, revenue is maximized by producing 410 T of the first type of brass and 4460 T of the second type.

PROBLEMS ON TRIGONOMETRIC FUNCTIONS

Find dy/dx:

1. $y = 2 \sin 3x$
2. $y = \cos^2 x$
3. $y = \sin x^2$
4. $y = \ln \cos x$

5. $y = \sqrt{1 - \sin x}$

6. $y = x^2 \sin x$

7. $y = x \cos \frac{1}{x}$

8. $y = \frac{\sin x}{x}$

9. $y = \frac{e^x \sin(2x + 1)}{\ln x + \cos x}$

10. $y = \frac{\tan(3x) + 5x}{x^2 + \cos x}$

11. $y = \sqrt{\frac{\cos^2 x + 1}{\sin^2 x + 1}}$

12. $y = \frac{e^{4x} \cos(3x)}{x^2 + 1}$

13. $e^x \cos y + x \sin y = 10$

14. $x = \sin y - y$

15. $\sin x \sin y + \cos x \cos y = 0$

16. Show that $f(x) = A \cos kx + B \sin kx$ satisfies the differential equation $f''(x) + k^2 f(x) = 0$.

17. Given that $f''(x) + 4f(x) = 0$, $f(0) = 1$, $f'(0) = 2$, find $f(x)$.

18. Show that $f(x) = e^x (A \cos x + B \sin x)$ satisfies the differential equation $f''(x) - 2f'(x) + 2f(x) = 0$.

19. Suppose that the grape farmer of problem 3, page 127, can sell the table grapes at 15¢ per lb., and the wine at 90¢ per gal. How much of each should he produce so as to maximize revenue?

11. PARTIAL DERIVATIVES.

Often we encounter situations in which a given quantity depends on more than one factor. For example, it could well be that the demand for beef depends not only on the price of beef but also on the price of pork and that the demand for pork also depends on both prices. Letting x be the demand for beef and y the demand for pork with p and q as the prices of beef and pork, respectively, we would describe this situation by equations

$$x = f(p, q), \; y = g(p, q).$$

Now, how about marginal demands? The significant idea here is the following. Suppose we hold the price of pork fixed and manipulate the price of beef. At what rate does this change the demand for beef or that for pork? On the other hand, we could fix the price of beef and play around with pork prices and thus introduce two other marginal demands.

These marginal quantities are illustrations of what are called *partial derivatives.* More generally, suppose

$$z = f(x, y).$$

If we hold y fixed, z depends only on x, and we can ask about the rate of change of z with respect to x. This follows the usual pattern for the idea of a derivative; the rate of change of z with respect to x for fixed y is given by

$$\lim_{h \to 0} \frac{f(x+h, y) - f(x, y)}{h}$$

Note what we are doing here. We do not change y at all, but we change x by an amount h, get the difference in values of z, divide by h, and take the limit. This is called the *partial derivative of z with respect to x* and is denoted by

$$\frac{\partial z}{\partial x}.$$

Here mathematical notation is inconsistent again. This looks like the Leibnitz notation for an ordinary derivative. Unlike dy/dx, the expression $\partial t/\partial x$ *cannot* be regarded as a fraction.

The other partial derivative is defined in a similar way:

$$\frac{\partial z}{\partial y} = \lim_{h \to 0} \frac{f(x, y+h) - f(x, y)}{h}.$$

Computation of partial derivatives is extremely simple. The partial derivative of z with respect to x means, "Hold y constant and differentiate with respect to x." This is exactly the way they are computed. For example, suppose

$$z = 2x^2 e^{xy};$$

we think of y as a constant and use the usual formulas:

5.11.1 $$\frac{\partial z}{\partial x} = 4x\, e^{xy} + 2x^2\, y\, e^{xy}.$$

The other partial derivative is even simpler in this example. If x is a constant,

5.11.2 $$\frac{\partial z}{\partial y} = 2x^3\, e^{xy}.$$

Partial derivatives have many uses. All we propose to do with them here is to apply them in Section 12 to multidimensional max-min problems. To that end we need to develop two items here. We shall do precisely that and stop, thereby leaving completely untouched most of the theory of partial derivatives.

The first thing we need is the expansion of the differential in multidimensional calculus. Given that

$$z = f(x, y)$$

it turns out that

5.11.3 $$dz = \frac{\partial z}{\partial x}\, dx + \frac{\partial z}{\partial y}\, dy.$$

This is all part of the theory of differentials, just as $dy = D_x y\, dx$ follows from $y = f(x)$. Thus, essentially, we are stating (5.11.3) without proof. The least we can do, however, is to point out that this formula is consistent with (5.3.1), (5.3.2) and (5.5.2) which we have previously given as the characteristic properties of the differential. Return to the example above in which

$$z = 2x^2\, e^{xy};$$

using (5.3.2) and (5.5.2), we have

$$\begin{aligned} dz &= 4x\, e^{xy}\, dx + 2x^2\, e^{xy}\, d(xy) \\ &= 4x\, e^{xy}\, dx + 2x^2\, e^{xy}(x\, dy + y\, dx) \\ &= (4x\, e^{xy} + 2x^2 y\, e^{xy})\, dx + 2x^3\, e^{xy}\, dy. \end{aligned}$$

A glance at (5.11.1) and (5.11.2) shows that this follows precisely (5.11.3).

The other item that we need is a rather deep theorem which we shall try to make plausible by considering a specific example. Look at the unit circle $x^2 + y^2 = 1$, and consider it as the locus of

$$u = 0$$

where

$$u = x^2 + y^2 - 1.$$

From $x^2 + y^2 - 1 = 0$ we get

5.11.4 $$2xdx + 2ydy = 0$$

and this is solvable for dy,

$$dy = -\frac{x\, dx}{y},$$

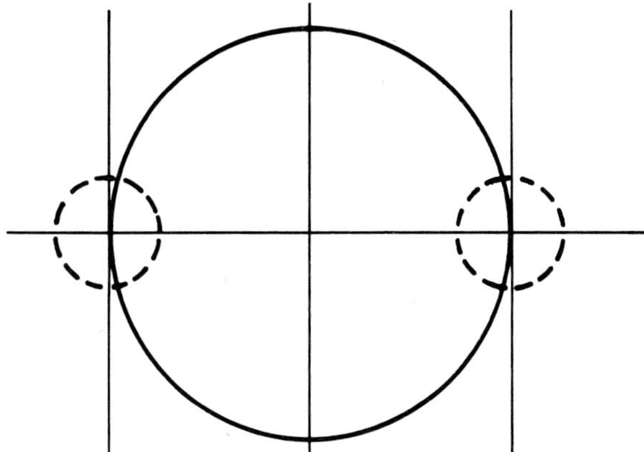

FIGURE V.11.1 Solution for y in terms of x not unique inside small circles.

provided $y \neq 0$. Now, look at Figure V.11.1. There are two points on the circle at which $y = 0$, and at each of these points we have the following features. (1) The tangent line is vertical and thus dx is identically zero. (2) Take any small circle around one of these points, and the part of the unit circle inside this small circle cannot be described by an equation $y = f(x)$ because for each value of x there are two values of y. On the other hand (Fig. V.11.2), at any point on the unit circle where $y \neq 0$ we can solve locally for y in terms of x and we have dx not identically zero. Thus, the points at which (5.11.4) is solvable for dy are those at which (1) the equation of the circle is solvable for y in terms of x and (2) dx is not identically zero.

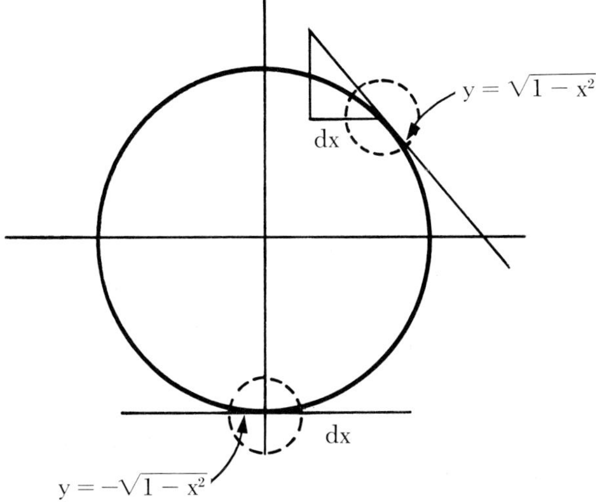

FIGURE V.11.2 Solution for y in terms of x is unique inside each small circle.

More generally, let $u=f(x, y)$ and consider the locus of $u=0$. Here

$$du=\frac{\partial u}{\partial x}\,dx+\frac{\partial u}{\partial y}\,dy=0,$$

and this is solvable for dy if

5.11.5
$$\frac{\partial u}{\partial y}\neq 0.$$

As in the special case of the circle, it turns out that at points at which (5.11.5) holds, the equation $u=0$ is locally solvable for y in terms of x; also, at these points dx is not identically zero.

The result just stated is known as the *implicit function theorem.* We also need to look at this theorem in higher dimensions. Let

$$u=f(x, y, z),$$
$$v=g(x, y, z),$$

and consider the locus of

$$u=0,\ v=0.$$

Here we have

$$\frac{\partial u}{\partial x}\,dx+\frac{\partial u}{\partial y}\,dy+\frac{\partial u}{\partial z}\,dz=0,$$
$$\frac{\partial v}{\partial x}\,dx+\frac{\partial v}{\partial y}\,dy+\frac{\partial v}{\partial z}\,dz=0.$$

This system of linear equations is solvable for dy and dz at points where

$$\begin{vmatrix}\dfrac{\partial u}{\partial y} & \dfrac{\partial u}{\partial z}\\[2ex] \dfrac{\partial v}{\partial y} & \dfrac{\partial v}{\partial z}\end{vmatrix}\neq 0,$$

and the implicit function theorem says that at these same points the nonlinear equations $u=v=0$ are solvable for y and z in terms of x. Furthermore, wherever this happens dx is not identically zero.

The third notion which we will develop here is that of maxima and minima of functions of several variables. Partial derivatives are used to obtain these, in a manner quite similar to that used for functions of a single variable. There is a slight complication due to the fact that, as the independent variables are in two or more dimensions, the function might vary in a different manner along each direction. Geometrically, we can think of a function as being represented by a surface; maxima will be "peaks", and minima will be "troughs." Apart from these, there may also be *saddle-points* ("mountain passes") which are maxima in one direction and minima in another. It is important to be able to recognize these.

The general rule is that, for either a maximum or a minimum, both partial derivatives must vanish:

5.11.6 $$\frac{\partial f}{\partial x} = \frac{\partial f}{\partial y} = 0$$

Condition (5.11.6) is, however, only a necessary (but not sufficient) condition. To determine whether a point satisfying (5.11.6) is indeed a maximum or a minimum, or something else, it is necessary to study the second-order partial derivatives:

$$\frac{\partial^2 f}{\partial x^2} = \frac{\partial}{\partial x}\left(\frac{\partial f}{\partial x}\right)$$

$$\frac{\partial^2 f}{\partial y^2} = \frac{\partial}{\partial y}\left(\frac{\partial f}{\partial y}\right)$$

$$\frac{\partial^2 f}{\partial x \partial y} = \frac{\partial}{\partial x}\left(\frac{\partial f}{\partial y}\right)$$

$$\frac{\partial^2 f}{\partial y \partial x} = \frac{\partial}{\partial y}\left(\frac{\partial f}{\partial x}\right)$$

(Note the difference between these last two derivatives. One, $\partial^2 f/\partial x \partial y$, is obtained by differentiating first with respect to y, and then with respect to x. The other, $\partial^2 f/\partial y \partial x$, differentiates first with respect to x, and then with respect to y. In most cases, and certainly in all cases considered in this book, it turns out that the order of differentiation is unimportant; $\partial^2 f/\partial x \partial y = \partial^2 f/\partial y \partial x$.)

The general rule for maxima and minima is that, when (5.11.6) holds, then f will have an extremum (i.e., either a maximum or a minimum) if the determinant

5.11.7 $$\Delta = \left(\frac{\partial^2 f}{\partial x^2}\right)\left(\frac{\partial^2 f}{\partial y^2}\right) - \left(\frac{\partial^2 f}{\partial x \partial y}\right)^2$$

is positive. If, on the other hand, $\Delta < 0$, then the point is a saddle-point. Finally, if $\Delta = 0$, then it is not clear whether the point is a maximum, a minimum, or a saddle-point.

V.11.1 *Example.* Find all maxima and minima of the function

$$f(x, y) = x^2 + y^2 - xy.$$

We have here

$$\frac{\partial f}{\partial x} = 2x - y \qquad \frac{\partial f}{\partial y} = 2y - x$$

 and setting these equal to zero we obtain $x = y = 0$. To check whether this is a maximum, minimum, or saddle point, we compute

$$\frac{\partial^2 f}{\partial x^2} = 2 \qquad \frac{\partial^2 f}{\partial y^2} = 2 \qquad \frac{\partial^2 f}{\partial x \partial y} = 1$$

which gives us $\Delta = 3$. Thus the point is an extremum, and, since the partials $\partial^2 f/\partial x^2$ and $\partial^2 f/\partial y^2$ are both positive, we conclude that (0,0) is a minimum.

V.11.2 Example. Find maxima and minima of

$$z = x^2 + y^2 + 4xy.$$

We have here

$$\frac{\partial z}{\partial x} = 2x + 4y \qquad \frac{\partial z}{\partial y} = 2y + 4x$$

and these vanish at $x = y = 0$. Checking the second-order derivatives, we have

$$\frac{\partial^2 z}{\partial x^2} = 2 \qquad \frac{\partial^2 z}{\partial y^2} = 2 \qquad \frac{\partial^2 z}{\partial x \partial y} = 4$$

and so $\Delta = -12$. We conclude that (0,0) must be a saddle-point for this function. Note that it is a minimum with respect to x (for fixed y) and also a minimum with respect to y (for fixed x), but that it is not a minimum of the function. (Consider its behavior along the line $x + y = 0$.)

V.11.3 Example. Find maxima and minima of

$$z = x^3 + 4x^2 - y^2 + 3xy - x + 7y.$$

We have here

$$\frac{\partial z}{\partial x} = 3x^2 + 8x + 3y - 1.$$

and

$$\frac{\partial z}{\partial y} = -2y + 3x + 7.$$

Setting both of these equal to zero, we obtain a system of two equations, which can be solved to give

$$x = -1, \; y = 2$$

and

$$x = -19/6,\ y = -5/4.$$

To determine the nature of these points, we check the second-order partial derivatives. These are

$$\frac{\partial^2 z}{\partial x^2} = 6x + 8 \qquad \frac{\partial^2 z}{\partial y^2} = -2 \qquad \frac{\partial^2 z}{\partial x \partial y} = 3$$

which gives

$$\Delta = -12x - 25.$$

At the point $(-1,2)$, we have $\Delta < 0$, and so this is a saddle-point. At $(-19/6, -5/4)$, $\Delta > 0$, which means that we have an extremum. Since $\partial^2 z/\partial x^2$ and $\partial^2 z/\partial y^2$ are both negative, this point is a maximum. In conclusion, this function has a maximum at $(-19/6, -5/4)$, a saddle-point at $(-1, 2)$, and no minima.

V.11.4 Example. By using x units of capital and y units of labor, the ABC company can produce goods worth z dollars, where

$$z = 15x + 18y + 2xy - x^2 - 2y^2.$$

If each unit of labor costs \$12, and each unit of capital costs \$10, what is the company's best policy?

We would like to maximize the profit function

$$p = z - 10x - 12y = 5x + 6y + 2xy - x^2 - 2y^2.$$

We therefore find the partial derivatives:

$$\frac{\partial p}{\partial x} = 5 + 2y - 2x \qquad \frac{\partial p}{\partial y} = 6 + 2x - 4y.$$

Setting both of these equal to zero, we obtain a system of simultaneous equations with solution $x = 8$, $y = 11/2$. Consideration of the second-order partial derivatives shows that this is, indeed, the maximum of the function, and we conclude that the company should use 8 units of capital and 11/2 units of labor.

PROBLEMS ON PARTIAL DERIVATIVES

Compute $\partial z/\partial x$ and $\partial z/\partial y$.

1. $z = x^2y - 2y^2x$

2. $z = 2x\,e^{-y}$
3. $z = x/y$
4. $z = y/x$
5. $z = \sqrt{x^2 + y^2}$
6. $z = 1/\sqrt{x^2 + y^2}$
7. $z = \ln(x^2 + y^2)$
8. $z = e^{-xy}$
9. $z = e^{-x/y}$
10. $z = x \ln y$

Compute dz by using (5.11.3); then compute dz by using (5.3.1), (5.3.2) and (5.5.2). Show that the two results are the same.

11. $z = e^{-xy}$
12. $z = \sqrt{x^2 - y^2}$
13. $z = \ln(x^2 + y^2)$
14. $z = x\,e^{-y}$
15. $z = x^2 + xy + y^2$
16. $z = x \ln y$
17. $z = e^{-x/y}$
18. $z = \ln \dfrac{x - y}{x + y}$

Find all maxima and minima of the following functions:

19. $z = x^3 + 3x^2 + 2xy + y^2 + 5x + 8y - 2.$
20. $z = e^x \cos y - 3x + 2y.$
21. $z = x + y - \sqrt{x^2 + y^2}.$

22. A company uses two types of raw materials, A and B, for its products. By using x units of A and y units of B, it obtains a production of z units, where

$$50z = 20x + 16y + 4xy - 3x^2 - 2y^2.$$

Each unit of A costs \$2, and each unit of B costs \$5. If each unit of the product can be sold for \$20, find the production policy which will optimize the company's profits.

23. A meat company finds that, if it markets x lb. of beef and y lb. of pork, then the price of beef is p¢ per lb., where

$$p = 90 - 0.02x - 0.005y$$

and the price of pork is q¢ per lb., where

$$q = 75 - 0.008x - 0.03y.$$

How much of each meat should the company market so as to maximize revenues?

12. LAGRANGE MULTIPLIERS.

We turn now to non-linear maximum-minimum problems with constraints. The constraint sets will take a variety of forms best explained by a few examples. The constraints might read

$$y = 1 - x^2,\ x \geq 0,\ y \geq 0,$$

in which case the constraint set is the arc of a parabola shown in Figure V.12.1. Or, we might have a constraint reading

$$x^2 + 2y^2 \leq 4,$$

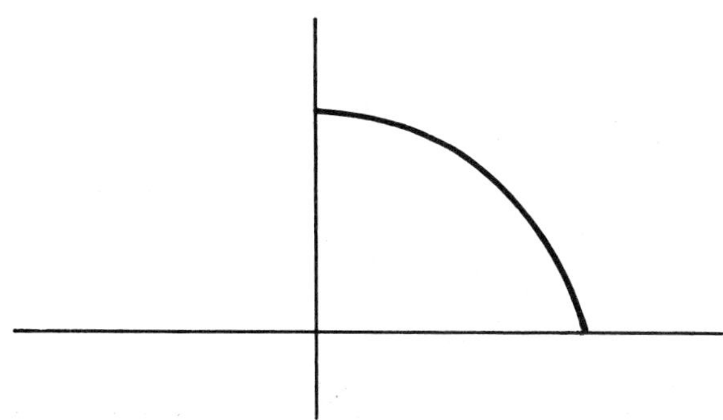

FIGURE V.12.1 $y = 1 - x^2,\ x \geq 0,\ y \geq 0.$

which describes the area inside an ellipse (Fig. V.12.2). As an example in three dimensions we might have

$$x + y + z = 1,\ x^2 + y^2 \leq 1.$$

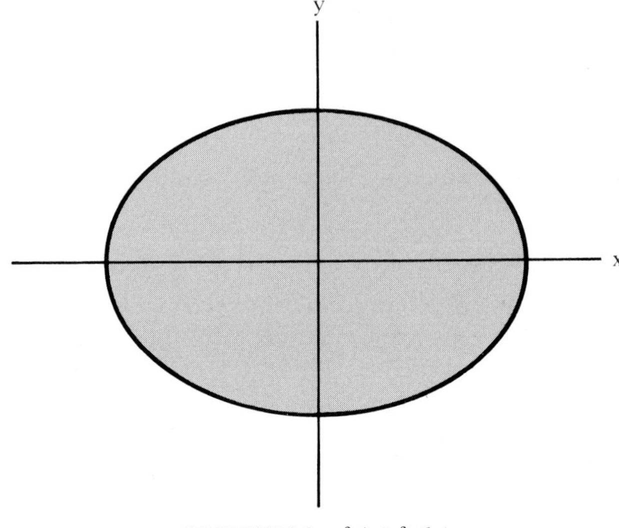

FIGURE V.12.2 $x^2 + 2y^2 \leq 4$.

The linear equation defines the tilted plane shown in Figure V.12.3, and the inequality restricts us to points inside or on the cylinder; so the constraint set is the shaded portion of the plane shown in Figure V.12.3.

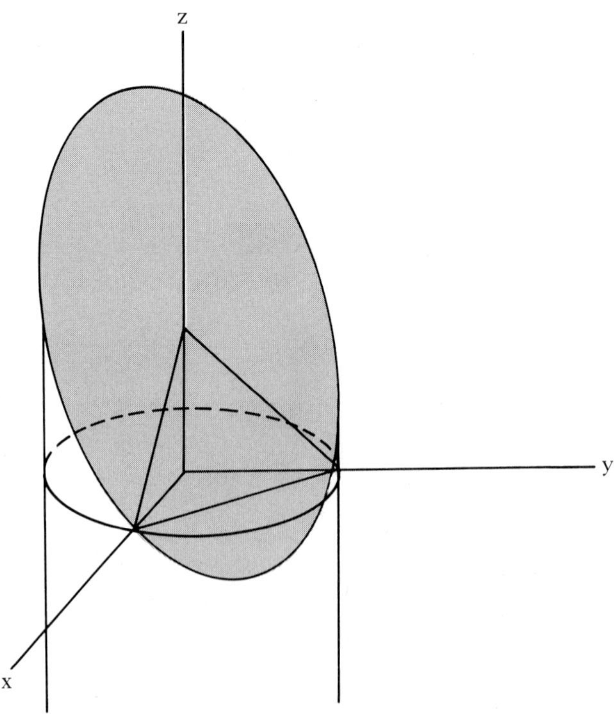

FIGURE V.12.3 $x + y + z = 1$, $x^2 + y^2 \leq 1$.

To solve the problems in this section it is, in general, not necessary to draw pictures of the constraint sets. We have sketched these three examples to illustrate the following points. The constraint sets have *interiors* and *boundaries*. In Figure V.12.1 the interior is the arc and the boundary consists of two points. In each of the other two examples the interior is a plane region and the boundary is an ellipse. More to the point is the following chart.

Figure	Equation satisfied by interior points	Additional equation satisfied by boundary points
V.12.1	$y = 1 - x^2$	$x = 0$ or $y = 0$
V.12.2	none	$x^2 + 2y^2 = 4$
V.12.3	$x + y + z = 1$	$x^2 + y^2 = 1$

Our constraints consisted of equations and inequalities. The given equations hold at interior points while the inequalities convert to additional equations that hold at the boundary points. We are going to develop here a procedure for locating maxima and minima of nonlinear functions subject to constraints in the form of equations. Thus, each problem comes in two parts. We go through our procedure with the *equations* given in the statement of the problem to find interior max-min points. Then, we start all over again with those equations plus others obtained from the *inequalities* given in the statement of the problem to find boundary max-min points.

So, it is now time to describe the method. Suppose

$$w = f(x, y, z),\ u = \phi(x, y, z),\ v = \psi(x, y, z),$$

and suppose we are looking for maxima and minima of w subject to the constraints

5.12.1 $$u = 0,\ v = 0.$$

At a maximum or minimum point for w, we will have $dw = 0$; so we seek a procedure that will locate all points at which $dw = 0$ (subject to the constraints) and then evaluate w at these points to see which are maxima and which are minima. Now, it follows from (5.12.1) that $du = dv = 0$; so at a $dw = 0$ point we have

5.12.2 $$dw + \lambda du + \mu dv = 0$$

for any numbers λ and μ. The variables λ and μ are called *Lagrange multipliers*. Note that we introduce one multipler for each constraining equation. We next expand (5.12.2) in terms of dx, dy and dz:

$$\frac{\partial w}{\partial x} dx + \frac{\partial w}{\partial y} dy + \frac{\partial w}{\partial z} dz + \lambda \left[\frac{\partial u}{\partial x} dx + \frac{\partial u}{\partial y} dy + \frac{\partial u}{\partial z} dz\right]$$
$$+ \mu \left[\frac{\partial v}{\partial x} dx + \frac{\partial v}{\partial y} dy + \frac{\partial v}{\partial z} dz\right] = 0;$$

 rearranging these terms we have

5.12.3 $$\left[\frac{\partial w}{\partial x}+\lambda\frac{\partial u}{\partial x}+\mu\frac{\partial v}{\partial x}\right]dx+\left[\frac{\partial w}{\partial y}+\lambda\frac{\partial u}{\partial y}+\mu\frac{\partial v}{\partial y}\right]dy$$
$$+\left[\frac{\partial w}{\partial z}+\lambda\frac{\partial u}{\partial z}+\mu\frac{\partial v}{\partial z}\right]dz=0.$$

The *Lagrange equations* for this problem are

5.12.4 $$\frac{\partial w}{\partial x}+\lambda\frac{\partial u}{\partial x}+\mu\frac{\partial v}{\partial x}=0,$$

5.12.5 $$\frac{\partial w}{\partial y}+\lambda\frac{\partial u}{\partial y}+\mu\frac{\partial v}{\partial y}=0,$$

5.12.6 $$\frac{\partial w}{\partial z}+\lambda\frac{\partial u}{\partial z}+\mu\frac{\partial v}{\partial z}=0.$$

Obviously, these last three equations imply (5.12.3), but the logic of our problem requires that we go the other way, because we want to say that if we find all solutions of the Lagrange equations then we have found all the possible max-min points.

To deduce the Lagrange equations from (5.12.3), we reason as follows. Recall that λ and μ can be anything; so let them be chosen so that (5.12.5) and (5.12.6) are satisfied. This is possible provided these two equations are solvable for λ and μ; that is, provided

5.12.7 $$\begin{vmatrix}\dfrac{\partial u}{\partial y} & \dfrac{\partial v}{\partial y}\\[2ex] \dfrac{\partial u}{\partial z} & \dfrac{\partial v}{\partial z}\end{vmatrix}\neq 0.$$

So, given (5.12.7), we can choose λ and μ so that (5.12.3) reduces to

5.12.8 $$\left[\frac{\partial w}{\partial x}+\lambda\frac{\partial u}{\partial x}+\mu\frac{\partial v}{\partial x}\right]dx=0.$$

However, in Section 11 we noted that given (5.12.7), dx is not identically zero; therefore in this case (5.12.8) implies (5.12.4).

What happens if (5.12.7) does not hold? Try another determinant of partial derivatives. If the determinant of partial derivatives of u and v with respect to x and z does not vanish, then λ and μ can be determined by (5.12.4) and (5.12.6); also in this case $dy \neq 0$ and (5.12.3) implies (5.12.5).

In this particular problem there are three different determinants of partial derivatives that can be used in the side condition. If we have a maximum or minimum point and any one of these determinants is different from zero, then the Lagrange equations must hold. So, to find all possible max-min points we find all solutions to the Lagrange equations plus all points at which all of the determinants that could be used in side conditions vanish.

Note one final point. We keep saying "solve the Lagrange equations." Actually, the system of equations to be solved consists of the Lagrange equations (5.12.4–5.12.6) together with the original constraint equations (5.12.1). Note that for this problem this gives five equations in x, y, z, λ, μ. This count always comes out right if we follow the rule of one Lagrange multiplier for each constraint equation.

V.12.1 Example. Find maxima and minima for

$$x^2 + y^2$$

subject to the constraints

$$y = 1 - x^2,\ x \geq 0,\ y \geq 0.$$

To follow the notation used above, we set

$$\begin{aligned} w &= x^2 + y^2, \\ u &= x^2 + y - 1. \end{aligned}$$

(Note that we ignore the inequalities in this first round.) Now,

$$\frac{\partial w}{\partial x} = 2x,\ \frac{\partial w}{\partial y} = 2y,\ \frac{\partial u}{\partial x} = 2x,\ \frac{\partial u}{\partial y} = 1;$$

so the equations to be solved are

$$\begin{aligned} 2x + 2\lambda x &= 0, \\ 2y + \lambda &= 0, \\ x^2 + y - 1 &= 0. \end{aligned}$$

From the second and third equations we have

$$\lambda = -2y = 2x^2 - 2$$

and substituting this in the first yields

$$\begin{aligned} x + x(2x^2 - 2) &= 0 \\ 2x^3 - x &= 0 \\ x(2x^2 - 1) &= 0 \\ x = 0,\ y &= 1; \\ x = 1/\sqrt{2},\ y &= 1/2. \end{aligned}$$

There is another solution, $x = -1/\sqrt{2}$, but this violates the constraint $x \geq 0$. The inequalities are not used in setting up the Lagrange equations for the interior, but they are used to reject solutions.

With only one multiplier, the "determinant" we have to check reduces to a single partial derivative. Since $\partial u/\partial y = 1$ (different from 0 everywhere), the Lagrange equations yield all the interior max-min points.

Turning to the boundary, we could set up Lagrange equations, but for a two-point boundary this is far too complicated. One of the boundary points has already appeared; so we add in the other one and tabulate w at three points.

x	y	w
0	1	1
$1/\sqrt{2}$	1/2	3/4
1	0	1

Thus, w has a minimum at $(1/\sqrt{2}, 1/2)$ and assumes a maximum value at each end point of the arc.

V.12.2 Example. Find maxima and minima for

$$x^3 + y^3$$

subject to the constraint

$$x^2 + 2y^2 \leq 4.$$

For the interior problem we have no constraint equation, therefore no multiplier, and the Lagrange equations are

$$3x^2 = 0,$$
$$3y^2 = 0.$$

These have the solution $x = y = 0$. Since there was no multiplier there is no side condition to be checked.

On the boundary we have the Lagrange equations

$$3x^2 + 2\lambda x = 0,$$
$$3y^2 + 4\lambda y = 0,$$
$$x^2 + 2y^2 - 4 = 0.$$

The first equation factors

$$x(3x + 2\lambda) = 0.$$

The solution $x = 0$ substituted in the third equation yields two points

$$(0, \pm\sqrt{2})$$

Similarly, we factor the second equation

$$y(3y + 4\lambda) = 0,$$

and the $y = 0$ solution yields

$$(\pm 2, 0).$$

Returning to the unused factors of the first two equations we have

$$-\frac{3x}{2} = \lambda = -\frac{3y}{4},$$
$$y = 2x.$$

Substituting this in the third equation we have

$$9x^2 = 4,$$
$$x = \pm 2/3.$$

Recalling that we want $y = 2x$, we have two more points

$$(2/3, 4/3), \ (-2/3, -4/3).$$

Turning to the side condition, we note that

$$\frac{\partial u}{\partial x} = 2x = 0$$

if $x = 0$. However, we have already explored the possibility that $x = 0$; so we have listed all the possible max-min points. Again, we tabulate.

x	y	w
0	0	0
0	$\sqrt{2}$	$\sqrt{8}$
0	$-\sqrt{2}$	$-\sqrt{8}$
2	0	8
−2	0	−8
2/3	4/3	8/3
−2/3	−4/3	−8/3

The maximum value is at (2, 0) and the minimum value is at (−2, 0).

V.12.3 Example. Find maxima and minima for

$$x^2 + y^2 + z^2$$

 subject to the constraints

$$x + y + z = 1,\ x^2 + y^2 \leq 1$$

In the interior we have

$$\frac{\partial u}{\partial x} = \frac{\partial u}{\partial y} = \frac{\partial u}{\partial z} = 1;$$

so all we have to do is solve the Lagrange equations

$$\begin{aligned} 2x + \lambda &= 0 \\ 2y + \lambda &= 0 \\ 2z + \lambda &= 0 \\ x + y + z &= 1. \end{aligned}$$

This system is easily solved:

$$x = y = z = \frac{-\lambda}{2},$$

$$x + y + z = \frac{-3\lambda}{2} = 1,$$

$$\lambda = -\frac{2}{3},$$

$$x = y = z = \frac{1}{3}.$$

Thus, we find one point

$$(1/3,\ 1/3,\ 1/3).$$

On the boundary we have two equations and so two multipliers. The Lagrange equations are

$$\begin{aligned} 2x + \lambda + 2x\,\mu &= 0, \\ 2y + \lambda + 2y\,\mu &= 0, \\ 2z + \lambda &= 0, \\ x + y + z &= 1, \\ x^2 + y^2 &= 1. \end{aligned}$$

Subtract the first equation from the second:

$$\begin{aligned} 2(y - x) + 2\mu(y - x) &= 0, \\ 2(y - x)(1 + \mu) &= 0. \end{aligned}$$

Thus,

$$y = x \quad \text{or} \quad \mu = -1.$$

Putting $y = x$ in the fifth equation gives

$$x = y = \frac{1}{\sqrt{2}} \quad \text{or} \quad x = y = \frac{-1}{\sqrt{2}}.$$

The fourth equation pairs these results with

$$z = 1 - \sqrt{2} \quad \text{and} \quad z = 1 + \sqrt{2},$$

respectively, and we have two points

$$\left(\frac{1}{\sqrt{2}}, \frac{1}{\sqrt{2}}, 1 - \sqrt{2}\right), \left(-\frac{1}{\sqrt{2}}, -\frac{1}{\sqrt{2}}, 1 + \sqrt{2}\right).$$

To pursue the possibility $\mu = -1$, substitute in the first equation to get

$$\lambda = 0;$$

then by the third equation

$$z = 0$$

and the fourth and fifth equations become

$$x + y = 1, \; x^2 + y^2 = 1.$$

This system has the solution

$$\begin{aligned} x^2 + (1 - x)^2 &= 1, \\ 2x^2 - 2x &= 0, \\ x &= 0 \text{ or } 1, \\ y &= 1 - x = 1 \text{ or } 0. \end{aligned}$$

Two more points appear:

$$(0, 1, 0), (1, 0, 0).$$

Now, let us look at determinants.

$$\begin{vmatrix} \dfrac{\partial u}{\partial x} & \dfrac{\partial v}{\partial x} \\ \\ \dfrac{\partial u}{\partial y} & \dfrac{\partial v}{\partial y} \end{vmatrix} = \begin{vmatrix} 1 & 2x \\ \\ 1 & 2y \end{vmatrix} = 2y - 2x,$$

and this is zero if $y=x$. However, we have already looked at the points where $y=x$; so we have all the possibilities and it is time to tabulate.

x	y	z	w
1/3	1/3	1/3	1/3
$1/\sqrt{2}$	$1/\sqrt{2}$	$1-\sqrt{2}$	$4-2\sqrt{2}$
$-1/\sqrt{2}$	$-1/\sqrt{2}$	$1+\sqrt{2}$	$4+2\sqrt{2}$
0	1	0	1
1	0	0	1

The minimum is at (1/3, 1/3, 1/3) and the maximum is at $(-1/\sqrt{2}, -1/\sqrt{2}, 1+\sqrt{2})$.

V.12.4 Example. A company has \$10,000 to spend on advertising in two cities, A and B. The potential market in these cities, as well as the total competitive advertising, are shown in Table V.12.1.

TABLE V.12.1

City	Market in dollars	Competitive Advertising
A	400,000	17,000
B	200,000	12,000

In each city, the market will be split among advertisers in direct proportion to their advertising there. If profit is 20 per cent of sales, minus advertising costs, how should the company allot its advertising budget?

If the company allots x thousand dollars to A, and y thousand to B, its sales (in thousands of dollars) will be

$$S = \frac{400x}{17+x} + \frac{200y}{12+y}$$

and profits will be

$$p = \frac{1}{5}S - x - y = \frac{80x}{17+x} + \frac{40y}{12+y} - x - y.$$

We must maximize p, subject to the constraints

$$x + y \leq 10 \qquad x \geq 0 \qquad y \geq 0.$$

We solve first, disregarding the constraints. We have

$$\frac{\partial p}{\partial x} = \frac{1360}{(17+x)^2} - 1$$

$$\frac{\partial p}{\partial y} = \frac{480}{(12+y)^2} - 1$$

and setting these equal to zero, we obtain (approximately)

$$x = 20 \qquad y = 10$$

which does not satisfy the first constraint. We therefore introduce a Lagrange multiplier for the first constraint, to obtain

$$F(x, y, \lambda) = \frac{80x}{17 + x} + \frac{40y}{12 + y} - x - y - \lambda(x + y - 10)$$

which has the partial derivatives

$$\frac{\partial F}{\partial x} = \frac{1360}{(17 + x)^2} - 1 - \lambda.$$

$$\frac{\partial F}{\partial y} = \frac{480}{(12 + y)^2} - 1 - \lambda.$$

Setting both of these equal to zero, we find

$$\frac{17}{(17 + x)^2} = \frac{6}{(12 + y)^2}$$

or

$$17(12 + y)^2 = 6(17 + x)^2$$

which, together with the original constraint

$$x + y = 10$$

gives

$$11x^2 - 952x + 6494 = 0$$

which has the solutions $x = 79$, $x = 7.5$. The first is clearly not valid, which leaves us with

$$x = 7.5, \; y = 2.5$$

as the only possible solution. To check whether this is indeed a maximum, we find the value $\lambda = 1.27$. For this (fixed) value of λ, a study of the second-order partials shows that (7.5, 2.5) does indeed give a maximum of the function $F(x, y, \lambda)$. We conclude that the company should spend \$7500 in city A and \$2500 in B.

V.12.5 Example. A man has \$10,000 to invest in 4 enterprises, A, B, C, and D. He calculates that an investment of x dollars in these enterprises will give a return of $\sqrt{x}$, $\sqrt{2x}$, $\sqrt{3x}$, and $\sqrt{4x}$ dollars, respectively. How much should he invest in each, to maximize his total return?

Suppose he invests x, y, z, and w dollars respectively. Then his return will be

$$r = \sqrt{x} + \sqrt{2y} + \sqrt{3z} + \sqrt{4w}$$

and we wish to maximize r, subject to the constraint

$$x + y + z + w \leq 10{,}000.$$

Clearly, the unconstrained problem has no maximum. We therefore form the Lagrangian

$$F = \sqrt{x} + \sqrt{2y} + \sqrt{3z} + \sqrt{4w} - \lambda(x + y + z + w - 10{,}000)$$

and take its derivatives

$$\frac{\partial F}{\partial x} = \frac{1}{2\sqrt{x}} - \lambda$$

$$\frac{\partial F}{\partial y} = \frac{\sqrt{2}}{2\sqrt{y}} - \lambda$$

$$\frac{\partial F}{\partial z} = \frac{\sqrt{3}}{2\sqrt{z}} - \lambda$$

$$\frac{\partial F}{\partial w} = \frac{\sqrt{4}}{2\sqrt{w}} - \lambda.$$

Setting all these equal to zero, we obtain

$$\sqrt{\frac{1}{x}} = \sqrt{\frac{2}{y}} = \sqrt{\frac{3}{z}} = \sqrt{\frac{4}{w}}$$

or, equivalently,

$$y = 2x,\ z = 3x,\ w = 4x$$

which, together with the original constraint

$$x + y + z + w = 10{,}000$$

gives

$$x = 1000,\ y = 2000,\ z = 3000,\ w = 4000$$

as the solution. Again, the reader may compute λ, and check that these values of x, y, z, and w do indeed maximize the function F.

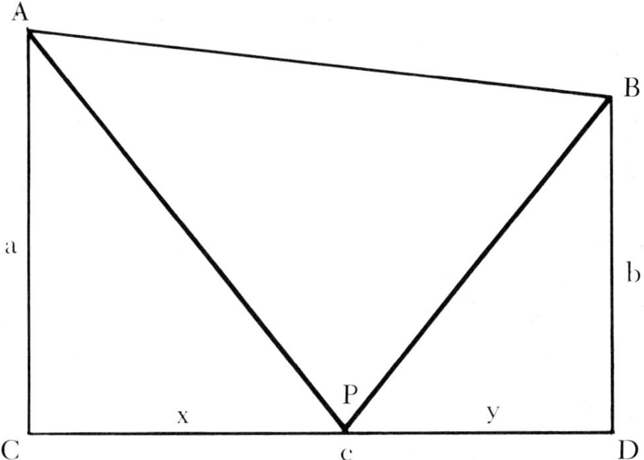

FIGURE V.12.4 Problems 1 to 6 refer to this figure.

PROBLEMS ON LAGRANGE MULTIPLIERS

In Figure V.12.4 find the proportion x/y to minimize, and then maximize, each of the following quantities.

1. $AP + PB$

2. $(AP)^2 + (PB)^2$

3. $(AP)^2 + 2(PB)^2$

4. $(AP)^2 - (PB)^2$

5. area ACP + area PDB

6. area APB

Find the maximum and minimum points for each of the following variables on the indicated figure.

7. $x + y$ on the sphere $x^2 + y^2 + z^2 = 1$

8. xyz on the curve $x^2 - y^2 = 1$, $y = z$

9. xy on the plane circle $x^2 + y^2 \leq 1$

10. $x + y + z$ on the solid sphere $x^2 + y^2 + z^2 \leq 1$

11. xyz on the solid sphere $x^2 + y^2 + z^2 \leq 1$

12. xyz on the plane triangle $x + y + z = 4$, $x \geq 1$, $y \geq 1$, $z \geq 1$

300 A box in the shape of a rectangular parallelepiped is to be made from a given amount of material. Find the proportions for maximum volume:

13. if it has a top.

14. if it has no top.

15. A figure consists of a right circular cylinder surmounted by a right circular cone. If the surface area (including base) is fixed, how should the height and radius of the cylinder and the vertex angle of the cone be related for maximum volume?

16. Consider a cross-section of the figure in Problem 15 (rectangle surmounted by a triangle). For a given perimeter, find the proportions for maximum area.

17. A manufacturer markets competing items and can sell u of the first and v of the second at prices x and y, respectively, where

$$u = 250(y - x), \quad v = 32{,}000 + 250(x - 2y).$$

Manufacturing costs for the first item are \$50 each and for the second, \$60 each. Find the selling prices for maximum profit.

18. To produce x units of a standard model and y units of a deluxe model, total costs are $x^2 + 2y^2$. If the total production of both models is to be at least 300 units, how many of each model should be produced for minimum total cost?

19. A company produces x tons of chemical A and in the process throws off x^2 pounds of waste that must be regarded as potential water pollutant. It produces y tons of chemical B with $4y^2$ pounds of waste and z tons of chemical C with $z^2 - 6z$ pounds of waste. They then mix the entire production of all three chemicals and have discovered that if they run the waste through the mixture $4xy$ pounds of it will be absorbed without hurting the final product. They want to produce at least 15 tons of the mixture. Find the proportions for minimal final water pollution.

CHAPTER VI

INTEGRAL CALCULUS

1. AREA UNDER A CURVE

A natural introduction to the concept of the *integral* is to consider the problem of finding the area of the shaded region in Figure VI.1.1.

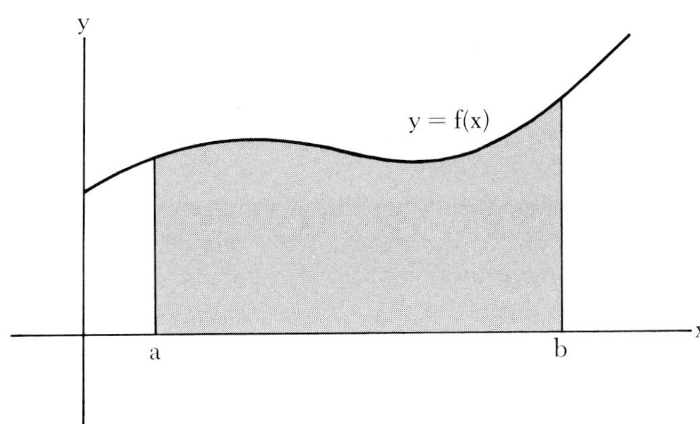

FIGURE VI.1.1 Area under a curve.

That is, if we know the equation, $y = f(x)$, of the top boundary, how can we compute the area of the shaded region? Figure VI.1.2. shows one way to approach this problem. The first rectangle has base $x_1 - x_0$ and altitude $f(x_0)$; therefore it has area

$$f(x_0)(x_1 - x_0).$$

The next rectangle has area $f(x_1)(x_2 - x_1)$, and similarly for each following rectangle. Thus, the sum of the areas of the rectangles is

$$f(x_0)(x_1 - x_0) + f(x_1)(x_2 - x_1) + \cdots + f(x_{n-1})(x_n - x_{n-1}).$$

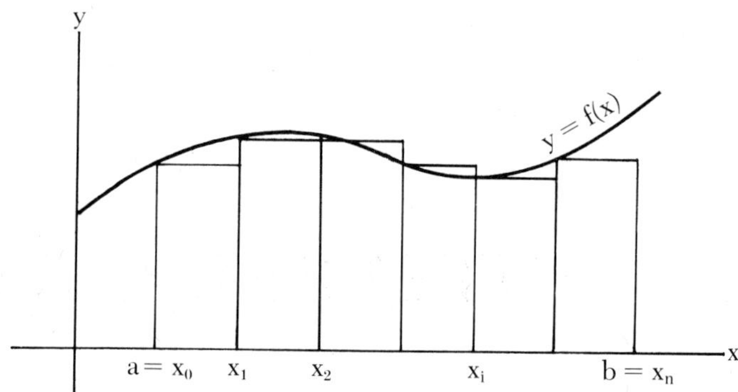

FIGURE VI.1.2 Approximation to area under a curve by rectangular areas.

Using the summation symbol (see Appendix), we can write this

6.1.1 $$\sum_{i=0}^{n-1} f(x_i)(x_{i+1} - x_i).$$

Now (except perhaps by coincidence) the sum of the areas of the rectangles in Figure VI.1.2 is not the area of the shaded region in Figure VI.1.1. However, if we continue to increase the number of rectangles and make them all thinner and thinner, it would appear that the sum (6.1.1) gets closer and closer to the area we are seeking. So, once again, a limit seems to be the answer to our problem. If we can discover a trend in the sums (6.1.1) as n increases, then the end result of that trend should be the answer to our area problem.

Before pursuing this idea further, let us rewrite (6.1.1) in slightly different notation. Recall that the differential dx measures horizontal projections of tangent line segments. Thus, we shall denote by dx_i the horizontal projection of the tangent line segment from $(x_i, f(x_i))$ over to the next partition line (Fig. VI.1.3).

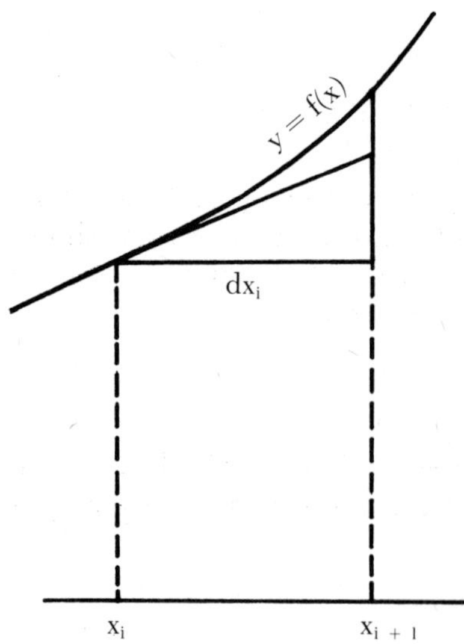

FIGURE VI.1.3 Geometric significance of dx_i.

The sum (6.1.1) may now be written

6.1.2 $$\sum_{i=0}^{n-1} f(x_i)\,dx_i,$$

and it is the limit of this expression that gives us the area under the curve.

We have been talking about areas in order to tie this discussion to something concrete. Actually, the process of forming the sum (6.1.2) and then taking its limit is merely a complicated operation to be performed on the function f. The result of all this procedure is called the *integral* of f from a to b. Notation for this result is

$$\int_a^b f(x)\,dx;$$

so we have as a definition that

6.1.3 $$\int_a^b f(x)\,dx = \lim_{n\to\infty} \sum_{i=0}^{n-1} f(x_i)\,dx_i.$$

We must pause now for a number of comments on terminology and notation.

(1) The elongated S on the left side of (6.1.3) is called an *integral sign*.

(2) The numbers a and b immediately following the integral sign are called the *limits of integration*. This terminology is unfortunate, but standard. The word "limit" used in this connection bears no relation to its use elsewhere in calculus.

(3) On the right side of (6.1.3) the symbol

$$\lim_{n\to\infty}$$

is read, "the limit as n tends to infinity." No philosophical discussion of the meaning of "infinity" is called for. The entire phrase means the end result of a trend as n continues to increase.

(4) The symbols a and b on the left side of (6.1.3) are understood to be connected with the symbols n, x_i and dx_i on the right by the following rules:

6.1.4 $$\begin{aligned} x_0 &= a, \\ x_n &= b, \\ dx_i &= x_{i+1} - x_i. \end{aligned}$$

VI.1.1 Example. Write as an integral

$$\lim_{n\to\infty} \sum_{i=0}^{n-1} \left(1 + \frac{2i}{n}\right)^2 \frac{2}{n}.$$

The idea is to find a, b and f so that the rules (6.1.4) are satisfied. As we shall see, this may be done in more than one way. Let us try setting

$$x_i = \frac{2i}{n}.$$

Then,

$$dx_i = x_{i+1} - x_i = \frac{2(i+1)}{n} - \frac{2i}{n} = \frac{2}{n},$$

and we want

$$f(x_i) = \left(1 + \frac{2i}{n}\right)^2 = (1 + x_i)^2$$

which is to say

$$f(x) = (1 + x)^2.$$

Finally,

$$a = x_0 = \frac{2 \cdot 0}{n} = 0, \quad b = x_n = \frac{2n}{n} = 2;$$

so the answer is

$$\int_0^2 (1 + x)^2 dx.$$

We could have tried

$$x_i = \frac{i}{n};$$

then $dx_i = 1/n$, and the 2 in the last factor must be put into the function f. That is,

$$f(x_i) = 2\left(1 + \frac{2i}{n}\right)^2 = 2(1 + 2x_i)^2.$$

In this analysis

$$a = x_0 = 0/n = 0, \quad b = x_n = n/n = 1,$$

and we have

$$\int_0^1 2(1 + 2x)^2 dx.$$

For a third solution, let

$$x_i = 1 + \frac{2i}{n};$$

then

$$dx_i = 2/n, \quad f(x_i) = \left(1 + \frac{2i}{n}\right)^2 = x_i^2,$$

and

$$a = x_0 = 1 + \frac{0}{n} = 1, \quad b = x_n = 1 + \frac{2n}{n} = 3.$$

Answer:

$$\int_1^3 x^2 dx.$$

The equality of these three seemingly different integrals follows from a general theorem on integrals which we shall look at later. As a prelude to that discussion we point out here how the three results spring from three different analyses of the definition of an integral.

VI.1.2 Example. Compute

$$\int_0^1 x^2 dx.$$

Let $x_i = i/n$; this yields $a = 0$, $b = 1$ and $dx_i = 1/n$ by (6.1.4). So we want

$$\lim_{n\to\infty} \sum_{i=0}^{n-1} \left(\frac{i}{n}\right)^2 \frac{1}{n}$$

$$= \lim_{n\to\infty} \frac{1}{n^3} \sum_{i=0}^{n-1} i^2.$$

To proceed any further we have to have a summation formula:

6.1.5 $$\sum_{i=0}^{n-1} i^2 = \frac{(n-1)n(2n-1)}{6}$$

Assuming this, we have

$$\int_0^1 x^2 dx = \lim_{n\to\infty} \frac{1}{n^3} \sum_{i=0}^{n-1} i^2$$

$$= \lim_{n\to\infty} \frac{(n-1)n(2n-1)}{6n^3}$$

$$= \lim_{n\to\infty} \frac{1}{6} \cdot \frac{n-1}{n} \cdot \frac{n}{n} \cdot \frac{2n-1}{n}$$

$$= \lim_{n\to\infty} \frac{1}{6} \left(1 - \frac{1}{n}\right) (1) \left(2 - \frac{1}{n}\right)$$

$$= \frac{1}{6} (1)(1)(2) = \frac{1}{3}.$$

We have worked out this example, and we give a few more like it as problems, in order to emphasize the definition of an integral. Obviously, in order to solve such a problem by direct computation you need a summation formula similar to (6.1.5). The following list will suffice for the problems that follow.

$$\sum_{i=0}^{n-1} i = \frac{n(n-1)}{2},$$

$$\sum_{i=0}^{n-1} i^3 = \frac{n^2(n-1)^2}{4},$$

$$\sum_{i=0}^{n-1} r^i = \frac{1-r^n}{1-r}.$$

We list these formulas for reference in connection with the problems below, but there is no point in memorizing the summation formulas. Happily, there are easier ways to evaluate integrals than by direct computation from the definition. This we will turn to in Section 2.

PROBLEMS ON AREA UNDER A CURVE

Write each of the following as an integral. Note: Answers are not unique (see Example VI.1.1).

1. $\lim_{n\to\infty} \sum_{i=1}^{n} \left(\frac{2i}{n} - \frac{3i^2}{n^2}\right)\frac{1}{n}$

2. $\lim_{n\to\infty} \sum_{i=1}^{n} \sqrt{\frac{i}{n}}\,\frac{1}{n}$

3. $\lim_{n\to\infty} \sum_{i=1}^{n} \sqrt{3+\frac{2i}{n}}\,\frac{2}{n}$

4. $\lim_{n\to\infty} \sum_{i=1}^{n} \frac{1}{1+\frac{2i}{n}}\,\frac{2}{n}$

5. $\lim_{n\to\infty} \sum_{i=1}^{n} \sqrt{\frac{10i}{n}}\,\frac{10}{n}$

6. $\lim_{n\to\infty} \sum_{i=1}^{n} \left(5+\frac{4i}{n}\right)^3 \frac{4}{n}$

7. $\lim_{n\to\infty} \sum_{i=1}^{n} \left(5+\frac{4i}{n}\right)^3 \frac{2}{n}$

8. $\lim_{n\to\infty} \sum_{i=1}^{n} 2^{i/n}(2^{i/n} - 2^{(i-1)/n})$

9. $\lim_{n\to\infty} \sum_{i=1}^{n} 2^{3i/n}(2^{i/n} - 2^{(i-1)/n})$

10. $\lim_{n\to\infty} \sum_{i=1}^{n} b^{i\alpha/n}(b^{i/n} - b^{(i-1)/n})$

Compute each of the following integrals.

11. $\int_0^1 x\,dx$

12. $\int_0^1 x^3\,dx$

13. $\int_2^3 x^2\,dx$

14. $\int_2^3 x^3\,dx$

15. $\int_a^b x^2\,dx$ Hint: Let $x_i = a + \dfrac{i(b-a)}{n}$

16. $\int_1^b x^k\,dx$ Hint: Let $x_i = b^{i/n} = (b^{1/n})^i$

17. $\int_a^b x^k\,dx$

2. THE FUNDAMENTAL THEOREM.

Archimedes was computing areas in the third century B.C. by methods similar to those introduced in Section 1. Specifically, he used the idea of approximating figures with curved sides by other figures with straight sides for which area formulas were well known. He then found the areas of the curve-sided figures by staring at the approximations until he saw what the limit was.

Problems 11 to 17 in Section 1 are designed primarily to show that this is not a very practical method of finding areas. So, while it might be said that Archimedes computed some integrals, no one else did for 1900 years after his death. Late in the 17th century A.D., Newton and Leibnitz discovered the idea of a derivative and then went on to discover that the idea of derivative is pertinent to the solution of the area problem. So, after 1900 years, integrals were "in" again.

One way to see this magical connection between integrals and derivatives is the following. Suppose we consider the idea of taking the integral of a derivative function,

$$\int_a^b f'(x)\,dx.$$

If we carry this back to the definition, we get

$$\lim_{x\to\infty} \sum_{i=0}^{n-1} f'(x_i)\,dx_i.$$

However,

$$f'(x_i)\,dx_i = dy_i$$

where dy_i is the vertical side of the right triangle formed by the tangent line, the horizontal and the next vertical partition line (Fig. VI.2.1). Thus, we have

$$\int_a^b f'(x)\,dx = \lim_{n\to\infty} \sum_{i=0}^{n-1} dy_i,$$

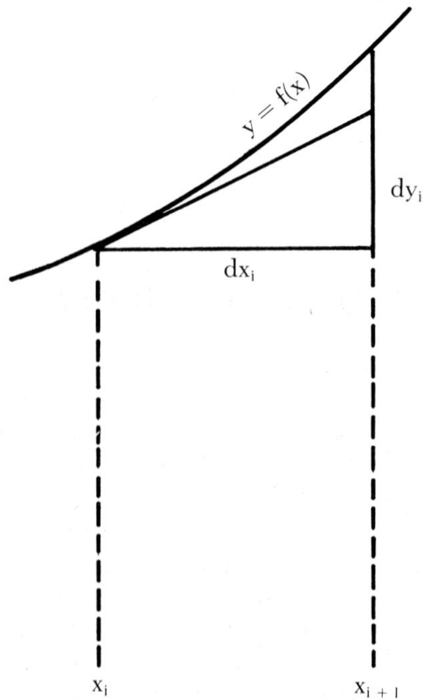

FIGURE VI.2.1 Geometric significance of dy_i.

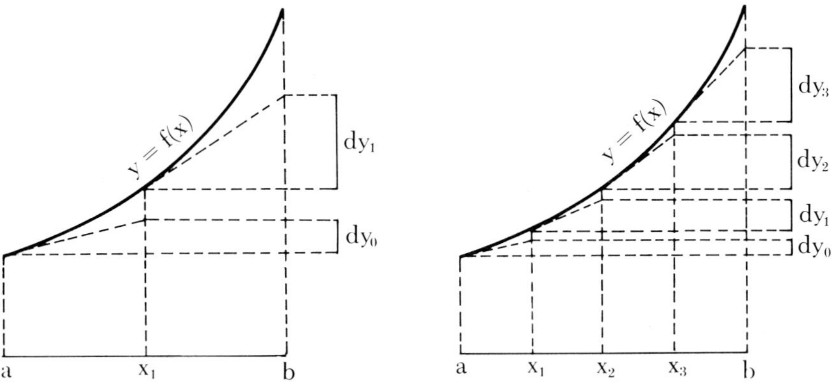

FIGURE VI.2.2 Approximations to $f(b) - f(a)$ by sums of dy_i.

and if we draw the graph of $y = f(x)$ we get a picture of the approximations to the integral of f', not as sums of areas but as sums of lengths of vertical line segments. Figure VI.2.2 shows two such approximations for the same curve, one with two subintervals, the other with four. These pictures suggest (but of course do not prove) that

$$\lim_{n\to\infty} \sum_{i=0}^{n-1} dy_i$$

is the length of the single vertical segment from the $f(a)$ level to the $f(b)$ level. This is, indeed, the case, and this fact is conveyed by the formula

6.2.1 $$\int_a^b f'(x)\,dx = f(b) - f(a).$$

The *fundamental theorem of calculus* asserts that if the limit defining the integral of f' exists, then formula 6.2.1 holds.

The fundamental theorem is the key to evaluating integrals. In words it says that we get the integral of a *derivative* by evaluating the *original* function at the limits of integration and subtracting. In Chapter V we developed a number of formulas for finding $f'(x)$ given $f(x)$. Now, to use the fundamental theorem we need to go the other way; given $f'(x)$, find $f(x)$. This process is quite naturally called *antidifferentiation;* and we now want to reverse some derivative formulas, giving what are called antiderivative formulas. First, let us introduce some slightly confusing but commonly used notation. We shall write

6.2.2. $$F(x) = \int f(x)\,dx$$

to mean that

6.2.3. $$F'(x) = f(x).$$

That is, the integral sign without limits of integration is interpreted to mean antiderivative. An antiderivative is not unique because if C is a constant, then $F(x)$ and $F(x)+C$ have the same derivative. However, in the fundamental theorem *any* antiderivative will do because

$$F(b) - F(a) = [F(b) + C] - [F(a) + C].$$

We shall write antiderivative formulas in the form (6.2.2). Note that whenever we assert that two functions F and f are related by (6.2.2) all we have to do to prove the assertion is to check that (6.2.3) holds.

Our first antiderivative formula is

$$\int x^n \, dx = \frac{x^{n+1}}{n+1} \qquad [n \neq -1].$$

To prove this, note that

$$D_x \frac{x^{n+1}}{n+1} = \frac{n+1}{n+1} x^{n+1-1} = x^n.$$

Others that come immediately from familiar derivative formulas are

$$\int e^x \, dx = e^x,$$

$$\int \frac{dx}{x} = \ln x,$$

$$\int \sin x \, dx = -\cos x,$$

$$\int \cos x \, dx = \sin x.$$

It should also be noted that because of the corresponding formula for derivatives we have

$$\int [a\, f(x) + b\, g(x)] \, dx = a \int f(x) \, dx + b \int g(x) \, dx.$$

VI.2.1 Example. Compute

$$\int_{-1}^{2} (x^3 - 5x^2 + 7) \, dx.$$

The antiderivative of $x^3 - 5x^2 + 7$ is

$$\frac{x^4}{4} - \frac{5x^3}{3} + 7x.$$

The following notational device is commonly employed in using the fundamental theorem of calculus to compute integrals.

$$\int_{-1}^{2} (x^3 - 5x^2 + 7)\, dx = \left(\frac{x^4}{4} - \frac{5x^3}{3} + 7x\right)\Bigg|_{-1}^{2}$$

$$= \left(4 - \frac{40}{3} + 14\right) - \left(\frac{1}{4} + \frac{5}{3} - 7\right)$$

$$= \frac{39}{4}.$$

In the first stage of the solution we write the antiderivative followed by a vertical bar with the limits of integration appended. This notation is interpreted to mean, "Continue with the fundamental theorem; substitute and subtract."

VI.2.2 Example.

$$\int_{1}^{2} \left(x^2 + e^x - \frac{1}{x}\right) dx = \left(\frac{x^3}{3} + e^x - \ln x\right)\Bigg|_{1}^{2}$$

$$= \left(\frac{8}{3} + e^2 - \ln 2\right) - \left(\frac{1}{3} + e - 0\right)$$

$$= \frac{7}{3} + e^2 - e - \ln 2.$$

Marginal and Total Costs. We saw in Chapter V that, if y is the total cost of producing x units, then its derivative $D_x y$ represents the marginal cost of producing one unit. It follows that total costs are an antiderivative of marginal costs. Thus, integration can be used to obtain total costs from marginal costs. (The same is true of profits, production, and many other quantities.)

VI.2.3 Example. A manufacturer has fixed costs of \$750. If he produces x pounds, his marginal cost per lb. is y dollars, where

$$y = 25 + 0.1x + 0.01x^{3/2}.$$

Find the total cost of producing 25 lb.

The total costs of producing x lb. are, here,

$$C(x) = \int (25 + 0.1x + 0.01x^{3/2})\,dx$$

$$= 25x + \frac{0.1x^2}{2} + \frac{0.01x^{5/2}}{5/2} + k$$

$$= 25x + 0.05x^2 + 0.004x^{5/2} + k$$

where k is a constant of integration. To find k, we notice that, for zero production, the costs will be precisely the fixed costs. Thus we must have $C(0) = 750$, and it follows that $k = 750$. Total costs, then, are

$$C(x) = 25x + 0.05x^2 + 0.004x^{5/2} + 750.$$

In particular, we find that $C(25) = 1418.75$. Thus, it will cost \$1418.75 to manufacture 25 lb.

VI.2.4 Example. An assembly line produces 200 items per hour. Because of errors, the proportion of defective items being produced after x hours is

$$p(x) = 0.20e^{-0.1x}.$$

How long will it take to produce 1000 good (i.e., non-defective) items?

The (marginal) number of good items produced per hour is clearly

$$f(x) = 200(1 - p) = 200 - 40e^{-0.1x}$$

and the total number produced after x hours is an antiderivative of f:

$$F(x) = \int (200 - 40e^{-0.1x})\,dx$$

$$= 200x - \frac{40e^{-0.1x}}{-0.1} + C$$

or

$$F(x) = 200x + 400e^{-0.1x} + C$$

where C is a constant. To evaluate this constant, we notice that production in 0 hours must be 0. Therefore $F(0) = 0$, and we must have $C = -400$. Thus

$$F(x) = 200x + 400e^{-0.1x} - 400.$$

We must now find x such that $F(x) = 1000$. Unfortunately, this equation cannot be solved exactly, but it may be checked that, approximately, $F(5.89) = 1000$. We conclude that it would take approximately 5 hours and 53 minutes to produce the first 1000 items. In a similar way, it may be seen that it would take about 52 hours to produce the first 10,000 items.

PROBLEMS ON THE FUNDAMENTAL THEOREM

Evaluate each of the following integrals:

1. $\int_{-1}^{1} (x^2 + x + 1)\,dx$

2. $\int_{-2}^{2} (4x^3 - 3x^2)\,dx$

3. $\int_{1}^{2} \left(x^2 + \frac{1}{x^2}\right) dx$

4. $\int_{-2}^{-1} \left(x^3 - \frac{1}{x^3}\right) dx$

5. $\int_{1}^{2} \left(\frac{2}{x^2} + \frac{1}{3x^3}\right) dx$

6. $\int_{1}^{2} (2\sqrt{x} + \sqrt{2x})\,dx$

7. $\int_{-1}^{1} (x^{1/3} + x^{2/3})\,dx$

8. $\int_{-8}^{-1} \left(x^{2/3} + \frac{1}{x^{1/3}}\right) dx$

9. $\int_{1}^{8} (x^{-4/3} + x^{-2/3})\,dx$

10. $\int_{1}^{2} \frac{(x^2 + 3x - 5)\,dx}{x^4}$

11. $\int_{1}^{4} \frac{(x^2 - 5x + 7)\,dx}{\sqrt{x}}$

12. $\int_{1}^{2} \frac{(x - \sqrt{x} + 1)\,dx}{\sqrt{x}}$

13. $\int_{1}^{2} \frac{2\,dx}{x}$

14. $\int_{0}^{\ln 3} 2\,e^x\,dx$

15. $\int_{1}^{2} \frac{(x + 1)\,dx}{x}$

16. $\int_{0}^{\pi/2} \sin x\,dx$

17. $\int_0^{\pi/4} \cos x \, dx$

18. $\int_{-\pi/2}^{\pi/2} (3 \cos x + 4 \sin x) dx$

19. A manufacturer has fixed costs of \$1200, plus marginal costs, per ton, of y dollars, where

$$y = 50 + 0.25\sqrt{x} + 0.1x$$

and x is the total amount produced, in tons. Find the total cost of producing 300 tons.

20. A metallurgist must bid on the production of 600 T. of a certain alloy. As workers learn, the number of man-hours spent on producing one ton will decrease; he calculates that it will be y, where

$$y = 20 + 10e^{-0.01x}$$

where x is the number of tons produced so far. How many man-hours will be required to produce the desired 600 tons?

3. INVERSE OF THE CHAIN RULE.

From the chain rule formula

$$D_x f\,[g(x)] = f'[g(x)]g'(x)$$

we get the antiderivative formula

$$\int f'[g(x)]g'(x)\,dx = f[g(x)].$$

To put it another way, if F is an antiderivative of f, then

$$\int f[g(x)]g'(x)\,dx = F[g(x)].$$

An even more suggestive way to write this is to note that

$$dg(x) = g'(x)\,dx$$

and so

6.3.1 $$\int f[g(x)]\,dg(x) = F[g(x)]$$

where F is an antiderivative of f.

Let us put this rule in words. If we have a composite function multiplied by the differential of the "inside" function, then we get an antiderivative by taking an antiderivative of the "outside" function *only*, and carrying the inside function along.

Formula (6.3.1) is applicable only to problems fitting a rather rigid pattern. Nevertheless, this pattern does occur sufficiently often that (6.3.1) is one of the more important calculus formulas. We commented several times in Chapter V that a mechanical hazard in the use of the chain rule is the diversity of notation for composite functions. As the following examples show, this same factor is still present as we reverse the chain rule to compute antiderivatives.

VI.3.1 Example.

$$\int_0^1 2x\sqrt{1+x^2}\,dx = \int_0^1 \sqrt{1+x^2}\,(2x\,dx) = \int_0^1 \sqrt{1+x^2}\,d(1+x^2)$$

$$= \frac{2}{3}(1+x^2)^{3/2}\Big|_0^1 = \frac{4\sqrt{2}-2}{3}.$$

VI.3.2 Example.

$$\int_0^1 x\sqrt{1-x^2}\,dx = -\frac{1}{2}\int_0^1 \sqrt{1-x^2}\,(-2x\,dx)$$

$$= -\frac{1}{2}\int_0^1 \sqrt{1-x^2}\,d(1-x^2)$$

$$= -\frac{1}{3}(1-x^2)^{3/2}\Big|_0^1 = \frac{1}{3}.$$

Note that in VI.3.1 we had outside the radical precisely the differential of the expression under the radical. In VI.3.2 we were off by a factor of -2. A constant factor can be adjusted, but *this can only be done with constants*. If we had had the wrong power of x outside the radical, the method we are using here would not have been applicable.

VI.3.3 Example.

$$\int_0^1 x\,e^{-x^2}\,dx = -\frac{1}{2}\int_0^1 e^{-x^2}(-2x\,dx) = -\frac{1}{2}\int_0^1 e^{-x^2}\,d(-x^2)$$

$$= -\frac{1}{2}e^{-x^2}\Big|_0^1 = \frac{1}{2}\left(1-\frac{1}{e}\right).$$

VI.3.4 Example.

$$\int_0^1 \frac{x^2\,dx}{2-x^3} = -\frac{1}{3}\int_0^1 \frac{-3x^2\,dx}{2-x^3} = -\frac{1}{3}\int_0^1 \frac{d(2-x^3)}{2-x^3}$$

$$= -\frac{1}{3}\ln(2-x^3)\Big|_0^1 = \frac{1}{3}\ln 2.$$

Note that the "inside" function has appeared in three different places so far: under the radical, in the exponent, and in the denominator.

VI.3.5 Example.

$$\int_0^1 e^{3x}dx = \frac{1}{3}\int_0^1 e^{3x}(3dx) = \int_0^1 e^{3x}d(3x) = \frac{1}{3}e^{3x}\Big|_0^1 = \frac{1}{3}(e^3 - 1).$$

VI.3.6 Example.

$$\int_0^1 \frac{e^x\,dx}{2+e^x} = \int_0^1 \frac{d(2+e^x)}{2+e^x} = \ln(2+e^x)\Big|_0^1$$

$$= \ln(2+e) - \ln 2 = \ln\left(1+\frac{e}{2}\right)$$

Trigonometric functions introduce additional variations in notation.

VI.3.7 Example.

$$\int_0^{\pi/4} \sin 2x\,dx = \frac{1}{2}\int_0^{\pi/4} \sin 2x(2\,dx) = \frac{1}{2}\int_0^{\pi/4} \sin 2x\,d(2x)$$

$$= -\frac{1}{2}\cos 2x\Big|_0^{\pi/4} = \frac{1}{2}.$$

VI.3.8 Example.

$$\int_0^{\pi/2} \cos^3 x \sin x\,dx = -\int_0^{\pi/2} \cos^3 x(-\sin x\,dx)$$

$$= -\int_0^{\pi/2} \cos^3 x\,d(\cos x)$$

$$= -\frac{1}{4}\cos^4 x\Big|_0^{\pi/2} = \frac{1}{4}.$$

VI.3.9 Example.

$$\int_0^{\pi/2} \cos^3 x\,dx = \int_0^{\pi/2} \cos^2 x(\cos x\,dx) = \int_0^{\pi/2} (1-\sin^2 x)(\cos x\,dx)$$

$$= \int_0^{\pi/2} (1-\sin^2 x)\,d(\sin x) = \left(\sin x - \frac{1}{3}\sin^3 x\right)\Big|_0^{\pi/2}$$

$$= \frac{2}{3}.$$

The technique used in VI.3.9 will furnish an antiderivative for $\cos^m x \sin^n x$ if either m or n is an odd integer.

VI.3.10 Example. A chemical firm produces a certain compound. Fixed costs are \$1000, while the marginal costs are y dollars per lb., where

$$y = x\sqrt{x^2 + 100}$$

and x is the amount produced, in pounds. Find the total cost of producing 20 lb. of this compound.

The total cost is, of course, an antiderivative:

$$C(x) = \int x\sqrt{x^2+100}\,dx$$

$$= \frac{1}{2}\int \sqrt{x^2+100}\,d(x^2+100)$$

$$= \frac{1}{2}\,\frac{(x^2+100)^{3/2}}{3/2} + k$$

or

$$C(x) = \frac{(x^2+100)^{3/2}}{3} + k.$$

We must now evaluate the constant of integration, k. We notice that $C(0) = 1000$, which means that $k = 2000/3$. Therefore,

$$C(x) = \frac{(x^2+100)^{3/2} + 2000}{3}$$

is the total cost of producing x pounds of the compound. For $x = 20$, we find

$$C(20) = \frac{500^{3/2} + 2000}{3} = \frac{13{,}680}{3}$$

and the costs will be approximately \$4560.

PROBLEMS ON INVERSE OF THE CHAIN RULE

Compute each of the following integrals.

1. $\int_0^1 x\sqrt{1+x^2}\,dx$

2. $\int_1^2 x\sqrt{4-x^2}\,dx$

3. $\int_0^1 \frac{x\,dx}{\sqrt[3]{2-x^2}}$

4. $\int_0^1 x^2\sqrt{1-x^3}\,dx$

5. $\int_0^1 x^4\sqrt{1-x^5}\,dx$

6. $\int_0^1 \frac{x^3\,dx}{(1+x^4)^{3/2}}$

7. $\int_0^1 \frac{3x\,dx}{(1+x^2)^{3/2}}$

8. $\int_0^2 \frac{4x^2\,dx}{(1+x^3)^{4/3}}$

9. $\int_0^1 2x^2(1-x^3)^{2/3}dx$

10. $\int_{-1}^1 5x\sqrt{1-x^2}\,dx$

11. $\int_{-1}^1 5x(1-x^2)^{2/3}dx$

12. $\int_{-1}^1 5x^2(1-x^3)^{2/3}dx$

13. $\int_{-1}^1 5x^2(1-x^3)^{1/3}dx$

14. $\int_{-1}^0 \frac{dx}{1-x}$

15. $\int_{-1}^0 \frac{dx}{\sqrt{1-x}}$

16. $\int_2^3 \frac{dx}{1-x}$

17. $\int_0^1 \frac{e^x\,dx}{1+e^x}$

18. $\int_1^2 \frac{e^x\,dx}{1-e^x}$

19. $\int_0^{\ln 2} e^{-x}\,dx$

20. $\int_0^1 \frac{e^x\,dx}{\sqrt{1+e^x}}$

21. $\int_0^3 \frac{x\,dx}{\sqrt{25-x^2}}$

22. $\int_0^3 \frac{x\,dx}{25-x^2}$

23. $\int_0^3 \frac{x\,dx}{x^2-25}$

24. $\int_0^1 x^2 e^{x^3}\,dx$

25. $\int_0^1 e^{3x}\,dx$

26. $\int_{-\pi}^{\pi} \cos^3 2x\,dx$

27. $\int_0^{\pi/3} \frac{\sin x\,dx}{\cos^2 x}$

28. $\int_0^{\pi} \cos x\sqrt{\sin x\,dx}$

29. $\int_0^{\pi/2} \sin^3 x\sqrt[3]{\cos x\,dx}$

30. $\int_0^{\pi/3} \frac{\sin x\,dx}{\sqrt{\cos x}}$

31. $\int_0^{\pi} \sin^2 x\cos x\,dx$

32. $\int_0^{\pi/2} \cos^3 x\sin^6 x\,dx$

33. $\int_0^{\pi/6} \frac{\sin 2x\,dx}{\cos^3 2x}$

34. $\int_0^{\pi/6} \frac{\sin 2x\,dx}{(\cos 2x)^{1/3}}$

35. $\int_0^{\pi} \frac{\cos x \, dx}{(2 - \sin x)^2}$

36. The daily rate of sales for a certain gadget, in dollars, is

$$y = 100x \, e^{-0.01x^2}$$

where x is the number of days that the gadget has been on the market. Find the total sales during the first 20 days that it is on the market.

37. The marginal cost of production of one ton of steel is y dollars, where

$$y = \frac{1000x}{(10{,}000 - x^2)^{3/2}}$$

and x is the amount produced, in tons. If fixed costs are $500, find the total cost of producing 30 tons of steel.

4. INTEGRATION BY SUBSTITUTION.

If we take the antiderivative formula (6.3.1) and apply it to integrals we have (assuming $F' = f$)

$$\int_a^b f[g(x)]g'(x)\,dx = F[g(x)]\Big|_a^b .$$

However,

$$F[g(x)]\Big|_a^b = F[g(b)] - F[g(a)] = F(u)\Big|_{g(a)}^{g(b)} = \int_{g(a)}^{g(b)} f(u)\,du.$$

Thus, we have

6.4.1 $$\int_a^b f[g(x)]g'(x)\,dx = \int_{g(a)}^{g(b)} f(u)\,du.$$

This is called the substitution formula for integrals. Note that if we set

6.4.2 $$u = g(x),$$

then we have

6.4.3 $$\begin{cases} du = g'(x)\,dx, \\ x = a \text{ when } u = g(a), \\ x = b \text{ when } u = g(b); \end{cases}$$

and these substitutions change the integral on the right in (6.4.1) into that on the left.

The technique introduced in Section 3 is frequently called integration by substitution. In a sense we do use the substitution formula (6.4.1) in Section 3, but we use it rather informally. Specifically, we recognize that an integral assumes precisely the form given on the left in (6.4.1); then we obtain the antiderivative by thinking of the form on the right in (6.4.1). However, we do not actually introduce the new variable u; we carry it as $g(x)$ all the way through. We shall refer to this process as informal substitution.

In the present section we want to consider what we will call formal substitution. In this we start with the right hand side of (6.4.1). This could be any integral at all; no special format is required. The process is actually to make the substitutions (6.4.2) and (6.4.3), thus ending up with the left side of (6.4.1).

VI.4.1 Example. Compute

$$\int_0^4 x^3\sqrt{16-x^2}\,dx.$$

We introduce a new variable u by setting

$$u^2 = 16 - x^2;$$

then

$$x^3 = (16-u^2)^{3/2},$$

and

$$2u\,du = -2x\,dx,$$

$$dx = -\frac{u\,du}{x} = \frac{-u\,du}{\sqrt{16-u^2}};$$

$u=4$ when $x=0$ and $u=0$ when $x=4$. Making these substitutions we have

$$\int_0^4 x^3\sqrt{16-x^2}\,dx = \int_4^0 (16-u^2)^{3/2}\,u\left(\frac{-u\,du}{\sqrt{16-u^2}}\right)$$

$$= \int_4^0 (u^4-16u^2)\,du = \left(\frac{u^5}{5}-\frac{16u^3}{3}\right)\Bigg|_4^0 = \frac{2048}{15}$$

This same integral could have been computed with another substitution process:

$$x = 4\sin u,$$
$$dx = 4\cos u\,du,$$
$$\sqrt{16-x^2} = 4\cos u;$$

$u = \pi/2$ when $x = 4$ and $u = 0$ when $x = 0$. Thus,

$$\int_0^4 x^3\sqrt{16 - x^2}\, dx = \int_0^{\pi/2} (64 \sin^3 u)(4 \cos u)(4 \cos u\, du)$$

$$= 1024 \int_0^{\pi/2} \sin^2 u \cos^2 u (\sin u\, du)$$

$$= -1024 \int_0^{\pi/2} (1 - \cos^2 u) \cos^2 u\, d(\cos u)$$

$$= 1024 \left(-\frac{1}{3} \cos^3 u + \frac{1}{5} \cos^5 u\right) \Big|_0^{\pi/2} = \frac{2048}{15}.$$

VI.4.2 Example. Compute

$$\int_1^2 \frac{dx}{x^3(9 - x^3)^{1/3}}.$$

In this case the following substitution will work wonders.

$$x^3u^3 = 9 - x^3,$$
$$3x^2u^3\, dx + 3x^3u^2\, du = -3x^2\, dx,$$
$$xu^2\, du = -(1 + u^3)\, dx,$$
$$dx = \frac{-xu^2\, du}{1 + u^3}.$$

From the first equation,

$$(9 - x^3)^{1/3} = xu;$$

$u = 2$ when $x = 1$ and $u = 1/2$ when $x = 2$. Thus,

$$\int_1^2 \frac{dx}{x^3(9 - x^3)^{1/3}} = \int_2^{1/2} \frac{1}{x^3(xu)} \cdot \frac{(-xu^2\, du)}{(1 + u^3)}$$

$$= \int_2^{1/2} \frac{-u\, du}{x^3 + x^3u^3}$$

$$= \int_2^{1/2} \frac{-u\, du}{9} = -\frac{u^2}{18}\Big|_2^{1/2} = \frac{5}{24}.$$

VI.4.3 Example. Compute

$$\int_0^2 x^2\sqrt{4 - x^2}\, dx$$

Here we substitute

$$x = 2 \sin u,$$
$$dx = 2 \cos u\, du,$$
$$\sqrt{4 - x^2} = 2 \cos u;$$

$u = 0$ when $x = 0$ and $u = \pi/2$ when $x = 2$. Thus,

$$\int_0^2 x^2\sqrt{4 - x^2}\, dx = \int_0^{\pi/2} (4 \sin^2 u)(2 \cos u)(2 \cos u\, du)$$

$$= 16 \int_0^{\pi/4} \sin^2 u \cos^2 u\, du.$$

If either sin or cos appears as an odd power, VI.3.9 shows how to compute the antiderivative. Here, however, we have a new form, and a new trick needs to be learned. We turn to the trigonometric identities.

6.4.4 $$\sin^2 u = \frac{1}{2}(1 - \cos 2u),$$

6.4.5 $$\cos^2 u = \frac{1}{2}(1 + \cos 2u).$$

Using these, we have

$$16 \int_0^{\pi/2} \sin^2 u \cos^2 u\, du = 4 \int_0^{\pi/2} (1 - \cos 2u)(1 + \cos 2u)\, du$$

$$= 4 \int_0^{\pi/2} du - 4 \int_0^{\pi/2} \cos^2 2u\, du$$

$$= 4 \int_0^{\pi/2} du - 2 \int_0^{\pi/2} (1 + \cos 4u)\, du$$

$$= \left(4u - 2u + \frac{1}{2} \sin 4u\right)\Big|_0^{\pi/2} = \pi.$$

These examples illustrate how integration by substitution works, but the obvious question is, "How do you know what substitution to make?" The answer to this question is to learn that for certain forms certain substitutions get the job done. The above examples illustrate the following rules.

The form

$$x^m(a + bx^n)^{r/s}\, dx$$

is frequently transformed into a rational form in u by one or the other of the substitutions

6.4.6 $$u^s = a + bx^n,$$

6.4.7 $$x^n u^s = a + bx^n.$$

Trial and error is probably as good a way as any to choose the proper substitution here, but it can be shown that (6.4.6) works if

$$\frac{m+1}{n}$$

is an integer, and (6.4.7) works if

$$\frac{m+1}{n}+\frac{r}{s}$$

is an integer.

As for trigonometric substitutions, given the form

$$x^m(a^2-x^2)^{n/2}\,dx,$$

the substitution

$$x = a \sin u$$

yields a product of powers of $\sin u$ and $\cos u$. If one of these is odd, VI.3.9 shows the way to proceed. If both are even, (6.4.4) and (6.4.5) may be used.

In a complete treatment of calculus methods, many other strategic formal substitutions are presented. We shall not pursue the matter in any greater detail here.

VI.4.4 Example. A farmer finds that his marginal harvest per acre is y bushels, where

$$y = \frac{500{,}000}{x^2+10{,}000}$$

where x is his total acreage under cultivation. How many bushels can he harvest from 100 acres?

When a factor of the form $x^2 + a^2$ appears, the substitution

$$x = a \tan u$$

is usually helpful. In this case, $a = 100$, so we set $x = 100 \tan u$, which gives us

$$dx = 100 \sec^2 u\, du.$$

Now, we know from elementary trigonometry that $\sec^2 u = \tan^2 u + 1$, and so

$$x^2 + 10{,}000 = 10{,}000 \sec^2 u$$

which gives us the integrand

$$\frac{500{,}000}{x^2 + 10{,}000}\,dx = 5000\,du.$$

Now, we know that total production is the integral of marginal production. Remembering that $\tan \pi/4 = 1$, we see that the total production on 100 acres is

$$\int_0^{100} \frac{500{,}000}{x^2 + 10{,}000}\,dx = \int_0^{\pi/4} 5000\,du = 1250\pi.$$

Production is, then, approximately 3940 bushels.

VI.4.5 Example. A wire manufacturer has marginal costs of y cents per yard of steel wire, where

$$y = \frac{0.01x^3}{\sqrt{1{,}000{,}000 - x^2}}$$

and x is the number of yards produced. If his fixed costs are \$100, find his total cost of producing 250 yards of the wire.

The variable costs here are, of course, the integral of marginal costs. If we set

$$x = 1000 \sin u$$

we will have

$$dx = 1000 \cos u\,du$$

and

$$\sqrt{1{,}000{,}000 - x^2} = 1000 \cos u.$$

Therefore,

$$\int_0^{250} \frac{0.01x^3}{\sqrt{1{,}000{,}000 - x^2}}\,dx = \int_0^{U} 10{,}000{,}000 \sin^3 u\,du$$

where U satisfies $\sin U = 0.25$, so that $\cos U = 0.968$. Thus variable costs are

$$10{,}000{,}000 \int_0^{U} \sin^3 u\,du = 10{,}000{,}000 \int_0^{U} (\sin u - \sin u \cos^2 u)du$$

$$= 10{,}000{,}000 \left(\frac{\cos^3 u}{3} - \cos u\right)\Big|_0^{U} = 9975.$$

Thus the variable costs of producing 250 yards of steel wire are \$99.75. Added to the \$100 in fixed costs, this gives total costs of \$199.75.

It should be noted that the substitution

$$v^2 = 1{,}000{,}000 - x^2$$

would also give this result, though with considerably more work.

PROBLEMS ON INTEGRATION BY SUBSTITUTION

Evaluate each of the following integrals.

1. $\int_0^1 x\sqrt{1-x}\,dx$

2. $\int_0^{2/3} x^2\sqrt{2-3x}\,dx$

3. $\int_0^{13} x(1+2x)^{2/3}\,dx$

4. $\int_0^1 \frac{x\,dx}{(3-2x)^{1/3}}$

5. $\int_0^1 x^8(1-x^3)^{1/3}\,dx$

6. $\int_0^1 \frac{x^3\,dx}{(1+x^2)^{2/3}}$

7. $\int_0^2 x^3\sqrt{4-x^2}\,dx$ using $x = 2\sin u$

8. $\int_0^2 x^3\sqrt{4-x^2}\,dx$ using $u^2 = 4-x^2$

9. $\int_1^2 \frac{dx}{x^3(9-x^3)^{1/3}}$ using $t^3 = \frac{9-x^3}{x^3}$

10. $\int_0^1 x^8(1-x^3)^{5/4}\,dx$ using $t^4 = 1-x^3$

11. $\int_0^2 \frac{x^5\,dx}{(1+x^3)^{3/2}}$ using $u^2 = 1+x^3$

12. $\int_{-1}^{1} \sqrt{2 - x^2}\, dx$

13. $\int_{-3}^{3} x^2\sqrt{9 - x^2}\, dx$

14. $\int_{0}^{1} x^3\sqrt{1 - x^2}\, dx$

15. $\int_{0}^{1} \frac{dx}{\sqrt{2 - x^2}}$

16. $\int_{0}^{1} \frac{x^2\, dx}{\sqrt{4 - x^2}}$

17. $\int_{-4}^{3} \frac{x^3\, dx}{\sqrt{25 - x^2}}$

18. A miner finds that he must spend at least \$100 if his mine is to produce anything. If he spends x dollars, where $x > 100$, his marginal production per dollar is y pounds of ore, where

$$y = \frac{400}{\sqrt{x^2 - 10{,}000}}.$$

What is his total production if he spends \$505? [Hint: let $x = 50(e^u + e^{-u})$. Remember that, for $x = 100$, total production is 0.]

19. A farmer's marginal production per acre of land is y bushels, where

$$y = \frac{4000}{\sqrt{x^2 + 10{,}000}}$$

and x is the number of acres under cultivation. How many bushels can he obtain from 495 acres? [Hint: let $x = 50(e^u - e^{-u})$.]

5. INTEGRATION BY PARTS.

The formula for the differential of a product may be written

6.5.1 $$d(uv) = u\, dv + v\, du.$$

By the fundamental theorem of calculus

$$\int_a^b d(uv) = uv \Big|_a^b;$$

 so we have from (6.5.1) that

6.5.2. $$\int_a^b u\,dv = uv\Big|_a^b - \int_a^b v\,du.$$

This is known as the *integration by parts* formula.

VI.5.1 Example. Compute

$$\int_0^1 x^3\sqrt{1-x^2}\,dx.$$

This could be done by trigonometric substitution, but integration by parts will also solve this problem. To fit this to the formula (6.5.2), we set

$$u = x^2$$
$$dv = x\sqrt{1-x^2}\,dx.$$

Then by informal substitution (Section 3) we have

$$v = \int x\sqrt{1-x^2}\,dx = -\frac{1}{2}\int \sqrt{1-x^2}\,d(1-x^2)$$
$$= -\frac{1}{2}\cdot\frac{2}{3}(1-x^2)^{3/2};$$

so by (6.5.2),

$$\int_0^1 x^3\sqrt{1-x^2}\,dx = (x^2)\left[-\frac{1}{3}(1-x^2)^{3/2}\right]\Bigg|_0^1 - \int_0^1 -\frac{1}{3}(1-x^2)^{3/2}(2x\,dx).$$

Now

$$-\frac{1}{3}(1-x^2)^{3/2}(2x\,dx) = \frac{1}{3}(1-x^2)^{3/2}\,d(1-x^2);$$

so

$$\int_0^1 x^3\sqrt{1-x^2}\,dx = -\frac{1}{3}x^2(1-x^2)^{3/2}\Big|_0^1 - \frac{1}{3}\cdot\frac{2}{5}(1-x^2)^{5/2}\Big|_0^1$$
$$= \frac{2}{15}.$$

As an aid to the computation of integrals, integration by parts has the effect of replacing the problem of integrating a given product by that of integrating a different product. In its most direct application in this connection, the secret of success lies in arranging things so that the second product has a more familiar antiderivative than the first. This involves strategic grouping of factors; where there are several options, it is a good idea to test them by visualizing the *form* of the new integral quickly without applying the formula in detail. Specifically, constants and signs are immaterial in this testing process. So forget them momentarily and think in terms like these: under differentiation $\sqrt{x} \to 1/\sqrt{x}$, $x^3 \to x^2$, and so on; under antidifferentiation $\sqrt{x} \to x^{3/2}$, $x^3 \to x^4$, and so on. Thinking in these terms, we could "invent" the procedure used in Example VI.5.1 as follows:

Under differentiation, $x^2 \to x$;

Under antidifferentiation, $x\sqrt{1-x^2} \to (1-x^2)^{3/2}$;

New integral form, $x(1-x^2)^{3/2}$.

Since this last form is one that we know how to handle, we put this plan into action.

VI.5.2 Example. Compute

$$\int_0^1 x\, e^{-x}\, dx.$$

Analysis:

Under differentiation, $x \to 1$;

Under antidifferentiation, $e^{-x} \to e^{-x}$;

New integral form, e^{-x}.

Details:

$$\int_0^1 x\, e^{-x}\, dx = x(-e^{-x})\Big|_0^1 - \int_0^1 -e^{-x}\, dx$$

$$= -x(e^{-x})\Big|_0^1 - e^{-x}\Big|_0^1 = 1 - \frac{2}{e}.$$

VI.5.3 Example. Compute

$$\int_1^2 \ln x\, dx.$$

Analysis:

Under differentiation, $\ln x \to 1/x$;

Under antidifferentiation, $1 \to x$

New integral form, $x(1/x) = 1$.

Details:

$$\int_1^2 \ln x \, dx = (\ln x)\, x \Big|_1^2 - \int_1^2 dx$$

$$= (x \ln x - x) \Big|_1^2 = 2 \ln 2 - 1.$$

VI.5.4 Example. Compute

$$\int_0^{\pi/2} e^x \sin x \, dx.$$

If e^x is differentiated and $\sin x$ antidifferentiated, the new integral involves $e^x \cos x$. Now, if we apply integration by parts to this new integral we come up with $e^x \sin x$ again. This opens the possibility of using the "circular technique" which is illustrated as follows.

$$\int_0^{\pi/2} e^x \sin x \, dx = e^x(-\cos x) \Big|_0^{\pi/2} - \int_0^{\pi/2} - e^x \cos x \, dx$$

$$= - e^x \cos x \Big|_0^{\pi/2} + e^x \sin x \Big|_0^{\pi/2} - \int_0^{\pi/2} e^x \sin x \, dx.$$

Now, we simply transpose the last integral on the right to the extreme left and thereby solve this equation for the desired integral:

$$2 \int_0^{\pi/2} e^x \sin x \, dx = (-e^x \cos x + e^x \sin x) \Big|_0^{\pi/2} = e^{\pi/2} + 1,$$

$$\int_0^{\pi/2} e^x \sin x \, dx = \frac{1}{2} (e^{\pi/2} + 1).$$

Discounted Future Earnings. We saw in Chapter III that, because of interest, a present sum of money will be worth more in the future. A corollary of this statement is the fact that the *present value* of a *future* sum of money is less than the future sum itself. (As an example, a borrower will sign a note for \$1000, payable one year later, and receive, in exchange, some \$900.) The fact is that future payments must be *discounted*, i.e., decreased by a factor which depends both on the interest rate and on the time which will elapse before repayment. From both Chapters III and IV, it seems reasonable that the discount factor should be an exponential function of time. Indeed, it is e^{-it}, where i is the interest rate (per year) and t is the number of years before repayment.

VI.5.5 Example. A manufacturer buys a new, labor-saving machine. It is estimated that the machine will have a useful life of 20 years, after which it can be sold for \$5000. It is also estimated that, after x years of use, the machine will occasion a saving of $20 - 0.02x^2$ dollars per day. Assuming 250 working days per year, and a discount (interest) rate of 10 per cent per annum, what is the effective present value of the machine?

Assuming 250 working days per year, the machine will give savings at the rate of

$$y = 250(20 - 0.02x^2) = 5000 - 5x^2$$

dollars per year at time x. This must be discounted, i.e., multiplied by the factor e^{-ix}. Since $i = 0.1$, the discounted rate of savings is

$$f(x) = (5000 - 5x^2)e^{-0.1x}$$

and the total (discounted) savings effected by the machine in 20 years will be

$$\int_0^{20} (5000 - 5x^2)e^{-0.1x}dx = \int_0^{20} 5000e^{-0.1x}dx - \int_0^{20} 5x^2e^{-0.1x}dx.$$

The first of these two integrals is easy to evaluate:

$$\int_0^{20} 5000e^{-0.1x}dx = -50{,}000e^{-0.1x}\Big|_0^{20} = 50{,}000(1 - e^{-2}) = 43{,}235$$

To evaluate the second integral, we use integration by parts, with

$$u = 5x^2$$
$$dv = e^{-0.1x}dx$$

which gives

$$du = 10x\,dx$$
$$v = -10e^{-0.1x}$$

so that

$$\int_0^{20} 5x^2e^{-0.1x}dx = -50x^2e^{-0.1x}\Big|_0^{20} + \int_0^{20} 100x\,e^{-0.1x}dx$$
$$= -20{,}000e^{-2} + \int_0^{20} 100x\,e^{-0.1x}dx.$$

This new integral must also be evaluated by the method of integration by parts. We now set

$$u = 100x$$
$$dv = e^{-0.1x}dx$$

so that

$$du = 100dx$$
$$v = -10e^{-0.1x}$$

and so

$$\int_0^{20} 100x\, e^{-0.1x}dx = -1000x\, e^{-0.1x}\Big|_0^{20} + \int_0^{20} 1000e^{-0.1x}dx$$
$$= -1000x\, e^{-0.1x}\Big|_0^{20} - 10{,}000e^{-0.1x}\Big|_0^{20}$$
$$= -20{,}000e^{-2} - 10{,}000e^{-2} + 10{,}000.$$

As a result we have

$$\int_0^{20} 5x^2e^{-0.1x}dx = -50{,}000e^{-2} + 10{,}000 = 3{,}235$$

and so

$$\int_0^{20} (5000 - 5x^2)\, e^{-0.1x}dx = 40{,}000.$$

Thus, the total discounted value of the savings effected by the machine is \$40,000. To this must be added the \$5000 salvage value, discounted also. The present value of \$5000 is $5000e^{-2} = 676.50$. Thus the effective present value of the machine (discounted future savings plus discounted salvage) is \$40,676.50.

PROBLEMS ON INTEGRATION BY PARTS

Compute each of the following integrals.

1. $\int_0^1 x\sqrt{1+x}\, dx$

2. $\int_0^1 \frac{x\, dx}{\sqrt{1+x}}$

3. $\int_0^1 \frac{(1-x)}{\sqrt{1+x}}\, dx$

4. $\int_0^{1/2} \frac{x^3\,dx}{\sqrt{1-x^2}}$

5. $\int_0^1 \frac{x^2\,dx}{\sqrt{1+x}}$

6. $\int_0^{\pi} x \sin x\,dx$

7. $\int_{-\pi}^{\pi} x \sin 2x\,dx$

8. $\int_0^{\pi/2} x^3 \cos 3x\,dx$

9. $\int_0^{\sqrt{\pi}} x^3 \sin(x^2)\,dx$

10. $\int_{-\pi}^{\pi} \sin\left(\frac{x}{2}\right) \cos x\,dx$

11. $\int_0^{\pi} \sin 3x \cos 5x\,dx$

12. $\int_0^1 xe^x\,dx$

13. $\int_0^1 x^2e^{2x}\,dx$

14. $\int_0^1 x^3e^{x^2}\,dx$

15. $\int_0^{\pi/2} e^{-x} \cos 2x\,dx$

16. $\int_0^{\pi/2} e^x \cos 2x\,dx$

17. $\int_0^{\pi} x^3 \sin x\,dx$

18. $\int_0^1 (x^2+2)e^{-2x}\,dx$

19. $\int_0^{\pi} \cos 3x \sin 4x\,dx$

20. $\int_0^{\pi} \cos 3x \ \cos 4x \, dx$

21. $\int_1^2 (\ln x)^2 dx$

22. $\int_1^t x^n \ln x \, dx$

23. The daily savings from a certain machine, after x years of use, are $20 - 0.4x$ dollars. After twenty years, the machine can be sold for a scrap value of $1000. Assuming an interest (discount) rate of 12 per cent per annum and 250 working days per year, find the effective present value of the machine.

24. A dairy company finds that its marginal profits per gallon of milk are y dollars, where

$$y = \frac{2000 \ln (x + 100)}{(x + 100)^2}$$

and x is total production in gallons. Find the firm's total profits, if it produces 400 gallons of milk.

25. A new chemical dye is brought on the market; daily sales are approximately y lb., where

$$y = 100x(1 - \cos \pi x)$$

and x is the time (in years) after the gadget was introduced. Assuming 300 sales days, find the total sales during the first year after the chemical is introduced.

6. APPLICATIONS

We introduced the notion of integral as the solution of the problem of the area under a curve. There are many other applications of integrals, but in this brief introduction to calculus we shall consider only problems tied rather closely to the notion of area.

First, let us look at a slightly more general picture of the area problem itself. If $y = f(x)$ is the top boundary of a plane region and $y = g(x)$ is the bottom boundary, then $f(x) - g(x)$ is the (positive) length of the vertical segment through the region at x. Note (Fig. VI.6.1) that this is true no matter where the axes are. At x_1, in Figure VI.6.1, we want

$$A - B = [-g(x_1)] - [-f(x_1)] = f(x_1) - g(x_1).$$

At x_2 we want

$$C + D = f(x_2) + [-g(x_2)] = f(x_2) - g(x_2).$$

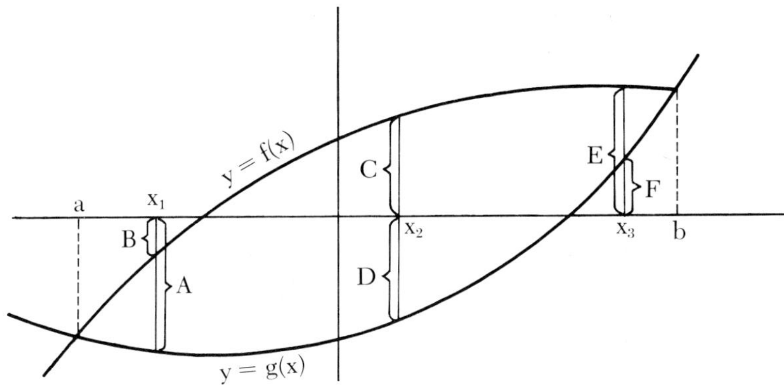

FIGURE VI.6.1 Positive distance across a figure.

At x_3 we want

$$E - F = f(x_3) - g(x_3).$$

Thus, the area of the region between the curves in Figure VI.6.1 is given by

$$\int_a^b [f(x) - g(x)]\,dx.$$

VI.6.1 Example. Find the area between the curves

$$y = x^2 \quad \text{and} \quad y = 2x.$$

A sketch is shown in Figure VI.6.2. We find the intersection points by solving the equations simultaneously:

$$\begin{aligned} x^2 &= 2x, \\ x(x-2) &= 0, \\ x &= 0, 2. \end{aligned}$$

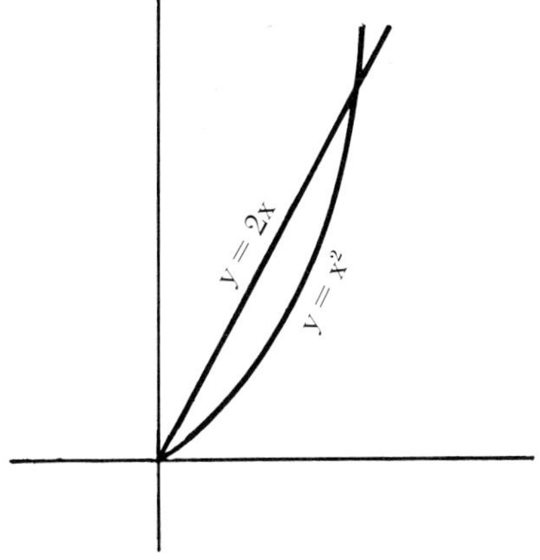

FIGURE VI.6.2 Sketch for Example VI.6.1.

 We see from the picture (and this is the principal purpose of drawing a sketch) that $y = 2x$ is the top and $y = x^2$ the bottom. Thus, the area is

$$\int_0^2 (2x - x^2)\,dx = \left(x^2 - \frac{1}{3}x^3\right)\Big|_0^2 = \frac{4}{3}.$$

We can equally well turn the area problem on its side and integrate with respect to y. Note Figure VI.6.3; the rectangle shown has area $[f(y_i) - g(y_i)]\,dy_i$, and the limit of the sum of such rectangular areas is

$$\int_c^d [f(y) - g(y)]\,dy.$$

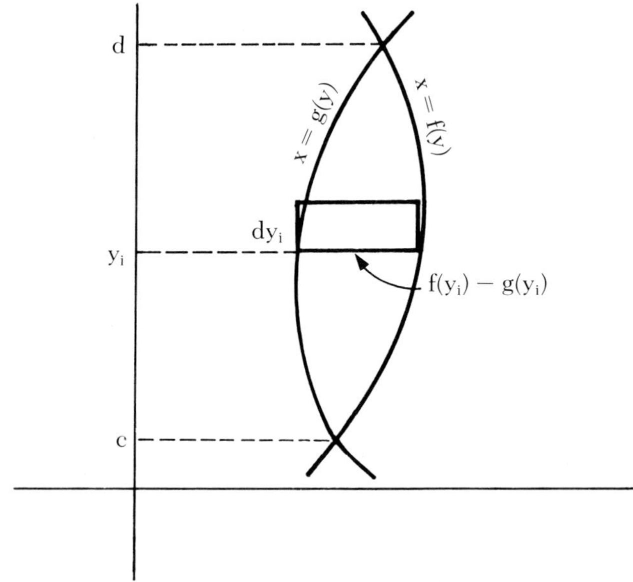

FIGURE VI.6.3 Analysis of area by integration with respect to y.

Let us return to Example VI.6.1. We see from Figure VI.6.2 that

$$x = \sqrt{y}$$

is the right side of the figure and

$$x = y/2$$

is the left side. In terms of y the figure runs from 0 to 4; so the area is

$$\int_0^4 \left(\sqrt{y} - \frac{y}{2}\right) dy = \left(\frac{2}{3}\,y^{3/2} - \frac{y^2}{4}\right)\Big|_0^4 = \frac{4}{3}.$$

VI.6.2 Example. Find the area bounded by

$$y = x^2 \quad \text{and } y = x + 2.$$

To find the intersection points we set

$$\begin{aligned} x^2 &= x + 2, \\ x^2 - x - 2 &= 0, \\ (x-2)(x+1) &= 0, \\ x &= 2, -1. \end{aligned}$$

A sketch is shown in Figure VI.6.4. The line is on the top and the parabola on the bottom; so the area is

$$\int_{-1}^{2} (x + 2 - x^2)\,dx = \left(\frac{x^2}{2} + 2x - \frac{x^3}{3}\right)\Bigg|_{-1}^{2} = \frac{9}{2}.$$

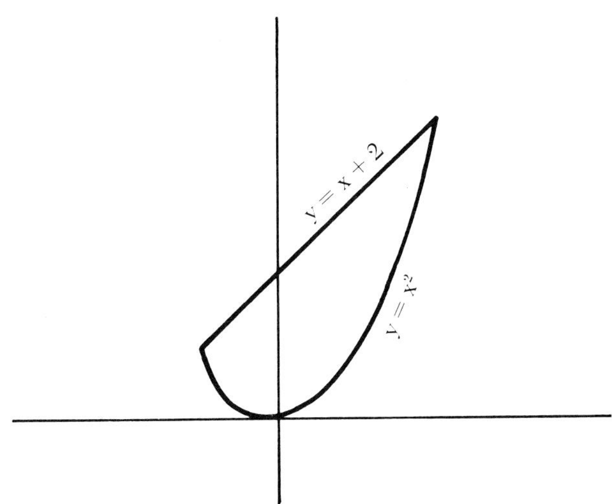

FIGURE VI.6.4 Figure for Example VI.6.2.

It is not practical to solve this problem by integrating with respect to y because (see Figure VI.6.5) the formula for the length of a horizontal rectangle changes at $y = 1$. This type of difficulty is sometimes referred to as "corner trouble."

Suppose it is known that you can sell x units of a given commodity at a price y where x and y are related by a "demand law" of the form

$$y = f(x).$$

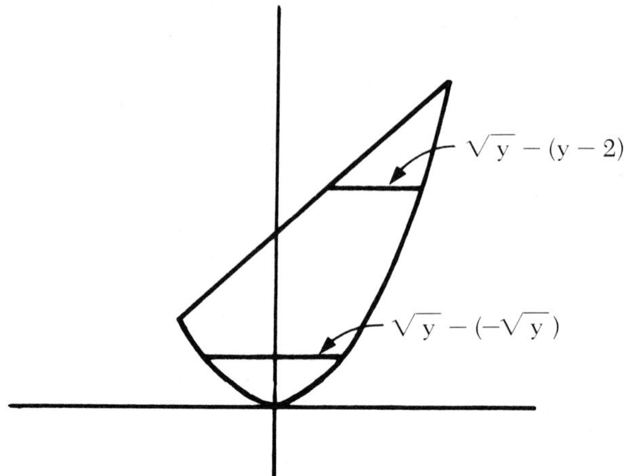

FIGURE VI.6.5 Obstacle to integrating with respect to y in Example VI.6.2.

We would expect the curve to look something like that shown in Figure VI.6.6. Increasing demand corresponds to decreasing price; there is a certain price at which you cannot sell any and there is a maximum demand even if you give them away. Now, suppose they are sold at a price y_0. The rectangle in Figure VI.6.7 indicates that dx_i consumers would have paid approximately $f(x_i) - y_0$ more than the market price; so they save a total of

$$[f(x_i) - y_0]\,dx_i.$$

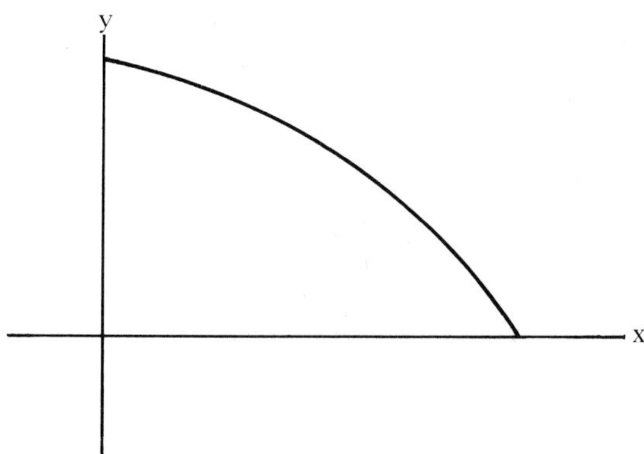

FIGURE VI.6.6 Typical demand curve.

Taking the limit of the sum of these savings, we define *consumers' surplus* as the area bounded by

$$x = 0,\ y = f(x) \quad \text{and} \quad y = y_0.$$

Similarly, suppose we are given a "supply law"

$$y = g(x)$$

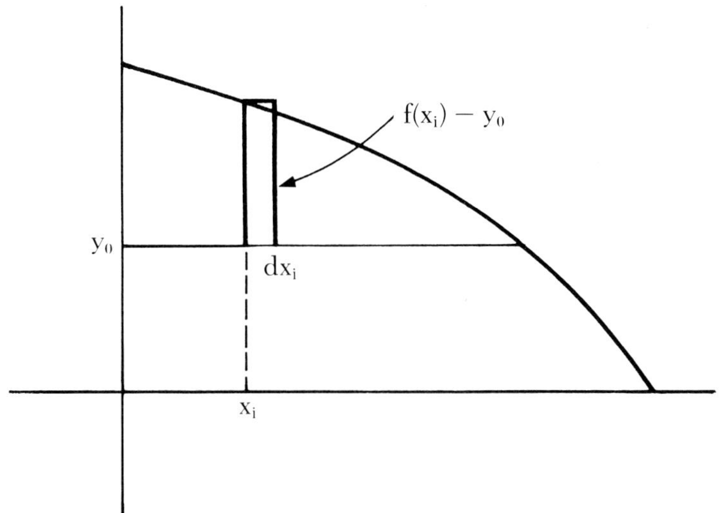

FIGURE VI.6.7 Consumers' surplus.

indicating that manufacturers will supply x items at a price y. We would expect this curve to have positive slope as in Figure VI.6.8. If, again, the price is set at y_0, there are some suppliers who would have sold at a lower price, and we define *producers' surplus* as the shaded area in Figure VI.6.8.

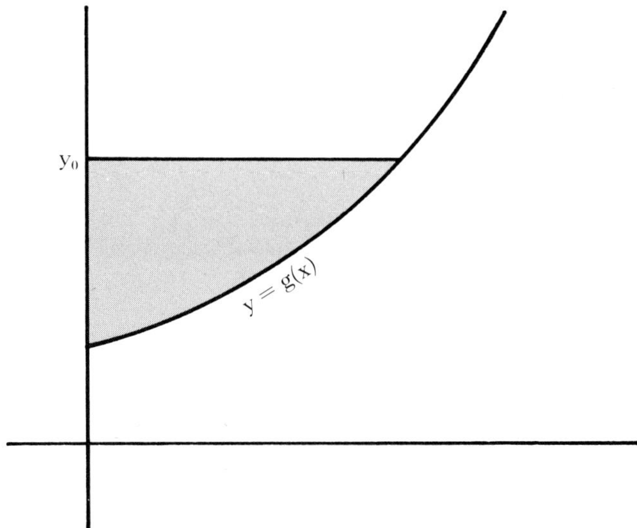

FIGURE VI.6.8 Producers' surplus.

"Pure competition" is defined as a situation in which the price is set so that supply equals demand. If we draw supply and demand curves on the same picture, their intersection point gives the pure

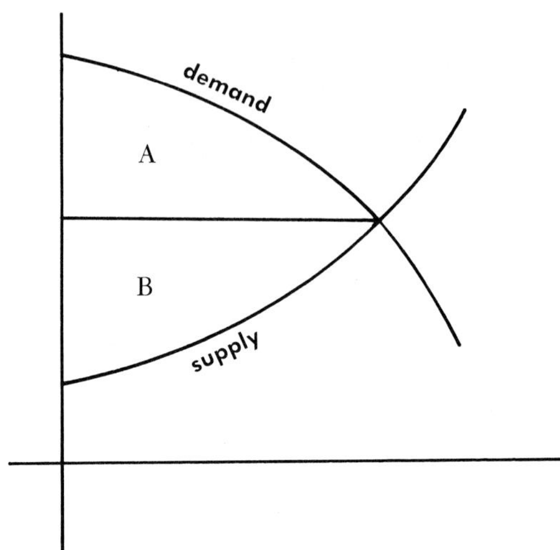

FIGURE VI.6.9 Consumers' and producers' surplus under pure competition.

competition price. In Figure VI.6.9, area A is consumers' surplus and area B is producers' surplus under pure competition.

PROBLEMS ON APPLICATIONS

Find the area of the region bounded by each of the following sets of curves.

1. $y = x^4$, $y = 8x$
2. $y = x^4 - 1$, $y = 5x + 5$
3. $y = x^3$, $y = 4x$
4. $y = x^4 - 4$, $y = 3x^2$
5. $x = y^2$, $x = y + 2$
6. $y = \sqrt{x - 1}$, $y = 5x - 5$
7. $y = 5 - x^2$, $y = 3 + x^2$
8. $x = y^2$, $5y = 6 - x^2$
9. $xy^2 = 1$, $y = 3 - 2\sqrt{x}$
10. $y = e^x$, $y = e$, $x = 0$
11. $y = \ln x$, $y = 0$, $x = 3$
12. $y = 1/x$, $2x + 2y = 5$

13. Suppose price y is related to demand x by

$$25y = 100 - x^2.$$

Find consumers' surplus if the price is 2.
What is consumers' surplus if the price is 4?
What is it if the product is given away?

14. Given that price y is related to supply x by

$$y = e^{x/10},$$

find producers' surplus if the price is e.

15. Given the demand curve

$$25y = 175 - x^2$$

and the supply curve

$$(x - 10)(y - 1) - 5,$$

find producers' and consumers' surpluses at the pure competition price.

7. IMPROPER INTEGRALS.

We turn in this section to the question of "area" for unbounded regions in the plane. Figure VI.7.1 shows the type of regions we have in mind. We want to discuss

$$\int_a^b f(x)\,dx$$

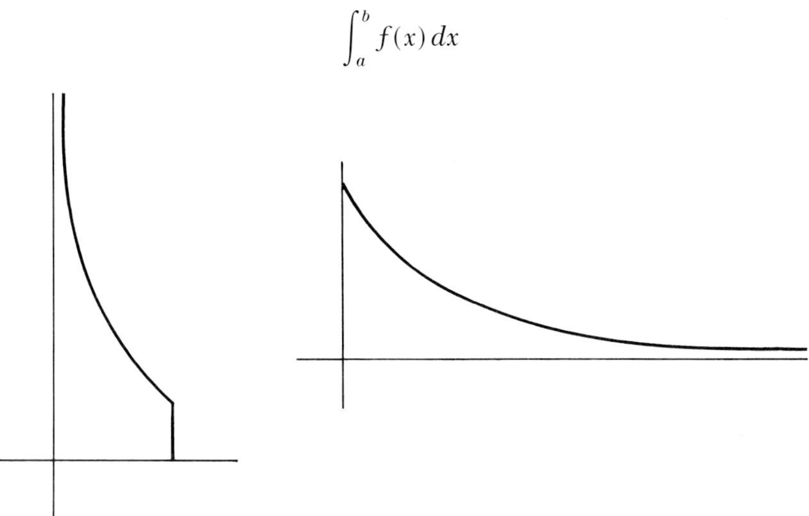

FIGURE VI.7.1 Unbounded areas.

 where $f(x) \to \infty$ as $x \to a$ (left hand picture in Figure VI.7.1) or

$$\int_a^\infty f(x)\,dx$$

(right hand picture in Figure VI.7.1).

The approach to this is to integrate over a bounded portion of the region and look for a limit of these integrals as the bounded portion expands. For example, Figure VI.7.2 shows a sketch of $y = 1/x^2$, and from this it appears that

$$\int_0^1 \frac{dx}{x^2}$$

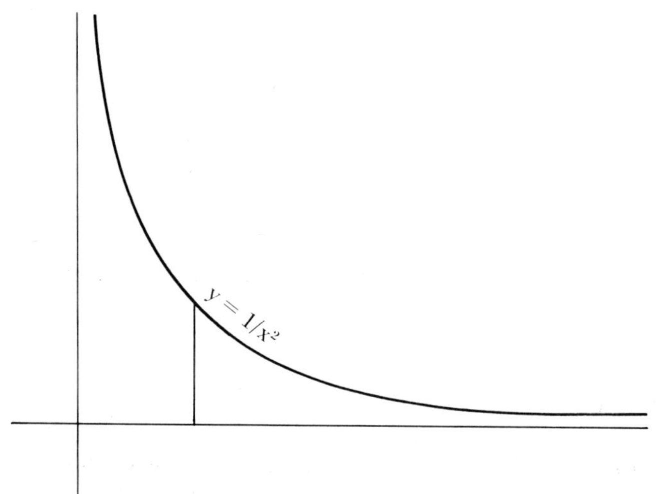

FIGURE VI.7.2 Figure for example of convergent and divergent improper integrals.

is an integral of the first type mentioned above and

$$\int_1^\infty \frac{dx}{x^2}$$

is one of the second type. Now, if we take $a > 0$, we have

$$\int_a^1 \frac{dx}{x^2} = \frac{-1}{x}\bigg|_a^1 = \frac{1}{a} - 1.$$

Clearly,

$$\lim_{a \to 0} \left(\frac{1}{a} - 1\right) = \infty;$$

so we say that

$$\int_0^1 \frac{dx}{x^2}$$

is a *divergent improper integral.* On the other hand,

$$\int_1^b \frac{dx}{x^2} = \frac{-1}{x}\bigg|_1^b = 1 - \frac{1}{b},$$

and

$$\lim_{b\to\infty}\left(1 - \frac{1}{b}\right) = 1;$$

so we say

$$\int_1^\infty \frac{dx}{x^2}$$

is a *convergent improper integral* and that it has the value 1.

In practice this discussion can be abbreviated as follows:

$$\int_0^1 \frac{dx}{x^2} = \frac{-1}{x}\bigg|_0^1 \text{ dvgt.}$$

$$\int_1^\infty \frac{dx}{x^2} = \frac{-1}{x}\bigg|_1^\infty = 1.$$

Though we are using familiar notation here, in the case of an improper integral it has a slightly different interpretation. One limit of integration is substituted into the antiderivative in the usual way. At the other limit of integration we want the limit of the antiderivative.

There occur integrals that might be called "multiply improper." A few sample pictures are shown in Figure VI.7.3. The important thing about these improper integrals is that they must be broken into "simple" cases, each of which must be convergent separately if the entire integral is to be classed as convergent.

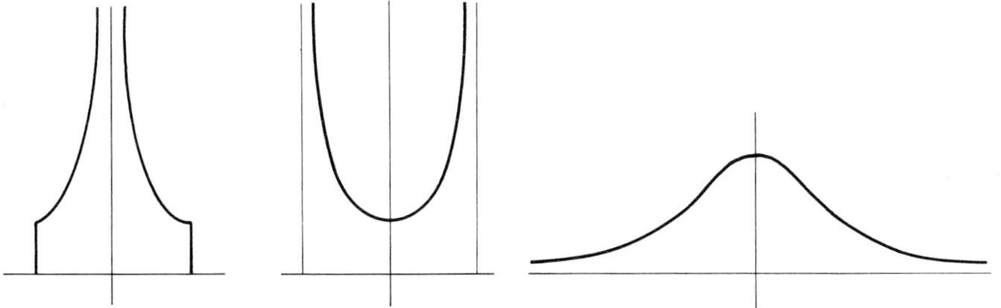

FIGURE VI.7.3 Figures illustrating multiply improper integrals.

Perhaps the left hand picture in Figure VI.7.3 is the most treacherous case. This is the case in which the function is unbounded at an interior point of the interval of integration. For example, if we ignore this feature, the following computation seems innocuous:

$$\int_{-1}^{1} \frac{dx}{x^2} = \frac{-1}{x}\Big|_{-1}^{1} = -2.$$

The "answer" -2 is suspicious because the curve is all above the x-axis. Actually, this is all wrong; we must write

$$\int_{-1}^{1} \frac{dx}{x^2} = \int_{-1}^{0} \frac{dx}{x^2} + \int_{0}^{1} \frac{dx}{x^2}$$

and each of these latter integrals is divergent.

The following facts about certain limits will be useful in connection with our study of improper integrals.

6.7.1 For any n, $\lim_{x\to\infty} x^n e^{-x} = 0$.

6.7.2 For $\alpha > 0$, $\lim_{x\to 0} x^\alpha \ln x = 0$.

In this brief treatment of calculus we do not propose to prove these statements. We merely list them here for reference. Another fact that we want to use but are not prepared to prove concerns an important improper integral:

6.7.3 $$\int_{-\infty}^{\infty} e^{-x^2/2\sigma^2} dx = \sigma\sqrt{2\pi}.$$

VI.7.1 *Example.* Evaluate $\int_0^1 \frac{dx}{\sqrt{1-x^2}}$. The substitution

$$x = \sin t,\ dx = \cos t\, dt$$

yields

$$\int_0^1 \frac{dx}{\sqrt{1-x^2}} = \int_0^{\pi/2} \frac{\cos t\, dt}{\cos t} = \int_0^{\pi/2} dt = \pi/2.$$

Here the substitution formula changes an improper integral into one that is not improper. This procedure is perfectly correct provided the only "trouble spot" is at the end of the interval of integration.

VI.7.2 *Example.* Integrating by parts we have

$$\int_0^\infty x e^{-x} dx = -x e^{-x}\Big|_0^\infty + \int_0^\infty e^{-x} dx$$

$$= (-x e^{-x} - e^{-x})\Big|_0^\infty = 1.$$

Here we used (6.7.1) to get the limit of $-x e^{-x}$ at ∞.

VI.7.3 Example. Evaluate

$$\int_{-\infty}^{\infty} e^x e^{-x^2/2} dx.$$

We write

$$e^x e^{-x^2/2} = e^{-(x^2-2x)/2}$$

and then complete the square in the exponent:

$$\begin{aligned} e^{-(x^2-2x)/2} &= e^{[-(x^2-2x+1)/2]+(1/2)} \\ &= e^{1/2}e^{-(x-1)^2/2}. \end{aligned}$$

Thus,

$$\begin{aligned} \int_{-\infty}^{\infty} e^x e^{-x^2/2} dx &= \int_{-\infty}^{\infty} e^{1/2}e^{-(x-1)^2/2} d(x-1) \\ &= e^{1/2}\sqrt{2\pi} = \sqrt{2\pi e} \end{aligned}$$

by (6.7.3).

VI.7.4 Example. A company makes an investment from which it expects to receive daily profits of y dollars, where

$$y = 10 + 0.1x$$

and x is the time elapsed (in years) since the investment. These profits may be expected to continue indefinitely. Assuming 250 working days per year, and an interest rate of 5 per cent per annum, find the present value of this investment.

Proceeding as in Example VI.5.4, we find that the discounted yearly rate of profits is

$$f(x) = (2500 + 25x)e^{-0.05x}.$$

The present value of all future earnings is, then, the improper integral

$$\int_0^{\infty} (2500 + 25x)e^{-0.05x} dx.$$

Integrating by parts, we have

$$\begin{aligned} \int_0^{\infty} (2500+25x)e^{-0.05x}dx &= -(50{,}000+500x)e^{-0.05x}\Big|_0^{\infty} + \int_0^{\infty} 500e^{-0.05x}dx \\ &= -(50{,}000+500x)e^{-0.05x}\Big|_0^{\infty} - 10{,}000e^{-0.05x}\Big|_0^{\infty} \\ &= 60{,}000. \end{aligned}$$

We see thus that the investment's present value is \$60,000—although it may be expected to produce an essentially infinite amount of revenues in the future.

VI.7.5 Example. A miner has marginal production of y lb. of ore per dollar, where

$$y = \frac{400{,}000}{x^2 + 10{,}000}$$

and x is his investment. What is the greatest possible amount that he can extract?

If the miner had a very large (essentially infinite) amount of capital, he could produce an amount of ore equal to the improper integral

$$\int_0^\infty \frac{400{,}000}{x^2 + 10{,}000}\, dx.$$

The substitution

$$x = 100 \tan u$$

gives the differential

$$dx = 100 \sec^2 u \; du = 100(\tan^2 u + 1) du$$

and changes the improper integral into a proper integral, as was done in Example VI.7.1. (Again, we point out that this is permissible, since the "trouble spot" is at the end of the interval of integration, ∞ in this case.) We have, then,

$$\int_0^\infty \frac{400{,}000}{x^2 + 10{,}000}\, dx = \int_0^{\pi/2} 4000 \; du = 2000\pi.$$

Thus, the total amount that the miner could produce, even if he had unlimited capital, is about 6283 pounds of ore. (This is possibly due to the fact that there is no more than this amount of ore in the mine.)

PROBLEMS ON IMPROPER INTEGRALS

Determine whether each of the following improper integrals is convergent or divergent, and find the value of each one that is convergent.

1. $\displaystyle\int_{-1}^{1} \frac{x\,dx}{\sqrt{1 - x^2}}$

2. $\int_{-1}^{1} \frac{dx}{\sqrt{1-x^2}}$

3. $\int_{-2}^{2} \frac{x\,dx}{1-x^2}$

4. $\int_{0}^{\infty} e^{-x}dx$

5. $\int_{-1}^{1} \frac{dx}{\sqrt[3]{x}}$

6. $\int_{-\infty}^{\infty} \frac{x^3\,dx}{1+x^4}$

7. $\int_{0}^{1} \frac{dx}{\sqrt[3]{2x-1}}$

8. $\int_{0}^{\infty} xe^{-x^2}dx$

9. $\int_{0}^{\infty} x^2e^{-x}dx$

10. $\int_{0}^{1} \frac{dx}{\sqrt{x}}$

11. $\int_{0}^{1} \ln x\,dx$

12. $\int_{0}^{1} (\ln x)^2dx$

13. Show that $\int_{-\infty}^{\infty} xe^{-x^2/2\sigma^2}\,dx = 0.$

14. Integrate by parts and use (6.7.3) and Problem 13 to show that

$$\frac{1}{\sigma\sqrt{2\pi}} \int_{-\infty}^{\infty} x^2e^{-x^2/2\sigma^2}dx = \sigma^2.$$

15. Follow the model of Example VI.7.3 to show that

$$\frac{1}{2\pi\sqrt{1-r^2}} \int_{-\infty}^{\infty} e^{-(x^2-2arx+a^2)/2(1-r^2)}dx = \frac{1}{\sqrt{2\pi}}\,e^{-a^2/2}$$

where a is any constant and r is a constant between -1 and 1.

16. A manufacturing company makes an investment which will bring daily profits of y dollars, where

$$y = 5 + 0.02\, x^2$$

and x is the number of years elapsed since making the investment. The investment can be continued indefinitely. Assuming a 20 per cent rate of interest, and assuming 250 working days per year, find the present value of all future profits obtained from the investment.

17. A miner has marginal production of y lb. of ore per dollar, where

$$y = \frac{400{,}000}{(x + 100)^2}$$

and x is his investment (in dollars). What is the greatest amount of ore that the miner can extract?

18. Suppose that the marginal production function of the miner mentioned in Problem 17 is

$$y = \frac{4000}{x + 100}.$$

How does this change the answer?

8. NUMERICAL INTEGRATION.

The fundamental theorem of calculus was indeed a momentous discovery, but it is not a panacea for all ills. The problem is that antiderivatives can be quite difficult (and sometimes impossible) to come by. In Sections 3, 4 and 5 of this chapter we have introduced some of the major tricks that are commonly used to compute antiderivatives. It should be apparent from a study of these sections that antidifferentiation is not as well organized as differentiation. The formulas in Chapter V will serve to find the derivative of any function that can be written in terms of powers, roots, exponentials, logarithms, and trigonometric functions. However, not every function that can be written in such terms has an antiderivative that can also be written in such terms. A well-known example of a function with a so-called "non-elementary" antiderivative is e^{-x^2}. In (6.7.3) we gave, without proof, the value of an integral of e^{-x^2}; but (6.7.3) is proved by a trick from multidimensional calculus, not by computing an antiderivative. Furthermore, the trick in question applies only to the integral from $-\infty$ to ∞. Integrals of e^{-x^2} with other limits of integration are quite a different matter.

Integrals of e^{-x^2} are sufficiently important in probability theory that they have been tabulated extensively. By this we mean, of course, that you can find tables giving decimal approximations to these integrals. Here we want to raise the larger question, "Suppose an antiderivative is difficult or impossible to come by. How can you get a decimal approximation to an integral without using antiderivatives?"

An integral is defined as the limit of a sequence of sums; so one of these sums is an obvious approximation to the integral. As a rule, however, this is not a very efficient approximation scheme, because to get a sum very close to the integral one must take a large number of partition points; and the more partition points are taken, the more work is involved in computing the sum. Figure VI.8.1 shows that, over a single subinterval, the rectangular area that appears in the sums defining the integral is a sort of approximation, that a trapezoidal area is probably closer, and that the area under a curved line is probably even better.

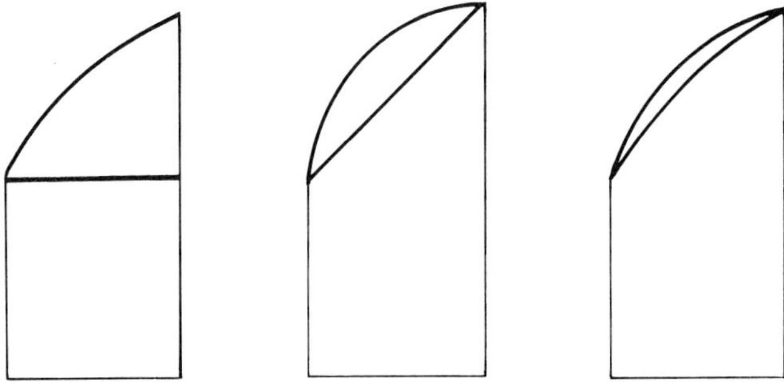

FIGURE VI.8.1 Approximations to area under a curve.

The most popular approximate integration formula, known as *Simpson's rule,* is based on the idea of using a parabolic arc across a pair of subintervals. Analytic details are as follows. The parabola

$$y = y_1 + \frac{y_2 - y_0}{2h}\,x + \frac{y_0 + y_2 - 2y_1}{2h^2}\,x^2$$

passes through the three points $(-h, y_0)$, $(0, y_1)$ and (h, y_2)—check this by direct substitution. Thus, the area under the parabola in Figure VI.8.2 is

6.8.1. $$y_1 x + \frac{y_2 - y_0}{4h}\,x^2 + \frac{y_0 + y_2 - 2y_1}{6h^2}\,x^3 \Big|_{-h}^{h}$$

$$= \frac{h}{3}\,(y_0 + 4y_1 + y_2).$$

Now, partition the interval $[a, b]$ into an even number of subintervals, each of length h. Let $y_i = f(x_i)$ where the x_i are the abscissas of the

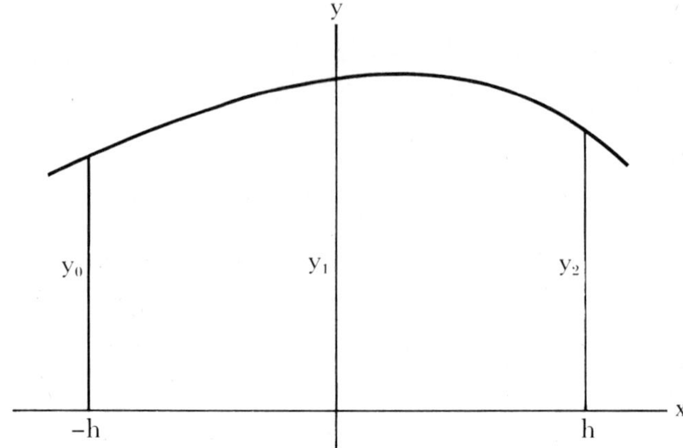

FIGURE VI.8.2 Parabolic arc through three points.

partition points. Through each triple of points (x_{2k}, y_{2k}), (x_{2k+1}, y_{2k+1}), (x_{2k+2}, y_{2k+2}) pass a parabola (see Figure VI.8.3); then use (6.8.1) to approximate $\int_a^b f$ by

$$\int_a^b f \approx \sum_{k=0}^{n-1} \frac{h}{3}(y_{2k} + 4y_{2k+1} + y_{2k+2}) = \frac{h}{3}(y_0 + 4y_1 + 2y_2 + 4y_3 + 2y_4 + \cdots + 2y_{2n-2} + 4y_{2n-1} + y_{2n}).$$

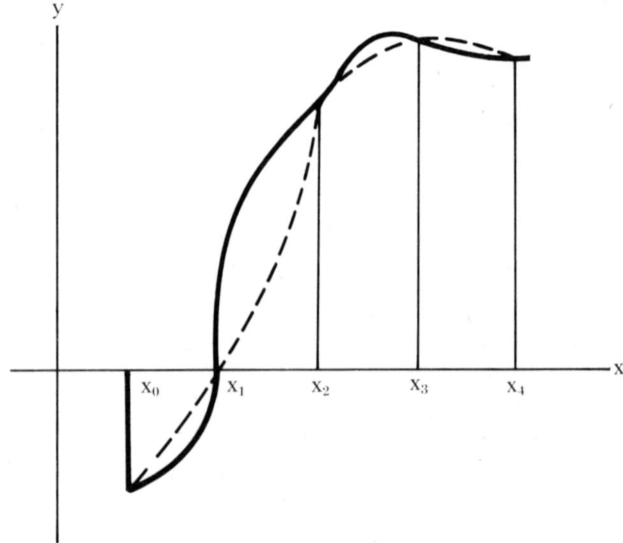

FIGURE VI.8.3 Graphical picture of Simpson's Rule.

Simpson's rule is a "natural" for a computer. The following very simple FORTRAN program is an all-purpose Simpson's rule program. To make it operative you fill in the FORMAT statements the way you want them and append a function subprogram that accepts X and returns F(X). Then, the data consists of the limits of integration a and b and the number n of subintervals to be used. The print-out repeats the data and adds the approximate value of

$$\int_a^b f(x)\,dx$$

computed by Simpson's rule with n subintervals.

```
1  READ (5,3) A, B, N
   EN = N
   H = (B - A)/EN
   Z = -1.
   K = N - 1
   S = F(A) + F(B)
   X = A
   DO 2 I = 1, K
   Z = -Z
   X = X + H
2  S = S + (Z + 3.) * F(X)
   ANS = H * S/3.
   WRITE (6,4) A, B, N, ANS
   STOP
3  FORMAT
4  FORMAT
   END
```

As would be expected, the accuracy of Simpson's rule increases as the number of subintervals is increased. However, Simpson's rule is often quite accurate with surprisingly few subintervals. This is worth remembering because there is a temptation to make the computer use a large number of subintervals. This costs nothing in human effort and very little in computer time, but the cost in round-off error can be disastrous. For example,

$$\int_0^1 \frac{4\,dx}{1+x^2} = \pi.$$

(To see this, substitute $x = \tan t$, $dx = \sec^2 t\,dt$.) Thus, approximate computations of this integral are easily checked for accuracy. Runs on an IBM 1620 (which carries computations to eight significant figures) gave this integral correct to seven significant figures using Simpson's rule with 16 subintervals. With 50 subintervals only six-figure accuracy was achieved, and with 150 subintervals only five figures were correct.

VI.8.1 Example. Use Simpson's rule with four subintervals to calculate

$$\ln 2 = \int_1^2 \frac{dx}{x}.$$

The numerical data may be tabulated as follows.

x	$1/x$	weight factor	product
1.00	1.00000	1	1.00000
1.25	0.80000	4	3.20000
1.50	0.66667	2	1.33333
1.75	0.57143	4	2.28572
2.00	0.50000	1	0.50000
			8.31905

This sum is to be multiplied by

$$\frac{h}{3} = \frac{1}{12};$$

so we get

$$\ln 2 \approx 0.69325.$$

The correct value to five decimal places is 0.69315.

PROBLEMS ON NUMERICAL INTEGRATION

Compute each of the following integrals by Simpson's rule, first with four subintervals, then with six. Find the exact value and compare.

1. $\int_0^\pi \sin x \, dx$

2. $\int_0^1 x^4 \, dx$

3. $\int_1^2 \frac{dx}{\sqrt{x}}$

4. $\int_{-1}^1 \sqrt{1 = x^2} \, dx$

5. Show that Simpson's rule with two subintervals gives the exact value of each of the following integrals:

$$\int_a^b x \, dx, \int_a^b x^2 \, dx, \int_a^b x^3 \, dx.$$

CHAPTER VII

VECTORS AND MATRICES

1. THE IDEA OF ABSTRACTION

Up to now, most of the mathematics that we have studied has dealt with the system of numbers—the natural numbers, the integers, the fractions, and finally, the real numbers—and its structure—its behavior with respect to the elementary operations of arithmetic, its order properties, and such. This is undoubtedly the natural process, for our mathematics is an outgrowth of the old sciences of arithmetic and geometry, most of whose concern was, precisely, with counting and measuring ideas that can only be explained in terms of the number system. Even today, most of the concrete applications of mathematics are, therefore, in terms of numbers.

It is the mathematician's desire, however, to look for ever more general mathematical structures, seeking always to understand the logical basis of mathematics—to uncover, as it were, the skeleton of the mathematical systems he knows—so as to adapt it to other related systems. This is the process of mathematical abstraction; in this manner the various branches of modern pure mathematics have been born.

Because mathematics evolved from arithmetic, the mathematical system that has been studied in greatest detail is that of the real numbers. It follows that most of the new branches of mathematics have been created by abstracting certain of the properties of the real number system and studying them in a sort of vacuum. Thus, the idea of "closeness" between numbers (or between points in Euclidean space, related to the numbers through the medium of analytic geometry) has been generalized—abstracted—to develop the science of *topology*. Alternatively, the ordering of the real numbers can be abstracted and generalized to give the complicated science of *cardinal* and *ordinal* numbers and *order-types*. It is also possible—

and this is our interest now—to abstract the idea of a *binary operation* (i.e., a rule for composing a pair of elements into a third element of the system), which appears amid the real numbers in the form of addition and the other fundamental arithmetic operations. This abstraction gives rise to the science of *modern algebra.*

We are interested in a branch of modern algebra, called *linear algebra,* that deals with certain generalizations of numbers, called *vectors* and *matrices.* The fact that they are generalizations means that some, but not all, of the properties of the number system will hold. We start by defining these concepts.

VII.1.1 Definition. A *matrix* is a rectangular array of numbers, arranged in m rows and n columns.

VII.1.2 Example. The following are all examples of matrices:

$$\text{(a)} \begin{pmatrix} 5 & 1 & 2 \\ 2 & 3 & 7 \end{pmatrix} \qquad \text{(b)} \begin{pmatrix} 1 \\ 5 \\ -1 \end{pmatrix} \qquad \text{(c)} \begin{pmatrix} 1 & 3 & 8 & 4 \\ 2 & 6 & -5 & 1 \\ -7 & 1 & 3 & 4 \end{pmatrix}$$

The numbers, m and n, of rows and columns in a matrix, are called the *dimensions* of the matrix. If a matrix has m rows and n columns it is called an $m \times n$ *matrix.* For instance, in Example II.1.2, (a) is a 2×3 matrix, (b) is 3×1, and (c) is 3×4.

The usual procedure, in dealing with matrices, is to represent all the numbers in the matrix, called *entries* of the matrix, by the same letter with different subscripts. Thus we have, as the "general" 2×3 matrix:

$$\begin{pmatrix} a_{11} & a_{12} & a_{13} \\ a_{21} & a_{22} & a_{23} \end{pmatrix}$$

or, more generally still, for the general $m \times n$ matrix:

7.1.1

$$A = \begin{pmatrix} a_{11} & a_{12} & \cdots & a_{1n} \\ a_{21} & a_{22} & \cdots & a_{2n} \\ \cdots & \cdots & \cdots & \cdots \\ a_{m1} & a_{m2} & \cdots & a_{mn} \end{pmatrix}$$

Note that the first subscript refers to the row, while the second subscript refers to the column. Thus a_{ij} is the entry in the ith row and jth column of the matrix A. The shorter, symbolic notation (a_{ij}) is also in common use for the matrix A given by (7.1.1).

VII.1.3 Definition. Two matrices, $A = (a_{ij})$ and $B = (b_{ij})$, are equal if they have the same dimensions and, for each i and j, $a_{ij} = b_{ij}$.

Equality of matrices is defined by saying that two matrices are equal only if they are identical in the sense of having the same shape and having corresponding entries equal. This means, among other things, that the single matrix equation $A = B$ is equivalent to the mn simultaneous equations $a_{ij} = b_{ij}$.

Of the possible dimensions for matrices, two cases of special interest arise.

VII.1.4 Definition. An $m \times n$ matrix is said to be *square* if $m = n$. An $n \times n$ matrix is also called an *nth-order* matrix.

VII.1.5 Definition. A $1 \times n$ matrix is called a *row vector,* or *row n-vector.* An $m \times 1$ matrix is called a *column vector,* or *column m-vector.*

A vector, as used here, is simply a special type of matrix. It may also be defined as an n-tuple of numbers, arranged either in a row, as, say

$$(5, 1, 4, 6, 2)$$

or in a column, as

$$\begin{pmatrix} 3 \\ 1 \\ 4 \\ 1 \end{pmatrix}$$

The entries in a vector are also called *components.* We shall usually denote a vector by a boldface letter, say $\mathbf{b} = (b_1, b_2, \ldots, b_m)$.

2. ADDITION AND SCALAR MULTIPLICATION

As already mentioned, matrices are an algebraic generalization of the concept of a number. This means that some operations can be performed on matrices. We introduce here two such operations: the first is that of addition.

VII.2.1 Definition. Let $A = (a_{ij})$ and $B = (b_{ij})$ be $m \times n$ matrices. Then by their sum, $A + B$, we mean the $m \times n$ matrix, $C = (c_{ij})$, where

7.2.1 $$c_{ij} = a_{ij} + b_{ij}$$

For example, we have the following additions:

$$\begin{pmatrix} 3 & 5 & 1 \\ 2 & 6 & 8 \end{pmatrix} + \begin{pmatrix} 1 & 4 & 6 \\ 2 & -2 & 3 \end{pmatrix} = \begin{pmatrix} 4 & 9 & 7 \\ 4 & 4 & 11 \end{pmatrix}$$

$$\begin{pmatrix} 1 & 3 \\ 1 & -4 \\ 2 & 5 \end{pmatrix} + \begin{pmatrix} -2 & 1 \\ 6 & 2 \\ 4 & -1 \end{pmatrix} = \begin{pmatrix} -1 & 4 \\ 7 & -2 \\ 6 & 4 \end{pmatrix}$$

$$(1, 5, 7, 2) + (2, 4, 6, 8) = (3, 9, 13, 10)$$

Addition is carried out in this way for matrices of equal dimensions by the simple expedient of adding corresponding entries in the two matrices. If two matrices have different dimensions, their sum is not defined, i.e., it is not possible to add them.

Thus, the sums

$$\begin{pmatrix} 1 & 3 & 5 \\ 2 & 6 & 4 \end{pmatrix} + \begin{pmatrix} -1 & 2 \\ 3 & 14 \\ 5 & -6 \end{pmatrix}$$

$$\begin{pmatrix} 1 & 3 \\ 5 & 8 \end{pmatrix} + \begin{pmatrix} 1 & -2 & 4 \\ 6 & 1 & -2 \end{pmatrix}$$

$$(1, 3, 1) + (2, 5)$$

are not defined; they *do not exist.*

The second operation to be considered is that of multiplication by a number. In the context of vectors and matrices, numbers are called *scalars,* so that this operation is called *scalar multiplication.*

VII.2.2 Definition. Let s be a scalar, and let $A = (a_{ij})$ be an $m \times n$ matrix. Then by their product sA we shall mean the $m \times n$ matrix $B = (b_{ij})$ given by

7.2.2 $$b_{ij} = sa_{ij}$$

Scalar multiplication is accomplished by multiplying each entry in the matrix by the scalar.

For example,

$$3\begin{pmatrix} 2 & 1 & 4 \\ 6 & -5 & 1 \end{pmatrix} = \begin{pmatrix} 6 & 3 & 12 \\ 18 & -15 & 3 \end{pmatrix}$$

$$-4\begin{pmatrix} 2 & 1 & -2 & 3 \\ -1 & 4 & -1 & 0 \end{pmatrix} = \begin{pmatrix} -8 & -4 & 8 & -12 \\ 4 & -16 & 4 & 0 \end{pmatrix}$$

$$2\begin{pmatrix} 1 \\ 4 \\ 5 \end{pmatrix} = \begin{pmatrix} 2 \\ 8 \\ 10 \end{pmatrix}$$

We shall use the symbol θ to represent a matrix all of whose entries are equal to zero. There are, of course, many such matrices—one for each pair of dimensions. This ambiguity is not very important, inasmuch as we will generally be able to tell, from the context, which of all these matrices θ is needed. For instance, if we see the expression $A + \theta$, we shall know that the θ here must have the same dimensions as A, since matrices cannot be added unless they have the same dimensions.

It is not too difficult to see that addition and scalar multiplication satisfy the following laws:

7.2.3 $$A + B = B + A$$

7.2.4 $$A + (B + C) = (A + B) + C$$

7.2.5 $$A + \theta = A$$

7.2.6 $$A + (-1)A = \theta$$

7.2.7 $$r(sA) = (rs)A$$

7.2.8 $$r(A + B) = rA + rB$$

7.2.9 $$(r + s)A = rA + sA$$

7.2.10 $$1A = A$$

The law (7.2.3) is called the *commutative law* for addition. Laws (7.2.4) and (7.2.7) are the *associative laws* for addition and scalar multiplication, respectively. Law (7.2.5) states that θ is an *additive identity,* while (7.2.6) says that $(-1)A$, which is more briefly written $-A$, is an *additive inverse* for A. The two laws (7.2.8) and (7.2.9) are the two *distributive laws;* finally, (7.2.10) states that the scalar 1 is an identity for scalar multiplication.

It is a well-known fact, of course, that the real numbers satisfy all these laws; it is, indeed, due to this fact that the matrices themselves satisfy them. Let us consider (7.2.4). This law states that, given three matrices A, B, and C (of the same dimensions, of course), we will obtain the same result if we first add A and B together and then add their sum to C as if we first add B and C together and then add A to *their* sum.

We prove this in the case of three 2×2 matrices, $A = (a_{ij})$, $B = (b_{ij})$, and $C = (c_{ij})$, as follows:

$$A + B = \begin{pmatrix} a_{11} & a_{12} \\ a_{21} & a_{22} \end{pmatrix} + \begin{pmatrix} b_{11} & b_{12} \\ b_{21} & b_{22} \end{pmatrix} = \begin{pmatrix} a_{11} + b_{11} & a_{12} + b_{12} \\ a_{21} + b_{21} & a_{22} + b_{22} \end{pmatrix}$$

so

$$\begin{aligned}(A + B) + C &= \begin{pmatrix} a_{11} + b_{11} & a_{12} + b_{12} \\ a_{21} + b_{21} & a_{22} + b_{22} \end{pmatrix} + \begin{pmatrix} c_{11} & c_{12} \\ c_{21} & c_{22} \end{pmatrix} \\ &= \begin{pmatrix} a_{11} + b_{11} + c_{11} & a_{12} + b_{12} + c_{12} \\ a_{21} + b_{21} + c_{21} & a_{22} + b_{22} + c_{22} \end{pmatrix}\end{aligned}$$

On the other hand,

$$B+C=\begin{pmatrix} b_{11} & b_{12} \\ b_{21} & b_{22} \end{pmatrix}+\begin{pmatrix} c_{11} & c_{12} \\ c_{21} & c_{22} \end{pmatrix}=\begin{pmatrix} b_{11}+c_{11} & b_{12}+c_{12} \\ b_{21}+c_{21} & b_{22}+c_{22} \end{pmatrix}$$

and so

$$\begin{aligned} A+(B+C) &= \begin{pmatrix} a_{11} & a_{12} \\ a_{21} & a_{22} \end{pmatrix}+\begin{pmatrix} b_{11}+c_{11} & b_{12}+c_{12} \\ b_{21}+c_{21} & b_{22}+c_{22} \end{pmatrix} \\ &= \begin{pmatrix} a_{11}+b_{11}+c_{11} & a_{12}+b_{12}+c_{12} \\ a_{21}+b_{21}+c_{21} & a_{22}+b_{22}+c_{22} \end{pmatrix}. \end{aligned}$$

We see that

$$(A+B)+C=A+(B+C)$$

For matrices of other dimensions, a similar proof using similar reasoning may be given. An alternative method, which the reader is invited to try, proves that the "typical" element (i.e., the element in the ith row and jth column) in each of the two matrices $(A+B)+C$ and $A+(B+C)$ is the same.

The other laws may all be proved in much the same way by using the fact that they hold for real numbers.

We may also, if we wish, define the operation of subtraction for matrices by analogy with the operation of subtraction for real numbers. In fact, if a and b are numbers, we know that $b-a$ is the number x, which satisfies the equation

$$x+a=b$$

In a similar way, for matrices A and B, the matrix $B-A$ would be an X such that

$$X+A=B$$

We notice first of all that this is possible only if A and B have the same dimensions. If this is the case, it is easy to see that $X=B+(-1)A$ satisfies the equation. In fact, the typical entry in $B+(-1)A$ is $b_{ij}-a_{ij}$; this means that the typical entry in $(B+(-1)A)+A$ is $(b_{ij}-a_{ij})+a_{ij}$, or b_{ij}.

We now have

$$B+(-1)A+A=B$$

Thus, we define subtraction by the natural expression

7.2.11 $$B-A=B+(-1)A$$

VII.2.3 Examples. Some examples of matrix addition and scalar multiplication are:

$$\begin{pmatrix}5 & 1\\2 & 4\end{pmatrix} - 2\begin{pmatrix}1 & -1\\2 & 6\end{pmatrix} + 3\begin{pmatrix}1 & 2\\-5 & 4\end{pmatrix}$$

$$= \begin{pmatrix}5-2(1)+3(1) & 1-2(-1)+3(2)\\2-2(2)+3(-5) & 4-2(6)\ +3(4)\end{pmatrix} = \begin{pmatrix}6 & 9\\-17 & 4\end{pmatrix}$$

$$\begin{pmatrix}1 & 3 & 5\\2 & 4 & 7\end{pmatrix} - \begin{pmatrix}6 & 1 & -8\\2 & 6 & 2\end{pmatrix} = \begin{pmatrix}-5 & 2 & 13\\0 & -2 & 5\end{pmatrix}$$

$$\begin{pmatrix}1\\2\\-1\end{pmatrix} + 5\begin{pmatrix}6\\-2\\3\end{pmatrix} = \begin{pmatrix}31\\-8\\14\end{pmatrix}$$

VII.2.4 Example. Trade among three countries, C_1, C_2, and C_3, in 1967 is given by the matrix $A = (a_{ij})$, where a_{ij} is the value (in millions of dollars) of exports from C_i to C_j.

$$A = \begin{pmatrix}0 & 31 & 7\\5 & 0 & 32\\11 & 16 & 0\end{pmatrix}.$$

Trade among the same countries in 1968, also in millions of dollars, is given by the matrix

$$B = \begin{pmatrix}0 & 25 & 10\\8 & 0 & 38\\13 & 12 & 0\end{pmatrix}.$$

What is the value of exports among these three countries for the two years 1967–68? What is its value in millions of West German marks (DM)?

The total exports, in millions of U.S. dollars, is clearly obtained by adding the matrices:

$$A + B = \begin{pmatrix}0 & 56 & 17\\13 & 0 & 70\\24 & 28 & 0\end{pmatrix}$$

To obtain the value in millions of DM, we use the fact that, in 1967–68, there was an official exchange rate of DM4 to \$1. Thus we must multiply every entry in the matrix by 4; equivalently, we multiply the entire matrix by 4:

$$4(A + B) = \begin{pmatrix}0 & 224 & 68\\52 & 0 & 280\\96 & 112 & 0\end{pmatrix}.$$

VII.2.5 Example. The population in town A is classified according to age and sex, obtaining the matrix

$$\begin{array}{l} \\ \text{Under 10} \\ \text{10–19} \\ \text{20–29} \\ \text{30–39} \\ \text{40–49} \\ \text{50–59} \\ \text{60 or over} \end{array} \begin{array}{c} \begin{array}{cc} \text{Male} & \text{Female} \end{array} \\ \begin{pmatrix} 40 & 37 \\ 38 & 36 \\ 35 & 36 \\ 29 & 31 \\ 23 & 24 \\ 12 & 18 \\ 5 & 11 \end{pmatrix} \end{array} = A.$$

A similar classification in town B yields the matrix

$$\begin{array}{l} \\ \text{Under 10} \\ \text{10–19} \\ \text{20–29} \\ \text{30–39} \\ \text{40–49} \\ \text{50–59} \\ \text{60 or over} \end{array} \begin{array}{c} \begin{array}{cc} \text{Male} & \text{Female} \end{array} \\ \begin{pmatrix} 53 & 49 \\ 51 & 47 \\ 45 & 44 \\ 38 & 41 \\ 33 & 38 \\ 25 & 31 \\ 17 & 28 \end{pmatrix} \end{array} = B.$$

Obtain a similar classification for the joint population of the two towns.

Clearly, this is obtained by adding corresponding entries in the two matrices, i.e., by adding the two matrices:

$$A + B = \begin{pmatrix} 93 & 86 \\ 89 & 83 \\ 80 & 80 \\ 67 & 72 \\ 56 & 62 \\ 37 & 49 \\ 22 & 39 \end{pmatrix}.$$

PROBLEMS ON ADDITION AND SCALAR MULTIPLICATION

1. Perform the indicated vector operations.

(a) $(3,5,1) + (2,8,-4)$

(b) $\begin{pmatrix} 5 \\ 1 \end{pmatrix} + \begin{pmatrix} 6 \\ -3 \end{pmatrix} - \begin{pmatrix} -2 \\ 4 \end{pmatrix}$

(c) $2(1,3,6,7) - 4(2,5,-2,2)$

(d) $\begin{pmatrix}3\\6\\1\end{pmatrix}+4\begin{pmatrix}1\\-2\\0\end{pmatrix}-3\begin{pmatrix}2\\-5\\-3\end{pmatrix}$

(e) $2(1,5)+4(-2,1)-3(1,-4)+(0,2)$

(f) $5(1,3,-5)+3(-2,6,2)-4(1,1,-1)$

(g) $(1,2,4)+3(1,6,2)-2(-2,4,1)$

(h) $\begin{pmatrix}3\\5\end{pmatrix}+\begin{pmatrix}2\\8\end{pmatrix}-3\begin{pmatrix}1\\-3\end{pmatrix}+4\begin{pmatrix}1\\-2\end{pmatrix}$

(i) $2\begin{pmatrix}1\\2\\4\\1\end{pmatrix}-3\begin{pmatrix}-2\\1\\-5\\6\end{pmatrix}+4\begin{pmatrix}1\\3\\-7\\2\end{pmatrix}$

(j) $(1,6,3)+2(-2,4,2)-3(1,5,-8)$

2. Solve the following vector equations.

(a) $2\begin{pmatrix}x\\y\end{pmatrix}+4\begin{pmatrix}1\\-2\end{pmatrix}=\begin{pmatrix}10\\2\end{pmatrix}$

(b) $5\begin{pmatrix}2x\\3y\end{pmatrix}+3\begin{pmatrix}4\\-2\end{pmatrix}=\begin{pmatrix}-8\\9\end{pmatrix}$

(c) $\begin{pmatrix}x\\2y\end{pmatrix}+\begin{pmatrix}-y\\3x\end{pmatrix}=\begin{pmatrix}1\\8\end{pmatrix}$

(d) $\begin{pmatrix}x\\3y\\2x\end{pmatrix}+\begin{pmatrix}2y\\-4x\\y\end{pmatrix}=\begin{pmatrix}4\\17\\-1\end{pmatrix}$

(e) $x(1,5)+y(-2,1)=(-5,-3)$

(f) $x\begin{pmatrix}3\\-2\end{pmatrix}-y\begin{pmatrix}2\\4\end{pmatrix}=\begin{pmatrix}-7\\-22\end{pmatrix}$

(g) $x\begin{pmatrix}2\\1\\4\end{pmatrix}+y\begin{pmatrix}1\\-2\\2\end{pmatrix}=\begin{pmatrix}7\\1\\14\end{pmatrix}$

(h) $x\begin{pmatrix}2\\1\\5\end{pmatrix}+y\begin{pmatrix}4\\2\\-3\end{pmatrix}=\begin{pmatrix}12\\7\\4\end{pmatrix}$

(i) $x(3,2,4)+y(6,4,1)=(9,6,-2)$

(j) $x\begin{pmatrix}2\\1\\4\end{pmatrix}+y\begin{pmatrix}1\\-3\\-2\end{pmatrix}+z\begin{pmatrix}1\\5\\1\end{pmatrix}=\begin{pmatrix}8\\14\\11\end{pmatrix}$

(k) $x\begin{pmatrix}1\\-1\\2\end{pmatrix}+y\begin{pmatrix}2\\1\\4\end{pmatrix}+z\begin{pmatrix}0\\-1\\1\end{pmatrix}=\begin{pmatrix}3\\3\\3\end{pmatrix}$

3. In each of the following problems, two vectors **a** and **b** are given. In which cases is it possible to express every vector **c** of the same dimensions in the form $x\mathbf{a}+y\mathbf{b}$? (In other words, in which cases will the vector equation $x\mathbf{a}+y\mathbf{b}=\mathbf{c}$ have solutions for all values of **c**?)

(a) $\mathbf{a}=\begin{pmatrix}1\\0\end{pmatrix}$ $\quad\mathbf{b}=\begin{pmatrix}0\\1\end{pmatrix}$

(b) $\mathbf{a}=\begin{pmatrix}2\\1\end{pmatrix}$ $\quad\mathbf{b}=\begin{pmatrix}3\\2\end{pmatrix}$

(c) $\mathbf{a}=(1,4)$ $\quad\mathbf{b}=(2,8)$

(d) $\mathbf{a}=(6,-4)$ $\quad\mathbf{b}=(-9,6)$

(e) $\mathbf{a}=(3,2)$ $\quad\mathbf{b}=(3,-2)$

4. Perform the following matrix operations.

(a) $\begin{pmatrix}2&1\\6&-3\end{pmatrix}+\begin{pmatrix}1&4\\1&8\end{pmatrix}-\begin{pmatrix}2&2\\-4&-1\end{pmatrix}$

(b) $5\begin{pmatrix}1&8\\6&-3\\2&-4\end{pmatrix}-3\begin{pmatrix}1&3\\2&-2\\-5&3\end{pmatrix}+4\begin{pmatrix}1&3\\2&0\\1&0\end{pmatrix}$

(c) $5\begin{pmatrix}3&4&1&2\\6&5&-4&0\\-5&3&0&1\end{pmatrix}+2\begin{pmatrix}1&-3&-5&7\\2&-4&6&-2\\-1&4&0&1\end{pmatrix}$

(d) $2\begin{pmatrix}1&5\\2&-2\\-3&6\\1&0\end{pmatrix}+3\begin{pmatrix}2&-3\\0&4\\-1&2\\5&1\end{pmatrix}-4\begin{pmatrix}1&5\\-2&4\\-6&7\\1&-3\end{pmatrix}$

(e) $-2\begin{pmatrix}1&5&8\\3&4&2\\2&-1&6\end{pmatrix}+3\begin{pmatrix}1&2&5\\6&-1&7\\-4&2&8\end{pmatrix}+\begin{pmatrix}1&3&0\\2&-4&2\\1&3&1\end{pmatrix}$

5. For the countries of Example VII.2.4, find the value of exports for 1967–68 (a) in Swiss Francs, given that one dollar is worth 4.3 SFr.; (b) in French Francs, given that one dollar was worth 4.9 FFr. in 1967, and 5.5 FFr. in 1968; (c) in pounds sterling, given that one dollar was worth £ 0.35 in 1967, and £ 0.42 in 1968.

6. Exports by countries D, E, and F can be classified as manufactured goods, inedible raw materials, foodstuffs, and miscellaneous. In 1967, their values in millions of U.S. dollars were as given by the table:

	Manufactured Goods	Industrial Raw Materials	Foodstuffs	Miscellaneous
D	12	37	29	12
E	38	20	10	17
F	45	5	11	4

whereas, in 1966, they were

D	7	40	27	17
E	35	18	12	14
F	47	7	10	5

Find the corresponding values for the two-year period, 1966–67, in millions of dollars. What would it be in millions of German marks ($1 = DM 4)?

7. A shoe manufacturer makes black and brown ladies' and men's shoes. Production at his New York plant is the following (in thousands of pairs):

	Men's	Ladies'
Black	27	31
Brown	48	17

Production at his Chicago plant (also in thousands of pairs) is:

	Men's	Ladies'
Black	32	29
Brown	55	14

How many of each type of shoe does he make at the two plants combined? He is thinking of opening a new plant in Denver, which will have twice the capacity of the New York plant. How many shoes will the Denver plant make? How many will all three plants, together, make?

3. SCALAR PRODUCTS OF VECTORS

Before considering any further binary operations on matrices, we introduce a *unitary* operation, a rule that assigns to each matrix another matrix. This is the operation of *transposition*.

VII.3.1 Definition. Let $A = (a_{ij})$ be an $m \times n$ matrix. Then the *transpose* of A is the $n \times m$ matrix $B = (b_{ij})$ defined by $b_{ij} = a_{ji}$.

We shall use the notation A^t for the transpose of A. Heuristically, the two matrices A and A^t are related in that each row of A is transformed into a column of A^t, and the converse. Examples of this follow.

VII.3.2 Example.

$$\begin{pmatrix} 2 & 5 \\ 1 & -1 \\ 3 & 2 \end{pmatrix}^t = \begin{pmatrix} 2 & 1 & 3 \\ 5 & -1 & 2 \end{pmatrix}$$

$$\begin{pmatrix} 1 & 3 & 1 \\ 2 & 6 & 7 \\ 4 & 1 & 5 \end{pmatrix}^t = \begin{pmatrix} 1 & 2 & 4 \\ 3 & 6 & 1 \\ 1 & 7 & 5 \end{pmatrix}$$

$$(1,5,6,2)^t = \begin{pmatrix} 1 \\ 5 \\ 6 \\ 2 \end{pmatrix}$$

VII.3.3 Example. Using the data of Example VII.2.4, what was the total amount of trade between each pair of countries in 1967?

Letting c_{ij} be the total amount of trade between C_i and C_j, it is easy to see that we must have

$$c_{ij} = a_{ij} + a_{ji}.$$

Therefore, the total-trade matrix is

$$C = A + A^t = \begin{pmatrix} 0 & 31 & 7 \\ 5 & 0 & 32 \\ 11 & 16 & 0 \end{pmatrix} + \begin{pmatrix} 0 & 5 & 11 \\ 31 & 0 & 16 \\ 7 & 32 & 0 \end{pmatrix}$$

or

$$C = \begin{pmatrix} 0 & 36 & 18 \\ 36 & 0 & 48 \\ 18 & 48 & 0 \end{pmatrix}.$$

Thus, for example, the trade between countries C_1 and C_2 was \$36 million. It is interesting to note that the matrix C is *symmetric;* i.e., $C = C^t$. This will always happen when we take the sum of a square matrix and its transpose.

It is easy to see that

7.3.1 $$(A+B)^t = A^t + B^t$$

7.3.2 $$(rA)^t = rA^t$$

7.3.3 $$A^{tt} = A$$

The two laws (7.3.1) and (7.3.2) state that the transposition preserves the operations of addition and scalar multiplication. Law (7.3.3) states that the transpose of the transpose is the original matrix.

The theory of matrices and their algebra was first developed in detail by two Englishmen, Arthur Cayley (1821-1895) and James Sylvester (1814-1897). It was their observation that certain physical processes could be described by matrices. Quite often it was desirable to consider the *resultant* of two such processes (i.e., the result of carrying these out one after the other). In obtaining the resultant, the following definition was found of considerable value.

VII.3.4 Definition. Let $\mathbf{x} = (x_1, \ldots, x_n)$ be a row vector, and $\mathbf{y} = (y_1, \ldots, y_n)^t$ be a column vector. Then the product $\mathbf{xy}$ is the scalar defined by

7.3.4 $$\mathbf{xy} = x_1y_1 + x_2y_2 + \ldots + x_ny_n$$

or, using the summation symbol (see Appendix),

7.3.5 $$\mathbf{xy} = \sum_{i=1}^{n} x_iy_i$$

Thus, the vectors $\mathbf{x}$ and $\mathbf{y}$ are multiplied by multiplying their corresponding components and adding the products obtained. We note that this is done only if the first vector is a row vector, the second is a column vector, and both have the same number of components. We add, further, that the product we have obtained here is $\mathbf{xy}$; it is *not* $\mathbf{yx}$. The product $\mathbf{yx}$ exists, but we will not define it as yet.

A possible heuristic interpretation would be as follows: let $\mathbf{x}$ be a "supply" vector, that is, let x_i represent the amount available of a certain commodity C_i. Let $\mathbf{y}$ be a "price" vector, that is, let y_i be the price of a unit amount of the commodity C_i. Then the total value of the supply $\mathbf{x} = (x_i, \ldots, x_n)$ is precisely the product $\mathbf{xy}$. We insist on this definition of product because it corresponds to the matrix multiplication which will be discussed below in Section 4 of this chapter.

VII.3.5 Example. We give, now, some examples of vector multiplication.

(a) $$(3,5,1)\begin{pmatrix} 2 \\ 4 \\ -2 \end{pmatrix} = 3(2) + 5(4) + 1(-2) = 24$$

(b) $$(6,1)\ (2,3)^t = 6(2) + 1(3) = 15$$

(c) $$(1,-2)\begin{pmatrix}2\\1\end{pmatrix} = 1(2) - 2(1) = 0$$

(d) $$(0,0,0)\begin{pmatrix}1\\4\\8\end{pmatrix} = 0(1) + 0(4) + 0(8) = 0$$

VII.3.6 Example. A pet store owner has 5 puppies, 3 kittens, 8 rabbits, 6 goldfish, and 2 parrots. The puppies sell for \$5 each; the kittens, for \$3; the rabbits, for \$2; the goldfish, for \$1; and the parrots for \$7 each. What is the total selling price of the man's stock?

We form here the product

$$(5,\ 3,\ 8,\ 6,\ 2)\begin{pmatrix}5\\3\\2\\1\\7\end{pmatrix} = 70$$

so that the stock will sell for \$70 in all.

VII.3.7 Example. The gross national product of the six European Economic Community countries in 1968 was: Belgium, 977 billion Francs; France, 613 billion Francs; German Federal Republic, 528.8 billion Marks; Italy, 45,122 billion Lire; Luxembourg, 38.3 billion Belgian Francs; Netherlands, 91.3 billion Guilders. Each Belgian Franc is worth \$0.02; each French Franc, \$0.18; each Mark, \$0.25; each Lira, \$0.0016, and each Guilder, \$0.275. Find the total GNP of the European Economic Community for 1968, in U.S. dollars.

We form here the product

$$(977,\ 613,\ 528.8,\ 45122,\ 38.3,\ 91.3)\begin{pmatrix}0.02\\0.18\\0.25\\0.0016\\0.02\\0.275\end{pmatrix} = 360.16$$

so the total GNP of the six countries was \$360,160 million.

It may be seen that the following laws hold for vector multiplication. Here, $\mathbf{x}$ and $\mathbf{x}'$ are row vectors, $\mathbf{y}$ and $\mathbf{y}'$ are column vectors, and s is a scalar:

7.3.6 $$(s\mathbf{x})\mathbf{y} = \mathbf{x}(s\mathbf{y}) = s(\mathbf{x}\mathbf{y})$$

7.3.7 $$(\mathbf{x} + \mathbf{x}')\mathbf{y} = \mathbf{x}\mathbf{y} + \mathbf{x}'\mathbf{y}$$

7.3.8 $$\mathbf{x}(\mathbf{y}+\mathbf{y}')=\mathbf{x}\mathbf{y}+\mathbf{x}\mathbf{y}'$$

7.3.9 $$\mathbf{x}\mathbf{y}=\mathbf{y}^t\mathbf{x}^t$$

The law (7.3.6) is called the homogeneity law; (7.3.7) and (7.3.8) are distributive laws, similar to (7.2.8) and (7.2.9). The law (7.3.9) is a modification of the commutative law $ab = ba$ which, we know, holds for multiplication of real numbers. Note that, while we have not, as yet, defined the product $\mathbf{y}\mathbf{x}$, the product $\mathbf{y}^t\mathbf{x}^t$ is well defined, since $\mathbf{y}^t$ is a row vector and $\mathbf{x}^t$ is a column vector.

Certain observations may be made concerning laws (7.3.6) to (7.3.9). We may, in (7.3.6), let $s=0$. In this case, we find that the product of any vector with the vector 0 (that is, the vector of which all the components are equal to zero) will, if defined, be the scalar 0. This corresponds to our usual idea that the product of any number by 0 is always equal to 0. This property of multiplication of real numbers carries over to the multiplication of vectors. On the other hand, the converse property of real numbers, that if the product ab of two numbers is equal to zero, then at least one of the two factors, a and b, must be equal to zero, does not hold for vectors. We can see this in Example VII.3.4(c), in which the product of $(1,-2)$ and $(2,1)^t$ is 0. Vectors such as these are called *orthogonal* vectors.

We note, finally, that the product (2.3.4) is quite generally called the *scalar product* or *inner product* of the two vectors, $\mathbf{x}$ and $\mathbf{y}$.

4. MATRIX MULTIPLICATION

We are now in a position to define multiplication of matrices. We shall do this in terms of the scalar products of vectors. In effect, to form the product AB of two matrices A and B, we would like to form the scalar products of all the rows of A with all the columns of B. These products (if they exist) can then be put into a matrix array in a very natural manner.

This, then, is the way in which we define the product of the two matrices, A and B. Each entry in the product matrix AB should be the scalar product of one of the rows of A with one of the columns of B. Now this is possible only if the rows of A and the columns of B have the same number of components. But the number of components in each row of A is equal to the number of columns in A, while the number of components in a column of B is equal to the number of rows in B. Thus, we find that *the product AB can be defined only if A has as many columns as B has rows.*

The reader may well ask: why do we multiply the rows of A with the columns of B? Why not the columns of A with the columns of B, or the rows of A with the rows of B? The answer is that a matrix is often thought of as a "black box", which takes input through the columns, and gives output through the rows. The product AB corresponds to the composition of the two mappings, which means that the output from A becomes the input into B. It follows that the rows of A must "mesh" with the columns of B. Thus, we multiply the rows of A with the columns of B.

VII.4.1 Definition. Let $A = (a_{ij})$ be an $m \times n$ matrix, and let $B = (b_{jk})$ be an $n \times p$ matrix. The product AB of these two matrices is the $m \times p$ matrix $C = (c_{ik})$, in which

7.4.1
$$c_{ik} = \sum_{j=1}^{n} a_{ij} b_{jk}$$

In other words, the entry in the ith row and kth column of C is the scalar product of the ith row of A and kth column of B.

VII.4.2 Examples. Some examples of matrix multiplication are:

(a)
$$\begin{pmatrix} 6 & 1 \\ 2 & 4 \\ -1 & 3 \end{pmatrix} \begin{pmatrix} 1 & 5 \\ 4 & 8 \end{pmatrix} = \begin{pmatrix} 6 \cdot 1 + 1 \cdot 4 & 6 \cdot 5 + 1 \cdot 8 \\ 2 \cdot 1 + 4 \cdot 4 & 2 \cdot 5 + 4 \cdot 8 \\ -1 \cdot 1 + 3 \cdot 4 & -1 \cdot 5 + 3 \cdot 8 \end{pmatrix} = \begin{pmatrix} 10 & 38 \\ 18 & 42 \\ 11 & 19 \end{pmatrix}$$

(b)
$$\begin{pmatrix} 3 & 5 \\ 1 & 4 \end{pmatrix} \begin{pmatrix} 2 & -3 \\ 1 & 8 \end{pmatrix} = \begin{pmatrix} 3 \cdot 2 + 5 \cdot 1 & 3(-3) + 5 \cdot 8 \\ 1 \cdot 2 + 4 \cdot 1 & 1(-3) + 4 \cdot 8 \end{pmatrix} = \begin{pmatrix} 11 & 31 \\ 6 & 29 \end{pmatrix}$$

(c)
$$\begin{pmatrix} 2 & -3 \\ 1 & 8 \end{pmatrix} \begin{pmatrix} 3 & 5 \\ 1 & 4 \end{pmatrix} = \begin{pmatrix} 2 \cdot 3 - 3 \cdot 1 & 2 \cdot 5 - 3 \cdot 4 \\ 1 \cdot 3 + 8 \cdot 1 & 1 \cdot 5 + 8 \cdot 4 \end{pmatrix} = \begin{pmatrix} 3 & -2 \\ 11 & 37 \end{pmatrix}$$

(d)
$$\begin{pmatrix} 1 & 5 & -1 \\ 2 & -4 & 3 \end{pmatrix} \begin{pmatrix} 6 & 2 \\ 1 & 4 \\ 1 & 8 \end{pmatrix} = \begin{pmatrix} 1 \cdot 6 + 5 \cdot 1 - 1 \cdot 1 & 1 \cdot 2 + 5 \cdot 4 - 1 \cdot 8 \\ 2 \cdot 6 - 4 \cdot 1 + 3 \cdot 1 & 2 \cdot 2 - 4 \cdot 4 + 3 \cdot 8 \end{pmatrix}$$
$$= \begin{pmatrix} 10 & 14 \\ 11 & 12 \end{pmatrix}$$

(e)
$$\begin{pmatrix} 2 & -1 & 8 \\ 6 & 2 & -1 \end{pmatrix} \begin{pmatrix} 1 \\ 5 \\ 2 \end{pmatrix} = \begin{pmatrix} 2 \cdot 1 - 1 \cdot 5 + 8 \cdot 2 \\ 6 \cdot 1 + 2 \cdot 5 - 1 \cdot 2 \end{pmatrix} = \begin{pmatrix} 13 \\ 14 \end{pmatrix}$$

VII.4.3 Example. A plumbing contractor has business in four different cities, A, B, C, and D. In each of these, his demand for copper pipe, steel pipe, and cast iron pipe are given in the table below (in hundreds of yards):

$$\begin{array}{l} \\ \text{Copper} \\ \text{Steel} \\ \text{Iron} \end{array} \begin{array}{c} \begin{array}{cccc} \;A\; & \;B\; & \;C\; & \;D\; \end{array} \\ \begin{pmatrix} 240 & 230 & 190 & 60 \\ 95 & 110 & 100 & 85 \\ 350 & 220 & 170 & 130 \end{pmatrix} \end{array} = M_1$$

In each of these four cities, the contractor has a choice of three suppliers, I, II, and III. He must buy all the pipe in one city from the same supplier, but he can use different suppliers in different cities. The suppliers quote the following prices (in cents per yard):

$$\begin{array}{c} \\ \text{I} \\ \text{II} \\ \text{III} \end{array} \quad \begin{array}{ccc} \text{Copper} & \text{Steel} & \text{Iron} \end{array} \atop \begin{pmatrix} 45 & 35 & 65 \\ 42 & 38 & 66 \\ 45 & 33 & 66 \end{pmatrix} = M_2.$$

Which suppliers should the contractor buy from, at each city? How would his expenses change if he could buy pipe from different contractors in the same city?

We must calculate the total cost of ordering the pipe from each supplier in each city. These are obtained by forming the product

$$M_2M_1 = \begin{pmatrix} 45 & 35 & 65 \\ 42 & 38 & 66 \\ 45 & 33 & 66 \end{pmatrix} \begin{pmatrix} 240 & 230 & 190 & 60 \\ 95 & 110 & 100 & 85 \\ 350 & 220 & 170 & 130 \end{pmatrix}$$

$$= \begin{pmatrix} 36875 & 28500 & 23100 & 14125 \\ 36790 & 28360 & 23000 & 14330 \\ 37035 & 28500 & 23070 & 14085 \end{pmatrix}.$$

From each column, now, we take the smallest entry, as representing the supplier whose costs are smallest for that city. Then, supplier II gives the lowest prices at A, B, and C, while supplier III gives lowest prices at D. If he orders in this manner, the contractor's costs for pipe will be

$$(36790, 28360, 23000, 14085).$$

If the contractor could order from different suppliers at the same city, then he should order the copper pipe from II, steel pipe from III, and cast iron pipe from I. In this case, his expenses at the different cities would be the product

$$(42, 33, 65) \begin{pmatrix} 240 & 230 & 190 & 60 \\ 95 & 110 & 100 & 85 \\ 350 & 220 & 170 & 130 \end{pmatrix} = (35965, 27590, 22330, 13775)$$

which represents a considerable saving (\$2675) over the previous cost vector.

VII.4.4 Example. A baker must decide how many cakes to bake. Each cake will cause expenses of \$2, and can be sold for \$6. He does not know how many customers will want to buy these cakes, but calculates that there is a 0.1 probability of 1 customer, 0.2 of 2 customers, 0.3 of 3 customers, 0.25 of 4 customers, and 0.15 of 5 customers. Any cakes left unsold at the end of the day must be thrown away. How many cakes should he prepare, so as to maximize his expected profit?

The baker's profits will depend, not only on his decision, but also on the demand for cakes. We may represent these by a matrix:

Number Baked \ Demand	1	2	3	4	5
1	4	4	4	4	4
2	2	8	8	8	8
3	0	6	12	12	12
4	−2	4	10	16	16
5	−4	2	8	14	20

$$\begin{pmatrix} 4 & 4 & 4 & 4 & 4 \\ 2 & 8 & 8 & 8 & 8 \\ 0 & 6 & 12 & 12 & 12 \\ -2 & 4 & 10 & 16 & 16 \\ -4 & 2 & 8 & 14 & 20 \end{pmatrix} = M$$

To obtain the expected profits, we must multiply this by the probability vector p:

$$p = (0.1, 0.2, 0.3, 0.25, 0.15)^t.$$

We have then

$$Mp = \begin{pmatrix} 4 & 4 & 4 & 4 & 4 \\ 2 & 8 & 8 & 8 & 8 \\ 0 & 6 & 12 & 12 & 12 \\ -2 & 4 & 10 & 16 & 16 \\ -4 & 2 & 8 & 14 & 20 \end{pmatrix} \begin{pmatrix} 0.1 \\ 0.2 \\ 0.3 \\ 0.25 \\ 0.15 \end{pmatrix} = \begin{pmatrix} 4.0 \\ 7.4 \\ 9.6 \\ 10.0 \\ 8.9 \end{pmatrix}$$

which shows that the baker will maximize his expected profits by baking 4 cakes; his expected profit will then be \$10.

The following rules may be seen to hold for matrix multiplication:

7.4.2 $$A(B + C) = AB + AC$$

7.4.3 $$(A + B)C = AC + BC$$

7.4.4 $$(sA)B = A(sB) = s(AB)$$

7.4.5 $$A(BC) = (AB)C$$

7.4.6 $$(AB)^t = B^tA^t$$

The laws (7.4.2) and (7.4.3) are distributive laws; (7.4.4) and (7.4.5) are associative laws for matrix multiplication. Law (7.4.6) is, like (7.3.9), a modification of the commutative law for multiplication. The commutative law itself, that is, $AB = BA$, does not, in general, hold for matrix multiplication (though in certain cases it may be true). A counterexample is provided by Example VII.4.2 (b) and (c).

Of the laws (7.4.2) to (7.4.6), the distributive laws are easily proved; they depend directly on the distributive laws for real numbers. The typical entry in the matrix AB is given by the expression

$$\sum_{j=1}^{n} a_{ij}b_{jk}$$

while that in AC is given by

$$\sum_{j=1}^{n} a_{ij}c_{jk}$$

Now, the typical entry in $A(B+C)$ is

$$\sum_{j=1}^{n} a_{ij}(b_{jk}+c_{jk})$$

and, by the distributive law for real numbers, it follows that the typical entry in $A(B+C)$ is the sum of the corresponding entries in AB and AC. Law (7.4.3) is proved similarly.

The law (7.4.6) is also easy to prove; it depends on the fact that the operation of transposition transforms each row into a column, and the converse. Thus the entry in the ith row and kth column of $(AB)^t$ is the same as the entry in the kth row and ith column of AB, and is therefore the scalar product of the kth row of A and ith column of B. But these are, respectively, the kth column of A^t and ith row of B^t. Application of (7.3.9) will then yield (7.4.6).

The associative law, (7.4.5), is somewhat less obvious. We shall prove this law in the case of 2×2 matrices $A = (a_{ij})$, $B = (b_{ij})$, and $C = (c_{ij})$. We have

$$AB = \begin{pmatrix} a_{11} & a_{12} \\ a_{21} & a_{22} \end{pmatrix} \begin{pmatrix} b_{11} & b_{12} \\ b_{21} & b_{22} \end{pmatrix} = \begin{pmatrix} a_{11}b_{11}+a_{12}b_{21} & a_{11}b_{12}+a_{12}b_{22} \\ a_{21}b_{11}+a_{22}b_{21} & a_{21}b_{12}+a_{22}b_{22} \end{pmatrix}$$

and so

$$\begin{aligned}(AB)C &= \begin{pmatrix} a_{11}b_{11}+a_{12}b_{21} & a_{11}b_{12}+a_{12}b_{22} \\ a_{21}b_{11}+a_{22}b_{21} & a_{21}b_{12}+a_{22}b_{22} \end{pmatrix} \begin{pmatrix} c_{11} & c_{12} \\ c_{21} & c_{22} \end{pmatrix} \\ &= \begin{pmatrix} (a_{11}b_{11}+a_{12}b_{21})c_{11}+(a_{11}b_{12}+a_{12}b_{22})c_{21} & (a_{11}b_{11}+a_{12}b_{21})c_{12}+(a_{11}b_{12}+a_{12}b_{22})c_{22} \\ (a_{21}b_{11}+a_{22}b_{21})c_{11}+(a_{21}b_{12}+a_{22}b_{22})c_{21} & (a_{21}b_{11}+a_{22}b_{21})c_{12}+(a_{21}b_{12}+a_{22}b_{22})c_{22} \end{pmatrix}\end{aligned}$$

On the other hand,

$$BC = \begin{pmatrix} b_{11} & b_{12} \\ b_{21} & b_{22} \end{pmatrix} \begin{pmatrix} c_{11} & c_{12} \\ c_{21} & c_{22} \end{pmatrix} = \begin{pmatrix} b_{11}c_{11}+b_{12}c_{21} & b_{11}c_{12}+b_{12}c_{22} \\ b_{21}c_{11}+b_{22}c_{21} & b_{21}c_{12}+b_{22}c_{22} \end{pmatrix}$$

and so

$$A(BC) = \begin{pmatrix} a_{11} & a_{12} \\ a_{21} & a_{22} \end{pmatrix} \begin{pmatrix} b_{11}c_{11} + b_{12}c_{21} & b_{11}c_{12} + b_{12}c_{22} \\ b_{21}c_{11} + b_{22}c_{21} & b_{21}c_{12} + b_{22}c_{22} \end{pmatrix}$$

$$= \begin{pmatrix} a_{11}(b_{11}c_{11} + b_{12}c_{21}) + a_{12}(b_{21}c_{11} + b_{22}c_{21}) & a_{11}(b_{11}c_{12} + b_{12}c_{22}) + a_{12}(b_{21}c_{12} + b_{22}c_{22}) \\ a_{21}(b_{11}c_{11} + b_{12}c_{21}) + a_{22}(b_{21}c_{11} + b_{22}c_{21}) & a_{21}(b_{11}c_{12} + b_{12}c_{22}) + a_{22}(b_{21}c_{12} + b_{22}c_{22}) \end{pmatrix}$$

It may now be seen directly (by expanding the products) that the two matrices $(AB)C$ and $A(BC)$ are equal. This being so, we will frequently dispense with the parentheses and write ABC instead of $(AB)C$ or $A(BC)$.

The proof for matrices of other dimensions is entirely similar (though of course somewhat longer), and the reader is invited to try it.

VII.4.3 Examples. The following examples illustrate these laws. Let

$$A = \begin{pmatrix} 5 & 1 \\ 3 & 2 \end{pmatrix} \quad B = \begin{pmatrix} -1 & 2 \\ 6 & 4 \end{pmatrix} \quad C = \begin{pmatrix} 2 & 3 \\ 1 & 0 \end{pmatrix}$$

Then

$$AB = \begin{pmatrix} 1 & 14 \\ 9 & 14 \end{pmatrix} \quad AC = \begin{pmatrix} 11 & 15 \\ 8 & 9 \end{pmatrix}$$

so that

$$AB + AC = \begin{pmatrix} 12 & 29 \\ 17 & 23 \end{pmatrix}$$

On the other hand,

$$B + C = \begin{pmatrix} 1 & 5 \\ 7 & 4 \end{pmatrix}$$

and so

$$A(B + C) = \begin{pmatrix} 5 & 1 \\ 3 & 2 \end{pmatrix} \begin{pmatrix} 1 & 5 \\ 7 & 4 \end{pmatrix} = \begin{pmatrix} 12 & 29 \\ 17 & 23 \end{pmatrix}$$

so that $A(B + C) = AB + AC$.

Similarly, we have

$$BC = \begin{pmatrix} 0 & -3 \\ 16 & 18 \end{pmatrix}$$

so that

$$A(BC) = \begin{pmatrix} 5 & 1 \\ 3 & 2 \end{pmatrix} \begin{pmatrix} 0 & -3 \\ 16 & 18 \end{pmatrix} = \begin{pmatrix} 16 & 3 \\ 32 & 27 \end{pmatrix}$$

 while

$$(AB)C = \begin{pmatrix} 1 & 14 \\ 9 & 14 \end{pmatrix} \begin{pmatrix} 2 & 3 \\ 1 & 0 \end{pmatrix} = \begin{pmatrix} 16 & 3 \\ 32 & 27 \end{pmatrix}$$

and so $A(BC) = (AB)C$.

Finally,

$$B^t A^t = \begin{pmatrix} -1 & 6 \\ 2 & 4 \end{pmatrix} \begin{pmatrix} 5 & 3 \\ 1 & 2 \end{pmatrix} = \begin{pmatrix} 1 & 9 \\ 14 & 14 \end{pmatrix}$$

and so $B^t A^t = (AB)^t$.

5. INVERSE MATRICES

We have defined (for certain cases) the operations of addition, subtraction, and multiplication of matrices. The question naturally arises whether some type of matrix division can be defined. In fact it cannot, but we shall see that the treatment of this question leads to very important results.

As usual, we proceed by analogy with the real number system. Where real numbers are concerned, the quotient a/b is defined as the number x, which is the solution of the equation

7.5.1 $$bx = a$$

or

7.5.2 $$xb = a$$

assuming that this solution is unique. We know that multiplication of real numbers satisfies the commutative law, so that (7.5.1) and (7.5.2) are equivalent equations; they will have a unique solution whenever $b \neq 0$. Thus, the quotient a/b is well defined except when the denominator b is equal to zero; i.e., division, except by zero, is a well-defined operation.

By analogy with this notion, we might try to define the quotient A/B as the solution of the equations

7.5.3 $$BX = A$$

or

7.5.4 $$XB = A$$

Two difficulties arise. One is that multiplication of matrices is not commutative, and so the two equations (7.5.4) and (7.5.3) will not, usually, be equivalent. The other difficulty is that, even if we consider only one of these two equations, there is no guarantee of a solution, much less of a unique solution. A deeper study of this question is necessary.

As we know, division by the real number b is equivalent to multiplication by its reciprocal, $1/b$. The reciprocal $1/b$ is in turn defined as the solution of the equation

$$bx = 1$$

We see that for real numbers, the number 1 plays an important role in the theory of multiplication and division. We look, therefore, for a matrix that will play a similar role. We find, in fact, that there are several such matrices: a unit matrix of order n for each value of n.

VII.5.1 Definition. The *unit matrix* of order n is the $n \times n$ matrix $I_n = (\delta_{ij})$, where

7.5.5 $$\delta_{ij} = \begin{cases} 0 & \text{if } i \neq j \\ 1 & \text{if } i = j \end{cases}$$

In other words, I_n is a square matrix that has 1's on the main diagonal, and zeros elsewhere:

$$I_4 = \begin{pmatrix} 1 & 0 & 0 & 0 \\ 0 & 1 & 0 & 0 \\ 0 & 0 & 1 & 0 \\ 0 & 0 & 0 & 1 \end{pmatrix}$$

and

$$I_2 = \begin{pmatrix} 1 & 0 \\ 0 & 1 \end{pmatrix}$$

There are, as we have said, an infinity of unit matrices, one for each value of n. When there is any ambiguity, a subscript is used to show the order of the matrix; in other cases, the order is clear from the context, and we omit the subscript.

As just mentioned, the unit matrices play a role analogous to that of the number 1 in ordinary multiplication. What this role is can best be seen from the following theorem.

VII.5.2 Theorem. Let A be an $m \times n$ matrix. Then

$$I_m A = A I_n = A$$

In other words, multiplication by a unit matrix, if it can be carried out at all, leaves any matrix unchanged.

Proof. We shall prove Theorem VII.5.2 for the case of 2×2 matrices. Consider $A = (a_{ij})$ and I_2. We have

$$AI_2 = \begin{pmatrix} a_{11} & a_{12} \\ a_{21} & a_{22} \end{pmatrix} \begin{pmatrix} 1 & 0 \\ 0 & 1 \end{pmatrix} = \begin{pmatrix} a_{11} \cdot 1 + a_{12} \cdot 0 & a_{11} \cdot 0 + a_{12} \cdot 1 \\ a_{21} \cdot 1 + a_{22} \cdot 0 & a_{21} \cdot 0 + a_{22} \cdot 1 \end{pmatrix}$$

$$= \begin{pmatrix} a_{11} & a_{12} \\ a_{21} & a_{21} \end{pmatrix} = A$$

Hence $AI = A$. The proof that $IA = A$ is quite similar.

The fundamental property of the unit matrices is that they serve as multiplicative identities, much as the matrices θ, considered earlier, serve as additive identities.

If we multiply the identity matrix I by a scalar s, we obtain a matrix, sI, all of whose entries are zero, except for the entries on the main diagonal, which are all equal to s. Such a matrix is called a *scalar matrix.*

If we apply (7.4.4) to a scalar matrix sI, we find that

$$(sI)A = s(IA) = sA$$

and

$$A(sI) = s(AI) = sA$$

so that multiplication, on either side, by the scalar matrix sI is equivalent to multiplication by the scalar s. This is, of course, the reason for the name.

We proceed now, to define the *inverse* of a matrix. By analogy with real numbers, we have

VII.5.3 Definition. By the *inverse* of a matrix A, we shall mean a matrix A^{-1}, satisfying the equations

$$AA^{-1} = A^{-1}A = I$$

If a matrix, A, has an inverse, we say that it is *invertible* or *non-singular.* A matrix that does not have an inverse is said to be *singular.*

It may be seen that the matrix A can have an inverse only if it is a square matrix; the matrix AA^{-1} will have as many rows as A, while $A^{-1}A$ will have as many columns as A. Hence A must have the same dimensions as the unit matrix I, which is, by definition, a square. It follows, then, that A^{-1} is also a square matrix of the same order as A. Thus, all non-singular matrices are square. The converse, however, is not true: a square matrix might be singular. The matrix θ, whatever its dimensions, will always be singular, since we know that $\theta X = \theta$ for any matrix X.

While we have no guarantee that a matrix will have an inverse, we see that, if the matrix does have an inverse, it will be unique. Suppose, for instance, that A had two inverses, B and C. Consider, then, the triple product $C(AB)$ or $(CA)B$. We have, by the associative law,

$$C(AB) = (CA)B$$

But, since $AB = I$,

$$C(AB) = CI = C$$

while, since $CA = I$,

$$(CA)B = IB = B$$

Therefore, $C = B$, and the inverse is unique. We are, hence, justified in talking about *the* inverse of A, and using the symbol A^{-1} to denote this inverse.

The following laws may be seen to hold for matrix inverses (we assume A,B are invertible).

7.5.4 $$(A^{-1})^{-1} = A$$

7.5.5 $$(sA)^{-1} = \frac{1}{s}A^{-1} \quad \text{if } s \neq 0$$

7.5.6 $$(A^t)^{-1} = (A^{-1})^t$$

7.5.7 $$(AB)^{-1} = B^{-1}A^{-1}$$

All these laws are quite easily proved; for instance, (7.5.4), which states that the inverse of the inverse is the original matrix, follows directly from the definition. Law (7.5.7) is proved by observing that

$$(AB)(B^{-1}A^{-1}) = A(BB^{-1})A^{-1} = AIA^{-1} = AA^{-1} = I$$

and

$$(B^{-1}A^{-1})(AB) = B^{-1}(A^{-1}A)B = B^{-1}IB = B^{-1}B = I$$

so that $B^{-1}A^{-1}$ is indeed the inverse of AB. Law (7.5.6) is proved by applying (7.4.6) and using the fact that $I^t = I$.

The actual problem of finding the inverse of a square matrix is quite complicated, and we shall defer it until a subsequent section of this chapter. We will, however, give some examples of matrix inverses; it is not difficult to check that a pair of matrices are inverses of each other, since all we have to do is perform a pair of matrix multiplications and see that the product is indeed an identity matrix.

VII.5.4 Example. Show that the two matrices

$$A = \begin{pmatrix} 5 & 1 \\ 4 & 1 \end{pmatrix} \quad B = \begin{pmatrix} 1 & -1 \\ -4 & 5 \end{pmatrix}$$

are inverses of each other.

This is done by constructing the two products, AB and BA. We have

$$AB = \begin{pmatrix} 5 \cdot 1 + 1(-4) & 5(-1) + 1 \cdot 5 \\ 4 \cdot 1 + 1(-4) & 4(-1) + 1 \cdot 5 \end{pmatrix} = \begin{pmatrix} 1 & 0 \\ 0 & 1 \end{pmatrix}$$

and

$$BA = \begin{pmatrix} 1 \cdot 5 - 1 \cdot 4 & 1 \cdot 1 - 1 \cdot 1 \\ -4 \cdot 5 + 5 \cdot 4 & -4 \cdot 1 + 5 \cdot 1 \end{pmatrix} = \begin{pmatrix} 1 & 0 \\ 0 & 1 \end{pmatrix}$$

so that $AB = BA = I$, and we find that A and B are indeed inverses.

VII.5.5 Example. Show that the two matrices

$$A = \begin{pmatrix} 1 & 1 & -3 \\ 2 & 5 & 1 \\ 1 & 3 & 2 \end{pmatrix} \quad B = \begin{pmatrix} 7 & -11 & 16 \\ -3 & 5 & -7 \\ 1 & -2 & 3 \end{pmatrix}$$

are inverses of each other.

Once again, we compute AB and BA:

$$AB = \begin{pmatrix} 1 \cdot 7 + 1(-3) - 3 \cdot 1 & 1(-11) + 1 \cdot 5 - 3(-2) & 1 \cdot 16 + 1(-7) + 2 \cdot 3 \\ 2 \cdot 7 + 5(-3) + 1 \cdot 1 & 2(-11) + 5 \cdot 5 + 1(-2) & 2 \cdot 16 + 5(-7) + 1 \cdot 3 \\ 1 \cdot 7 + 3(-3) + 2 \cdot 1 & 1(-11) + 3 \cdot 5 + 2(-2) & 1 \cdot 16 + 3(-7) + 2 \cdot 3 \end{pmatrix}$$

$$= \begin{pmatrix} 1 & 0 & 0 \\ 0 & 1 & 0 \\ 0 & 0 & 1 \end{pmatrix}$$

and

$$BA = \begin{pmatrix} 7 \cdot 1 - 11 \cdot 2 + 16 \cdot 1 & 7 \cdot 1 - 11 \cdot 5 + 16 \cdot 3 & 7(-3) - 11 \cdot 1 + 16 \cdot 2 \\ -3 \cdot 1 + 5 \cdot 2 - 7 \cdot 1 & -3 \cdot 1 + 5 \cdot 5 - 7 \cdot 3 & -3(-3) + 5 \cdot 1 - 7 \cdot 2 \\ 1 \cdot 1 - 2 \cdot 2 + 3 \cdot 1 & 1 \cdot 1 - 2 \cdot 5 + 3 \cdot 3 & 1(-3) - 2 \cdot 1 + 3 \cdot 2 \end{pmatrix}$$

$$= \begin{pmatrix} 1 & 0 & 0 \\ 0 & 1 & 0 \\ 0 & 0 & 1 \end{pmatrix}$$

and A and B are inverses.

VII.5.6 Example. Find the inverse of the matrix

$$A = \begin{pmatrix} a & b \\ c & d \end{pmatrix}$$

if it has one.

We are looking, here, for a matrix

$$X = \begin{pmatrix} x & y \\ z & w \end{pmatrix}$$

that will satisfy $AX = XA = I$. Let us consider the equation $AX = I$. We have

$$AX = \begin{pmatrix} ax + bz & ay + bw \\ cx + dz & cy + dw \end{pmatrix} = \begin{pmatrix} 1 & 0 \\ 0 & 1 \end{pmatrix}$$

The two matrices will be equal if, and only if, corresponding entries are equal. We therefore obtain a system of four equations in the unknowns x, y, z, w:

$$\begin{aligned} ax + bz &= 1 \\ cx + dz &= 0 \\ ay + bw &= 0 \\ cy + dw &= 1 \end{aligned}$$

This system can be split into two systems of two equations each: the first two equations, with the variables x and z only,

$$\begin{aligned} ax + bz &= 1 \\ cx + dz &= 0 \end{aligned}$$

can be solved by the methods discussed in Chapter I, giving

$$x = \frac{d}{ad - bc}, \quad z = \frac{-c}{ad - bc}$$

as solution, assuming, of course, that $ad - bc \neq 0$. The last two equations, in the variables y and w, will have a solution

$$y = \frac{-b}{ad - bc}, \quad w = \frac{a}{ad - bc}$$

Let us write $\Delta = ad - bc$. This quantity is called the *determinant* of the matrix A, and is often denoted by the symbol $|A|$, or det A. We find, thus, that the matrix equation $AX = I$ has a solution if $\Delta \neq 0$; this solution is

$$X = \begin{pmatrix} \frac{d}{\Delta} & \frac{-b}{\Delta} \\ \frac{-c}{\Delta} & \frac{a}{\Delta} \end{pmatrix}$$

 Before stating that X is the inverse of A, we must still check that $XA = I$. We have

$$XA = \begin{pmatrix} \dfrac{ad-bc}{\Delta} & \dfrac{bd-bd}{\Delta} \\ \dfrac{-ac+ac}{\Delta} & \dfrac{-bc+ad}{\Delta} \end{pmatrix} = \begin{pmatrix} 1 & 0 \\ 0 & 1 \end{pmatrix}$$

and so we can write $X = A^{-1}$.

VII.5.7 *Example.* Find the inverses of the matrices

$$A = \begin{pmatrix} 5 & 2 \\ 3 & 1 \end{pmatrix} \quad B = \begin{pmatrix} 1 & 4 \\ 2 & 8 \end{pmatrix} \quad C = \begin{pmatrix} 2 & 3 \\ 5 & 1 \end{pmatrix}$$

We can apply Example VII.5.6. For matrix A, we find

$$\Delta = 5 \cdot 1 - 3 \cdot 2 = -1$$

and so we will have

$$A^{-1} = \begin{pmatrix} \dfrac{1}{-1} & \dfrac{-2}{-1} \\ \dfrac{-3}{-1} & \dfrac{5}{-1} \end{pmatrix} = \begin{pmatrix} -1 & 2 \\ 3 & -5 \end{pmatrix}$$

For B, we find that $\Delta = 1 \cdot 8 - 2 \cdot 4 = 0$; this means that B is singular and does not have an inverse.

For C, we have

$$\Delta = 2 \cdot 1 - 5 \cdot 3 = -13$$

and so

$$C^{-1} = \begin{pmatrix} \dfrac{-1}{13} & \dfrac{3}{13} \\ \dfrac{5}{13} & -\dfrac{2}{13} \end{pmatrix}$$

It was mentioned that not all square matrices are invertible, i.e., some, such as the matrix θ, are singular. The problem of deciding which are invertible is solved in the next two sections of this chapter; here we give the following negative criterion:

VII.5.8 Theorem. If $AB = \theta$, but $B \neq \theta$, then A is singular.

Proof. Suppose A is invertible. Then

$$A^{-1}(AB) = A^{-1}\theta = \theta$$

But

$$(A^{-1}A)B = IB = B$$

By the associative law, $A^{-1}(AB) = (A^{-1}A)B$, meaning that $B = \theta$. We had assumed, however, that $B \neq \theta$. This contradiction proves that A^{-1} cannot exist.

VII.5.9 Example. Find the inverses of

$$A = \begin{pmatrix} 1 & 3 & 1 \\ 2 & 6 & 2 \\ 5 & 1 & 4 \end{pmatrix} \quad B = \begin{pmatrix} 5 & 1 \\ 10 & 2 \end{pmatrix}$$

if they exist.

We can apply Theorem VII.5.8 here. In fact, we can see that

$$\begin{pmatrix} 1 & 3 & 1 \\ 2 & 6 & 2 \\ 5 & 1 & 4 \end{pmatrix} \begin{pmatrix} 11 \\ 1 \\ -14 \end{pmatrix} = \begin{pmatrix} 0 \\ 0 \\ 0 \end{pmatrix}$$

so that we have $AS = \theta$ with $S \neq \theta$. Similarly,

$$\begin{pmatrix} 5 & 1 \\ 10 & 2 \end{pmatrix} \begin{pmatrix} 1 \\ -5 \end{pmatrix} = \begin{pmatrix} 0 \\ 0 \end{pmatrix}$$

and we have $BT = \theta$ with $T \neq \theta$. It follows that both A and B are singular matrices: their inverses do not exist.

Leontief Models. We consider here a model, introduced by W. W. Leontief, of an economy with n basic commodities. Each commodity is produced through a process which uses (as input) some (or all) of the other commodities in the economy. More precisely, the ith process $(i = 1, \ldots, n)$, when operating at unit intensity, uses as input a_{ij} units of the jth commodity $(j = 1, \ldots, n)$, and produces as output 1 unit of the ith commodity.

The economy can be represented by the technological matrix $A = (a_{ij})$. It is easy to see that, if the ith process operates at intensity x_i, then the total (gross) production of the economy is given by the vector

$$x = (x_1, x_2, \ldots, x_n),$$

whereas the total consumption (for production purposes) is given by the vector xA. The net production, therefore, is given by

$$y = x - xA = x(I - A).$$

Now there is no guarantee that the matrix $I - A$ will be invertible. It can be shown, however, that, if the sum of the entries in each row of

A is smaller than 1, then the inverse matrix $(I-A)^{-1}$ exists. (This is a reasonable assumption, meaning simply that, for the units chosen, production of one unit requires the consumption of less than one unit.) In this case, we see that we have

$$x = y(I-A)^{-1}$$

so that, given the desired net production y, we can find the gross production x which will be necessary for this.

VII.5.10 Example. Consider an economy with the two fundamental products of steel and wheat. Production of one ton of steel requires the consumption of 1/6 T. of steel (to prime the mills) and 1/4 bushel of wheat. In turn, production of one bushel of wheat requires the consumption of 1/12 T. of steel (wear and tear on the tractors) and 1/8 bu. of wheat (for seed). Find the gross production necessary to obtain a net production of 510 T. of steel and 680 bu. of wheat. How should this be changed if we wish to increase net wheat production to 1020 bu.?

In this case, we have the technological matrix

$$A = \begin{pmatrix} 1/6 & 1/4 \\ 1/12 & 1/8 \end{pmatrix}$$

and so

$$I - A = \begin{pmatrix} 5/6 & -1/4 \\ -1/12 & 7/8 \end{pmatrix}.$$

By the method described above, we calculate the inverse of this matrix:

$$(I-A)^{-1} = \begin{pmatrix} 21/17 & 6/17 \\ 2/17 & 20/17 \end{pmatrix}.$$

The desired net production, now, is $y = (510, 680)$. This gives us

$$x = (510, 680)\begin{pmatrix} 21/17 & 6/17 \\ 2/17 & 20/17 \end{pmatrix} = (710, 980)$$

so that gross production would be 710 T. of steel and 980 bu. of wheat. If we wish, instead, 510 T. of steel and 1020 bu. of wheat, we have the new vector

$$x = (510, 1020)\begin{pmatrix} 21/17 & 6/17 \\ 2/17 & 20/17 \end{pmatrix} = (750, 1380)$$

so that gross steel production must be increased to 750 T., and gross wheat production to 1380 bu. In general, we see from the matrix $(I-A)^{-1}$ that net production of each ton of steel requires gross production of 21/17 T. and 6/17 bu., while net production of each bushel of wheat requires a gross production of 2/17 T. and 20/17 bu.

PROBLEMS ON INVERSE MATRICES

1. Perform the following operations.

(a) $(3,1,5)\,(2,4,6)^t$

(b) $(2,4)\begin{pmatrix}1\\-2\end{pmatrix}$

(c) $(1,5,3,6)\begin{pmatrix}2\\6\\1\\-8\end{pmatrix}$

(d) $\begin{pmatrix}5\\-1\\3\end{pmatrix}^t\begin{pmatrix}2\\6\\5\end{pmatrix}$

(e) $\begin{pmatrix}2&6&3\\3&-1&8\end{pmatrix}\begin{pmatrix}1&5&4\\-3&1&2\\0&-6&3\end{pmatrix}$

(f) $\begin{pmatrix}1&0&3\\0&1&0\\0&0&1\end{pmatrix}\begin{pmatrix}1&5\\2&4\\6&-3\end{pmatrix}$

(g) $\begin{pmatrix}2&1&3\\4&6&-2\\5&3&3\end{pmatrix}\begin{pmatrix}1&4&-2\\6&0&1\\2&-5&0\end{pmatrix}$

(h) $\begin{pmatrix}1&4&-2\\6&0&1\\2&-5&0\end{pmatrix}\begin{pmatrix}2&1&3\\4&6&-2\\5&3&3\end{pmatrix}$

(i) $\begin{pmatrix}1&6&2\\4&0&-5\\-2&1&0\end{pmatrix}\begin{pmatrix}2&4&5\\1&6&3\\3&-2&3\end{pmatrix}$

(j) $\begin{pmatrix}-1&3&5\\2&-1&-2\\0&5&8\end{pmatrix}\begin{pmatrix}1&3&-1\\-8&-24&8\\5&15&-5\end{pmatrix}$

(k) $\begin{pmatrix}1&3&6&4\\2&-4&1&3\\5&1&8&4\end{pmatrix}\begin{pmatrix}1\\2\\-4\\6\end{pmatrix}$

(l) $\begin{pmatrix}2&5&8\\6&1&2\end{pmatrix}\begin{pmatrix}1&3&1\\-2&4&2\end{pmatrix}^t$

(m) $\begin{pmatrix}2&5&8\\6&1&2\end{pmatrix}^t\begin{pmatrix}1&3&1\\-2&4&2\end{pmatrix}$

(n) $\begin{pmatrix}5&1&6\\3&-5&1\end{pmatrix}\begin{pmatrix}1&4&3&1\\2&-4&3&2\\0&5&6&8\end{pmatrix}$

(o) $\begin{pmatrix} 1 & 3 \\ 2 & 4 \end{pmatrix} \begin{pmatrix} 1 & 5 \\ 6 & -1 \end{pmatrix}$

(p) $\begin{pmatrix} 2 & 6 \\ 3 & 1 \end{pmatrix} \begin{pmatrix} 1 & 3 \\ 4 & 8 \end{pmatrix}$

(q) $\begin{pmatrix} -1 & 5 \\ 2 & 6 \end{pmatrix} \begin{pmatrix} 1 & 4 \\ 6 & 2 \end{pmatrix}$

(r) $\begin{pmatrix} 1 & 5 \\ -2 & 4 \\ 1 & -6 \end{pmatrix} \begin{pmatrix} 1 & 4 \\ 1 & 6 \end{pmatrix}$

(s) $(1,5,3)\ (1,6,-8)^t$

(t) $\begin{pmatrix} 1 \\ 4 \\ 2 \end{pmatrix} (1,-3,4)$

2. Verify the associative law, $A(BC) = (AB)C$, by computing the following products in two different ways.

(a) $\begin{pmatrix} 1 & 5 \\ 2 & -4 \end{pmatrix} \begin{pmatrix} 3 \\ -2 \end{pmatrix} (1,3,1)$

(b) $\begin{pmatrix} 1 & -2 \\ 6 & 4 \end{pmatrix} \begin{pmatrix} 0 & 5 \\ 1 & -8 \end{pmatrix} \begin{pmatrix} -3 & 1 \\ 2 & 4 \end{pmatrix}$

(c) $\begin{pmatrix} 1 & 3 \\ 2 & 4 \\ 1 & -6 \end{pmatrix} \begin{pmatrix} 1 & 5 & 1 & 8 \\ 2 & -6 & -1 & 4 \end{pmatrix} \begin{pmatrix} 1 & -3 & 2 \\ 6 & 1 & -5 \\ 2 & 1 & 2 \\ 1 & 0 & 1 \end{pmatrix}$

(d) $\begin{pmatrix} 2 & 5 \\ 1 & 4 \end{pmatrix} \begin{pmatrix} 2 & 6 & 3 \\ 1 & -7 & 4 \end{pmatrix} \begin{pmatrix} 3 & 1 \\ 2 & -4 \\ 1 & 8 \end{pmatrix}$

(e) $\begin{pmatrix} 2 & 4 \\ 1 & 6 \end{pmatrix} \begin{pmatrix} 2 & 1 \\ 1 & -8 \end{pmatrix} \begin{pmatrix} 3 & 2 & 5 \\ 1 & 6 & 8 \end{pmatrix}$

(f) $(1,3,4) \begin{pmatrix} 1 & 6 & -2 \\ 2 & 4 & -3 \\ -1 & 0 & 2 \end{pmatrix} \begin{pmatrix} 1 \\ 5 \\ -2 \end{pmatrix}$

3. Solve the following equations.

(a) $\begin{pmatrix} 3 & 1 \\ 1 & 5 \end{pmatrix} \begin{pmatrix} x \\ y \end{pmatrix} = \begin{pmatrix} 7 \\ 7 \end{pmatrix}$

(b) $\begin{pmatrix} 4 & 2 \\ 6 & 0 \end{pmatrix} \begin{pmatrix} x \\ y \end{pmatrix} = \begin{pmatrix} 2 \\ 3 \end{pmatrix}$

(c) $\begin{pmatrix} 1 & 5 \\ 2 & 4 \\ 4 & 2 \end{pmatrix} \begin{pmatrix} x \\ y \end{pmatrix} = \begin{pmatrix} 3 \\ 0 \\ -6 \end{pmatrix}$

(d) $\begin{pmatrix} 2 & 1 \\ 3 & 5 \end{pmatrix} \begin{pmatrix} x & y \\ z & w \end{pmatrix} = \begin{pmatrix} 7 & 0 \\ 0 & 7 \end{pmatrix}$

(e) $\begin{pmatrix} 1 & 2 \\ 2 & 3 \end{pmatrix} \begin{pmatrix} x & y \\ z & w \end{pmatrix} = \begin{pmatrix} 3 & 5 \\ 1 & 4 \end{pmatrix}$

(f) $\begin{pmatrix} 1 & 3 \\ 2 & 5 \end{pmatrix} \begin{pmatrix} x & y \\ z & w \end{pmatrix} = \begin{pmatrix} 2 & 4 \\ 3 & 1 \end{pmatrix}$

4. A farmer divides his acreage in the following manner: wheat, 78 acres; corn, 43 A; alfalfa, 21 A; soybeans, 8 A. He calculates that his yearly profits will be \$70 per acre on the wheat, \$120 per A on the corn, \$50 per A on the alfalfa, and \$110 per A on the soybeans. How much profit does the farmer expect to make?

5. The 1960 areas (in square miles) and population densities (in people per sq. mi.) of the six Central American countries are given in the table below:

	Area	Population density
Belize	8,867	10.2
Costa Rica	19,695	54.7
El Salvador	8,165	303.2
Guatemala	42,042	90.9
Honduras	43,277	43.5
Nicaragua	53,668	27.0

Find the total population of Central America in 1960.

6. Consider a set of n persons, labeled $P_1, P_2, \ldots, P_n$, some of whom (but not necessarily all) know each other. Form a matrix $A = (a_{ij})$, by

$$a_{ij} = \begin{cases} 1, \text{ if } P_i \text{ knows } P_j \text{ and } i \neq j \\ 0, \text{ otherwise} \end{cases}$$

and call this the *acquaintance matrix* of the set. Form the product $A^2 = AA = (b_{ij})$. Show that, for $i \neq j$, b_{ij} is the number of people (in the set) whom P_i and P_j both know. What is b_{ii}?

7. Suppose that 5 people have the acquaintance matrix

$$A = \begin{pmatrix} 0 & 1 & 0 & 1 & 0 \\ 1 & 0 & 1 & 1 & 0 \\ 0 & 1 & 0 & 0 & 0 \\ 1 & 1 & 0 & 0 & 0 \\ 0 & 0 & 0 & 0 & 0 \end{pmatrix}.$$

Form the product A^2. How many individuals in this set do both P_1 and P_2 know? How many does P_3 know?

8. An investor is trying to decide whether to buy stock in company A, B, or C. Company A has recently executed a merger, and hopes to execute a second one shortly. Unfortunately, the Justice Department may block the forthcoming merger, and may even force A to divest itself of its recently acquired subsidiary. The investor calculates that, if both mergers are allowed, stock A will give him a gain of \$20,000, stock B a loss of \$25,000, and stock C a gain of \$12,000. If both mergers are disallowed, stock A will give a loss of \$6000, stock B, a loss of \$10,000, and stock C, a gain of \$12,000. Finally, if only the past merger is allowed, stock A will give a gain of \$5000, stock B, a gain of \$30,000, and stock C, a loss of \$4000. He feels that there is a 20 per cent probability that both mergers will be allowed, 30 per cent that both will be disallowed, and 50 per cent that the past merger will be allowed, but the future one blocked. What stock should our investor buy, so as to maximize expected profits?

9. A company has five factories. At each factory, there are four types of workers: managers, foremen, journeymen, and apprentices. Total number at each factory are given by the table below:

Factory	1	2	3	4	5
Managers	2	1	1	2	1
Foremen	7	3	5	9	4
Journeymen	45	20	25	55	18
Apprentices	30	12	18	37	15

Managers are paid \$300 per week; foremen, \$250; journeymen, \$200; and apprentices, \$125. What is the total weekly payroll at each factory?

10. A grinding machine can produce either ground steak, or hamburger. To produce one pound of ground steak, it uses 14 ounces of meat and 2 ounces of scraps; to produce one lb. of hamburger, it uses 8 ounces of meat and 8 ounces of scraps. If the machine is given x ounces of meat and y ounces of scraps, how many pounds of ground steak and hamburger can it produce (assuming that none of the meat or scraps is thrown away)? In particular, how many pounds of ground steak and hamburger can it produce from 20 lb. of meat and 10 lb. of scrap? (One lb. = 16 oz.)

11. Consider an economy with two basic products: transportation and fuel. The fuel must be taken from its natural deposit (mine, or oil well) to a processing plant, whereas the transportation needs fuel to run. Assume that each ton of fuel requires 100 T.-mi. of transportation, while each ton-mile of transportation consumes 2 lb. of fuel. Choosing suitable units, construct a technological matrix for this economy. What intensity vector should be used, if there is an outside demand for 1000 T. of fuel and 25,000 T.-mi. of transportation?

12. In a Leontief economy, let p_i be the price of 1 unit of the ith commodity, and q_i be the total value added in producing 1 unit of that commodity.

(a) Show that the two vectors

$$\mathbf{p} = \begin{pmatrix} p_1 \\ p_2 \\ \vdots \\ p_n \end{pmatrix} \quad \mathbf{q} = \begin{pmatrix} q_1 \\ q_2 \\ \vdots \\ q_n \end{pmatrix}$$

are related by the equation

$$(I - A)\mathbf{p} = \mathbf{q}$$

(b) In the foregoing example, what should the price of fuel and the freight rates be, given that the value added by the fuel industry is 5¢ per pound, while the value added by transportation is 10¢ per T.-mi.?

6. ROW OPERATIONS AND THE SOLUTION OF SYSTEMS OF LINEAR EQUATIONS

Let us consider a system of two linear equations in three unknowns:

$$\begin{aligned} 2x + 5y - 4z &= 6 \\ x - 2y + 2z &= 3 \end{aligned}$$

It may be seen that this can be rewritten in matrix notation as

$$\begin{pmatrix} 2 & 5 & -4 \\ 1 & -2 & 2 \end{pmatrix} \begin{pmatrix} x \\ y \\ z \end{pmatrix} = \begin{pmatrix} 6 \\ 3 \end{pmatrix}$$

More generally, the system of m linear equations in n unknowns,

7.6.1
$$\begin{aligned} a_{11}x_1 + a_{12}x_2 + \ldots + a_{1n}x_n &= c_1 \\ a_{21}x_1 + a_{22}x_2 + \ldots + a_{2n}x_n &= c_n \\ \cdots\cdots\cdots\cdots\cdots\cdots & \\ a_{m1}x_1 + a_{m2}x_2 + \ldots + a_{mn}x_n &= c_m \end{aligned}$$

can be rewritten in the form

7.6.2
$$\begin{pmatrix} a_{11} & a_{12} & \ldots & a_{1n} \\ a_{21} & a_{22} & \ldots & a_{2n} \\ \cdots & \cdots & \cdots & \cdots \\ a_{m1} & a_{m2} & \ldots & a_{mn} \end{pmatrix} \begin{pmatrix} x_1 \\ x_2 \\ \vdots \\ x_n \end{pmatrix} = \begin{pmatrix} c_1 \\ c_2 \\ \vdots \\ c_m \end{pmatrix}$$

 or, more simply, in the form

7.6.3 $$A\,\mathbf{x} = \mathbf{c}$$

where A is the matrix of coefficients, $\mathbf{x}$ is the column vector of variables, and $\mathbf{c}$ is the vector of constant terms. Let K be any $m \times m$ matrix. If we multiply both sides of (7.6.3) by K, we obtain, by the associative law,

7.6.4 $$(KA)\,\mathbf{x} = K\mathbf{c}$$

and it is clear that (7.6.3) implies (7.6.4), i.e., any vector $\mathbf{x}$ that satisfies (7.6.3) will also satisfy (7.6.4). The converse is not, generally, true: the solutions of (7.6.4) need not be solutions to (7.6.3). If, however, K is invertible, we may multiply both sides of (7.6.4) by K^{-1}:

$$K^{-1}(KA)\mathbf{x} = K^{-1}K\mathbf{c}$$

which reduces to (7.6.3). Thus, in this case, (7.6.4) implies (7.6.3), and so the two matrix equations, (7.6.3), and (7.6.4), are equivalent: any vector that solves the one will also solve the other. Now, it must be admitted that (7.6.4) has approximately the same form as (7.6.3): for an arbitrary, non-singular matrix K, we have no reason to believe that solution of the matrix equation (7.6.4), will be any easier than that of (7.6.3). We shall see, however, that the matrix K can be chosen in such a way that the equation (7.6.4) is either trivially easy to solve or else shows clearly that no solution is possible. More exactly, we shall show the existence of sequence, $K_1, K_2, \ldots, K_p$, of non-singular matrices such that the equation

$$K_p \ldots K_2\,K_1\,A\,\mathbf{x} = K_p \ldots K_2\,K_1\,\mathbf{c}$$

can be easily solved if it has any solution at all.

Let us consider the following operations that may be performed on any matrix:

7.6.5 Multiply every entry in the kth row by the non-zero scalar, r; i.e., replace a_{kj} by ra_{kj}.

7.6.6 Multiply each entry in the lth row by r and add to the corresponding entry in the kth row; i.e., replace a_{kj} by $a_{kj} + ra_{lj}$ (where $k \neq l$).

These two operations are called *fundamental row* operations; examples of them are:

VII.6.1 Examples

(a) $\begin{pmatrix} 1 & 5 & 2 \\ -1 & 4 & 6 \\ 2 & 1 & 8 \end{pmatrix} \rightarrow \begin{pmatrix} 1 & 5 & 2 \\ 3(-1) & 3\cdot 4 & 3\cdot 6 \\ 2 & 1 & 8 \end{pmatrix} = \begin{pmatrix} 1 & 5 & 2 \\ -3 & 12 & 18 \\ 2 & 1 & 8 \end{pmatrix}$

In this example, the second row was multiplied by the scalar 3.

(b) $\begin{pmatrix} 1 & -1 & 6 \\ 2 & 4 & 5 \\ 3 & 1 & 8 \end{pmatrix} \rightarrow \begin{pmatrix} 1+2\cdot 2 & -1+2\cdot 4 & 6+2\cdot 5 \\ 2 & 4 & 5 \\ 3 & 1 & 8 \end{pmatrix} = \begin{pmatrix} 3 & 7 & 16 \\ 2 & 4 & 5 \\ 3 & 1 & 8 \end{pmatrix}$

Here, the second row was multiplied by 2, and added to the first row.

(c) $\begin{pmatrix} 1 & 4 & 7 \\ 1 & 4 & 6 \\ 2 & 1 & 8 \end{pmatrix} \rightarrow \begin{pmatrix} 1 & 4 & 7 \\ 1-1 & 4-4 & 6-7 \\ 2 & 1 & 8 \end{pmatrix} = \begin{pmatrix} 1 & 4 & 7 \\ 0 & 0 & -1 \\ 2 & 1 & 8 \end{pmatrix}$

In this example, the first row was multiplied by -1, and added to the second row, i.e., the first row was subtracted from the second.

It is our desire to use these row operations to solve systems of linear equations. To do this, we shall show, first, that these operations are always equivalent to multiplication on the left by a non-singular matrix K; and second, that the matrix K depends only on the form of the operation, and not on the particular matrix A upon which we operate. This second property means that multiplication by K will perform the same operation on the vector **c** as on the matrix A.

Consider, then, operation (7.6.5). It may be seen that the corresponding matrix is $T = (t_{ij})$, defined by

7.6.7 $$t_{ij} = \begin{cases} r & \text{if } i = j = k \\ 1 & \text{if } i = j \neq k \\ 0 & \text{if } i \neq j \end{cases}$$

(T is almost an identity matrix, the only difference being that its (k,k)th entry is r instead of 1.) The matrix T is non-singular; its inverse is $T^{-1} = (t'_{ij})$, where

7.6.8 $$t'_{ij} = \begin{cases} 1/r & \text{if } i = j = k \\ 1 & \text{if } i = j \neq k \\ 0 & \text{if } i \neq j \end{cases}$$

That T will, in fact, cause the operation (7.6.5) can be proved in a manner similar to the proof, given earlier, that the matrix I is a multiplicative identity. That (7.6.8) does, in fact, define the inverse matrix is most easily seen when we consider that it induces the inverse operation, i.e., it causes the kth row to be *divided* by the scalar r.

Next let us consider operation (7.6.6). It may be seen that this is induced by the matrix $I + S$, in which I is the identity matrix, and $S = (s_{ij})$ is defined by

7.6.9
$$s_{ij} = \begin{cases} r & \text{if } i = k, j = l \\ 0 & \text{otherwise} \end{cases}$$

(S has all entries equal to zero except for the (k,l)th entry, which is equal to r.) It is not difficult to see, since $k \neq l$, that $SS = \theta$, and so

$$\begin{aligned}(I + S)(I - S) &= II + SI - IS - SS \\ &= I + S - S + \theta \\ &= I\end{aligned}$$

Similarly, we can see that $(I - S)(I + S) = I$, so that $I - S = (I + S)^{-1}$, and $I + S$ is non-singular.

We are now in a position to apply these elementary row operations to the solution of systems of linear equations. The general idea is to perform the operations on the matrix of coefficients until a very simple type of matrix is obtained; we try to change A, if possible, into an identity matrix. If A is singular, this will not be feasible, so we have to content ourselves with something that looks as much like an identity matrix as possible.

We have seen that each of these row operations is equivalent to multiplication by a non-singular matrix K. The vector of constant terms, **c**, must simultaneously be multiplied by the same K. But K induces the same row operation on all matrices (so long as they have the proper number of rows so that multiplication is possible). Hence we need only perform the same row operations on the vector **c** as we do on the matrix A.

Since the same row operations are to be performed on both A and **c**, it is generally more convenient to deal with the *augmented* matrix

7.6.10
$$(A \mid \mathbf{c}) = \left(\begin{array}{cccc|c} a_{11} & a_{12} & \cdots & a_{1n} & c_1 \\ a_{21} & a_{22} & \cdots & a_{2n} & c_2 \\ \cdots & \cdots & \cdots & \cdots & \cdots \\ a_{m1} & a_{m2} & \cdots & a_{mn} & c_m \end{array}\right)$$

It is then simply a question of performing the row operations on $(A \mid \mathbf{c})$; this guarantees that they be performed simultaneously on A and **c**.

VII.6.2. Example. Solve the system of equations

$$\begin{aligned} x + 2y - z + w &= 5 \\ 2x - y \phantom{{}- z} + w &= 6 \\ x - y + z + 3w &= 9 \\ 4y - z - 2w &= 1 \end{aligned}$$

This system may be written, using matrix notation, in the form

$$\begin{pmatrix} 1 & 2 & -1 & 1 \\ 2 & -1 & 0 & 1 \\ 1 & -1 & 1 & 3 \\ 0 & 4 & -1 & -2 \end{pmatrix}\begin{pmatrix} x \\ y \\ z \\ w \end{pmatrix} = \begin{pmatrix} 5 \\ 6 \\ 9 \\ 1 \end{pmatrix}$$

We obtain, thus, the augmented matrix

$$\left(\begin{array}{cccc|c} 1 & 2 & -1 & 1 & 5 \\ 2 & -1 & 0 & 1 & 6 \\ 1 & -1 & 1 & 3 & 9 \\ 0 & 4 & -1 & -2 & 1 \end{array}\right)$$

We would like, if possible, to change this matrix, by means of row operations, until the first four columns form an identity matrix. The best method is, generally, to take the columns one at a time, changing the "main diagonal" entry to a 1 and getting rid of any other non-zero entries in the column. In the first column, here, we find that the first entry is, as desired, a 1. The entries in the second and third rows are not zero, so we must eliminate them. To eliminate the 2 in the second row, we can multiply the first row by -2 and add to the second row; we also subtract the first row from the third, obtaining

$$\left(\begin{array}{cccc|c} 1 & 2 & -1 & 1 & 5 \\ 2-2 & -1-4 & 0+2 & 1-2 & 6-10 \\ 1-1 & -1-2 & 1+1 & 3-1 & 9-5 \\ 0 & 4 & -1 & -2 & 1 \end{array}\right) = \left(\begin{array}{cccc|c} 1 & 2 & -1 & 1 & 5 \\ 0 & -5 & 2 & -1 & -4 \\ 0 & -3 & 2 & 2 & 4 \\ 0 & 4 & -1 & -2 & 1 \end{array}\right)$$

We have, thus, reduced the first column to the required form. We now attack the second column: we must obtain here a 1 in the second row and zeros elsewhere, *being careful meanwhile that the first column remains as it is.* (This means that we cannot, say, multiply the first row by 3 and add to the second row; while this would accomplish the feat of putting a 1 in the second row, second column, it would also put a 3 in the second row, first column.)

There are many ways of putting a 1 in the second row, second column. One possibility is to multiply the third row by -2 and add to the second row. Another possibility is to multiply the second row by $-1/5$. The former procedure has the advantage of avoiding fractions, at least for the moment. The latter has the advantage of being somewhat more systematic. Let us, then, carry out this latter procedure; we obtain the matrix

$$\left(\begin{array}{cccc|c} 1 & 2 & -1 & 1 & 5 \\ 0 & 1 & -2/5 & 1/5 & 4/5 \\ 0 & -3 & 2 & 2 & 4 \\ 0 & 4 & -1 & -2 & 1 \end{array}\right)$$

392 We now use the second row to remove all the other non-zero entries in this column. We can multiply the second row by -2, and add to the first row; by 3, and add to the third row; by -4, and add to the fourth row. Carrying out all these operations, we will have

$$\left(\begin{array}{cccc|c} 1 & 2-2 & -1+4/5 & 1-2/5 & 5-8/5 \\ 0 & 1 & -2/5 & 1/5 & 4/5 \\ 0 & -3+3 & 2-6/5 & 2+3/5 & 4+12/5 \\ 0 & 4-4 & -1+8/5 & -2-4/5 & 1-16/5 \end{array}\right) = \left(\begin{array}{cccc|c} 1 & 0 & -1/5 & 3/5 & 17/5 \\ 0 & 1 & -2/5 & 1/5 & 4/5 \\ 0 & 0 & 4/5 & 13/5 & 32/5 \\ 0 & 0 & 3/5 & -14/5 & -11/5 \end{array}\right)$$

We proceed to the third column. We can multiply the third row by 5/4:

$$\left(\begin{array}{cccc|c} 1 & 0 & -1/5 & 3/5 & 17/5 \\ 0 & 1 & -2/5 & 1/5 & 4/5 \\ 0 & 0 & 1 & 13/4 & 8 \\ 0 & 0 & 3/5 & -14/5 & -11/5 \end{array}\right)$$

and now we use the third row to remove the other non-zero entries in the third column, obtaining

$$\left(\begin{array}{cccc|c} 1 & 0 & -1/5+1/5 & 3/5+13/20 & 17/5+8/5 \\ 0 & 1 & -2/5+2/5 & 1/5+3/10 & 4/5+16/5 \\ 0 & 0 & 1 & 13/4 & 8 \\ 0 & 0 & 3/5-3/5 & -14/5-39/20 & -11/5-24/5 \end{array}\right) = \left(\begin{array}{cccc|c} 1 & 0 & 0 & 5/4 & 5 \\ 0 & 1 & 0 & 3/2 & 4 \\ 0 & 0 & 1 & 13/4 & 8 \\ 0 & 0 & 0 & -19/4 & -7 \end{array}\right)$$

We next multiply the bottom row by $-4/19$:

$$\left(\begin{array}{cccc|c} 1 & 0 & 0 & 5/4 & 5 \\ 0 & 1 & 0 & 3/2 & 4 \\ 0 & 0 & 1 & 13/4 & 8 \\ 0 & 0 & 0 & 1 & 28/19 \end{array}\right)$$

and use this row to eliminate the other fourth-column entries:

$$\left(\begin{array}{cccc|c} 1 & 0 & 0 & 0 & 60/19 \\ 0 & 1 & 0 & 0 & 34/19 \\ 0 & 0 & 1 & 0 & 61/19 \\ 0 & 0 & 0 & 1 & 28/19 \end{array}\right)$$

We have now accomplished what we set out to do: the matrix of coefficients has become an identity matrix, and the original matrix equation, $\mathbf{Ax} = \mathbf{c}$, has been reduced to the form

$$I\begin{pmatrix} x \\ y \\ z \\ w \end{pmatrix} = \begin{pmatrix} 60/19 \\ 34/19 \\ 61/19 \\ 28/19 \end{pmatrix}$$

But the matrix I will not change anything. This last equation gives us, directly, the solution of the original system. It is, precisely, $(x, y, z, w) = (60/19, 34/19, 61/19, 28/19)$. That this satisfies the original system may be checked directly.

VII.6.3 Example. Solve the system

$$\begin{aligned} x - 2y + z + 2w &= 18 \\ 2x + y + 4z - w &= 16 \\ x + y - z + w &= 4 \\ x + 3y - 2z + w &= 1 \end{aligned}$$

As before, we form the augmented matrix

$$\left(\begin{array}{cccc|c} 1 & -2 & 1 & 2 & 18 \\ 2 & 1 & 4 & -1 & 16 \\ 1 & 1 & -1 & 1 & 4 \\ 1 & 3 & -2 & 1 & 1 \end{array}\right)$$

We use the first row to remove the other non-zero entries in the first column:

$$\left(\begin{array}{cccc|c} 1 & -2 & 1 & 2 & 18 \\ 0 & 5 & 2 & -5 & -20 \\ 0 & 3 & -2 & -1 & -14 \\ 0 & 5 & -3 & -1 & -17 \end{array}\right)$$

Next, we divide the second row by 5 and use this row to remove the non-zero entries in the second column:

$$\left(\begin{array}{cccc|c} 1 & 0 & 9/5 & 0 & 10 \\ 0 & 1 & 2/5 & -1 & -4 \\ 0 & 0 & -16/5 & 2 & -2 \\ 0 & 0 & -5 & 4 & 3 \end{array}\right)$$

We multiply the third row by $-5/16$ and use it to clear the third column:

$$\left(\begin{array}{cccc|c} 1 & 0 & 0 & 9/8 & 71/8 \\ 0 & 1 & 0 & -3/4 & -17/4 \\ 0 & 0 & 1 & -5/8 & 5/8 \\ 0 & 0 & 0 & 7/8 & 49/8 \end{array}\right)$$

Finally, we multiply the last row by 8/7 and clear the fourth column,

$$\left(\begin{array}{cccc|c} 1 & 0 & 0 & 0 & 1 \\ 0 & 1 & 0 & 0 & 1 \\ 0 & 0 & 1 & 0 & 5 \\ 0 & 0 & 0 & 1 & 7 \end{array}\right)$$

This gives us the solution; it is $(x, y, z, w) = (1, 1, 5, 7)$. Again, we can check directly by substituting these values in the original equations.

VII.6.4 Example. A mixture must be made of food pellets of types A, B, and C. The fat, carbohydrate, and protein contents of each of the pellets are given in Table VII.6.1 (in grams):

TABLE VII.6.1

	A	B	C
Fat	1	3	1
Carbohydrate	2	1	0
Protein	1	0	4

It is desired to make a mixture containing exactly 300 gm. of fat, 269 gm. of carbohydrate, and 240 gm. of protein. How should this be done?

The vector of requirements can be adjoined to the coefficient matrix, obtaining the augmented matrix

$$\left(\begin{array}{ccc|c} 1 & 3 & 1 & 300 \\ 2 & 1 & 0 & 269 \\ 1 & 0 & 4 & 240 \end{array}\right)$$

We perform elementary row operations on this matrix, as shown below:

$$\left(\begin{array}{ccc|c} 1 & 3 & 1 & 300 \\ 0 & -5 & -2 & -331 \\ 0 & -3 & 3 & -60 \end{array}\right)$$

$$\left(\begin{array}{ccc|c} 1 & 0 & -1/5 & 507/5 \\ 0 & 1 & 2/5 & 331/5 \\ 0 & 0 & 21/5 & 693/5 \end{array}\right)$$

$$\left(\begin{array}{ccc|c} 1 & 0 & 0 & 108 \\ 0 & 1 & 0 & 53 \\ 0 & 0 & 1 & 33 \end{array}\right).$$

We conclude from this last matrix that the desired quantities of nutrients can be obtained by using 108 pellets of type A, 53 of type B, and 33 of type C.

7. SOLUTION OF GENERAL $m \times n$ SYSTEMS OF EQUATIONS

Naturally, it is not always possible to solve systems of equations as we did in Examples VII.6.2 and VII.6.3. Even if the matrix of coefficients is square, we have no guarantee that it will be invertible. We can see what happens in such cases from the following examples.

VII.7.1 Example. Solve the system

$$\begin{aligned} x + 2y - z &= 6 \\ 3x - y + z &= 8 \\ 5x + 3y - z &= 20 \end{aligned}$$

In this case, we form the augmented matrix

$$\left(\begin{array}{ccc|c} 1 & 2 & -1 & 6 \\ 3 & -1 & 1 & 8 \\ 5 & 3 & -1 & 20 \end{array}\right)$$

and use the first row to clear the first column:

$$\left(\begin{array}{ccc|c} 1 & 2 & -1 & 6 \\ 0 & -7 & 4 & -10 \\ 0 & -7 & 4 & -10 \end{array}\right)$$

It may be seen that the second and third rows have become identical; this means that one of the equations is redundant, so that, with only two independent equations, we cannot hope to have a unique solution. If we continue with our procedure, multiplying the second row by $-1/7$ and clearing the second column, we obtain

$$\left(\begin{array}{ccc|c} 1 & 0 & 1/7 & 22/7 \\ 0 & 1 & -4/7 & 10/7 \\ 0 & 0 & 0 & 0 \end{array}\right)$$

Consider this new matrix: we would like to get a 1 in the third row, third column. This cannot be done by multiplying the third row, since we now have a zero in that position. We could do this by using one of the *other* rows, say, by multiplying the first row by 7 and adding to the third row. If we do this, however, we obtain a 7 in the third row, first column, thus undoing some of our previous work. We find, in fact, that we can go no further in simplifying the coefficient matrix. Our final result is the system

$$\begin{aligned} x + 1/7z &= 22/7 \\ y - 4/7z &= 10/7 \end{aligned}$$

 The third equation, of course, reduces to the identity $0 = 0$, and can be dispensed with. We can, then, solve for x and y in terms of z:

$$\begin{aligned} x &= 22/7 - 1/7z \\ y &= 10/7 + 4/7z \end{aligned}$$

The variable z is arbitrary. This is the general solution, or, more exactly, one form of the general solution. We could, of course, solve for any two of the variables in terms of the third. Let us suppose that we want to solve for y and z in terms of x. We can do this by obtaining a second-order identity matrix in the columns corresponding to y and z, rather than in those corresponding to x and y. Starting from the last of the foregoing matrices, we can multiply the first row by 7:

$$\left(\begin{array}{ccc|c} 7 & 0 & 1 & 22 \\ 0 & 1 & -4/7 & 10/7 \\ 0 & 0 & 0 & 0 \end{array}\right)$$

We then clear the third column by means of the first row:

$$\left(\begin{array}{ccc|c} 7 & 0 & 1 & 22 \\ 4 & 1 & 0 & 14 \\ 0 & 0 & 0 & 0 \end{array}\right)$$

We could, of course, have discarded the third row a long time ago, reducing to the 2×4 matrix

$$\left(\begin{array}{ccc|c} 7 & 0 & 1 & 22 \\ 4 & 1 & 0 & 14 \end{array}\right)$$

Now, the two columns corresponding to y and z do not form an identity matrix, but this is not too important. In fact, these two columns form what is called a *permutation* matrix, a matrix that can be transformed into an identity matrix by simply interchanging some of the rows. As such, the foregoing matrix is sufficient for our purposes. Indeed, the equations we now have are

$$\begin{aligned} 7x \qquad + z &= 22 \\ 4x + y \qquad &= 14 \end{aligned}$$

or, solving for y and z in terms of x,

$$\begin{aligned} y &= 14 - 4x \\ z &= 22 - 7x \end{aligned}$$

which, for arbitrary x, is another form of the general solution to the system.

VII.7.2 Example. Solve the system

$$\begin{aligned} x + 2y - z &= 6 \\ 3x - y + z &= 8 \\ 5x + 3y - z &= 22 \end{aligned}$$

Again we form the augmented matrix

$$\left(\begin{array}{rrr|r} 1 & 2 & -1 & 6 \\ 3 & -1 & 1 & 8 \\ 5 & 3 & -1 & 22 \end{array}\right)$$

and use the first row to clear the first column:

$$\left(\begin{array}{rrr|r} 1 & 2 & -1 & 6 \\ 0 & -7 & 4 & -10 \\ 0 & -7 & 4 & -8 \end{array}\right)$$

We then divide the second row by -7, and clear the second column, obtaining

$$\left(\begin{array}{rrr|r} 1 & 0 & 1/7 & 22/7 \\ 0 & 1 & -4/7 & 10/7 \\ 0 & 0 & 0 & 2 \end{array}\right)$$

Consider, now, the third row: it represents the contradiction $0 = 2$. This means that the system has no solution. In general, a system will be infeasible (i.e., have no solution) if a row is obtained in which all the entries except the last are zero. On the other hand, as we saw from Example VII.7.1, a row that has all its entries (including the last one) equal to zero may be discarded and does not affect the system.

For systems that have fewer equations than variables, the procedure will generally be as in Example VII.7.1: we try to obtain an identity matrix (or a permutation matrix) in the columns corresponding to some of the variables; this solves the system for the corresponding variables in terms of the others.

VII.7.3 Example. Solve the system

$$\begin{aligned} 2x + y - 4z &= 10 \\ x - y + 2z &= 6 \end{aligned}$$

We can, here, attempt to solve the system for any two of the variables in terms of the third. (In this case, our attempts will be successful, but in some cases they need not be.) Let us, then, solve for x and z in terms of y. We will work on the matrix in such a way that the columns corresponding to x and z form an identity (or permutation) matrix. The augmented matrix is

$$\left(\begin{array}{rrr|r} 2 & 1 & -4 & 10 \\ 1 & -1 & 2 & 6 \end{array}\right)$$

We can divide the first row by 2 and clear the first column:

$$\left(\begin{array}{ccc|c} 1 & 1/2 & -2 & 5 \\ 0 & -3/2 & 4 & 1 \end{array}\right)$$

Next we divide the second row by 4 and clear the third column to obtain

$$\left(\begin{array}{ccc|c} 1 & -1/4 & 0 & 11/2 \\ 0 & -3/8 & 1 & 1/4 \end{array}\right)$$

so that the desired solution is

$$\begin{aligned} x &= 11/2 + 1/4y \\ z &= \ \ 1/4 + 3/8y \end{aligned}$$

Let us suppose, now that we want to solve for y and z in terms of x. In this last matrix, we multiply the first row by -4:

$$\left(\begin{array}{ccc|c} -4 & 1 & 0 & -22 \\ 0 & -3/8 & 1 & 1/4 \end{array}\right)$$

and use the first row to clear the second column:

$$\left(\begin{array}{ccc|c} -4 & 1 & 0 & -22 \\ -3/2 & 0 & 1 & -8 \end{array}\right)$$

thus obtaining the solution

$$\begin{aligned} y &= -22 + \ \ 4x \\ z &= -8 \ \ + 3/2x \end{aligned}$$

VII.7.4 Example. Solve the system

$$\begin{aligned} 2x + y + 4z &= 10 \\ x - y + 2z &= \ \ 6 \end{aligned}$$

Suppose we want to solve this system for x and z in terms of y. We take the augmented matrix

$$\left(\begin{array}{ccc|c} 2 & 1 & 4 & 10 \\ 1 & -1 & 2 & 6 \end{array}\right)$$

and, as before, divide the first row by 2, and clear the first column:

$$\left(\begin{array}{ccc|c} 1 & 1/2 & -2 & 5 \\ 0 & -3/2 & 0 & 1 \end{array}\right)$$

We would like to clear the third column now, but it has become impossible. In fact, we cannot solve for x and z in terms of y. The reason for this is best seen if we note that the second row represents the equation

$$-\frac{3}{2}y = 1$$

which tells us that $y = -2/3$. In effect, we have found that y is not arbitrary; if we were to give x and z in terms of y, we would get only one solution. But the problem has an infinity of solutions.

While it is not possible, here, to solve for x and z in terms of y, we may, if we wish, solve for x and y in terms of z. In the last matrix given, we may multiply the second row by $-2/3$, and clear the second column:

$$\left(\begin{array}{ccc|c} 1 & 0 & -2 & 16/3 \\ 0 & 1 & 0 & -2/3 \end{array}\right)$$

which gives us

$$x = \frac{16}{3} + 2z$$

$$y = \frac{-2}{3}$$

Note that x is given in terms of z; y, however, has only one value.

In case we have more equations than variables, the usual procedure is to try to obtain an identity matrix in the first n rows and n columns; below this, we try to get nothing but zeros. When this has been done, we might find that the last $m - n$ rows contain zeros only; if so, we have obtained the solution. On the other hand, it may be that one of the last $m - n$ rows has a non-zero entry in the right-hand column, but all zeros otherwise. In this case, the problem has no solution, as this row represents the contradiction $0 = 1$.

VII.7.5 Example. Solve the system

$$\begin{aligned} x + y + z &= 9 \\ 2x - y + 2z &= 15 \\ x - y + 2z &= 12 \\ x + 2y - z &= 0 \end{aligned}$$

We take the augmented matrix

$$\left(\begin{array}{ccc|c} 1 & 1 & 1 & 9 \\ 2 & -1 & 2 & 15 \\ 1 & -1 & 2 & 12 \\ 1 & -2 & -1 & 0 \end{array}\right)$$

and use the first row to clear the first column:

$$\left(\begin{array}{rrr|r} 1 & 1 & 1 & 9 \\ 0 & -3 & 0 & -3 \\ 0 & -2 & 1 & 3 \\ 0 & 1 & -2 & -9 \end{array}\right)$$

We divide the second row by -3, and use it to clear the second column:

$$\left(\begin{array}{rrr|r} 1 & 0 & 1 & 8 \\ 0 & 1 & 0 & 1 \\ 0 & 0 & 1 & 5 \\ 0 & 0 & -2 & -10 \end{array}\right)$$

Finally, we can use the third row to clear the second column, obtaining the matrix

$$\left(\begin{array}{rrr|r} 1 & 0 & 0 & 3 \\ 0 & 1 & 0 & 1 \\ 0 & 0 & 1 & 5 \\ 0 & 0 & 0 & 0 \end{array}\right)$$

which gives us the solution

$$\begin{aligned} x &= 3 \\ y &= 1 \\ z &= 5 \end{aligned}$$

Note that the last row consists only of zeros: it may be disregarded.

VII.7.6 Example. Solve the system

$$\begin{aligned} x + y - z &= 1 \\ 2x + y + z &= 9 \\ x - 2y + z &= 1 \\ x + 2y - 2z &= 0 \\ x - y + z &= 2 \end{aligned}$$

Let us once again take the augmented matrix

$$\left(\begin{array}{rrr|r} 1 & 1 & -1 & 1 \\ 2 & 1 & 1 & 9 \\ 1 & -2 & 1 & 1 \\ 1 & 2 & -2 & 0 \\ 1 & -1 & 1 & 2 \end{array}\right)$$

Once again, we use the first row to clear the first column:

$$\left(\begin{array}{ccc|c} 1 & 1 & -1 & 1 \\ 0 & -1 & 3 & 7 \\ 0 & -3 & 2 & 0 \\ 0 & 1 & -1 & -1 \\ 0 & -2 & 2 & 1 \end{array}\right)$$

We multiply the second row by -1, and clear the second column:

$$\left(\begin{array}{ccc|c} 1 & 0 & 2 & 8 \\ 0 & 1 & -3 & -7 \\ 0 & 0 & -7 & -21 \\ 0 & 0 & 2 & 6 \\ 0 & 0 & -4 & -13 \end{array}\right)$$

Finally, we divide the third row by -7, and clear the third column:

$$\left(\begin{array}{ccc|c} 1 & 0 & 0 & 2 \\ 0 & 1 & 0 & 2 \\ 0 & 0 & 1 & 3 \\ 0 & 0 & 0 & 0 \\ 0 & 0 & 0 & 1 \end{array}\right)$$

Consider this matrix: the first three rows seem to give us the solution $(x, y, z) = (2, 2, 3)$. The last row, however, represents the contradiction $0 = 1$. We conclude that the system is infeasible, since the unique solution to the first three equations fails to satisfy the fifth equation.

VII.7.7 Example. A grocer has three types of mixed nuts. Mixture A contains 60 per cent cashews and 20 per cent peanuts (and 20 per cent other nuts) and costs \$1.50 per pound. Type B contains 20 per cent cashews and 20 per cent peanuts, and costs \$1.20 per lb. Type C contains 40 per cent cashews and 40 per cent peanuts, and costs 80¢ per lb. The grocer would like to make 100 lb. of a mixture which will contain 46 per cent cashews and 26 per cent peanuts, to sell at \$1.23 per lb. How much of each mixture should he use?

Letting x, y, and z be the number of pounds of mixtures A, B, and C respectively, we obtain the system

$$\begin{aligned} x + y + z &= 100 \\ 0.6x + 0.2y + 0.4z &= 46 \\ 0.2x + 0.2y + 0.4z &= 26 \\ 1.5x + 1.2y + 0.8z &= 123 \end{aligned}$$

 which is represented by the augmented matrix

$$\left(\begin{array}{ccc|c} 1 & 1 & 1 & 100 \\ 0.6 & 0.2 & 0.4 & 46 \\ 0.2 & 0.2 & 0.4 & 26 \\ 1.5 & 1.2 & 0.8 & 123 \end{array}\right).$$

By means of elementary row operations, we will (eventually) obtain the matrix

$$\left(\begin{array}{ccc|c} 1 & 0 & 0 & 50 \\ 0 & 1 & 0 & 20 \\ 0 & 0 & 1 & 30 \\ 0 & 0 & 0 & 0 \end{array}\right)$$

which tells us that the system is indeed feasible; the grocer should use 50 lb. of mixture A, 20 lb. of B, and 30 lb. of C.

VII.7.8 Example. A chemist is given three solutions. Solution A contains 20 per cent alcohol and 6 per cent salt, and costs \$1 per gallon. Solution B contains 40 per cent alcohol and 4 per cent salt, and costs \$1.50 per gallon. Solution C contains 50 per cent alcohol and 3 per cent salt, and costs \$1.90 per gallon. The chemist needs 40 gallons of a mixture which will contain 35 per cent alcohol and 4½ per cent salt. Is it possible to make such a mixture, costing \$58?

Letting x, y, and z be the amounts of mixtures A, B, and C, respectively, we have the equations

$$\begin{aligned} x + y + z &= 40 \\ 0.2x + 0.4y + 0.5z &= 14 \\ 0.06x + 0.04y + 0.03z &= 1.8 \\ x + 1.6y + 1.9z &= 58 \end{aligned}$$

which gives us the matrix

$$\left(\begin{array}{ccc|c} 1 & 1 & 1 & 40 \\ 0.2 & 0.4 & 0.5 & 14 \\ 0.06 & 0.04 & 0.03 & 1.8 \\ 1 & 1.6 & 1.9 & 58 \end{array}\right).$$

After several row operations, we obtain

$$\left(\begin{array}{ccc|c} 1 & 0 & -0.5 & 10 \\ 0 & 1 & 1.5 & 30 \\ 0 & 0 & 0 & 0 \\ 0 & 0 & 0 & 0 \end{array}\right)$$

which means that the system has the general solution

$$\begin{aligned}&z \text{ arbitrary}\\ &x = 10 + 0.5z\\ &y = 30 - 1.5z.\end{aligned}$$

We remember, however, that the variables x, y, and z must all be non-negative. Thus z is not entirely arbitrary; it must satisfy $z \geqslant 0$, and also $z \leqslant 20$, so that $y \geqslant 0$. Thus the chemist can use any amount z, up to 20 gal., of C. The amounts of A and B are then $10 + z/2$ and $30 - 3z/2$, respectively.

PROBLEMS ON SOLUTION OF GENERAL $m \times n$ SYSTEMS OF EQUATIONS

1. Solve the following systems of simultaneous equations.

(a) $$\begin{aligned} 3x + 5y &= 6\\ x - 2y &= 2\end{aligned}$$

(b) $$\begin{aligned} 3x + 5y &= 4\\ x + 2y &= 7\end{aligned}$$

(c) $$\begin{aligned} 5x + 2y + 4z &= 7\\ 3x - 2y + z &= 6\\ x + y + z &= 4\end{aligned}$$

(d) $$\begin{aligned} 2x - 6y + z &= -11\\ x + 2y - z &= 2\\ -x + y + 2z &= 12\end{aligned}$$

(e) $$\begin{aligned} x - 2y + 3z - w &= 12\\ 2x + y + z - 2w &= -6\\ -x + y + 2z + w &= 2\\ -y + z + w &= 1\end{aligned}$$

(f) $$\begin{aligned} 3x + y + z &= 2\\ x - 2y + 2z &= 5\\ 2x - y - z &= -7\\ 5x + 3y + z &= 2\end{aligned}$$

(g) $$\begin{aligned} x + 2y + 4z + w &= 6\\ 3x + 2y - 3z - w &= 5\\ -x + 6y + 2z + 3w &= 8\end{aligned}$$

(h) $$\begin{aligned} 2x + 3y + z - 2w &= 0\\ x - 2y + 2z - w &= 9\\ 3x - y + 3z + w &= 6\\ -x + y - z + 3w &= 2\end{aligned}$$

(i)
$$\begin{aligned} 3x + 2y + 6z + w &= 12 \\ x - 2y + 3z + 2w &= 4 \\ -2x + 4y - 2z - w &= -1 \\ x + y - 2z + 3w &= 3 \end{aligned}$$

(j)
$$\begin{aligned} 2x + 4y + 2z - w &= 18 \\ x + 3y - 2z + w &= 13 \\ -x + 2y \qquad + 2w &= 28 \\ 3x + y + 4z \qquad &= 1 \end{aligned}$$

(k)
$$\begin{aligned} x + 2y + 6z - w &= 5 \\ 3x + y + 4z - 2w &= 7 \\ x - y + 7z + w &= 9 \\ -x + 2y + 3z - 2w &= 5 \end{aligned}$$

(l)
$$\begin{aligned} x - 3y + 5z - w &= -3 \\ 2x + y + 4z + w &= 11 \\ -x + y + z - 3w &= -7 \\ 3x + 6y + 2z - 4w &= 5 \\ 2x + 4y + z - 7w &= -10 \\ x + y + z - w &= 1 \end{aligned}$$

(m)
$$\begin{aligned} x - 2y + 3z + w &= 6 \\ 2x + y + 6z - 2w &= 1 \\ 5x + 2y - 3z - 2w &= -4 \\ 4x + y - 2z - w &= -1 \\ 6x + 2y + z + w &= 10 \end{aligned}$$

(n)
$$\begin{aligned} 2x + 3y + z - 5w &= 8 \\ -x + 3y + 7z + 5w &= 30 \\ x - 2y + 4z - 6w &= 7 \\ -2x + y + z + 8w &= 11 \end{aligned}$$

(o)
$$\begin{aligned} 3x + y + 2z - w &= 6 \\ -x - y + 2z - 2w &= 5 \\ x + 4y + 6z - 3w &= 8 \end{aligned}$$

2. Suppose that one gram of food X contains 25 units of vitamin A, 12 units of vitamin B, and 15 units of vitamin C. One gram of food y contains 11 units of vitamin A, 15 units of vitamin B, and 12 units of vitamin C. Finally, one gram of food Z contains 5 units of vitamin A and 25 units of vitamin C. Is it possible to obtain a mixture containing exactly 276 units of vitamin A, 186 units of B, and 242 units of C? Is it possible to make such a mixture, weighing exactly 16 grams? [Hint: If the fractions are difficult to work with, try interchanging the rows of the matrix.]

3. A mixture is to be made of three solutions. Solution X contains 20 per cent alcohol and 12 per cent sugar. Solution Y contains 16 per cent alcohol and 13 per cent sugar, while solution Z contains 32 per cent alcohol and 9 per cent sugar. It is desired to make 20 gallons of a

mixture containing 28 per cent alcohol and 10 per cent sugar. How can this be done?

4. Suppose that, in Problem 3 above, solution X costs 50¢ per gallon, Y costs 40¢ per gal., and Z costs 65¢ per gal. Is it possible to make the desired solution, costing 56¢ per gallon? Costing 60¢ per gallon? Costing 64¢ per gallon?

5. A food company makes three blends of coffee. Each blend contains different ratios of Colombian, Brazilian, and African coffees, as shown in the table below (in ounces per pound of the blend):

	Blend X	Y	Z
Colombian	15	6	2
Brazilian	1	4	8
African	0	6	6

If the company has 400 lb. of Colombian coffee, 800 lb. of Brazilian, and 600 lb. of African, how many pounds of each blend can it make (so as not to waste any coffee)?

8. COMPUTATION OF THE INVERSE MATRIX

In Section 4 of this chapter, we saw that we can, in effect, solve a matrix equation $A\mathbf{x} = \mathbf{c}$ by using the inverse matrix; the solution is $\mathbf{x} = A^{-1}\mathbf{c}$. On the other hand, it may be noticed that, while all the row operations gave us this solution, the matrix A^{-1} was never actually computed. It is true, of course, that this was not necessary in the context of such problems. We may, however, conceive of a situation in which computation of A^{-1} may actually help us to avoid a great amount of work.

Let us suppose, for instance, that we are given several systems of equations:

7.8.1 $$A\mathbf{x} = \mathbf{c}$$

7.8.2 $$A\mathbf{x} = \mathbf{c}'$$

7.8.3 $$A\mathbf{x} = \mathbf{c}''$$

. . . .

7.8.4 $$A\mathbf{x} = \mathbf{c}^{(p)}$$

in which, in each case, the matrix of coefficient A, remains the same, but the vector of constant terms varies. This may happen, in a physical case, where the matrix A represents the physical characteristics of a system, while the vectors $\mathbf{c}$, $\mathbf{c}'$, $\mathbf{c}''$, . . . , $\mathbf{c}^{(p)}$ represent the several outputs that one may wish to obtain from the system. The variables $\mathbf{x}$ would then be the inputs necessary to obtain the desired outputs.

Now, we may solve each of the systems (7.8.1), (7.8.2), (7.8.3), . . . , (7.8.4) by the method of Section VII.6, if we wish. On the other hand, if there are many of them, this will mean carrying out the same set of row operations several times over. It becomes much more practical, in this case, to compute the matrix A^{-1}; it is then quite simple to generate the solutions $A^{-1}\mathbf{c}, A^{-1}\mathbf{c}', \ldots, A^{-1}\mathbf{c}^{(p)}$ of the several systems. This procedure also has the advantage of allowing us to see how the solution vector, $\mathbf{x}$, depends on the vector of constant terms, $\mathbf{c}$.

We wish, then, a procedure for inverting matrices. Our method will once again employ the fundamental row operations (7.6.5) and (7.6.6). As we have seen, a non-singular matrix A may, by means of these operations, be transformed into the identity matrix I. But each row operation corresponds to multiplication by a matrix K. What these operations do, in effect, is to provide us with a sequence of matrices, $K_1, K_2, \ldots, K_q$, such that

7.8.5 $$K_q \ldots K_2 K_1 A = I$$

But this means that

7.8.6 $$K_q \ldots K_2 K_1 = A^{-1}$$

Now, precisely because I is the multiplicative identity, we have

7.8.7 $$A^{-1} = K_q \ldots K_2 K_1 I$$

This means that the inverse matrix, A^{-1}, may be obtained by subjecting I to those row operations that changed A into I.

For those readers who are mechanically inclined, we give here a program for computing the inverse of a matrix A. The reader may translate it into any programming language (e.g., Fortran) that he is familiar with. We assume that we start with a square matrix; i.e., we are given a set of numbers a_{ij}, for $i,j = 1,2, \ldots, n$. Then:

[1] Define $b_{ij} = 1$ if $i = j$, $b_{ij} = 0$ if $i \neq j$, for $i,j = 1,2, \ldots, n$. Proceed to [2].

[2] Let $k = 1$. *Proceed to* [3].

[3] For each $i \neq k$, and for each j, replace a_{ij} by

$$a'_{ij} = a_{ij} - \frac{a_{ik} a_{kj}}{a_{kk}}$$

and replace b_{ij} by

$$b'_{ij} = b_{ij} - \frac{a_{ik} b_{kj}}{a_{kk}}.$$

Proceed to $\boxed{4}$.

$\boxed{4}$ For each j, replace a_{kj} by $a'_{kj} = a_{kj}/a_{kk}$, and replace b_{kj} by $b'_{kj} = b_{kj}/a_{kk}$. Proceed to $\boxed{5}$.

$\boxed{5}$ Is $k = n$? If NO, proceed to $\boxed{6}$. If YES, then STOP, the numbers b_{ij} are the answer.

$\boxed{6}$ Replace k by $k+1$. Proceed to $\boxed{3}$.

It should be clear, of course, that the program can abort, and will do so if ever one of the main diagonal entries a_{kk} is equal to zero. This will certainly happen if A is singular, but may happen even if A is invertible. The reader may wish to modify the program to take this into account. Hint: if $a_{kk} = 0$, but A is invertible, then some a_{ik}, with $i > k$, must be different from zero. The ith and k^{th} rows can then be interchanged.

VII.8.1 Example. Invert the matrix

$$A = \begin{pmatrix} 3 & 1 \\ 5 & 2 \end{pmatrix}$$

We could, of course, invert the matrix directly by using the results of Section VII.5. Let us, however, invert it by the method of row operations. We form the double matrix

$$(A \mid I) = \left(\begin{array}{cc|cc} 3 & 1 & 1 & 0 \\ 5 & 2 & 0 & 1 \end{array}\right)$$

In this matrix, we divide the first row by 3:

$$\left(\begin{array}{cc|cc} 1 & 1/3 & 1/3 & 0 \\ 5 & 2 & 0 & 1 \end{array}\right)$$

and use this row to clear the first column:

$$\left(\begin{array}{cc|cc} 1 & 1/3 & 1/3 & 0 \\ 0 & 1/3 & -5/3 & 1 \end{array}\right)$$

Next we multiply the second row by 3, and use it to clear the second column:

$$\left(\begin{array}{cc|cc} 1 & 0 & 2 & -1 \\ 0 & 1 & -5 & 3 \end{array}\right)$$

We have now, by means of row operations, reduced the double matrix, $(A \mid I)$ to the form $(I \mid B)$. But this B is precisely the desired inverse matrix. In other words,

$$A^{-1} = \begin{pmatrix} 2 & -1 \\ -5 & 3 \end{pmatrix}$$

which, of course, coincides with the results of Section VII.5.

VII.8.2 Example. Invert the matrix

$$A = \begin{pmatrix} 1 & -2 & 1 \\ 2 & 1 & 4 \\ 1 & 1 & -1 \end{pmatrix}$$

Once again, we form the double matrix

$$\left(\begin{array}{ccc|ccc} 1 & -2 & 1 & 1 & 0 & 0 \\ 2 & 1 & 4 & 0 & 1 & 0 \\ 1 & 1 & -1 & 0 & 0 & 1 \end{array}\right)$$

We use the first row to clear the first column:

$$\left(\begin{array}{ccc|ccc} 1 & -2 & 1 & 1 & 0 & 0 \\ 0 & 5 & 2 & -2 & 1 & 0 \\ 0 & 3 & -2 & -1 & 0 & 1 \end{array}\right)$$

We then divide the second row by 5, and clear the second column:

$$\left(\begin{array}{ccc|ccc} 1 & 0 & 9/5 & 1/5 & 2/5 & 0 \\ 0 & 1 & 2/5 & -2/5 & 1/5 & 0 \\ 0 & 0 & -16/5 & 1/5 & -3/5 & 1 \end{array}\right)$$

Finally, we multiply the third row by $-5/16$, and use this row to clear the third column:

$$\left(\begin{array}{ccc|ccc} 1 & 0 & 0 & 5/16 & 1/16 & 9/16 \\ 0 & 1 & 0 & -3/8 & 1/8 & 1/8 \\ 0 & 0 & 1 & -1/16 & 3/16 & -5/16 \end{array}\right)$$

This gives us, then, the inverse matrix; it is the right half of this double matrix, or, more simply,

$$A^{-1} = \frac{1}{16}\begin{pmatrix} 5 & 1 & 9 \\ -6 & 2 & 2 \\ -1 & 3 & -5 \end{pmatrix}$$

It may be checked directly that this is, indeed, the inverse matrix.

VII.8.3 Example. Find the solution of the system

$$\begin{aligned} x + 3y - z + 4w &= a \\ -x + 2y + z - w &= b \\ 2x - y + w &= c \\ x + 2z + 3w &= d \end{aligned}$$

in which a, b, c, d are *parameters*, i.e., constants that may be changed at the option of the person handling the system.

Since a solution is wanted in terms of general values for the parameters, we invert the matrix of coefficients. Once again, we take the double matrix

$$\left(\begin{array}{cccc|cccc} 1 & 3 & -1 & 4 & 1 & 0 & 0 & 0 \\ -1 & 2 & 1 & -1 & 0 & 1 & 0 & 0 \\ 2 & -1 & 0 & 1 & 0 & 0 & 1 & 0 \\ 1 & 0 & 2 & 3 & 0 & 0 & 0 & 1 \end{array}\right)$$

and we use the first row to clear the first column. The intermediate steps follow without further comment:

$$\left(\begin{array}{cccc|cccc} 1 & 3 & -1 & 4 & 1 & 0 & 0 & 0 \\ 0 & 5 & 0 & 3 & 1 & 1 & 0 & 0 \\ 0 & -7 & 2 & -7 & -2 & 0 & 1 & 0 \\ 0 & -3 & 3 & -1 & -1 & 0 & 0 & 1 \end{array}\right)$$

$$\left(\begin{array}{cccc|cccc} 1 & 0 & -1 & 11/5 & 2/5 & -3/5 & 0 & 0 \\ 0 & 1 & 0 & 3/5 & 1/5 & 1/5 & 0 & 0 \\ 0 & 0 & 2 & -15/4 & -3/5 & 7/5 & 1 & 0 \\ 0 & 0 & 3 & 4/5 & -2/5 & 3/5 & 0 & 1 \end{array}\right)$$

$$\left(\begin{array}{cccc|cccc} 1 & 0 & 0 & 4/5 & 1/10 & 1/10 & 1/2 & 0 \\ 0 & 1 & 0 & 3/5 & 1/5 & 1/5 & 0 & 0 \\ 0 & 0 & 1 & -7/5 & -3/10 & 7/10 & 1/2 & 0 \\ 0 & 0 & 0 & 5 & 1/2 & -3/2 & -3/2 & 1 \end{array}\right)$$

$$\left(\begin{array}{cccc|cccc} 1 & 0 & 0 & 0 & 1/50 & 17/50 & 37/50 & -4/25 \\ 0 & 1 & 0 & 0 & 7/50 & 19/50 & 9/50 & -3/25 \\ 0 & 0 & 1 & 0 & -4/25 & 7/25 & 2/25 & 7/25 \\ 0 & 0 & 0 & 1 & 1/10 & -3/10 & -3/10 & 1/5 \end{array}\right)$$

The right side of this last matrix gives us the inverse matrix; the solution to the problem is then simply given by

$$\begin{pmatrix} x \\ y \\ z \\ w \end{pmatrix} = \frac{1}{50}\begin{pmatrix} 1 & 17 & 37 & -8 \\ 7 & 19 & 9 & -6 \\ -8 & 14 & 2 & 14 \\ 5 & -15 & -15 & 10 \end{pmatrix}\begin{pmatrix} a \\ b \\ c \\ d \end{pmatrix}$$

or, equivalently,

$$\begin{aligned} x &= 1/50a + 17/50b + 37/50c - 4/25d \\ y &= 7/50a + 19/50b + 9/50c - 3/25d \\ z &= -4/25a + 7/25b + 1/25c + 7/25d \\ w &= 1/10a - 3/10b - 3/10c + 1/5d \end{aligned}$$

VII.8.4 Example. An economy depends on three fundamental goods: wheat, steel, and transportation. The (gross) production of one bushel of wheat requires 0.2 bu. of wheat, 0.2 lb. of steel, and 0.1 T.-mi. of transportation. One pound of steel requires 0.1 bu. of wheat, 0.3 lb. of steel, and 0.2 T.-mi. of transportation. Finally, one ton-mile of transportation requires 0.1 bu. of wheat and 0.5 lb. of steel. What gross production is necessary for the net production of 1 bu. of wheat? Of 1 lb. of steel? Of 1 T.-mi. of transportation?

The technological matrix here is

$$A = \begin{pmatrix} 0.2 & 0.2 & 0.1 \\ 0.1 & 0.3 & 0.2 \\ 0.1 & 0.5 & 0 \end{pmatrix}$$

and so

$$I - A = \begin{pmatrix} 0.8 & -0.2 & -0.1 \\ -0.1 & 0.7 & -0.2 \\ -0.1 & -0.5 & 1 \end{pmatrix}.$$

We invert this matrix, to obtain

$$(I - A)^{-1} = \begin{pmatrix} 50/37 & 125/222 & 55/222 \\ 10/37 & 395/222 & 85/222 \\ 10/37 & 35/37 & 45/37 \end{pmatrix}.$$

Each row of the matrix gives us the gross production necessary to obtain a net production of one unit. Thus, the net production of one bushel of wheat requires the gross production of 50/37 bu., 125/222 lb., and 55/222 T.-mi. One lb. of steel requires 10/37 bu., 395/222 lb., and 85/222 T.-mi. Finally, one T.-mi. of transportation (net) requires 10/37 bu., 35/37 lb., and 45/37 T.-mi. If, for example, we desired a net production of 444 bu. of wheat, 666 lb. of steel, and 222 T.-mi. of transportation, the required gross production would be

$$x = (444, 666, 222)(I - A)^{-1} = (840, 1470, 635)$$

or, 840 bu. of wheat, 1470 lb. of steel, and 635 T.-mi. of transportation.

VII.8.5 Example. As mentioned in problem 12, page 387, the matrix $(I - A)^{-1}$ can also be used to obtain relationships among prices and values added by the several industries. Using the data of Example VII.8.4, let us suppose that the total value added in the gross production of one bushel of wheat is 37¢; in the production of one lb. of steel, 11.1¢, and, in the production of one T.-mi. of transportation, 22.2¢. What are the "natural" prices for the three goods?

We saw above that the relationship between the column vector, $\mathbf{p}$, of prices, and the vector, $\mathbf{q}$, of value added, is $\mathbf{p} = (I - A)^{-1}\mathbf{q}$. In this case, we will have

$$\mathbf{p} = (I - A)^{-1}\begin{pmatrix} 0.37 \\ 0.111 \\ 0.222 \end{pmatrix} = \begin{pmatrix} 0.6175 \\ 0.3825 \\ 0.4750 \end{pmatrix}.$$

Thus, the "natural" prices will be $61\frac{3}{4}$¢ per bushel of wheat, $38\frac{1}{4}$¢ per pound of steel, and $47\frac{1}{2}$¢ per ton-mile of transportation.

PROBLEMS ON COMPUTATION OF INVERSE MATRICES

1. Find the inverses of the following matrices.

(a) $\begin{pmatrix} 2 & 5 \\ 6 & 1 \end{pmatrix}$

(b) $\begin{pmatrix} 3 & 4 \\ 7 & 8 \end{pmatrix}$

(c) $\begin{pmatrix} 1 & -3 \\ 2 & 5 \end{pmatrix}$

(d) $\begin{pmatrix} 1 & 6 \\ 1 & 5 \end{pmatrix}$

(e) $\begin{pmatrix} -1 & 3 \\ -2 & 4 \end{pmatrix}$

(f) $\begin{pmatrix} 3 & 1 & -5 \\ 2 & 1 & 7 \\ 8 & 3 & -2 \end{pmatrix}$

(g) $\begin{pmatrix} 3 & 6 & 1 \\ 2 & 4 & 8 \\ 5 & 1 & 9 \end{pmatrix}$

(h) $\begin{pmatrix} 6 & 4 & 2 \\ 3 & 1 & 5 \\ 3 & 3 & -3 \end{pmatrix}$ (j) $\begin{pmatrix} 5 & 1 & 4 & 2 \\ 3 & 2 & 7 & 1 \\ 2 & -1 & -3 & 4 \\ 1 & -4 & -13 & 2 \end{pmatrix}$

(i) $\begin{pmatrix} 2 & 4 & 7 & 1 \\ 6 & 3 & -2 & 4 \\ -5 & 4 & 8 & 1 \\ 2 & 6 & 3 & -7 \end{pmatrix}$

2. The company described in problem 5, page 405, is given a pounds of African coffee, b lb. of Brazilian, and c lb. of Colombian. How many pounds of each blend can it make (so as not to waste any coffee)?

3. An economy depends on three fundamental goods: steel, fuel, and transportation. Production of one pound of steel requires 0.1 lb. of fuel and 0.2 T.-mi. of transportation. One pound of fuel requires 0.2 lb. of steel and 0.05 T.-mi. of transportation. One ton-mile of transportation requires 0.05 lb. of steel and 0.3 lb. of fuel. What gross production is necessary in order to meet a net outside demand of one lb. of steel? Of one lb. of fuel? Of one T.-mi. of transportation? What is the necessary production if the net outside demand is 500 lb. of steel, 2000 lb. of fuel, and 1000 T.-mi. of transportation?

4. In the economy described in Problem 3, above, the net value added in producing one lb. of steel is 7¢; one lb. of fuel, 5¢; and one T.-mi. of transportation, 4¢. What are the prices for the three goods?

5. A company makes scaffoldings. Each scaffolding is obtained by taking three assemblies of type A and one assembly of type B, and joining them by 6 bolts and 3 coils of wire. Each assembly A consists of two sub-assemblies C, joined by 2 bolts. Each assembly B consists of one sub-assembly C and one sub-assembly D, joined by 1 bolt and 2 coils of wire. Each sub-assembly C consists of two rods joined by two bolts, and each sub-assembly D consists of three rods joined by 2 bolts and 2 coils of wire. How many rods, bolts, and coils of wire are needed to construct a scaffolding? (It is of course possible to solve this problem by direct enumeration. The reader should, however, construct the technological matrix A, and compute the associated matrix $(I-A)^{-1}$.)

CHAPTER VIII

LINEAR PROGRAMMING

1. LINEAR PROGRAMS

In Chapter II, we studied systems of linear inequalities and saw that, in general, such systems do not have unique solutions. Nor is it generally possible to characterize their solutions as simply as those of systems of linear equations. This is not, however, an important consideration, for it is seldom desired to find *all* the solutions to such a system. It is usually only necessary to find *one* solution, i.e., a point that satisfies the constraints (inequalities) of the system. In a few cases, any solution will do, but normally, a special solution is required —the one that is best in a particular sense, say, that maximizes profits or minimizes costs. We can see this best from the following example.

VIII.1.1 Example. A dietitian must prepare a mixture of foods A, B, and C. Each unit of food A contains 3 oz. protein, 2 oz. carbohydrate, and 6 oz. fat, and costs 60¢. Each unit of food B contains 5 oz. protein, 6 oz. carbohydrate, and 2 oz. fat, and costs $1.25. Each unit of food C contains 3 oz. protein, 4 oz. carbohydrate, and 5 oz. fat, and costs 50¢. The mixture must contain at least 20 oz. protein, 18 oz. carbohydrate, and 30 oz. fat. It is desired to find the mixture of minimal total cost, subject to these constraints.

Letting x, y, and z be the amounts of foods A, B, and C, respectively, and letting w represent the total cost of the mixture, we can summarize the data of the problem in a table (Table VIII.1.1).

TABLE VIII.1.1

Food	Amount	Protein	Carbohydrate	Fat	Cost ($)
A	x	3	2	6	0.60
B	y	5	6	2	1.25
C	z	3	4	5	0.50
Total		≥ 20	≥ 18	≥ 30	

We may express the problem in the following form:
Minimize

$$w = 0.60x + 1.25y + 0.50z$$

Subject to

$$\begin{aligned} 3x + 5y + 3z &\geq 20 \\ 2x + 6y + 4z &\geq 18 \\ 6x + 2y + 5z &\geq 30 \\ x,\ y,\ z &\geq 0 \end{aligned}$$

A problem such as this, of maximizing or minimizing a linear function of the variables, subject to linear constraints (equations or, more commonly, inequalities), is called a *linear program.* It is not possible, at this point, to solve the example since we have not yet developed the necessary mathematical tools. We will, therefore, give a strict mathematical analysis of the subject and come back to the example later on.

The science of linear programming is a relatively new branch of mathematics; it was among the many studies that were given impetus by the interest of military planners during World War II. (Prior to this war, only trial-and-error and approximation techniques were available for the solution of linear programs.) The most commonly used method nowadays, the *simplex algorithm,* was developed by G. B. Dantzig during the late 1940's; certain refinements in technique and notation as well as the theory of duality were later introduced by A. W. Tucker.

The standard form of the linear programming problem may be given as

8.1.1 Maximize

$$w = c_1x_1 + c_2x_2 + \ldots + c_nx_n$$

Subject to

8.1.2

$$\begin{aligned} a_{11}x_1 + a_{12}x_2 + \ldots + a_{1n}x_n &\leq b_1 \\ a_{21}x_1 + a_{22}x_2 + \ldots + a_{2n}x_n &\leq b_2 \\ \cdots\cdots\cdots\cdots\cdots\cdots\cdots\cdots \\ a_{m1}x_1 + a_{m2}x_2 + \ldots + a_{mn}x_n &\leq b_m \end{aligned}$$

8.1.3 $$x_1, x_2, \ldots, x_n \geq 0$$

or, in matrix notation:

8.1.4 Maximize

$$w = \mathbf{c}^t \mathbf{x}$$

Subject to

8.1.5 $$A\mathbf{x} \leq \mathbf{b}$$

8.1.6 $$\mathbf{x} \geq 0$$

where $A = (a_{ij})$ is the matrix of coefficients, **b**, and **c** are constant column vectors, **x** is the column vector of variables, and 0 is the zero column vector. (We add that vector inequalities hold whenever the inequalities hold, component by component.)

We give some definitions. The variable w, given by (8.1.1) or (8.1.4), is called the *objective function.* The inequalities (8.1.2) and (8.1.3) or (8.1.5) and (8.1.6) are the *constraints* of the problem. The set of points satisfying the constraints is called the *constraint set.* Such points are *feasible* points.

Let us suppose that there is a vector, $\mathbf{x}^*$, that satisfies the constraints and such that, if **x** is any vector in the constraint set, $c^t\mathbf{x} \leq c^t\mathbf{x}^*$. In this case we say that $\mathbf{x}^*$ is the *solution* of the program (8.1.1) through (8.1.3) or (8.1.4) through (8.1.6); while $c^t\mathbf{x}^*$ is the *value* of the program.

Strictly speaking, the program (8.1.1) through (8.1.3) is not the most general type of linear program. It may be, for example, that the objective function is to be minimized rather than maximized; that some of the constraints have opposite sign, $\geq$ instead of $\leq$; or even that some of the constraints may be equations instead of inequalities. It may also happen that some of the variables are allowed to have negative values. We can always, however, reduce a program to the standard form of (8.1.1) through (8.1.3). In fact, a function can be minimized by maximizing its negative. If an inequality has the form $\geq$, we can reverse it through the simple expedient of multiplying by -1. In turn, an equation $\alpha = \beta$, may be replaced by the two inequalities $\alpha \leq \beta$, and $-\alpha \leq -\beta$. Finally, an unrestricted variable (i.e., one that is allowed to have negative values) may be expressed as the difference of two non-negative variables. The program is, however, as general as we desire; we are concerned, mainly, with programs that have this form, though occasionally problems of other forms may be considered without reduction to the standard form.

Let us analyze the problem by looking at the geometry of the situation. The first thing to be pointed out is that the constraint set is *convex.* Geometrically, a set, S, is convex if, whenever the points P and Q both lie in S, so does the entire line-segment determined by these two points (i.e., every point that lies on the line PQ, *between* P and Q, lies in the set S). Figures VIII.1.1 (a) and (b) show examples of convex sets. On the other hand Figure VIII.1.1 (c) shows an example of a set that is not convex.

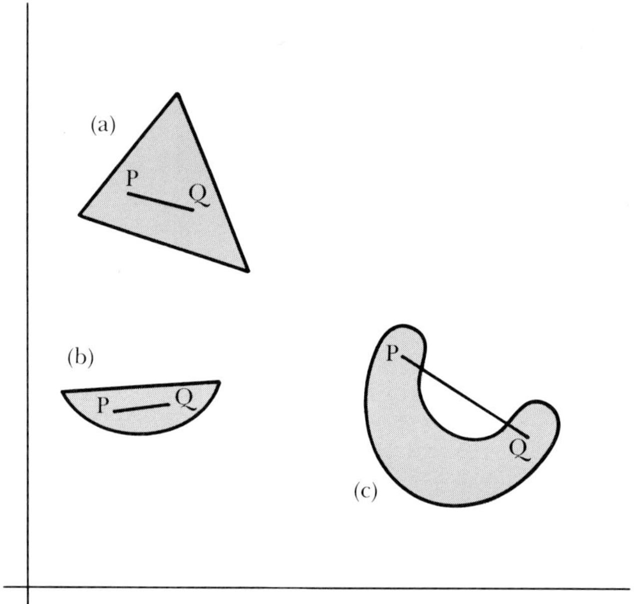

FIGURE VIII.1.1 (a) and (b) are convex; (c) is not convex.

To prove that the constraint set for a linear program is convex, we recall that each of the constraints (8.1.2) or (8.1.3) is satisfied by the set of points lying on one side of a hyperplane. It is clear that such a set is convex, since, if two points lie above a plane, then so does any point between them. Suppose that P and Q both satisfy all the constraints, and R is any point between them (Figure VIII.1.2). It is clear that R will lie on the "correct" side of each of the hyperplanes, since P and Q do. But this means that R lies in the constraint set. Thus, the constraint set is convex.

The second thing to be pointed out is that the objective function is linear:

$$w = c_1x_1 + c_2x_2 + \ldots + c_nx_n$$

Now, consider any line in n-dimensional space. As we know from previous study, such a line can be expressed by making one of the variables, say x_1, arbitrary and expressing all the other variables in terms of it:

8.1.7 $$x_j = \alpha_j x_1 + \beta_j$$

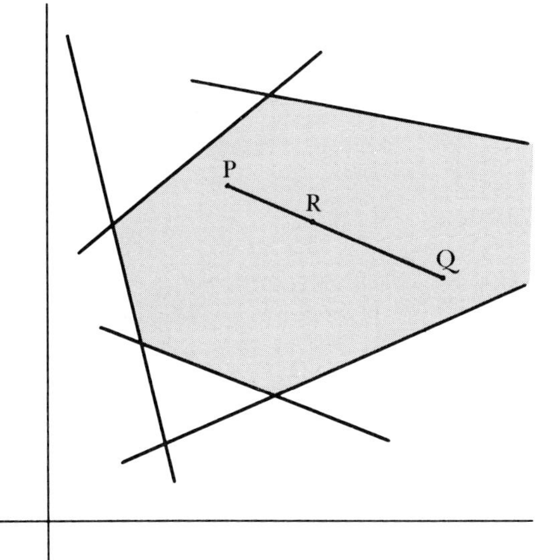

FIGURE VIII.1.2 The set of points that satisfy a collection of linear inequalities is convex.

where the α_j and β_j are constants, for each value of $j = 2,3, \ldots, n$. If we substitute (8.1.7) into (8.1.1), we obtain

$$w = \left(c_1 + \sum_{j=2}^{n} c_j\ \alpha_j\right) x_1 + \sum_{j=2}^{n} \beta_j$$

or, more compactly,

8.1.8 $$w = \alpha x_1 + \beta$$

where α and β depend only on the particular line considered.

Considering Figure VIII.1.2 once again, we see that, if the value of the coordinate x is greater at Q than at P, then it must increase steadily from P to Q. Now, consider the objective function, w, which, as we have seen, can be expressed *along the line PQ* by (8.1.8). If $\alpha = 0$, then w is constant, equal to β, along this line. If $\alpha > 0$, then w increases whenever x_1 increases, and so it either increases steadily or decreases steadily along PQ. If $\alpha < 0$, then w decreases whenever x_1 increases, and once again, it either decreases or increases steadily along PQ. The important thing is that, along any given line, the function w can do any one of these three things: remain constant, increase steadily, or decrease steadily. It cannot do anything else: it cannot, say, increase from P to R and then decrease from R to Q, or be constant from P to R and then increase from R to Q.

The importance of these observations can best be seen from the following theorem. We already know, from Chapter II, that the constraint set is a polyhedral set; we have seen how its vertices, or extreme points, can be found.

VIII.1.2 Theorem. The maximum of the objective function (8.1.1) over the constraint set, if such a maximum exists, will be attained at one of the extreme points of the constraint set.

We will not give a strict mathematical proof of this theorem here. The general idea of the proof runs as follows: first, for any point, P, we shall let $w(P)$ denote the value of the objective function, w, at the point P. Now, suppose Q is the point that maximizes the function, i.e., Q lies in the constraint set, and, for any P in the constraint set, $w(P) \leq w(Q)$. If Q is an extreme point, the theorem is proved. If not, then there will be an extreme point, P, of the constraint set, such that the line PQ has points, such as R, which lie "beyond" Q (i.e., Q lies between P and R) but still inside the constraint set (Figure VIII.1.3).

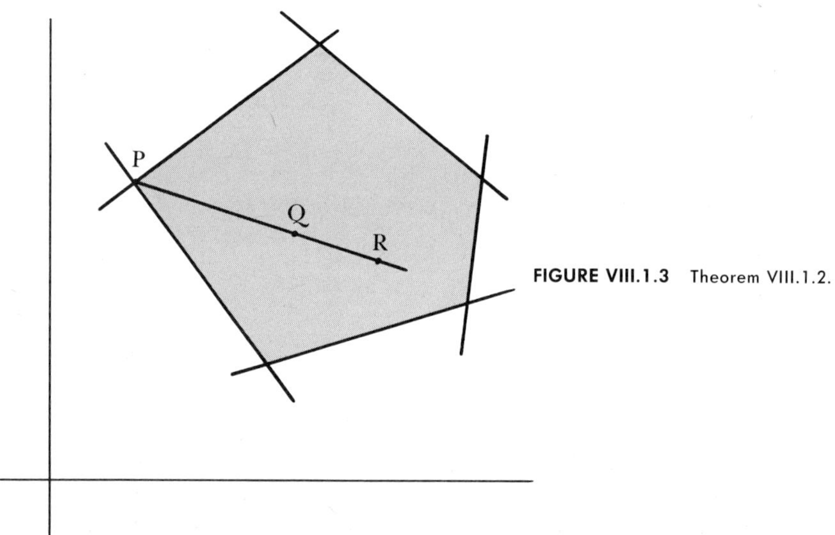

FIGURE VIII.1.3 Theorem VIII.1.2.

Suppose, now, that $w(P) < w(Q)$. By the linearity of w, we know that if w increases from P to Q, it must continue to increase from Q to R, and so $w(Q) < w(R)$. But this contradicts the assumption that Q gave the maximum value of w. It follows that we cannot have $w(P) < w(Q)$; we must have $w(P) = w(Q)$. Thus, P also gives the maximum value for w. Since P is extreme, this proves the theorem.

In a sense, Theorem VIII.1.2 gives us a method for solving the linear program (8.1.1) to (8.1.3). Since the maximum is to be found at an extreme point, all that we have to do is consider all the extreme points, find the value of the objective function at each of them, and take the one that gives the best value. As there are only a finite number of extreme points, it is clear that this method will give us the solution in a finite number of steps.

There are, unfortunately, two difficulties with this. The first is conceptual: we have, generally, no guarantee that the maximum exists, and, if it does not, Theorem VIII.1.2 will not tell us so. The second difficulty is practical: the number of extreme points, though finite, can be quite large, and finding them all might be beyond the scope of even the largest computers in use today. Nevertheless, we give examples of this method of solution of linear programs. We repeat that it is useful only in very small programs (i.e., those with only a few variables and constraints).

VIII.1.3 Example

Maximize

$$w = 2x + y + 3z$$

Subject to

$$\begin{aligned} x + 2y + z &\leq 25 && (i_1) \\ 3x + 2y + 2z &\leq 30 && (i_2) \\ x &\geq 0 && (i_3) \\ y &\geq 0 && (i_4) \\ z &\geq 0 && (i_5) \end{aligned}$$

We shall, as in Chapter II, look for the extreme points of the constraint set. (It is for this reason that we have numbered the constraints.) As before, we take the constraints, three at a time, and solve them as equations; taking constraints (i_1, i_2, i_3), we obtain the system

$$\begin{aligned} x + 2y + z &= 25 \\ 3x + 2y + 2z &= 30 \\ x &= 0 \end{aligned}$$

which has the solution (0,10,5). We check, then, that the remaining constraints are satisfied; since they are, we evaluate the objective function, which has the value of $2 \cdot 0 + 10 + 3 \cdot 5 = 25$ at this point. Continuing in this way with each combination of three inequalities, we obtain Table VIII.1.2.

TABLE VIII.1.2

Constraints	Point	Check	w
1,2,3	(0,10,5)	Yes	25
1,2,4	(−20,0,45)	No	
1,2,5	(5/2,45/4,0)	Yes	65/4
1,4,5	(25,0,0)	No	
2,3,4	(0,0,15)	Yes	45
2,3,5	(0,15,0)	No	
1,3,4	(0,0,25)	No	
1,3,5	(0,25/2,0)	Yes	25/2
2,4,5	(10,0,0)	Yes	20
3,4,5	(0,0,0)	Yes	0

We see from the table that, of the six extreme points of the constraint set, the point (0,0,15) gives the best value for the objective function, $w = 45$. We conclude that, if the program has a solution, it is this point.

The program does have a solution. It can be seen, multiplying the first constraint by 3, that $3x + 6y + 3z \leq 75$. Since the variables x,y are non-negative, it follows that $w \leq 75$, i.e., w cannot become arbitrarily large. From this and the fact that the constraint set contains its boundary points, it can be shown that the function w will have a maximum. We have, thus, solved the program: the solution is

$$\begin{aligned} x &= 0 \\ y &= 0 \\ z &= 15 \\ w &= 45 \end{aligned}$$

VIII.1.4 Example

Maximize

$$w = x + 2y$$

Subject to

$$\begin{aligned} -2x + y &\leq 10 && (i_1) \\ x - 2y &\leq 6 && (i_2) \\ x &\geq 0 && (i_3) \\ y &\geq 0 && (i_4) \end{aligned}$$

Once again, we take the constraints two at a time and solve as equations. We obtain Table VIII.1.3.

TABLE VIII.1.3

Constraints	Point	Check	w
1,2	(−26/3,−22/3)	No	
1,3	(0.10)	Yes	20
1,4	(−5,0)	No	
2,3	(0,−3)	No	
2,4	(6,0)	Yes	6
3,4	(0,0)	Yes	0

We see from the table that, if the program has a solution, it must be the point (0,10), which gives a value of 20 for w. Let us, however, look at Figure VIII.1.4, which shows the constraint set of the program. We see here that the constraint set is not bounded on the upper righthand side; in fact, for any positive value of x, the point (x,x) lies in the constraint set. But this means that $w = x + 2y$ can be made arbitrarily large; the program fails to have a solution. In this case, our method of looking at the extreme points fails us.

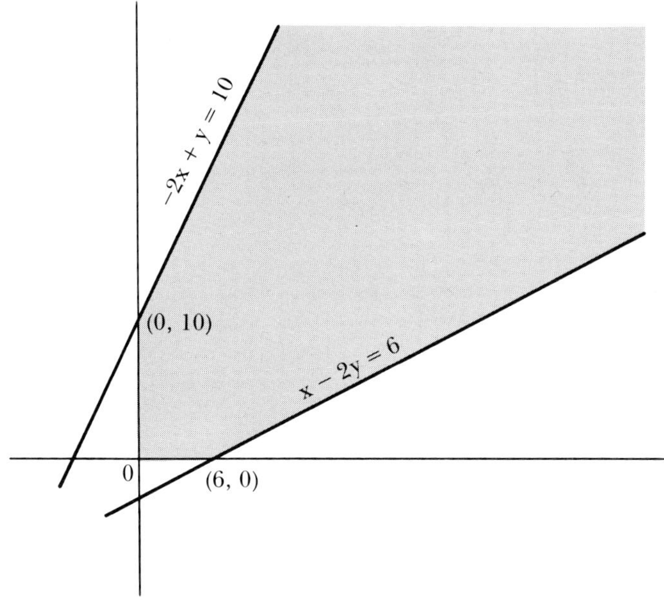

FIGURE VIII.1.4 Constraint set for Example VIII.1.4.

PROBLEMS ON LINEAR PROGRAMMING

1. Make a sketch of the constraint set in each of these problems.

(a) Maximize

$$3x+5y$$

Subject to

$$\begin{aligned} x + 3y &\leq 12 \\ 2x + y &\leq 10 \\ x + y &\leq 4 \\ x &\geq 0 \\ y &\geq 0 \end{aligned}$$

(b) Maximize

$$2x + y$$

Subject to

$$\begin{aligned} x + 3y &\leq 12 \\ 2x + y &\leq 10 \\ x + y &\leq 4 \\ x &\geq 0 \\ y &\geq 0 \end{aligned}$$

(c) Minimize

$$3x + 2y$$

Subject to

$$\begin{aligned} x + 3y &\geq 6 \\ 2x + y &\geq 3 \\ x &\geq 0 \\ y &\geq 0 \end{aligned}$$

(d) Maximize

$$3x - y$$

Subject to

$$\begin{aligned} x + 2y &\geq 4 \\ x + y &\leq 12 \\ 2x + 4y &\leq 30 \\ x &\geq 0 \\ y &\geq 0 \end{aligned}$$

2. A dietitian must prepare a mixture of two foods A and B. Each unit of food A contains 10 gm. of protein and 20 gm. of carbohydrate, and costs 15¢. Each unit of food B contains 20 gm. of protein and 25 gm. of carbohydrate, and costs 20¢. What is the cheapest mixture that can be obtained, subject to the constraint that it must contain at least 500 gm. of protein and 800 gm. of carbohydrate?

3. A nut company has 600 lb. of peanuts and 400 lb. of walnuts. It can sell the peanuts alone at 20¢ per lb.; it can also mix the peanuts and walnuts in a ratio of three parts peanuts and one part walnuts, or in a ratio of one part peanuts and two parts walnuts. The first mixture sells at 35¢ per lb., and the third at 50¢ per lb. How much of each mixture should it produce to maximize sales revenue?

4. A food processing company has 1000 lb. of African coffee, 2000 lb. of Brazilian coffee, and 500 lb. of Colombian coffee. It produces two grades of coffee. Grade A is a mixture of equal parts of African and Brazilian coffees, and sells for 60¢ per lb. Grade B is a mixture of three parts Brazilian and one part Colombian coffee, and sells for 85¢ per lb. How much of each should it produce to maximize sales revenue?

2. THE SIMPLEX ALGORITHM: SLACK VARIABLES

As we have seen from the examples in Section VIII.1, as a method of solving a linear program, the process of enumerating vertices is both impractical and uncertain. It is uncertain because we generally have no guarantee that the program does, indeed, have a solution; it may just possibly be unbounded, and the method of enumeration will not tell us when this happens. It is impractical because, for a large program, the process of enumerating the extreme points can be extremely long.

To take an example, a program with 6 variables and 10 constraints (including the non-negativity constraints) would require us to solve a total of $\binom{10}{6} = 120$ systems of 6 equations in 6 unknowns.

Even if we could be certain of having the answer when we finished, the amount of work required would necessitate using a medium-sized computer to obtain the solution with reasonable speed. The trouble is that enumeration of the combinations of constraints, while it may be systematic, is not to the point. A need arises, therefore, for a method of enumeration that will be both systematic *and* to the point (i.e., that will discard as many vertices as possible without more than a passing glance, if that much). The *simplex algorithm* is such a method.

To understand more fully the point of the simplex algorithm, let us return to the geometric study of the program. As we have pointed out several times, the constraint set for a linear program will be a convex polyhedral set. We shall say that two vertices (extreme points) of this set are *adjacent extreme points* if they are joined by one of the edges of the polyhedron. In Figure VIII.2.1, the vertices A and B are adjacent, as are A and C, or C and D. The vertices B and C, however, are not adjacent; neither are D and E, nor C and F.

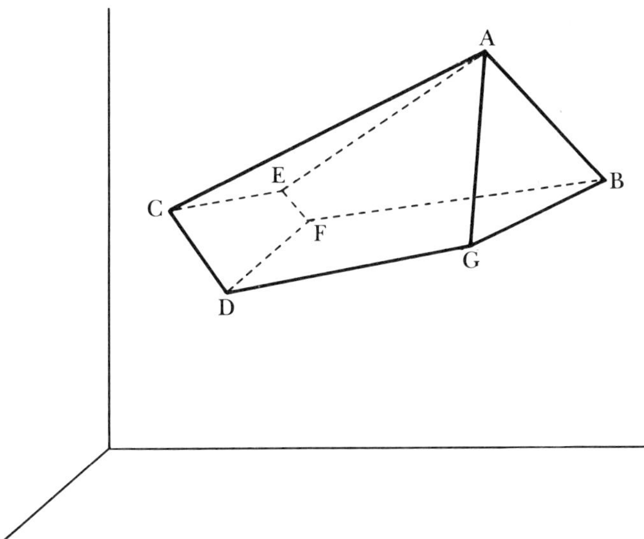

FIGURE VIII.2.1 Adjacent and non-adjacent vertices in a polyhedron.

The simplex algorithm, as we shall see, is a procedure that allows us to move from one extreme point of the constraint set to an *adjacent* extreme point, in such a way that the value of the objective function is improved at each step, until we find the solution of the program (or until we see that it has no solution). As such, it is a very effective method of implicit enumeration of the extreme points—it automatically discards a great many points before they are even considered.

The mathematical justification for the simplex algorithm is given by the following theorem:

VIII.2.1 Theorem. Let P be an extreme point of the constraint set of the program (8.1.1) through (8.1.3). Then, if P does not maximize the objective function w, there is an edge of the constraint set, starting at P, along which the objective function increases.

Again, we will not give a strict proof. The general idea of the proof, for a two-dimensional set, is as follows. Since P does not maximize w, there is a point in the constraint set, say Q, such that $w(Q) > w(P)$ (see Figure VIII.2.2). By the convexity of the constraint set, and the linearity of the objective function, we know that if R is on the line-segment PQ, then R is in the constraint set, and $w(R) > w(P)$. Through R, a line may be drawn cutting the two edges that begin at P at the points S and S'.

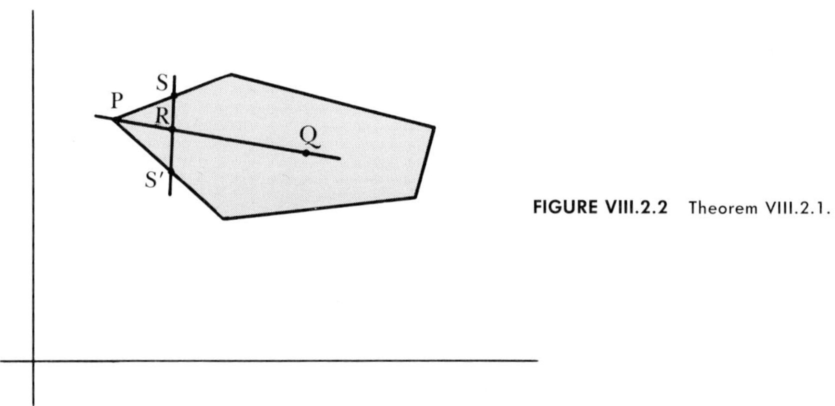

FIGURE VIII.2.2 Theorem VIII.2.1.

Now, if $w(S) \geq w(R)$, we have $w(S) > w(P)$, and so w increases along the edge PS. If, on the other hand, $w(S) < w(R)$, then we must (by linearity) have $w(S') > w(R)$, and so $w(S') > w(P)$, so that w increases along the edge PS'. Thus w will increase along one, at least, of the two edges that begin at P.

For higher dimensions, the proof is similar, though somewhat greater care must be observed in the arguments. The theorem is true in any case.

Theorem VIII.2.1 tells us that the procedure just outlined, considering only extreme points adjacent to those that have already been obtained, will always bring us to the maximum if it exists. At the same time, if the objective function is not bounded, we will always find an edge along which the function increases without bound.

One of the difficulties in dealing with linear programs is that constraints of the type (8.1.2) are generally difficult to handle; this is something that we saw in Chapter II. On the other hand, equations are comparatively easy to handle, as are the non-negativity constraints (8.1.3). It follows that the program will become that much easier to solve if we can replace the constraints (8.1.2) by equations or by constraints of the type (8.1.3). This can be effected by introducing so-called *slack variables.*

Let us consider one of the constraints (8.1.2):

8.2.1 $$a_{i1}x_1 + a_{i2}x_2 + \ldots + a_{in}x_n \leq b_i$$

This can be restated in the form

$$a_{i1}x_1 + a_{i2}x_2 + \ldots + a_{in}x_n - b_i \leq 0$$

Since the left-hand side of this inequality is negative or zero, we can write

8.2.2 $$a_{i1}x_1 + a_{i2}x_2 + \ldots + a_{in}x_n - b_i = -u_i$$

where

8.2.3 $$u_i \geq 0$$

Thus we have replaced the single constraint (8.2.1) by the equation (8.2.2) and the non-negativity constraint (8.2.3). This is what we desired to do. Proceeding in this manner with each of the constraints (8.1.2), we find that the program (8.1.1) through (8.1.3) can be rewritten in the form

8.2.4 Maximize

$$w = c_1x_1 + \ldots + c_nx_n$$

Subject to

8.2.5
$$\begin{array}{l} a_{11}x_1 + a_{12}x_2 + \ldots + a_{1n}x_n - b_1 = -u_1 \\ a_{21}x_1 + a_{22}x_2 + \ldots + a_{2n}x_n - b_2 = -u_2 \\ \cdots\cdots\cdots\cdots\cdots\cdots\cdots\cdots \\ \cdots\cdots\cdots\cdots\cdots\cdots\cdots\cdots \\ a_{m1}x_1 + a_{m2}x_2 + \ldots + a_{mn}x_n - b_m = -u_m \end{array}$$

8.2.6
$$\begin{array}{l} x_1, x_2, \ldots, x_n \geq 0 \\ u_1, u_2, \ldots, u_m \geq 0 \end{array}$$

or, in matrix notation,

8.2.7 Maximize

$$w = \mathbf{c}'\mathbf{x}$$

Subject to

8.2.8
$$\begin{array}{c} \mathbf{A}\mathbf{x} - \mathbf{b} = -\mathbf{u} \\ \mathbf{x}, \mathbf{u} \geq 0 \end{array}$$

3. THE SIMPLEX TABLEAU

We introduce next, a system of notation that simplifies the computational procedure for the simplex algorithm to a considerable extent. This is the *simplex tableau* or *schema.* In a sense, it is only a shorthand notation for a system of linear equations. The schema

8.3.1

x_1	x_2	$\dots$	x_n	1	
a_{11}	a_{12}	$\dots$	a_{1n}	$-b_1$	$=-u_1$
a_{21}	a_{22}	$\dots$	a_{2n}	$-b_2$	$=-u_2$
$\dots$	$\dots$	$\dots$	$\dots$	$\dots$	$\dots$
$\dots$	$\dots$	$\dots$	$\dots$	$\dots$	$\dots$
$\dots$	$\dots$	$\dots$	$\dots$	$\dots$	$\dots$
a_{m1}	a_{m2}	$\dots$	a_{mn}	$-b_m$	$=-u_m$
c_1	c_2	$\dots$	c_n	0	$=w$

represents the equations (8.2.4) and (8.2.5). Generally speaking, each of the rows inside the box represents a linear equation, namely, the equation obtained by multiplying each entry in the row by the corresponding entry in the top row, and then adding and setting this sum equal to the entry in the right-hand column (outside the box). Thus the first row represents the equation

$$a_{11}x_1 + a_{12}x_2 + \dots + a_{1n}x_n - b_1 = -u_1$$

which is precisely the first of the equations (8.2.5), and so on for the other rows in the tableau. The bottom row in the tableau is nothing other than the definition of the objective function, w, according to (8.2.4).

We point out that, in most linear programming texts, a different form of the simplex tableau is used. The present form is due to A. W. Tucker. We prefer it for several reasons, not the least of which is the fact that this form allows us to make use of the mathematical duality theory (discussed in Section 11, below) to the fullest. Any reader who may be familiar with other forms of the simplex tableau is asked to bear with us.

Let us, for the present, ignore the non-negativity constraints (8.2.6). The equation constraints (8.2.5) are, of course, a system of m equations in the $m + n$ unknowns x_j, u_i. The system is "solved" for the u_i in terms of the x_j, but, as we know from Chapters II and VII, it is normally possible to solve such a system for any m variables in terms of the remaining n (though in practice there may be certain combinations of m variables for which this is not possible). What we are looking for now is a method of solving successively for different combinations of the variables.

To see how this is done, let us suppose that we have, at some time, solved for the variables $r_1, r_2, \ldots, r_m$ in terms of the variables $s_1, s_2, \ldots, s_n$. We have a tableau very similar to (8.3.1):

8.3.2

s_1	s_2	$\ldots$	s_n	1	
a_{11}	a_{12}	$\ldots$	a_{1n}	$-b_1$	$=-r_1$
a_{21}	a_{22}	$\ldots$	a_{2n}	$-b_2$	$=-r_2$
$\cdots$	$\cdots$	$\cdots$	$\cdots$	$\cdots$	$\cdot$
$\cdots$	$\cdots$	$\cdots$	$\cdots$	$\cdots$	$\cdot$
$\cdots$	$\cdots$	$\cdots$	$\cdots$	$\cdots$	$\cdot$
a_{m1}	a_{m2}	$\ldots$	a_{mn}	$-b_m$	$=-r_m$
c_1	c_2	$\ldots$	c_n	δ	$= \; w$

(where, of course, the entries are different from those in (8.3.1)). For such a tableau, the m variables $r_1, \ldots, r_m$, for which we have solved, are called *basic variables*. The variables $s_1, \ldots, s_n$, in terms of which we have solved, are called *non-basic variables*. Let us suppose, next, that we want to interchange the roles of the basic variable r_i and the non-basic variable s_j, i.e., we wish to solve for $m-1$ of the r's, and s_j, in terms of $n-1$ of the s's, and r_i. Consider, then, the ith row of the tableau (8.3.2):

$$a_{i1}s_1 + a_{i2}s_2 + \ldots + a_{ij}s_j + \ldots + a_{in}s_n - b_i = -r_i$$

We can solve this for s_j, if $a_{ij} \neq 0$, obtaining

8.3.3 $$\frac{a_{i1}}{a_{ij}}s_1 + \frac{a_{i2}}{a_{ij}}s_2 + \ldots + \frac{1}{a_{ij}}r_i + \ldots + \frac{a_{in}}{a_{ij}}s_n - \frac{b_i}{a_{ij}} = -s_j$$

We must now substitute this value of s_j in the remaining equations that (8.3.2) represents. Let us take any other equation, say the kth (in which $k \neq i$):

$$a_{k1}s_1 + \ldots + a_{kj}s_j + \ldots + a_{kn}s_n - b_k = -r_k$$

Substituting (8.3.3), we obtain

$$a_{k1}s_1 + \ldots - a_{kj}\left(\frac{a_{i1}}{a_{ij}}s_1 + \ldots + \frac{1}{a_{ij}}r_1 + \ldots - \frac{b_i}{a_{ij}}\right) + \ldots + a_{kn}s_n - b_k = -r_k$$

or, collecting terms,

8.3.4 $$\left(a_{k1} - \frac{a_{kj}a_{i1}}{a_{ij}}\right)s_1 + \ldots - \frac{a_{kj}}{a_{ij}}r_i$$
$$+ \ldots + \left(a_{kn} - \frac{a_{kj}a_{in}}{a_{ij}}\right)s_n - \left(b_k - \frac{b_i a_{kj}}{a_{ij}}\right) = -r_k$$

Equations (8.3.3) and (8.3.4) tell us, now, how the simplex tableau (8.3.2) is to be changed. We can summarize them as follows:

8.3.5 The variables r_i and s_j are interchanged.

8.3.6 The entry a_{ij}, called the *pivot* of the transformation, is replaced by its reciprocal, $1/a_{ij}$.

8.3.7 The other entries a_{il}, or $-b_i$ in the pivot row, are replaced by a_{il}/a_{ij}, or $-b_i/a_{ij}$, respectively.

8.3.8 The other entries a_{kj}, or c_j in the pivot column, are replaced by $-a_{kj}/a_{ij}$ or $-c_j/a_{ij}$ respectively.

8.3.9 The other entries in the tableau, a_{kl}, $-b_k$, c_l, or δ, are replaced by $a_{kl} - a_{kj}a_{il}/a_{ij}$, $-(b_k - b_i a_{kj}/a_{ij})$, $c_l - c_j a_{il}/a_{ij}$, or $\delta + b_i c_j/a_{ij}$, respectively.

A transformation such as that outlined in (8.3.5) to (8.3.9) is called a *pivot step*, or *pivot transformation*. We give the rules in the schematic form

8.3.10
$$\begin{array}{|cc|}\hline p & q \\ & \\ r & s \\ \hline\end{array} \rightarrow \begin{array}{|cc|}\hline \dfrac{1}{p} & \dfrac{q}{p} \\ & \\ -\dfrac{r}{p} & s - \dfrac{qr}{p} \\ \hline\end{array}$$

The schema (8.3.10) is to be interpreted as follows: p is the pivot, q is any other entry in the pivot row, r is any other entry in the pivot column, and s is the entry in the row of r and column of q. The arrow shows how the pivot step affects these entries: the pivot is replaced by its reciprocal; other entries in the pivot row are divided by the pivot; other entries in the pivot column are divided by the negative of the pivot. The entry s is replaced by $s - qr/p$, where p is the pivot and q and r are the two entries that form a rectangle with s and p (i.e., q is in the row of p and column of s, while r is in the column of p and row of s).

An example of such a pivot step follows.

VIII.3.1 *Example.* Solve the system

$$\begin{aligned} r &= x + 2y - 3 \\ s &= 2x + 5y + 1 \end{aligned}$$

for x and y in terms of r and s.

There are actually two ways of doing this: one is by means of row operations on the matrix of coefficients; the other, by pivot steps. We shall use the latter method.

We form the tableau

$$\begin{array}{ccc|l} x & y & 1 & \\ \hline -1^* & -2 & 3 & = -r \\ -2 & -5 & -1 & = -s \end{array}$$

We wish to interchange the roles of x and y with those of r and s. This means we must carry out two pivot steps. In the first, we pivot on the starred entry to interchange the variables x and r, thus obtaining

$$\begin{array}{ccc|l} r & y & 1 & \\ \hline 1/-1 & -2/-1 & 3/-1 & = -x \\ -(-2/-1) & -5 - \dfrac{(-2)(-2)}{-1} & -1 - \dfrac{3(-2)}{-1} & = -s \end{array}$$

or, carrying out the operations,

$$\begin{array}{ccc|l} r & y & 1 & \\ \hline -1 & 2 & -3 & = -x \\ -2 & -1^* & -7 & = -s \end{array}$$

This gives us x and s in terms of r and y. To interchange s and y, we pivot on the starred entry to obtain

$$\begin{array}{ccc|l} r & s & 1 & \\ \hline -5 & 2 & -17 & = -x \\ 2 & -1 & 7 & = -y \end{array}$$

or, in equation form,

$$\begin{aligned} x &= 5r - 2s + 17 \\ y &= -2r + s - 7 \end{aligned}$$

This could also have been done by means of row operations, as in Chapter VII: we have the system

$$\begin{aligned} -x - 2y + r \qquad &= -3 \\ -2x - 5y \qquad + s &= 1 \end{aligned}$$

which gives us the augmented matrix

$$\left(\begin{array}{cccc|c} -1 & -2 & 1 & 0 & -3 \\ -2 & -5 & 0 & 1 & 1 \end{array}\right)$$

 We multiply the first row by −1, and clear the first column:

$$\left(\begin{array}{cccc|c} 1 & 2 & -1 & 0 & 3 \\ 0 & -1 & -2 & 1 & 7 \end{array}\right)$$

Now we multiply the second row by −1, and clear the second column, obtaining

$$\left(\begin{array}{cccc|c} 1 & 0 & -5 & 2 & 17 \\ 0 & 1 & 2 & -1 & -7 \end{array}\right)$$

which gives us the desired solution. It may be checked that, as the row operations are carried out, the matrices obtained at each step contain an identity matrix formed by two columns. The remaining columns are the same, except for a possible interchange of columns, as in the simplex tableaux obtained previously (the only difference being in the signs of the last column). In fact, the pivot steps are nothing other than a condensed format for the row operations.

VIII.3.2 Example. Interchange the roles of s and y in the tableau

x	y	1	
1	4	−1	$= -r$
5	-9^*	2	$= -s$
6	−5	−3	$= -t$

In this case, the pivot is the entry −9, in the y-column and s-row. The new tableau will be

x	s	1	
$1 - 4 \cdot 5/-9$	$-4/-9$	$-1 - 4 \cdot 2/-9$	$= -r$
$5/-9$	$1/-9$	$2/-9$	$= -y$
$6 - 5(-5)/-9$	$-(-5)/-9$	$-3 - 2(-5)/-9$	$= -t$

or

x	s	1	
$\frac{29}{9}$	$\frac{4}{9}$	$\frac{-1}{9}$	$= -r$
$\frac{-5}{9}$	$\frac{-1}{9}$	$\frac{-2}{9}$	$= -y$
$\frac{29}{9}$	$\frac{-5}{9}$	$\frac{-37}{9}$	$= -t$

PROBLEMS ON PIVOT TRANSFORMATIONS

1. In each of the following problems, solve the given system of equations for x, y, and z in terms of the other variables.

(a)
$$\begin{aligned} 3s + 2y - z + 4 &= -x \\ s - 2y + 4z - 2 &= -t \\ 4s + 3y - 2z + 1 &= -u \end{aligned}$$

(b)
$$\begin{aligned} 2x + 3y - 5 &= -z \\ x + 2y - 3 &= -s \\ 3x + 7y + 1 &= -t \end{aligned}$$

(c)
$$\begin{aligned} x - 2s + 5z - 1 &= -y \\ 3x + s - 5z + 3 &= -t \\ -x + 4s + 2z + 1 &= -r \end{aligned}$$

(d)
$$\begin{aligned} 4x + 3y - 2z - 5 &= -r \\ x + 2y - 3z + 3 &= -s \\ 2x + y + 2z + 6 &= -t \end{aligned}$$

(e)
$$\begin{aligned} 3x + 2y + 6s + 1 &= -r \\ x + y + 2s + 4 &= -z \\ -2x - y + 3s - 5 &= -t \end{aligned}$$

(f)
$$\begin{aligned} 5r + x - 2y + 4 &= -z \\ 3r + x + y - 6 &= -s \\ r - 5x + 2y + 7 &= -t \end{aligned}$$

(g)
$$\begin{aligned} x + y + 3z - 2 &= -s \\ 2x + y - 2z + 4 &= -t \\ 5x + 2y - z + 6 &= -u \end{aligned}$$

4. THE SIMPLEX ALGORITHM: OBJECTIVES

Now that we have seen what a pivot step is, we have to decide what to do with it. We must remember that, essentially, it is nothing other than a method of rewriting the system of equations (8.2.5) to obtain a different but equivalent system. Let us suppose that, after one or several pivot steps, a tableau such as (8.3.2) is reached, which has the property that all the entries in the right-hand column, except possibly the bottom entry, are non-positive; that is, for each value of i, $b_i \geqslant 0$. Since (8.3.2) is just a "solution" of the equations (8.2.5) for the basic variables $r_1, \ldots, r_m$ in terms of the non-basic variables $s_1, \ldots, s_n$, we may assign values arbitrarily to the non-basic variables. We might let each of the non-basic variables equal zero. This gives us the values

8.4.1 $$s_j = 0 \quad \text{for } j = 1, \ldots, n$$

8.4.2 $$r_i = b_i \quad \text{for } i = 1, \ldots, m$$

as a particular solution to the system (8.2.5). But we are assuming that $b_i \geqslant 0$. It follows that these values of the variables satisfy the non-negativity constraints (8.2.6) as well as the equations (8.2.5) and thus give us a point in the constraint set of the program (8.2.4) to (8.2.6).

Let us suppose, further, that each entry in the bottom row, except possibly the right-hand entry, is non-positive, i.e., for each value of j, $c_j \leqslant 0$. The bottom row of (8.3.2) says simply that

8.4.3 $$w = c_1 s_1 + c_2 s_2 + \ldots + c_n s_n + \delta$$

We know that a feasible point is obtained by letting each s_j equal zero. If we do so, we find that $w = \delta$, i.e., this feasible point gives a value of δ for the objective function. On the other hand, each term on the right side of (8.4.3) except the last is the product of a non-positive constant c_j and a variable s_j that is non-negative for any feasible point. Thus, for any feasible point, the right side of (8.4.3) is the sum of n non-positive terms $c_j s_j$, and δ. It follows that we must have $w \leq \delta$. But the point (8.4.1) and (8.4.2) gives a value of δ, and is thus the *solution* of the program (8.2.4) to (8.2.6); it gives the maximum of the objective function.

We now know what we want to accomplish by means of the pivot steps: we wish to make every entry in the right-hand column (except possibly the bottom entry) non-positive; we wish to make every entry in the bottom row (except possibly the right-hand entry) non-positive. Of course, if we wish to minimize the objective function, we will try to make the entries in the bottom row non-negative; except for this, there is no great difference between maximization and minimization problems. To see how this can best be done, we shall look at the geometric situation. First, however, we give some definitions.

VIII.4.1 Definition. A simplex tableau is associated with a point in $(m+n)$-dimensional space, namely, the point obtained by setting the n non-basic variables equal to zero. Such a point is called a *basic point* of the program. If all the entries in the right-hand column (except possibly the bottom entry) are non-positive, then the point is a *basic feasible point* (b.f.p). Two basic points are said to be *adjacent* if the corresponding simplex tableaux can be obtained from each other by a single pivot step.

Geometrically, the situation is as follows: each of the planes $x_j = 0$ or $u_i = 0$ is one of the bounding hyperplanes of the constraint set. The intersection of n of these determines a basic point of the program. If the remaining constraints are all satisfied, we have (as we saw in Chapter VII) a vertex of the constraint set, so that a b.f.p. is simply one of the vertices of the constraint set. An edge of the constraint set is determined by setting $n-1$ of the variables equal to zero, so that two vertices will lie on a common edge if they have $n-1$ of their non-basic variables in common, i.e., if their tableaux can be obtained from each other by a single pivot step. The idea of adjacent b.f.p.'s corresponds to the geometric idea of adjacent vertices.

5. THE SIMPLEX ALGORITHM: CHOICE OF PIVOTS

The process of solving a linear program through the simplex method generally can be thought of as consisting of two stages. In Stage I, the entries in the right-hand column are made non-negative to obtain a basic feasible point. In Stage II, while care is exercised to maintain a b.f.p. at all times, the entries in the bottom row are made non-positive (or non-negative in case of a minimization problem) to obtain the solution. The two stages have to be considered separately since the objectives in the two cases are distinct. Let us consider Stage II first; the rules for dealing with Stage I can be adapted from those for Stage II.

For Stage II, Figure VIII.5.1 gives a two-dimensional illustration of what we are trying to do. The vertex A lies at the intersection of $x=0$ and $r=0$; i.e., its non-basic variables are r and x. From this point, suppose we see that the objective function increases along the edge AB, i.e., by letting x increase while r remains at zero. We improve the value of the function, then, if we interchange x with one of the basic variables, y,s,t. But which one? Clearly not y, since this would cause

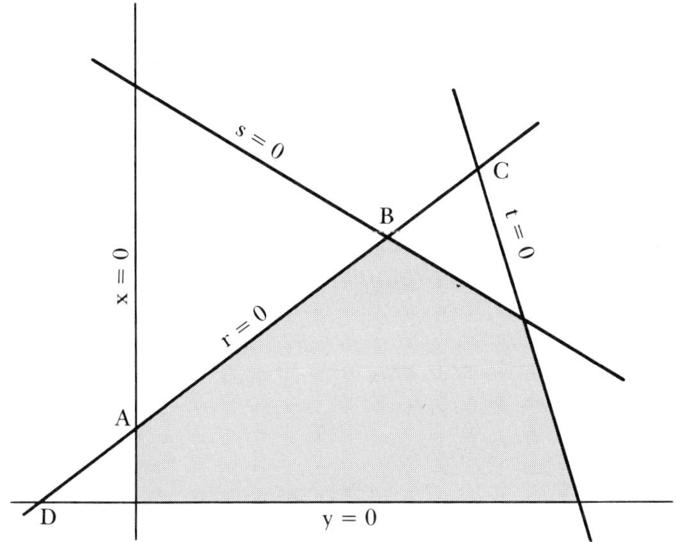

FIGURE VIII.5.1 Movement from *A* to *B*.

x to decrease (giving the point D). Not t either, since this would cause x to increase too much; it would give us the point C, at which s is negative. We conclude that we should interchange x and s, since, of those interchanges that cause x to increase, this is the one that causes the smallest increase. We thus obtain the adjacent vertex, B, and continue from B in the same manner.

Accordingly, we obtain the following rules:

VIII.5.1 Rules for Choosing a Pivot (Stage II). Let c_j be any positive entry (other than the right-hand entry) in the bottom row of the tableau. This gives us the pivot column. For each *positive* entry a_{ij} in this column, form the quotient b_i/a_{ij} (obtained by dividing a_{ij} into the negative of the corresponding entry in the right-hand column). Let b_k/a_{kj} be the smallest of these quotients. Then a_{kj} will be the pivot. (If there is a tie for the smallest quotient, the pivot may be chosen arbitrarily from among those that tie.)

The rules VIII.5.1 are for a maximization problem. If we are required to minimize the objective function, the procedure is changed only in that we take c_j to be a negative entry; otherwise, the choice of pivot is the same.

We shall not, at this moment, prove that the rules VIII.5.1 work in the sense that they will eventually give us the solution of the program if such a solution exists. We give, instead, a numerical example.

VIII.5.2 Example. A company produces three types of toys, A, B, and C, by using three machines called a forge, a press, and a painter.

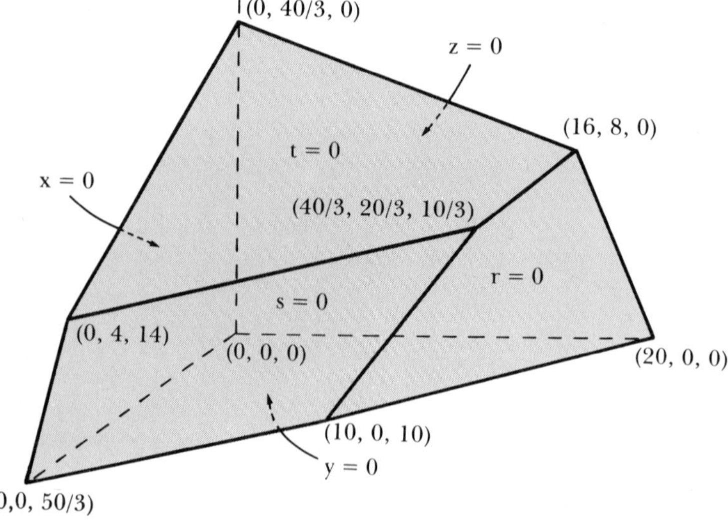

FIGURE VIII.5.2 Constraint set for Example VIII.5.2. Note that each bounding plane corresponds to one of the variables.

Each toy of type A must spend 4 hr. in the forge, 2 hr. in the press, and 1 hr. in the painter, and yields a profit of \$5. Each toy of type B must spend 2 hr. in the forge, 2 hr. in the press, and 3 hr. in the painter and yields a profit of \$3. Each toy of type C spends 4 hr. in the forge, 3 hr. in the press, and 2 hr. in the painter and yields a profit of \$4. In turn, the forge may be used up to 80 hr. per week; the press, up to 50 hr., and the painter up to 40 hr. How many of each type of toy should be made in order to maximize the company's profits?

Letting x, y, and z be the amounts produced of toys A, B, and C, respectively, we obtain the linear program:

Maximize

$$5x + 3y + 4z = w$$

Subject to

$$\begin{aligned} 4x + 2y + 4z &\leq 80 \\ 2x + 2y + 3z &\leq 50 \\ x + 3y + 2z &\leq 40 \\ x,y,z &\geq 0 \end{aligned}$$

Figure VIII.5.2 shows the constraint set for this problem. Introducing the slack variables r, s, and t, we obtain the tableau

x	y	z	1	
4^*	2	4	-80	$= -r$
2	2	3	-50	$= -s$
1	3	2	-40	$= -t$
5	3	4	0	$= w$

We see that we have a basic feasible point, since the entries in the right-hand column are negative. On the other hand, there are three positive entries in the bottom row (see also Figure VIII.5.3). Let us

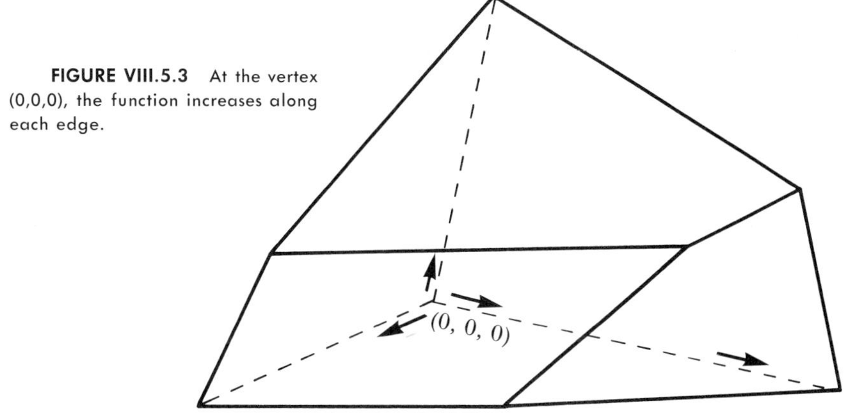

FIGURE VIII.5.3 At the vertex (0,0,0), the function increases along each edge.

choose the first column, which has one of these positive entries, as the pivot column. According to the rules VIII.5.1, we divide each of the positive entries in this column into the corresponding b_i. The corresponding quotients b_i/a_{ij} are then:

$$b_1/a_{11} = 80/4 = 20$$
$$b_2/a_{21} = 50/2 = 25$$
$$b_3/a_{31} = 40/1 = 40$$

The first row gives the smallest quotient: the entry 4 (starred) will be the pivot. This gives us the new tableau

r	y	z	1	
1/4	1/2	1	−20	$= -x$
−1/2	1	1	−10	$= -s$
−1/4	5/2*	1	−20	$= -t$
−5/4	1/2	−1	100	$= \quad w$

This tableau gives an improved b.f.p.: the profit here is $100. It is still not the maximum, since there is a positive entry in the bottom row, second column (see also Figure VIII.5.4). Once again, we consider the quotients b_i/a_{ij} for each positive a_{ij} in the second column. These quotients are, respectively, $20/(1/2) = 40$, $10/1 = 10$, and $20/(5/2) = 8$. Since the starred entry, 5/2, gives the smallest quotient, we use it as pivot. We then obtain the tableau

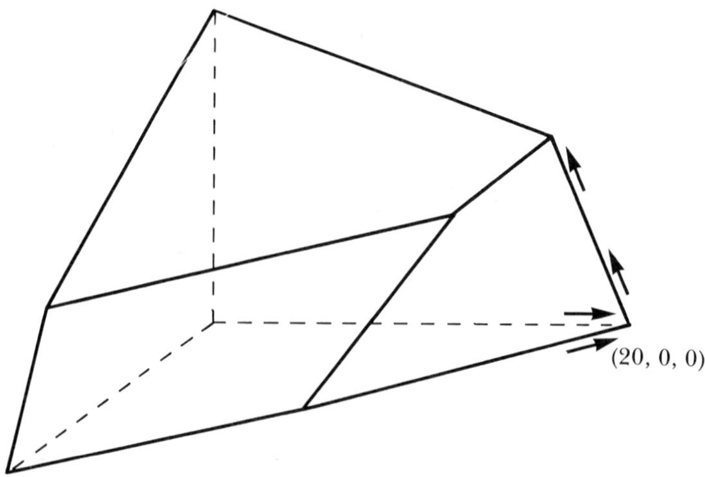

FIGURE VIII.5.4 At (20,0,0), the function increases along one of the three edges.

r	t	z	1	
3/10	−1/5	4/5	−16	$= -x$
−2/5	−2/5	3/5	−2	$= -s$
−1/10	2/5	2/5	−8	$= -y$
−6/5	−1/5	−6/5	104	$= \ w$

(see Figure VIII.5.5). This gives us the solution: it is obtained by setting the non-basic variables equal to zero, and so

$$\begin{aligned} x &= 16 \\ y &= 8 \\ z &= 0 \\ w &= 104 \end{aligned}$$

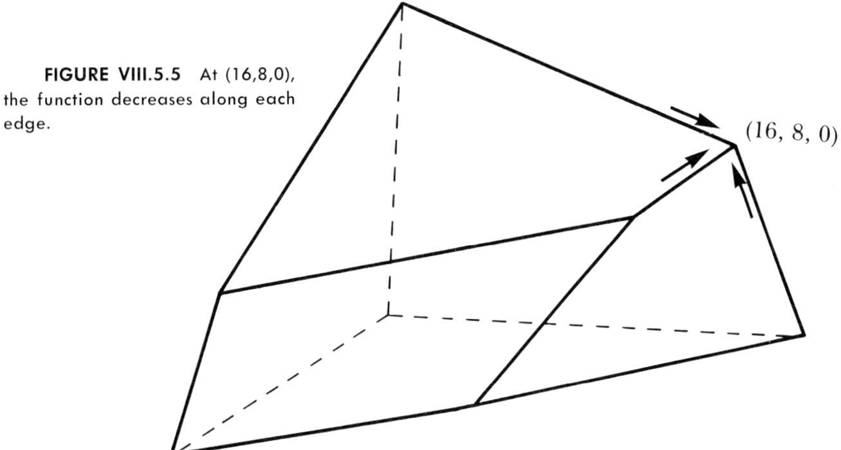

FIGURE VIII.5.5 At (16,8,0), the function decreases along each edge.

The company should produce 16 type A toys and 8 type B toys every week. Type C toys will not be produced. It may be checked directly that the constraints are satisfied; it is not possible, at least yet, to check that this is indeed the solution. Note that the press is idle 2 hr. a week; this corresponds to the value $s = 2$ for the related slack variable.

VIII.5.3 Example

Maximize

$$w = 2x + y + 3z$$

Subject to

$$\begin{aligned} x + 2y + \ z &\leq 25 \\ 3x + 2y + 2z &\leq 30 \\ x, y, z &\geq 0 \end{aligned}$$

This is, of course, the same as Example VIII.1.3. We propose to solve it, now by the simplex algorithm. Introducing slack variables we have the b.f.p.

x	y	z	1	
1	2	1	-25	$=-r$
3	2	2^*	-30	$=-s$
2	1	3	0	$=w$

We may choose any one of the columns as the pivot column. Say we choose the third column. The corresponding quotients are $25/1=25$ and $30/2=15$. Since 15 is the smaller quotient, we choose the starred entry as pivot. We obtain

x	y	s	1	
$-1/2$	1	$-1/2$	-10	$=-r$
$3/2$	1	$1/2$	-15	$=-z$
$-5/2$	-2	$-3/2$	45	$=w$

which gives the solution $x=0$, $y=0$, $z=15$, $w=45$. This is, of course, the same as was obtained by other means earlier in the chapter.

6. THE SIMPLEX ALGORITHM: RULES FOR STAGE I

In Stage I, the situation is, as we mentioned, somewhat different. The general rule is to take the several entries $-b_i$ in the right-hand column and work on them, one at a time, until all have become non-positive. Some care must be observed, of course, that, while one $-b_i$ becomes non-positive, others do not simultaneously become positive. The method for "working" on a positive entry $-b_i$, then, is as follows: if it can be made non-positive immediately without disturbing the non-positivity of the other entries in the column, we do so. If not, then we treat the $-b_i$ as a sort of "secondary objective": we try to decrease it as much as possible while keeping the other entries in the column non-positive. We have the following rules.

VIII.6.1 Rules for Choosing a Pivot (Stage I). Let $-b_l$ be the lowest positive entry (other than the bottom entry) in the right-hand column. Let a_{lj} be any negative entry in this row; this gives us the pivot column. Now, for a_{lj}, and for each *positive* entry a_{ij} *below* a_{lj} in this column, form the quotient b_i/a_{ij}. Let b_k/a_{kj} be the smallest of these quotients. Then a_{kj} is the pivot. (Once again, in case of ties, any one of those entries that tie to give the smallest quotient may be chosen as pivot.)

We shall leave the proof that these rules work, i.e., that they eventually give us a b.f.p., until later. We are now in a position to solve Example VIII.1.1, which follows.

VIII.6.2 Example

Minimize

$$w = 0.60x + 1.25y + 0.50z$$

Subject to

$$\begin{aligned} 3x + 5y + 3z &\geq 20 \\ 2x + 6y + 4z &\geq 18 \\ 6x + 2y + 5z &\geq 30 \\ x, y, z &\geq 0 \end{aligned}$$

Introducing slack variables, we have the tableau

x	y	z	1	
−3	−5	−3	20	$= -r$
−2	−6	−4	18	$= -s$
−6	−2	−5*	30	$= -t$
3/5	5/4	1/2	0	$= w$

which does not represent a b.f.p., since there are several positive entries in the right-hand column. The lowest such entry is in the third row. We take a negative entry in this row, say the −5 in the third column. According to the rules VIII.6.1, we must consider this entry and any positive entry below it except that in the bottom row. This reduces the choice to no choice at all, of course; and the −5 will be the pivot. We then obtain

x	y	t	1	
3/5	−19/5	−3/5*	2	$= -r$
14/5	−22/5	−4/5	−6	$= -s$
6/5	2/5	−1/5	−6	$= -z$
0	21/20	1/10	3	$= w$

which is still not a b.f.p. We now work on the top row: there are two negative entries here, and either one can be used to obtain the pivot column. Let us take the third column. There are no positive entries in this column, other than the bottom entry, so that the starred entry, −3/5, will be the pivot. We then obtain

x	y	r	1	
−1	19/3	−5/3	−10/3	$=-t$
2	2/3	−4/3	−26/3	$=-s$
1	5/3	−1/3	−20/3	$=-z$
1/10	5/12	1/6	10/3	$=w$

which is not only a b.f.p., but actually the solution, since all the entries in the bottom row are positive (and we want to minimize the objective function). We have then

$$\begin{aligned} x &= 0 \\ y &= 0 \\ z &= 20/3 \\ w &= 10/3 \end{aligned}$$

and the mixture desired in Example VIII.1.1 consists of 20/3 units of food of type C, and nothing else, for a total cost of \$3.33.

VIII.6.3 Example

Maximize

$$w = x + 2y + z$$

Subject to

$$\begin{aligned} x - 10y - 4z &\le -20 \\ 3x + \quad y + \quad z &\le 3 \\ x,\, y,\, z &\ge 0 \end{aligned}$$

We obtain, in this case, the tableau

x	y	z	1	
1	−10	−4	20	$=-r$
3	1	1*	−3	$=-s$
1	2	1	0	$=w$

There is a positive entry at the top of the right-hand column. We can choose the pivot column to be either the second or third column, since these are the negative entries in the top row. Let us choose the third. The pivot is then chosen from between the negative entry −4 and the positive entry below it (the bottom row does not count, of course). The corresponding quotients are −20/−4 = 5 and 3/1 = 3. Since 3 is smaller, we choose the starred entry as pivot. This gives us

x	y	s	1	
13	-6^*	4	8	$=-r$
3	1	1	-3	$=-z$
-2	1	-1	3	$=w$

which is still not a b.f.p. Note how the right-hand entry in the top row has been decreased, while the right-hand entry in the second row remains non-positive.

The only negative entry in the first row is in the second column, which is, therefore, the pivot column. The choice is between the negative entry in the top row and the positive entry below it; the corresponding quotients are $-8/-6=4/3$ and $3/1=3$, and the pivot is the starred entry. We then get the tableau

x	r	s	1	
$-13/6$	$-1/6$	$-2/3$	$-4/3$	$=-y$
$31/6$	$1/6^*$	$5/3$	$-5/3$	$=-z$
$1/6$	$1/6$	$-1/3$	$13/3$	$=w$

which is a b.f.p., but not the solution, since there are two positive entries in the bottom row. Let us choose the second column as the pivot column; the pivot will be the only positive entry in this column (other than the bottom entry) which is starred. We then get the tableau

x	z	s	1	
3	1	1	-3	$=-y$
31	6	10	-10	$=-r$
-5	-1	-2	6	$=w$

which gives us the solution: it is

$$x=0$$
$$y=3$$
$$z=0$$
$$w=6$$

7. THE SIMPLEX ALGORITHM: PROOF OF CONVERGENCE

Now that we have illustrated the two sets of rules, VIII.5.1 and VIII.6.1, for choosing pivots, we shall show that they actually work in the sense of giving us the solution if this exists and telling us that there is no solution if such is the case. We shall make, at first, an assumption of *non-degeneracy:* no more than n of the bounding hyperplanes of the original program (8.1.1) to (8.1.3) pass through a single point. This is the general case and will "almost always" happen, unless the program is of a very special type. In practice, this means that there will be no zeros in the right-hand column (except, possibly, in the bottom row).

VIII.7.1 Proof for Stage II.

In Stage II, the constant term in the pivot row, $-b_k$, is replaced by

8.7.1
$$-b'_k = -\frac{b_k}{a_{kj}}$$

By hypothesis, $-b_k \leq 0$, and $a_{kj} > 0$. Hence $-b'_k \leq 0$.

The other terms $-b_i$ in the right-hand column are replaced by

8.7.2
$$-b'_i = -b_i + \frac{b_k a_{ij}}{a_{kj}}$$

We consider two possibilities: $a_{ij} \leq 0$ and $a_{ij} > 0$. If $a_{ij} \leq 0$, then, since $b_k \geq 0$ and $a_{kj} > 0$, it will follow that the second term on the right side of (8.7.2) is non-positive. In this case, $-b'_i \leqslant -b_i$. But, by hypothesis, $-b_i \leq 0$. Hence $-b'_i \leq 0$.

If $a_{ij} > 0$, then (8.7.2) can be rewritten

8.7.3
$$-b'_i = a_{ij}\left(\frac{b_k}{a_{kj}} - \frac{b_i}{a_{ij}}\right)$$

Now, $a_{ij} > 0$, which means that it was among the terms considered as possible pivots. The entry a_{kj} was chosen as pivot, which means (by rules VIII.5.1) that

$$\frac{b_k}{a_{kj}} \leq \frac{b_i}{a_{ij}}$$

But this means that the parenthesis on the right side of (8.7.3) is non-positive. Since $a_{ij} > 0$, it follows that $-b'_i \leq 0$.

Finally, the term δ, in the lower right-hand corner of the tableau is replaced by

8.7.4
$$\delta' = \delta + \frac{b_i c_j}{a_{ij}}$$

Now, according to the rules, c_j and a_{ij} are both positive. By hypothesis, $b_i \geq 0$; under the non-degeneracy assumption, however, we cannot have $b_i = 0$, so that $b_i > 0$. This means that the second term on the right side of (8.7.4) is positive, and so $\delta' > \delta$.

We have seen, then, that following the rules VIII.5.1 will give us a new tableau that represents a b.f.p. (since all the $-b_i'$ are non-positive). Moreover, this new b.f.p. will (under the non-degeneracy assumption) give an improved value for the objective function. Thus, every time that we follow the rules, we obtain a new vertex that is an improvement over the previous one. This means that we cannot obtain the same vertex twice. As there are only a finite number of vertices, it follows that we must eventually get a tableau at which the rules VIII.5.1 cannot be followed.

Suppose, then, that it is impossible to follow the rules. This may be for either of two reasons. One is that there are no positive entries (other than the right-hand entry) in the bottom row. But this means that we have the solution.

The other possibility is that there may be a column (other than the right-hand column) with a positive entry in the bottom row, and no other positive entries, i.e., for some j, $c_j > 0$, but $a_{ij} \leq 0$ for each i. If we let all the non-basic variables, except s_j, vanish, we have

$$\begin{aligned} r_i &= b_i - a_{ij} s_j \quad \text{for each } i \\ w &= \delta + c_j s_j \end{aligned}$$

Now, we can see that, for any positive value of s_j, we will have $r_i \geq b_i$ (since $a_{ij} \leq 0$). But $b_i \geq 0$ (by hypothesis, we have a b.f.p.), and so $r_i \geq 0$. On the other hand, since $c_j > 0$, it follows that we can make w as large as desired simply by making s_j large enough. But this means that the program has no solution: it is unbounded.

We have seen that, starting from any b.f.p., the rules VIII.5.1 will eventually *either* give us the solution to the program *or* tell us that the program is unbounded. An example of the latter possibility follows.

VIII.7.2 Example

Maximize

$$w = 2x + y + z$$

Subject to

$$\begin{aligned} x - 2y - 4z &\leq -20 \\ 3x - 5y + z &\leq 3 \\ x, y, z &\geq 0 \end{aligned}$$

We obtain, in this case, the tableau

x	y	z	1	
1	−2	−4	20	$= -r$
3	−5	1*	−3	$= -s$
1	2	1	0	$= w$

We have a positive entry in the right-hand column. Taking the first column as our pivot column, the rules VIII.6.1 tell us to pivot on the starred entry, giving us the tableau

x	y	s	1	
13	−22*	4	8	$=-r$
3	−5	1	−3	$=-z$
−2	7	−1	3	$=w$

which is still not a b.f.p. We pivot on the starred entry (which is the only negative entry in the top row) obtaining

x	r	s	1	
−13/22	−1/22	−2/11	−4/11	$=-y$
1/22*	−5/22	1/11	−53/11	$=-z$
47/22	7/22	3/11	61/11	$=w$

which is a b.f.p., but not the solution, as there are several positive entries in the bottom row. If we pivot on the starred entry, we obtain the tableau

z	r	s	1	
13	−3	1	−63	$=-y$
22	−5	2	−106	$=-x$
−47	11	−4	232	$=w$

This tableau is not the solution, as there is still a positive entry in the bottom row, second column. There are, however, no other positive entries in the second column, so the rules VIII.5.1 cannot be followed. We conclude that the program does not have a solution: it is unbounded.

In fact, we may check that, for any positive value of r, the point

$$\begin{aligned} x &= 106 + 5r \\ y &= 63 + 3r \\ z &= 0 \\ w &= 232 + 11r \end{aligned}$$

satisfies the constraints. By making r large enough, w can be made as large as desired. If, for example, we want w to be greater than 1,000,000, we simply let $r = 100{,}000$; then

$$\begin{aligned} x &= 500{,}106 \\ y &= 300{,}063 \end{aligned}$$

$$z = 0$$
$$w = 1{,}100{,}232$$

is a feasible point of the program, as may be checked directly.

We return to the rules VIII.6.1. Once again, we make the non-degeneracy assumption: no zeros appear in the right-hand column (except possibly in the bottom row).

VIII.7.3 Proof for Stage I.

In Stage I, the constant term in the pivot row, $-b_k$, and the other terms in the right-hand column, $-b_i$, are replaced by $-b'_k$ and $-b'_i$, respectively, according to equations (8.7.1) and (8.7.2). It can be shown, exactly as in VIII.7.1, that, for $i > l, -b'_i \leq 0$ (i.e., the entries in the right-hand column below the lth row remain non-positive.)

As for the entry $-b_l$, it is replaced by $-b'_l$, which will be given by (8.7.1) or (8.7.2), depending on whether the pivot is in the lth row ($k = l$) or in another row ($k \neq l$).

In case the pivot is in the lth row we have

8.7.5 $$-b'_l = -\frac{b_l}{a_{il}}$$

Now, by assumption, b_l and a_{il} are both negative. This means that $-b'_l < 0$.

If the pivot is in another row, we will have

8.7.6 $$-b'_l = -b_l + \frac{b_k a_{il}}{a_{ik}}.$$

By the rules, $a_{il} < 0$ while $a_{ik} > 0$. By the non-degeneracy assumption, $b_k > 0$, since $-b_k$ is a term in the right-hand column, below $-b_l$. It follows that the second term on the right side of (8.7.6) is negative, and so $-b'_l < -b_l$.

We see, thus, that, by following the rules VIII.6.1, we obtain a new tableau that preserves the non-positivity of the terms in the right-hand column *below* the lth row. Under the non-degeneracy assumption, the term $-b_l$ (the lowest positive entry in the right-hand column) will be decreased. Using once again the fact that any program can give only a finite number of distinct tableaux, we conclude that, so long as we follow the rules VIII.6.1, the term $-b_l$ will eventually become negative. We then work on the lowest remaining positive term in the right-hand column, and it follows that, eventually, *either* a b.f.p. will be reached *or* a tableau will appear at which it is impossible to follow rules VIII.6.1.

Suppose then, that we find it impossible to follow the rules VIII.6.1, although we do not yet have a b.f.p. This will happen if a row has a positive entry in the right-hand column, and only non-negative entries otherwise, i.e., if for an i, $-b_i > 0$, and $a_{ij} \geq 0$ for all j. The ith row then represents the equation

8.7.7 $$r_i = b_i - (a_{i1}s_1 + a_{i2}s_2 + \ldots + a_{in}s_n)$$

Now, each term inside the parenthesis in the right side of (8.7.7) is the product of a non-negative constant a_{ij} and a non-negative variable s_j. It follows that the expression inside the parenthesis is non-negative, and so $r_i \leq b_i$. But $b_i < 0$, and so $r_i < 0$. But r_i must be non-negative. We have, thus, a contradiction. It means that the program is *infeasible*: the constraints cannot be satisfied.

We see, then, that the rules for Stage I will either lead us to a b.f.p. or tell us that the program is infeasible. An example of this possibility is:

VIII.7.4 Example

Maximize

$$w = x - 4y$$

Subject to

$$\begin{aligned} x + 3y &\geq 2 \\ 2x + 5y &\leq 1 \\ x + y &\leq 4 \\ x, y &\geq 0 \end{aligned}$$

We have here the tableau

x	y	1	
−1	−3	2	$= -r$
2	5*	−1	$= -s$
1	1	−4	$= -t$
1	−4	0	$= w$

which has a positive entry in the right-hand column, top row. There are two negative entries in this row: let us choose the second column as the pivot column. The negative entry in the top row and both positive entries below it have to be considered in choosing a pivot. The corresponding quotients are 2/3, 1/5, and 4. The smallest corresponds to the second row; we pivot, therefore, on the starred entry, to obtain the tableau

x	s	1	
1/5	3/5	7/5	$= -r$
2/5	1/5	−1/5	$= -y$
3/5	−1/5	−19/5	$= -t$
13/5	4/5	−4/5	$= w$

This is not a b.f.p., since there is still a positive entry (albeit smaller) in the right-hand column. In fact, however, all the entries in the top row are positive. We conclude that the program is infeasible.

8. EQUATION CONSTRAINTS

In some cases, some of the constraints in a program may be equations instead of inequalities. As we mentioned earlier, an equation may be replaced by two opposite inequalities, so that it is always possible to reduce the program to standard form. On the other hand, it should be clear that if we replace an equation by two inequalities, and then introduce two slack variables to make each of these inequalities into an equation, we are merely multiplying work for ourselves. It is better to leave the equations as they are; there is then, of course, no slack variable corresponding to the constraint. The technique is slightly different, though the same general rules apply. The following example shows this.

VIII.8.1 Example. A steel company has two warehouses, W_1 and W_2, and three distributors, D_1, D_2, and D_3. The two warehouses hold, respectively, 50,000 and 60,000 tons of steel. Of these, 30,000 tons must be shipped to D_1, 40,000 tons to D_2, and 40,000 tons to D_3. The transportation costs, in dollars per ton, between warehouses and distributors, are given in Table VIII.8.1.

TABLE VIII.8.1

From Warehouse	To Distributor		
	D_1	D_2	D_3
W_1	1	4	1
W_2	5	9	2

The required amounts must be transported in such a way as to minimize the total costs.

Letting x_{ij} represent the amount shipped from warehouse W_i to distributor D_j, we have the program

Minimize

$$w = x_{11} + 4x_{12} + x_{13} + 5x_{21} + 9x_{22} + 2x_{23}$$

Subject to

$$
\begin{array}{l}
x_{11} + x_{12} + x_{13} \qquad\qquad\qquad\qquad\qquad\qquad = 50 \\
\qquad\qquad\qquad\qquad x_{21} + x_{22} + x_{23} = 60 \\
x_{11} \qquad\qquad\qquad + x_{21} \qquad\qquad\qquad = 30 \\
\qquad x_{12} \qquad\qquad\qquad + x_{22} \qquad\quad = 40 \\
\qquad\qquad x_{13} \qquad\qquad\qquad + x_{23} = 40 \\
\qquad\qquad\qquad\qquad\qquad\qquad\qquad x_{ij} \geq 0
\end{array}
$$

where, for example, the first constraint means that the amounts shipped from W_1 to the three distributors must add up to 50, and so on for the other constraints (we are measuring in units of a thousand). Of the five equation constraints, one can be dispensed with: the fifth constraint is equal to the sum of the first two, minus the sum of the third and fourth equations. (This is, of course, because the total amounts available at the warehouses are equal to the total requirements by the distributors. If D_1 and D_2 receive what they require, then the amount left for D_3 is eactly what he requires.) Discarding the fifth equation, we get the tableau

x_{11}	x_{12}	x_{13}	x_{21}	x_{22}	x_{23}	1	
1	1	1	0	0	0	−50	= 0
0	0	0	1	1	1*	−60	= 0
1	0	0	1	0	0	−30	= 0
0	1	0	0	1	0	−40	= 0
1	4	1	5	9	2	0	$= w$

It is clear that zeros have no business in the space reserved for basic variables. To remove them, we pivot in the usual manner, say, on the starred entry. This will bring one of the zeros into the row of non-basic variables:

x_{11}	x_{12}	x_{13}	x_{21}	x_{22}	0	1	
1	1	1	0	0	0	−50	= 0
0	0	0	1	1	1	−60	$= -x_{23}$
1	0	0	1	0	0	−30	= 0
0	1	0	0	1*	0	−40	= 0
1	4	1	3	7	−2	120	$= w$

We cross out the column corresponding to the zero, since the elements in that column would merely be multiplied by zero. Pivoting on the next starred entry, we get

x_{11}	x_{12}	x_{13}	x_{21}	0	1	
1	1	1	0	0	−50	$=0$
0	−1	0	1	−1	−20	$=-x_{23}$
1	0	0	1*	0	−30	$=0$
0	1	0	0	1	−40	$=-x_{22}$
1	−3	1	3	−7	400	$=w$

Again, we cross out the column corresponding to zero, and pivot on the starred entry

x_{11}	x_{12}	x_{13}	0	1	
1	1	1*	0	−50	$=0$
−1	−1	0	−1	10	$=-x_{23}$
1	0	0	1	−30	$=-x_{21}$
0	1	0	0	−40	$=-x_{22}$
−2	−3	1	−3	490	$=w$

We repeat the procedure:

x_{11}	x_{12}	0	1	
1	1	1	−50	$=-x_{13}$
−1	−1*	0	10	$=-x_{23}$
1	0	0	−30	$=-x_{21}$
0	1	0	−40	$=-x_{22}$
−3	−4	−1	540	$=w$

Having reduced the tableau to the desired form, we now proceed to solve the problem by applying rules VIII.5.1 and VIII.6.1. We show the pivots, in each case, by asterisks.

x_{11}	x_{23}	1	
0	1	−40	$=-x_{13}$
1	−1	−10	$=-x_{12}$
1	0	−30	$=-x_{21}$
−1	1*	−30	$=-x_{22}$
1	−4	500	$=w$

x_{11}	x_{22}	1	
1^*	-1	-10	$=-x_{13}$
0	1	-40	$=-x_{12}$
1	0	-30	$=-x_{21}$
-1	1	-30	$=-x_{23}$
-3	4	380	$=w$

x_{13}	x_{22}	1	
1	-1	-10	$=-x_{11}$
0	1	-40	$=-x_{12}$
-1	1	-20	$=-x_{21}$
1	0	-40	$=-x_{23}$
3	1	350	$=w$

This last tableau gives us the minimum cost solution: it is

$$x_{11}=10 \qquad x_{12}=40 \qquad x_{13}=0$$
$$x_{21}=20 \qquad x_{22}=0 \qquad x_{23}=40$$

with a total cost of 350. (Measured in thousands, this is really \$350,000.)

Such problems, called *transportation problems*, can be solved by a considerably simpler method; we shall see this in a later section of this chapter.

9. DEGENERACY PROCEDURES

In case of degeneracy, i.e., assuming that some zeros appear in the right-hand column of the simplex tableau, the foregoing arguments do not hold. The main difficulty is that, in Stage II, for instance, the new tableau obtained need not give a better value for the objective function. But it is precisely the improvement in the objective function that guarantees that the process will terminate; in a case of degeneracy we might, with bad luck, find that the process "cycles"—it brings us back to a previous position. Similar difficulties arise in Stage I.

From the geometric point of view, degeneracy arises when more than n of the bounding planes of the constraint set pass through a point (n, here, is the dimension of the space). Figure VIII.9.1 illustrates this: it shows the set satisfying

8.9.1
$$\begin{aligned} x \quad\;\; + z &\le 1 \\ y + z &\le 1 \\ x,y,z &\ge 0 \end{aligned}$$

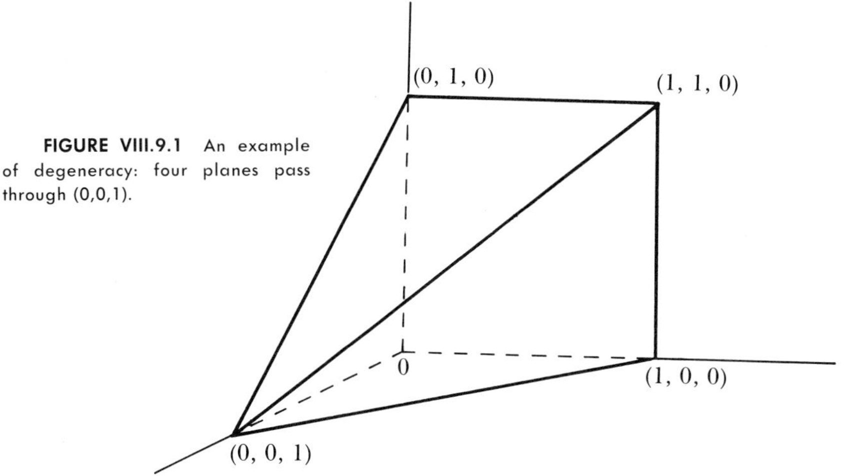

FIGURE VIII.9.1 An example of degeneracy: four planes pass through (0,0,1).

As may be seen, the planes $x + z = 1$, $y + z = 1$, $x = 0$, and $y = 0$ all pass through the point (0,0,1). This is not normal: in three-dimensional space, four planes will usually "meet," three at a time, at four distinct points.

Precisely because this situation is not normal, we can change it by what is called a *perturbation*. Let us consider the slightly different set

8.9.2
$$\begin{aligned} x \quad + z &\le 11/10 \\ y + z &\le 1 \\ x,y,z &\ge 0 \end{aligned}$$

which differs from the previous system in that the first inequality has been slightly "perturbed," i.e., the right side has been changed by 1/10. Figure VIII.9.2 shows the new constraint set: the vertex at (0,0,1) has been split into two vertices, one of which is now at (1/10,0,1). Two of the other vertices have been slightly perturbed, the others not at all.

Suppose, now, that we are given a program with (8.9.1) as constraint set. We replace the constraint set by (8.9.2) and solve. The solution of the new program will be at one of the vertices of Figure VIII.9.2. Each of these vertices corresponds to a *unique* vertex of Figure VIII.9.1: for instance, (0,1,0) corresponds to (0,1,0), while (11/10,0,0) corresponds to (1,0,0). (The converse is not true: a vertex of Figure VIII.9.1 may correspond to more than one vertex of Figure VIII.9.2.) That vertex of Figure VIII.9.1 that corresponds to the solution of the program (8.9.2) will *probably* be the solution of the program (8.9.1); if (8.9.2) has its solution at (1/10,0,1), then (8.9.1) will *probably* have its solution at (0,0,1). We say probably, because it may be that if the perturbation has been too large the solution of the perturbed problem (8.9.2) does not correspond to the solution of (8.9.1). In that

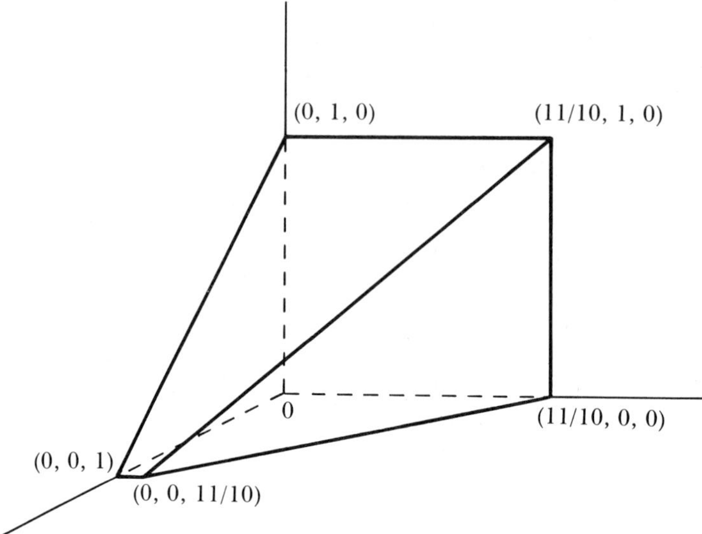

FIGURE VIII.9.2 A slight perturbation of Figure VIII.9.1.

case, the perturbation should have been smaller: the first constraint may be changed, say, to $x + z \leq 101/100$.

This, then, is the usual technique for perturbation. Each time a zero appears in the right-hand column, we perturb slightly, replacing the zero by a small number. Since we do not know how small is small enough to preserve the position of the solution, we generally use the letter $-\epsilon$, representing a very small but negative amount. The second time that we obtain a zero in the right-hand column, we replace it by $-\epsilon^2$, and so on. The solution, when obtained, will be in terms of ϵ. We then obtain the solution to the unperturbed problem by letting ϵ be equal to zero.

In a sense, this technique is needlessly complicated; careful selection of the pivot will generally make it unnecessary. The procedure is described here only because it helps to complete the proof of convergence for the simplex algorithm, using rules VIII.5.1 and VIII.6.1. Normally, "cycling" occurs only when there is a very systematic choice of the pivot. A recommended procedure when using computers for the solution of linear programs is to instruct the computer to choose its pivots at random whenever the rules allow it latitude for such choice. This makes certain that the pivot is not chosen systematically, and cycling is avoided.

Although the perturbation method is not generally recommended, we give the following example.

VIII.9.1 Example

Maximize

$$w = x + y + 3z$$

Subject to

$$\begin{aligned} x \qquad + z &\le 1 \\ y + z &\le 1 \\ x, y, z &\ge 0 \end{aligned}$$

We have here the tableau

x	y	z	1	
1	0	1	−1	$=-r$
0	1	1*	−1	$=-s$
1	1	3	0	$=w$

Let us assume we take our pivot from the third column; the two quotients are equal, and so we may pivot on either one of the entries in this column. *It is this situation that gives rise to degeneracy.* Pivoting on the starred entry, we get the tableau

x	y	s	1	
1	−1	−1	0	$=-r$
0	1	1	−1	$=-z$
1	−2	−3	3	$=w$

This shows degeneracy: we therefore perturb the tableau to get

x	y	s	1	
1*	−1	−1	$-\epsilon$	$=-r$
0	1	−1	−1	$=-z$
1	−2	−3	3	$=w$

Pivoting on the starred entry, we then have

r	y	s	1	
1	−1	−1	$-\epsilon$	$=-x$
0	1	1	−1	$=-z$
−1	−1	−2	$3+\epsilon$	$=w$

Thus the solution to the perturbed problem is $x = \epsilon$, $y = 0$, $z = 1$, with a value $w = 3 + \epsilon$ for the objective function. We conclude that the unperturbed problem has the solution $x=0$, $y=0$, $z=1$, and $w=3$.

10. SOME PRACTICAL COMMENTS

The number of pivot steps necessary to solve an $m \times n$ program (i.e., a program with m inequalities in n variables, not counting non-negativity constraints) is generally of the order of magnitude of $m + n$. This means that, in most cases, it will be between, say, $(m + n)/2$ and $2(m + n)$. This is certainly a tremendous gain over the "brute force" method described in Section VIII.1. For instance, if $m = n = 10$, we expect that the simplex algorithm will use between 10 and 40 pivot steps, whereas inspecting all the intersections of 10 of the planes would imply solving $\binom{20}{10} = 184{,}756$ systems of 10 equations in 10 unknowns! Nevertheless, for large programs, it is always desirable to decrease even the number of steps that the simplex method uses. This can be done by judicious choice of the pivot. In fact, rules VIII.5.1 and VIII.6.1 often give us great latitude for choice; for instance, there may be many positive entries in the bottom row. Much work has been done in perfecting the choice of pivot so as to cut the number of steps considerably, in some cases by as much as one half, over indiscriminate choices. Unfortunately, we cannot, within the scope of this book, study all these systems. Interested students should read more advanced treatises on linear programming.

PROBLEMS ON THE SIMPLEX ALGORITHM

1. Maximize

$$w = 2x + 3y + z$$

Subject to

$$\begin{aligned} x + 4y - 2z &\le 10 \\ 3y + z &\le 7 \\ 2x - 3y + 4z &\le 12 \\ x &\ge 0 \\ y &\ge 0 \\ z &\ge 0 \end{aligned}$$

2. Minimize

$$w = x + 2y + z$$

Subject to

$$\begin{aligned} 3x + y + 2z &\ge 20 \\ x - 3y + 4z &\ge 15 \\ -x + y + z &\ge 8 \\ x &\ge 0 \\ y &\ge 0 \\ z &\ge 0 \end{aligned}$$

3. Maximize

$$w = x - 2y + 4z$$

Subject to

$$\begin{aligned} x + y + 2z &\le 20 \\ x - y + z &\le 5 \\ 2x + 3y + 2z &\le 35 \\ x &\ge 0 \\ y &\ge 0 \\ z &\ge 0 \end{aligned}$$

4. Minimize

$$w = 5x + 2y + 3z$$

Subject to

$$\begin{aligned} 3x - 2y + 6z &\ge 25 \\ x + y + 2z &\ge 15 \\ 3x + y - 3z &\ge 10 \\ x &\ge 0 \\ y &\ge 0 \\ z &\ge 0 \end{aligned}$$

5. A dietitian is given three foods. Each unit of food A contains 10 oz. protein, 4 oz. fat, and 6 oz. carbohydrate, and costs \$1. A unit of food B has 4 oz. protein, 6 oz. fat, and 8 oz. carbohydrate, and costs 50¢. A unit of food C contains 2 oz. protein, 12 oz. fat, and 8 oz. carbohydrate, and costs 30¢. A mixture of these three foods must be made, containing at least 50 oz. protein, 60 oz. fat, and 65 oz. carbohydrate. What is the cheapest such mixture possible?

6. A toy company makes three types of toy, which must be processed through three machines called a twister, a bender, and a painter. Toy A requires 1 hr. in the twister, 2 hr. in the bender, and 1 hr. in the painter, and sells for \$4. Toy B requires 2 hr. in the twister, 1 hr. in the bender, and 3 hr. in the painter, and sells for \$5. Toy C requires 3 hr. in the twister, 2 hr. in the bender, and 1 hr. in the painter, and sells for \$9. The twister can work at most 50 hr. per week, the bender, at most 40 hr.; and the painter, at most 60 hr. How many toys of each type should the company produce to maximize revenues?

7. A furniture firm manufactures chairs, tables, and beds, and has three workers. The time in hours that each employee must spend to produce one of these, and the corresponding profit, are given in the table:

	A	B	C	Profit ($)
Chair	1	3	2	10
Table	3	5	4	25
Bed	5	4	8	35

Employee A can work only 20 hr. per week, while B and C can work 50 hr. each. How many chairs, beds, and tables should be made so as to maximize profits?

8. A food processing company has 10,000 lb. of African coffee, 12,000 lb. of Brazilian coffee, and 7000 lb. of Colombian coffee. It sells four blends of coffee, whose contents (in ounces of the coffee per pound of blend) and prices are given in the table:

	African	Brazilian	Colombian	Price (¢)
A	0	0	16	90
B	0	12	4	70
C	6	8	2	60
D	10	6	0	50

How many pounds of each blend should be produced to maximize revenues?

9. A candy manufacturer has 500 lb. of chocolate, 100 lb. of nuts, and 50 lb. of fruit in inventory. He produces three types of candy. A box of type A uses 3 lb. chocolate, 1 lb. nuts, and 1 lb. fruit, and sells for $10. A box of type B uses 4 lb. chocolate and 1/2 lb. nuts, and sells for $6. A box of type C contains 5 lb. chocolate, and sells for $4. How many boxes of each type should be made to maximize revenues?

10. A metal company needs 50 tons of iron, 5 tons of copper, and 1 ton of nickel to fulfill a contract. It is offered three types of ore. Type A contains 10 per cent iron and 1 per cent copper, and costs $5 per ton. Type B contains 5 per cent copper and 2 per cent nickel, and sells for $15 per ton. Type C contains 3 per cent iron, 1 per cent copper, and 2 per cent nickel, and sells for $6 per ton. How many tons of each ore should be bought to minimize costs?

11. DUALITY

Let us consider the two linear programs

8.11.1 Maximize

$$w = c_1x_1 + c_2x_2 + \ldots + c_nx_n$$

Subject to

8.11.2
$$\begin{cases} a_{11}x_1 + a_{12}x_2 + \ldots + a_{1n}x_n \le b_1 \\ a_{21}x_1 + a_{22}x_2 + \ldots + a_{2n}x_n \le b_2 \\ \cdots\cdots\cdots\cdots\cdots\cdots \\ \cdots\cdots\cdots\cdots\cdots\cdots \\ a_{m1}x_1 + a_{m2}x_2 + \ldots + a_{mn}x_n \le b_m \end{cases}$$

8.11.3
$$x_1, x_2, \ldots, x_n \ge 0$$

and

8.11.4 Minimize

$$z = b_1y_1 + b_2y_2 + \ldots + b_my_m$$

Subject to

8.11.5
$$\begin{cases} a_{11}y_1 + a_{21}y_2 + \ldots + a_{m1}y_m \ge c_1 \\ a_{12}y_1 + a_{22}y_2 + \ldots + a_{m2}y_m \ge c_2 \\ \cdots\cdots\cdots\cdots\cdots\cdots \\ \cdots\cdots\cdots\cdots\cdots\cdots \\ a_{1n}y_1 + a_{2n}y_n + \ldots + a_{mn}y_m \ge c_m \end{cases}$$

8.11.6
$$y_1, y_2, \ldots, y_m \ge 0$$

In matrix notation, these programs are

8.11.7 Maximize

$$w = \mathbf{c}^t\mathbf{x}$$

Subject to

8.11.8
$$A\mathbf{x} \le \mathbf{b}$$

8.11.9
$$\mathbf{x} \ge 0$$

and

8.11.10 Minimize

$$z = \mathbf{b}^t\mathbf{y}$$

Subject to

$$A^t\mathbf{y} \geq \mathbf{c} \tag{8.11.11}$$

$$\mathbf{y} \geq 0 \tag{8.11.12}$$

where the matrix A and the vectors $\mathbf{b}$ and $\mathbf{c}$ are the same for both programs. We shall call the program (8.11.10) to (8.11.12)—or (8.11.4) to (8.11.6)—the *derived program* of program (8.11.7) to (8.11.9)—or (8.11.1) to (8.11.3).

Let us see whether we can find the derived program of (8.11.10) to (8.11.12). We notice, of course, that (8.11.10) to (8.11.12) is not the same type of program as (8.11.7) to (8.11.9), since it is a minimization problem, and, moreover, the inequalities (8.11.11) go the wrong way. We dispose of these difficulties by multiplying both the objective function and the constraints (8.11.11) by -1; this gives us a restatement of (8.11.10) to (8.11.12) in the form

Maximize

$$-z = (-\mathbf{b})^t\mathbf{y}$$

Subject to

$$(-A^t)\mathbf{y} \leq -\mathbf{c}$$
$$\mathbf{y} \geq 0$$

We can now form the derived program of this; it is done by taking the transpose of the coefficient matrix and interchanging the two vectors $-\mathbf{b}$ and $-\mathbf{c}$. We then obtain

8.11.13 Minimize

$$r = (-\mathbf{c})^t s$$

Subject to

$$(-A^t)^t\mathbf{s} \geq -\mathbf{b} \tag{8.11.14}$$

$$\mathbf{s} \geq 0 \tag{8.11.15}$$

We can, however, do the same to this program as we just did to (8.11.10) to (8.11.12). Using the fact that $A^{tt} = A$, and multiplying through by -1, we have the equivalent program

8.11.16 Maximize

$$-r = \mathbf{c}^t\mathbf{s}$$

Subject to

$$A\mathbf{s} \leq \mathbf{b} \tag{8.11.17}$$

$$\mathbf{s} \geq 0 \tag{8.11.18}$$

It may be seen that this program (8.11.16) to (8.11.18) is the same as (8.11.7) to (8.11.9): in fact, the only difference between the two lies in the fact that the variable vector is **x** in one case, and **s** in the other; in other words, the only difference lies in the labeling of the variables. But this is no difference at all. We find, thus, that the derived of the derived program (8.11.10) to (8.11.12) is the original program (8.11.7) to (8.11.9). When a relation is of this sort (in which each of two problems may be obtained from the other by the same set of rules) it is generally called a *duality* relation: the two programs, in this case, are said to be *dual* to each other.

VIII.11.1 Definition. The two programs (8.11.1) to (8.11.3) and (8.11.4) to (8.11.6) are *dual* programs. Each of these two programs is the *dual* of the other.

From a purely theoretical point of view, each of the two programs (8.11.1) to (8.11.3) and (8.11.4) to (8.11.6) is the dual of the other; hence neither enjoys any primacy over the other. From the practical point of view, of course, this is not so. The decision-maker is given a practical problem, which he expresses (if possible) in the form of a linear program. The dual program has, in general, no practical interpretation (though we shall see that it may be given one in terms of so-called *shadow prices*). Thus, the original practical program does enjoy a definite primacy and will, therefore, be called the *primal*. The other problem will be called the *dual*.

The relation between two dual linear programs is extremely strong, as is evidenced by the theorems that follow.

VIII.11.2 Theorem. Let the vectors $\mathbf{x} = (x_1, \ldots, x_n)^t$ and $\mathbf{y} = (y_1, \ldots, y_m)^t$ satisfy the constraints (8.11.8) and (8.11.9) and (8.11.11) and (8.11.12), respectively. Then

$$\mathbf{c}^t\mathbf{x} \leq \mathbf{b}^t\mathbf{y}$$

Proof. We prove this theorem by the following sequence of relations:

By the commutative law for vector multiplication,

$$\mathbf{c}^t\mathbf{x} = \mathbf{x}^t\mathbf{c}$$

Now, $\mathbf{c} \leq A^t\mathbf{y}$, and $\mathbf{x}^t \geq 0$, and so

$$\mathbf{x}^t\mathbf{c} \leq \mathbf{x}^t(A^t\mathbf{y})$$

By the associative law for matrix multiplication,

$$\mathbf{x}^t(A^t\mathbf{y}) = (\mathbf{x}^tA^t)\mathbf{y}$$

Now, $A\mathbf{x} \leq \mathbf{b}$, so $\mathbf{x}^tA^t \leq \mathbf{b}^t$; since $\mathbf{y} \geq 0$, then

$$(\mathbf{x}^tA^t)\mathbf{y} \leq \mathbf{b}^t\mathbf{y}$$

This gives us the chain of relations

$$\mathbf{c}^t\mathbf{x} = \mathbf{x}^t\mathbf{c} \leq \mathbf{x}^t(A^t\mathbf{y}) = (\mathbf{x}^tA^t)\mathbf{y} \leq \mathbf{b}^t\mathbf{y}$$

which proves the theorem.

VIII.11.3 Corollary. Suppose $\mathbf{x}^*$ and $\mathbf{y}^*$ satisfy the constraints (8.11.8) to (8.11.9) and (8.11.11) to (8.11.12) respectively, and, moreover,

$$\mathbf{c}^t\mathbf{x}^* = \mathbf{b}^t\mathbf{y}^*$$

Then $\mathbf{x}^*$ and $\mathbf{y}^*$ are the solutions of the programs (8.11.7) to (8.11.9) and (8.11.10) to (8.11.12) respectively.

Proof. This is clear since, if $\mathbf{x}$ is any vector satisfying (8.11.8) to (8.11.9), then, by Theorem VIII.11.2,

$$\mathbf{c}^t\mathbf{x} \leq \mathbf{b}^t\mathbf{y}^* = \mathbf{c}^t\mathbf{x}^*$$

Hence, $\mathbf{x}^*$ solves (8.11.7) to (8.11.9), and, similarly, $\mathbf{y}^*$ solves (8.11.10) to (8.11.12).

Corollary VIII.11.3 gives us a method of checking that a given pair of vectors, $\mathbf{x}^*$ and $\mathbf{y}^*$, solve the two dual programs (8.11.7) to (8.11.9) and (8.11.10) to (8.11.12), respectively. In fact, if they satisfy the constraints of the programs, and give the same values for the objective functions, then VIII.11.3 confirms that they are solutions. The converse of VIII.11.3 is also true; if $\mathbf{x}^*$ and $\mathbf{y}^*$ are solutions to the two mutually dual programs (8.11.7) to (8.11.9) and (8.11.10) to (8.11.12), then $\mathbf{c}^t\mathbf{x}^* = \mathbf{b}^t\mathbf{y}^*$. This equality of values is both necessary and sufficient for the vectors $\mathbf{x}^*$ and $\mathbf{y}^*$ to solve the two programs. (We shall prove this statement later.) We give an example of this property now.

VIII.11.4 Example. Consider the two programs

Maximize

$$w = 3x_1 + 2x_2 + 4x_3$$

Subject to

$$\begin{aligned} x_1 + 3x_2 + 2x_3 &\leq 10 \\ 2x_1 + x_2 + x_3 &\leq 8 \\ x_1, x_2, x_3 &\geq 0 \end{aligned}$$

and

Minimize

$$z = 10y_1 + 8y_2$$

Subject to

$$\begin{aligned} y_1 + 2y_2 &\geq 3 \\ 3y_1 + \quad y_2 &\geq 2 \\ 2y_1 + \quad y_2 &\geq 4 \\ y_1, y_2 &\geq 0 \end{aligned}$$

The maximizing problem gives us the tableau

x_1	x_2	x_3	1	
1	3	2*	−10	$= -u_1$
2	1	1	−8	$= -u_2$
3	2	4	0	$= w$

Pivoting on the starred entries, we obtain the tableaux

x_1	x_2	u_1	1	
1/2	3/2	1/2	−5	$= -x_3$
3/2*	−1/2	−1/2	−3	$= -u_2$
1	−4	−2	20	$= w$

and

u_2	x_2	u_1	1	
−1/3	5/3	2/3	−4	$= -x_3$
2/3	−1/3	−1/3	−2	$= -x_1$
−2/3	−11/3	−5/3	22	$= w$

which gives us the solution; it is $x_1 = 2$, $x_2 = 0$, $x_3 = 4$, and gives a value $w = 22$.

Consider, now, the minimizing problem. Its tableau is

y_1	y_2	1	
−1	−2	3	$= -v_1$
−3	−1	2	$= -v_2$
−2*	−1	4	$= -v_3$
10	8	0	$= z$

 We pivot on the starred entries to obtain

v_3	y_2	1	
$-1/2$	$-3/2^*$	1	$= -v_1$
$-3/2$	$1/2$	-4	$= -v_2$
$-1/2$	$1/2$	-2	$= -y_1$
5	3	20	$= z$

and

v_3	v_1	1	
$1/3$	$-2/3$	$-2/3$	$= -y_2$
$-5/3$	$1/3$	$-11/3$	$= -v_2$
$-2/3$	$1/3$	$-5/3$	$= -y_1$
4	2	22	$= z$

which gives the solution: $y_1 = 5/3$, $y_2 = 2/3$ with a value $z = 22$. It is now possible to check that the given vectors are, indeed, the solutions to the two programs. In fact, they satisfy the constraints, and give the same value for the objective functions.

Looking over the solutions of the pair of dual programs, we can see that the implications of duality are far greater than the simple fact that the values of the two programs (i.e., the maximum and minimum values, respectively, of their objective functions) are equal. In fact, it is not only the objective functions that give the same value: the corresponding tableaux for the two programs are, at each step, dually related. Each tableau is "almost" the negative transpose of the corresponding tableau for the other problem. That this is so for any pair of dual programs is immensely important.

Let us, by the introduction of slack variables, rewrite (8.11.7) to (8.11.9) in the form

8.11.19 Maximize

$$w = \mathbf{c}^t\mathbf{x}$$

Subject to

8.11.20
$$A\mathbf{x} - \mathbf{b} = -\mathbf{u}$$

8.11.21
$$\mathbf{x},\mathbf{u} \geq 0$$

Similarly, we may rewrite (8.11.10) to (8.11.12) in the form

8.11.22 Minimize

$$z = \mathbf{b}^t\mathbf{y}$$

Subject to

8.11.23 $$A^t\mathbf{y} - \mathbf{c} = \mathbf{v}$$

8.11.24 $$\mathbf{y},\mathbf{v} \geq 0$$

We may write these two programs simultaneously in the tableau

8.11.25

	x_1	x_2	. . . x_n	1	
y_1	a_{11}	a_{12}	. . . a_{1n}	$-b_1$	$=-u_1$
y_2	a_{21}	a_{22}	. . . a_{2n}	$-b_2$	$=-u_2$
. . .	. . .	. . .	. . .	. . .	. . .
. . .	. . .	. . .	. . .	. . .	. . .
. . .	. . .	. . .	. . .	. . .	. . .
y_m	a_{m1}	a_{m2}	. . . a_{mn}	$-b_m$	$=-u_m$
-1	c_1	c_2	. . . c_n	0	$=w$
	$=v_1$	$=v_2$	. . . v_n	$=-z$	

The rows of (8.11.25) represent, as usual, the equations (8.11.19) and (8.11.20); the columns, now, represent the equations (8.11.22) and (8.11.23). Thus, the single tableau does double duty as a representation of the pair of problems simultaneously. The pivot steps of the simplex algorithm were devised precisely to maintain the validity of the row equations. What we will show is that they have the property of preserving the column equations as well. Let us, then, consider the column system of a tableau:

8.11.26

q_1	a_{11}	a_{12}	. . . a_{1n}	$-b_1$
q_2	a_{21}	a_{22}	. . . a_{2n}	$-b_2$
. . .	. . .	. . .	. . .	. . .
. . .	. . .	. . .	. . .	. . .
. . .	. . .	. . .	. . .	. . .
q_m	a_{m1}	a_{m2}	. . . a_{mn}	$-b_m$
-1	c_1	c_2	. . . c_n	δ
	$=p_1$	$=p_2$	. . . $=p_n$	$=-z$

In (8.11.26), the p_j are the basic variables; the q_i are non-basic. Suppose that we wish to interchange p_j and q_i. The jth column represents the equation

8.11.27 $$a_{1j}q_1 + a_{2j}q_2 + \ldots + a_{ij}q_i + \ldots - c_j = p_j$$

We can solve (8.11.27) for q_i, obtaining

8.11.28 $$-\frac{a_{1j}}{a_{ij}}q_1 - \frac{a_{2j}}{a_{ij}}q_2 - \ldots + \frac{1}{a_{ij}}p_j - \ldots + \frac{c_j}{a_{ij}} = q_i$$

 In turn, the lth column represents

8.11.29 $$a_{1l}q_1 + a_{2l}q_2 + \ldots + a_{il}q_i + \ldots - c_l = p_l$$

Let us, now, substitute (8.11.28) in (8.11.29). This gives us

8.11.30 $$\left(a_{1l} - \frac{a_{il}a_{1j}}{a_{ij}}\right) q_1 + \left(a_{2l} - \frac{a_{il}a_{2j}}{a_{ij}}\right) q_2 + \ldots + \frac{a_{il}}{a_{ij}} p_j + \ldots - \left(c_l - \frac{a_{il}c_j}{a_{ij}}\right) = p_l$$

The changes in the coefficients can be represented by the schema

8.11.31 $$\boxed{\begin{matrix} p & q \\ r & s \end{matrix}} \rightarrow \boxed{\begin{matrix} \dfrac{1}{p} & \dfrac{q}{p} \\ -\dfrac{r}{p} & s - \dfrac{qr}{p} \end{matrix}}$$

in which, as in Section VIII.2, p represents the pivot, q an entry in the pivot row, r an entry in the pivot column, and s the entry in the row of r and column of q. But (8.11.31) is exactly the same as (8.3.10). Thus, *the pivot transformations that preserve the row equations also preserve the column equations.*

Let us suppose that the program (8.11.1) to (8.11.3) has a solution. This solution is obtained, as we know, from a tableau in which all the entries in the right-hand column and in the bottom row (except possibly the bottom right-hand entry) are non-positive. Consider then, the column equations in this tableau:

8.11.32

q_1	a_{11}	a_{12}	...	a_{1n}	$-b_1$
q_2	a_{21}	a_{22}	...	a_{2n}	$-b_2$
...	...	...	...	...	...
...	...	...	...	...	...
...	...	...	...	...	...
q_m	a_{m1}	a_{m2}	...	a_{mn}	$-b_m$
-1	c_1	c_2	...	c_n	δ
	$=p_1$	$=p_2$	...	$=p_n$	$=-z$

It may be seen from (8.11.32) that, if we set the non-basic variables $q_1, \ldots, q_m$ equal to zero, each of the basic variables, $p_1, \ldots, p_n$ will be equal to the negative of the corresponding entry in the bottom row: $p_j = -c_j$. But, by hypothesis, $c_j \leq 0$. Hence $p_j \geq 0$. Thus the tableau (8.11.32) gives a b.f.p. for the dual program (8.11.4) to (8.11.6).

Consider, next, the right-hand column of (8.11.32). It represents the equation

8.11.33 $$z = b_1q_1 + b_2q_2 + \ldots + b_mq_m + \delta$$

Now, by hypothesis, each of the terms b_iq_i is non-negative (being the product of a non-negative constant b_i and a non-negative variable q_i). It follows that, for every point in the constraint set, $z \geq \delta$. We have here, however, a point that gives a value $z = \delta$ for the objective function. It follows that the b.f.p. of (8.11.32) is the solution of (8.11.4) to (8.11.6); *the solutions to both programs (8.11.1) to (8.11.3) and (8.11.4) to (8.11.6) are given by the same tableau (8.11.32).* Therefore, the optimal values of the objective functions must be the same for both programs, i.e., δ. We obtain the following theorem, known as the *fundamental theorem of linear programming.*

VIII.11.5 Theorem. If either one of the two programs (8.11.1) to (8.11.3) and (8.11.4) to (8.11.6) has a solution, then so does the other, and the solutions give the same value to the objective functions. If both programs are feasible, then both have solutions. If one of the two programs is feasible but unbounded, then the other is infeasible. Finally, if one of the two programs is infeasible, then the other is either infeasible or unbounded.

Proof. The first statement of this theorem has just been proved. The other statements depend on the fact that a program has a solution if it is bounded and feasible (seen in our development of the simplex algorithm), as well as on Theorem VIII.11.2.

That the same tableau solves both of a pair of dual problems is additionally advantageous because very often the first tableau obtained is a b.f.p., not for the primal problem, but for the dual. By attacking the dual, rather than the primal, we are able to go into Stage II immediately, avoiding perhaps half of the pivot steps normally necessary.

VIII.11.6 Example. A mixture is to be made of foods, A, B, and C. Each unit of food A contains 3 oz. protein and 4 oz. carbohydrate, and costs \$1. Each unit of food B contains 5 oz. protein and 3 oz. carbohydrate, and costs \$3. Each unit of food C contains 1 oz. protein and 4 oz. carbohydrate, and costs 50c. The mixture is to contain at least 15 oz. protein and 10 oz. carbohydrate. Find the mixture of minimal cost.

The program here is:

Minimize

$$z = y_1 + 3y_2 + \frac{1}{2}y_3$$

Subject to

$$\begin{aligned} 3y_1 + 5y_2 + y_3 &\geq 15 \\ 4y_1 + 3y_2 + 4y_3 &\geq 10 \\ y_1, y_2, y_3 &\geq 0 \end{aligned}$$

The first tableau obtained for this program will not be a b.f.p.; y = 0 is not a feasible vector. On the other hand, we may write the tableau that represents this program as the column system: the dual will then be the row system:

	x_1	x_2	1	
y_1	3*	4	−1	$= -u_1$
y_2	5	3	−3	$= -u_2$
y_3	1	4	$-\frac{1}{2}$	$= -u_3$
−1	15	10	0	$= w$
	$= v_1$	$= v_2$	$= -z$	

It may be seen that this is a b.f.p. for the dual problem. We pivot on the starred element, obtaining

	u_1	x_2	1	
v_1	1/3	4/3	−1/3	$= -x_1$
y_2	−5/3	−11/3	−4/3	$= -u_2$
y_3	−1/3	8/3	−1/6	$= -u_3$
−1	−5	−10	5	$= w$
	$= y_1$	$= v_2$	$= -z$	

which gives the solution: $y_1 = 5$, $y_2 = 0$, $y_3 = 0$ with a value $z = 5$. Note how Stage I was avoided by solving the dual program.

In a pair of mutally dual programs, generally one of the pair has a clear practical application: it has been formulated, precisely, to solve a practical problem. Yet, what about the dual? Is it possible to obtain a "practical" interpretation of the dual program?

Let us consider the dual of Example VIII.11.6:

Maximize

$$15x_1 + 10x_2$$

Subject to

$$\begin{aligned} 3x_1 + 4x_2 &\leq 1 \\ 5x_1 + 3x_2 &\leq 3 \\ x_1 + 4x_2 &\leq \frac{1}{2} \\ x_1, x_2 &\geq 0 \end{aligned}$$

and see whether some interpretation may be obtained. Each of the two variables, x_1 and x_2, corresponds to one of the constraints. But these constraints correspond to the two nutrients, protein and carbohydrate. Thus, x_1 and x_2 are numbers assigned, respectively, to protein and carbohydrate; the type of number can be seen from the fact that the constants in the constraints correspond to costs—the costs of the different foods, A, B, and C. It follows that we may consider x_1 and x_2 as representing some sort of value, in dollars per ounce, for the two nutrients. We are assigning such values, known as *shadow prices*, to these nutrients. The constraints represent, in a sense, a "profitless" system: each of three foods costs at least as much as the values of the nutrients it contains. The program seeks to maximize the total shadow prices of the required diet.

The solution of the dual problem is given by the tableau: $x_1 = 1/3$, $x_2 = 0$, with a value $w = 5$. Note that this seems to make the carbohydrate valueless. This is not to say that the carbohydrate is useless; rather, it represents the well-known economic fact that surplus goods lose all commercial value, and in this case the solution to the primal problem gives a surplus of carbohydrate. Note also that the solution of the primal does not use any of foods B and C, coincident with the fact that the costs of these foods are higher than the shadow prices of the nutrients contained (according to the dual solution).

Generally speaking, the dual may be interpreted in terms of shadow prices whenever the primal problem deals with costs or profits.

VIII.11.7 Example. A company makes three products, A, B, and C, which must be processed by three employees, D, E, and F. Each unit of product A must be processed 1 hr. by employee D, 2 hr. by E, and 3 hr. by F, and yields a profit of \$50. Each unit of B must be processed 4 hr. by D, 1 hr. by E, and 1 hr. by F, and yields a profit of \$30. Each unit of C must be processed 1 hr. by D, 3 hr. by E, and 2 hr. by F, and yields a profit of \$20. Employee D can work at most 40 hr. per week; E, at most 50 hr., and F, at most 30 hr. It is required to maximize the profit.

The program here is:

Maximize

$$w = 50x_1 + 30x_2 + 20x_3$$

Subject to

$$\begin{aligned} x_1 + 4x_2 + x_3 &\le 40 \\ 2x_1 + x_2 + 3x_3 &\le 50 \\ 3x_1 + x_2 + 2x_3 &\le 30 \\ x_1, x_2, x_3 &\ge 0 \end{aligned}$$

The *dual* program is:

Minimize

$$z = 40y_1 + 50y_2 + 30y_3$$

Subject to

$$\begin{aligned} y_1 + 2y_2 + 3y_3 &\geq 50 \\ 4y_1 + y_2 + y_3 &\geq 30 \\ y_1 + 3y_2 + 2y_3 &\geq 20 \\ y_1, y_2, y_3 &\geq 0 \end{aligned}$$

which may be interpreted, in one sense, as the assignation of "natural" wages to the three employees. It may be checked that the solutions to the two problems are

$$x_1 = \frac{80}{11},\ x_2 = \frac{90}{11},\ x_3 = 0,\ w = \frac{6700}{11}$$

for the primal, and

$$y_1 = \frac{40}{11},\ y_2 = 0,\ y_3 = \frac{170}{11},\ z = \frac{6700}{11}$$

for the dual. This should not be taken to mean that employee E should receive no wages; it does mean, however, that he should be encouraged to work a shorter week, since he is clearly idle more than half the time. Conversely, this does not suggest that employee F should be paid over $15 per hour; it does say that his week should be lengthened if extra hours for him will cost less than $15.45 per hour. The "shadow wages" tend to represent the *marginal productivity* of the three employees concerned. (The classic economist would explain that employee F's wages would seek the high level causing him to work longer hours until an equilibrium was found.)

PROBLEMS ON DUALITY

1. Give the duals of the following linear programs. Obtain solutions to each pair of programs.

(a) Maximize

$$2x + 4y + 7z = w$$

Subject to

$$\begin{aligned} 3x + y + 2z &\le 25 \\ x + 3y - z &\le 10 \\ 4y + z &\le 30 \\ x &\ge 0 \\ y &\ge 0 \\ z &\ge 0 \end{aligned}$$

(b) Minimize

$$6x + 2y + 3z = w$$

Subject to

$$\begin{aligned} x + 4y - 3z &\ge 15 \\ 2x + 3y + z &\ge 25 \\ -x + 4y + 2z &\ge 12 \\ x &\ge 0 \\ y &\ge 0 \\ z &\ge 0 \end{aligned}$$

(c) Minimize

$$2x + y + 2z = w$$

Subject to

$$\begin{aligned} x + 3y + z &\ge 30 \\ x - y &\ge -10 \\ 3x + y + z &\ge 40 \\ x &\ge 0 \\ y &\ge 0 \\ z &\ge 0 \end{aligned}$$

(d) Maximize

$$3x + 2y + z = w$$

Subject to

$$\begin{aligned} x + y + 3z &\le 25 \\ x - y &\le -10 \\ y + 2z &\le 15 \\ x &\ge 0 \\ y &\ge 0 \\ z &\ge 0 \end{aligned}$$

2. Give the duals of problems 5 to 10 on pages 455–456. Solve these duals and give a heuristic interpretation (shadow prices).

3. Show that, to an equation constraint in the primal problem, there corresponds an unrestricted (i.e., not necessarily non-negative) variable in the dual, and vice versa.

470 12. TRANSPORTATION PROBLEMS

The simplex algorithm is probably the most efficient method available for the solution of linear programs in general. This does not mean, however, that it will be *uniformly* the most efficient method. The reason is that there are often special types of problem that, precisely because of their special type, can be handled more efficiently by a method other than the simplex algorithm. Much of the present research in linear programming consists precisely in the search for new methods of solving special types of linear programs—generally those that the researcher (an applied mathematician or operations analyst) has encountered in practical experience. Some of these new methods are tremendously efficient—for the particular type of problem concerned, generally considerably larger than could be solved with the simplex algorithm.

We consider here a special type of linear program: the *network* program. Such a program was seen earlier as Example VIII.8.1. Its characteristic property is that, as may be seen from inspection of the several tableaux in VIII.8.1, the entries in the inner part of the tableau (i.e., other than on the bottom row or right-hand column) are always 0, 1, or -1. The importance of this fact is that the pivot is always 1 or -1; in effect, this means that no multiplications or divisions will ever be necessary—the simpler operations of addition and subtraction will always be sufficient for solution.

Example VIII.8.1 is a typical *transportation problem.* In the general case, we might picture a company with a total of m warehouses, each containing a certain amount of the company's product, and n distributors, each having a demand for a certain amount of the product. The problem is to transport the desired amounts from the warehouses to the distributors in such a way as to minimize shipping costs.

More precisely, let a_i be the availability at the ith warehouse; let b_j be the requirement at the jth distributor. We make the assumption:

8.12.1 $$\sum_{i=1}^{m} a_i = \sum_{j=1}^{n} b_j$$

which means that the total availabilities at the m warehouses are exactly equal to the total requirements at the n distributors. Let c_{ij} be the cost of transporting a unit from the ith warehouse to the jth distributor. The problem can then be expressed as a linear program:

8.12.2 Minimize

$$\sum_{i=1}^{m} \sum_{j=1}^{n} c_{ij} x_{ij}$$

Subject to

8.12.3 $$\sum_{j=1}^{n} x_{ij} = a_i \text{ for each } i$$

8.12.4 $$\sum_{i=1}^{m} x_{ij} = b_j \text{ for each } j$$

8.12.5 $$x_{ij} \geq 0 \quad \text{for each } i,j$$

The first thing to notice is that there is no need for slack variables, since all the constraints (8.12.3) and (8.12.4) are equations rather than inequalities. This means that the "flows" x_{ij} are the only variables. Since each b.f.p. of a program is determined by the non-basic variables, to determine a b.f.p. of this program, it is sufficient to state which of the variables x_{ij} are equal to zero. This being so, it follows that we can represent each b.f.p. by a schematic graph.

Technically speaking, a graph is a collection of points called *nodes* and segments of curved lines called *arcs* that connect some of the nodes. (This is a *topological* graph, as distinguished from the graph of a function, which we saw in Chapter II.) To each b.f.p. we assign a graph as follows: there is a node W_i for each of the warehouses and a node D_j for each of the distributors; there will be an arc joining W_i to D_j if x_{ij} is positive, and there will be no other arcs. Thus, the tableau for the first b.f.p. reached in the solution of Example VIII.8.1 can be represented either by a graph or by a matrix:

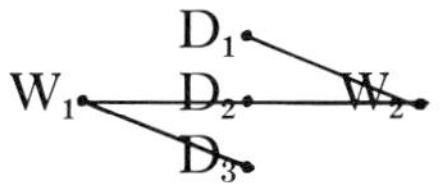

D_1
W_1 D_2 W_2
D_3

$$\begin{array}{c} \\ W_1 \\ W_2 \end{array} \begin{array}{c} \begin{array}{ccc} D_1 & D_2 & D_3 \end{array} \\ \begin{pmatrix} 0 & 10 & 40 \\ 30 & 30 & 0 \end{pmatrix} \end{array}$$

We repeat that the graph determines the b.f.p. entirely: in fact, the "missing" arcs, i.e., from W_1 to D_1 and from W_2 to D_3 in this example, correspond to the non-basic variables. These determine the values of the basic variables entirely.

The following theorem is a fundamental part of the method of solution of transportation problems. We say a graph is connected if any two of the nodes can be joined by a sequence of arcs from the graph.

VIII.12.1 Theorem. The graph corresponding to a b.f.p. will not contain any closed loops. Moreover, assuming non-degeneracy of the program, the graph will be connected, and so the addition of any arc will close a loop.

Proof. Let us suppose that the graph corresponding to a b.f.p. does contain a closed loop. This loop must have an even number of arcs, since each arc connects a warehouse to a distributor. Let α be the smallest flow along any of the arcs in the loop. If we number these arcs consecutively, it may be seen that we can increase the flow along the odd-numbered arcs by $\alpha/2$ and decrease it along the even-numbered arcs by the same amount without contradicting any of the constraints. This gives us a different point that has the same arcs "missing" from its graph, which contradicts the fact that the non-basic variables entirely determine a b.f.p. The contradiction proves the first statement of the theorem.

To prove the second statement, we point out that the graph can be disconnected only if the availabilities at a subcollection of the warehouses are equal to the requirements at a subcollection of the distributors. This is, precisely, degeneracy. Assuming that the graph is connected, any new arc will close a loop, since it will connect two nodes that were already connected in some other manner.

Now that we have replaced the simplex tableau by a graph, we must see what form the pivot-step takes. We know that a pivot-step has the property of increasing the value of one of the non-basic variables, and it will, thus, correspond to adding one of the missing arcs to the graph. This will cause one of the arcs presently on the graph to be removed. Let us see which one.

In the preceding graph, there are two missing arcs: W_1D_1 and W_2D_3 (corresponding to $x_{11}=x_{23}=0$). We can make x_{11} positive by adding the arc W_1D_1 to the graph: this closes a loop, which we denote as $W_1D_1W_2D_2W_1$. Suppose, that we do increase the variable x_{11}: this means that because a smaller quantity of the product is left at W_1, we must decrease either x_{12} or x_{13} (the flows from W_1 to the other two D_j). We cannot decrease x_{13}, as this would entail an increase in x_{23}, and we want to keep x_{23} fixed at zero (the fundamental idea of pivot-steps). Instead we decrease x_{12}. Now the requirement at D_2 is not being met, and this deficiency must be made up by increasing x_{22}. In turn, this causes x_{21} to decrease. The decrease in x_{21} will be exactly enough to meet the increase in x_{11}, keeping the required quantity at D_1. Thus, an increase in x_{11} will cause an equal increase in x_{22}, and corresponding equal decreases in x_{12} and x_{21}. We increase x_{11}, then, until either x_{12} or x_{21} has decreased to zero. Since $x_{12}=10$ and $x_{21}=30$, it is clear that x_{12} will be the first to vanish: adding the arc W_1D_1 to the graph has caused us to remove the arc W_1D_2. We obtain the new b.f.p.

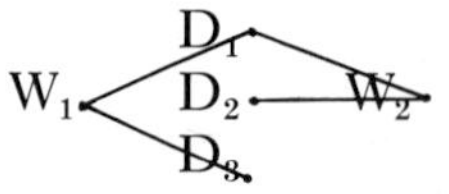

W_1 — D_1, D_2, D_3; D_1 — W_2; D_2 — W_2

$$\begin{array}{c} \\ W_1 \\ W_2 \end{array} \begin{array}{c} \begin{array}{ccc} D_1 & D_2 & D_3 \end{array} \\ \begin{pmatrix} 10 & 0 & 40 \\ 20 & 40 & 0 \end{pmatrix} \end{array}$$

It remains to be seen whether this new b.f.p. is an improvement over the previous one. This can actually be calculated directly; if we do, we see that the total costs for this b.f.p. are 510, whereas, for the previous b.f.p., they were only 500. Since we wish to minimize costs, this is not an improvement. On the other hand, we could also have seen that the change would increase the costs in the following manner: each unit increase in x_{11} causes an equal increase in x_{22}, and equal decreases in x_{21} and x_{12}. The total change in costs for a unit change in x_{11} is

$$c_{11} - c_{12} + c_{22} - c_{21} = 1 - 4 + 9 - 5 = 1$$

Since this is positive, we would conclude that the change would cause an increase in total costs, and therefore reject it.

This describes, then, the nature of the fundamental step for the transportation problem. Under the assumption of non-degeneracy, Theorem VIII.12.1 tells us that each missing arc will, when added to the graph, close a loop. If we number the arcs in this loop consecutively, starting with the newly included arc, we see that an increase along the first arc will cause equal increases along the odd-numbered arcs, and decreases along the even-numbered arcs. To see whether the change will be an improvement, we add the unit costs along the odd arcs, and subtract from these the unit costs along the even arcs. If the result is negative the change is an improvement (it will decrease costs). The new arc is then added to the diagram, while one of the even arcs in the loop (the one with the lowest flow) is removed.

With this method explained, we proceed to solve the problem of Example VIII.8.1.

VIII.12.2 Example. Solve the 2×3 transportation problem with availabilities

$$a_1 = 50 \qquad a_2 = 60$$

requirements

$$b_1 = 30 \qquad b_2 = 40 \qquad b_3 = 40$$

and transportation costs given by the table:

TABLE VIII.12.1

From Warehouse To Distributor

	D_1	D_2	D_3
W_1	1	4	1
W_2	5	9	2

We already have obtained a b.f.p., given by

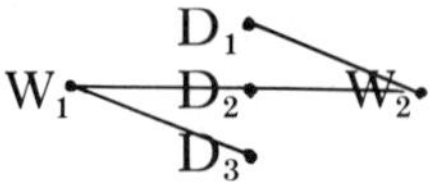

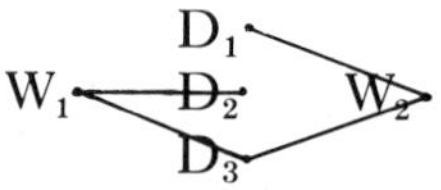

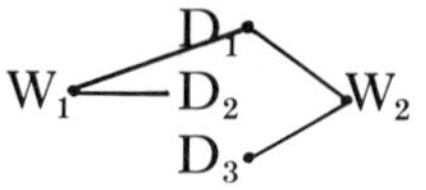

$$\begin{array}{c} \\ W_1 \\ W_2 \end{array}\begin{array}{ccc} D_1 & D_2 & D_3 \\ \end{array}$$
$$\begin{array}{c|ccc} & D_1 & D_2 & D_3 \\ \hline W_1 & 0 & 10 & 40 \\ W_2 & 30 & 30 & 0 \end{array}$$

of the two missing arcs, we have already seen that W_1D_1 will not cause a favorable change. Consider, then, the arc W_2D_3. This will close the loop $W_2D_3W_1D_2W_2$. The change in cost per unit is given by

$$c_{23} - c_{13} + c_{12} - c_{22} = 2 - 1 + 4 - 9 = -4$$

Since this is negative, the change is a favorable one. We therefore add the arc W_2D_3 to the graph. Of the even arcs in the loop, W_1D_3 and W_2D_2, we see that W_2D_2 has the smaller flow: $x_{13} = 40$ and $x_{22} = 30$. Hence, we remove W_2D_2 from the graph; x_{23} and x_{12} increase by 30, while x_{13} and x_{22} decrease by 30.

$$\begin{array}{c|ccc} & D_1 & D_2 & D_3 \\ \hline W_1 & 0 & 40 & 10 \\ W_2 & 30 & 0 & 30 \end{array}$$

We repeat the procedure. Of the two arcs missing in the new graph, W_2D_2 will clearly not give any improvement; we would simply be replacing the arc we removed in the previous step. Consider, then, the arc W_1D_1. This closes the loop $W_1D_1W_2D_3W_1$. We calculate

$$c_{11} - c_{21} + c_{23} - c_{13} = 1 - 5 + 2 - 1 = -3$$

This also is negative, so W_1D_1 is added to the graph. We see, next, that $x_{21} = 30$, while $x_{13} = 10$. The arc W_1D_3 should be removed. We increase x_{11} and x_{23} by 10, while decreasing x_{21} and x_{13} by the same amount. This gives us

$$\begin{array}{c|ccc} & D_1 & D_2 & D_3 \\ \hline W_1 & 10 & 40 & 0 \\ W_2 & 20 & 0 & 40 \end{array}$$

This new scheme is the solution. Of the two missing arcs, W_1D_3, which was just removed, can give no improvement; as for W_2D_2, it closes the loop $W_2D_2W_1D_1W_2$. We calculate

$$c_{22} - c_{12} + c_{11} - c_{21} = 9 - 4 + 1 - 5 = 1$$

Since this is also positive, it follows that this arc gives no improvement. There are no other arcs. We conclude that this is the solution.

We have seen the simplification of Stage II of the simplex algorithm for the transportation problem. Let us consider next, Stage I: obtaining a b.f.p. A very easy method of obtaining a b.f.p.

for these problems is called the *northwest corner* rule; it can be illustrated by referring once again to Example VIII.8.2.

We wish to obtain a b.f.p. for the problem with availabilities

$$a_1 = 50 \qquad a_2 = 60$$

and requirements

$$b_1 = 30 \qquad b_2 = 40 \qquad b_3 = 40$$

We start at the upper left-hand (i.e., northwest) corner. We have to get 30 units to D_1. Since W_1, has all this (and more) available, we let $x_{11} = 30$. There are still 20 units at W_1. Now, D_2 requires 40 units. We give it the 20 remaining units at W_1, and 20 others from W_2, obtaining $x_{12} = 20$, $x_{22} = 20$. There are still 40 units left at W_2; these are needed at D_3, so we let $x_{23} = 40$. We obtain the scheme

$$\begin{array}{c} \\ W_1 \\ W_2 \end{array} \begin{array}{c} \begin{array}{ccc} D_1 & D_2 & D_3 \end{array} \\ \begin{pmatrix} 30 & 20 & 0 \\ 0 & 20 & 40 \end{pmatrix} \end{array}$$

which may be seen to be a b.f.p.

This, then, is the general idea behind the northwest corner rule: as many of the D_j as possible are serviced (on a first-come, first-served basis) from W_1; the remainder are serviced, in turn, from W_2,W_3, and so on. Mathematically, the description is as follows: let x_{11} be the smaller of a_1 and b_1. If $x_{11} < a_1$, then let x_{12} be the smaller of $a_1 - x_{11}$ and b_2. If, on the other hand, $x_{11} < b_1$, then let x_{21} be the smaller of a_2 and $b_1 - x_{11}$. Note that each step, in a sense, reduces the problem to that of finding a b.f.p. for a smaller transportation problem. The same process can be continued until all the availabilities and requirements have been exhausted. It is not obvious, but this scheme will actually give a b.f.p. (that is, a vertex of the constraint set, rather than an arbitrary point). The reason is that each of the variables x_{ij} has been found by solving an equation: thus, the point is an intersection of sufficiently many of the bounding planes of the constraint set.

We illustrate this procedure with an example:

VIII.12.3 Example. Find a b.f.p. for the 6×10 transportation program with availabilities

$$a = (40, 30, 50, 20, 80, 30)$$

and requirements

$$b = (30, 20, 10, 40, 20, 20, 30, 10, 10, 60)$$

We check first that $\Sigma a_i = \Sigma b_j = 250$. Using the first warehouse, we service as many of the distributors as possible, starting with D_1. Thus, $a_1 = 40$, and $b_1 = 30$, so we let $x_{11} = 30$. This leaves 10 units at W_1. Since $b_2 = 20$, we set $x_{12} = 10$, and continue to service D_2 from W_2. We get $x_{22} = 10$. There are still 20 units left at W_2; we service D_3 by letting $x_{23} = 10$. The remaining 10 units at W_2 go to D_4, and $x_{24} = 10$. Then we set $x_{34} = 30$ to complete the requirements at D_4. The remaining 20 units at W_3 will go to D_5. We continue in this manner until we obtain the scheme

$$\begin{array}{c} \\ W_1 \\ W_2 \\ W_3 \\ W_4 \\ W_5 \\ W_6 \end{array}
\begin{array}{c}
\begin{array}{cccccccccc} D_1 & D_2 & D_3 & D_4 & D_5 & D_6 & D_7 & D_8 & D_9 & D_{10} \end{array} \\
\begin{pmatrix}
30 & 10 & 0 & 0 & 0 & 0 & 0 & 0 & 0 & 0 \\
0 & 10 & 10 & 10 & 0 & 0 & 0 & 0 & 0 & 0 \\
0 & 0 & 0 & 30 & 20 & 0 & 0 & 0 & 0 & 0 \\
0 & 0 & 0 & 0 & 0 & 20 & 0 & 0 & 0 & 0 \\
0 & 0 & 0 & 0 & 0 & 0 & 30 & 10 & 10 & 30 \\
0 & 0 & 0 & 0 & 0 & 0 & 0 & 0 & 0 & 30
\end{pmatrix}
\end{array}$$

It may be checked directly that this scheme satisfies the constraints of the example.

VIII.12.4 Example. Solve the 3×5 transportation problem with availabilities

$$a = (50, 45, 65)$$

requirements

$$b = (20, 40, 40, 35, 25)$$

and unit costs given by the table:

TABLE VIII.12.2

From Warehouse	To Distributor D_1	D_2	D_3	D_4	D_5
W_1	2	6	5	3	5
W_2	8	9	7	9	3
W_3	2	3	8	4	6

First of all, we obtain a b.f.p. by the northwest corner method. Thus, W_1 supplies all 20 units for D_1 and 30 units for D_2. W_2 supplies the remaining 10 units for D_2 and 35 units for D_3. W_3 supplies the remainder. We obtain the scheme

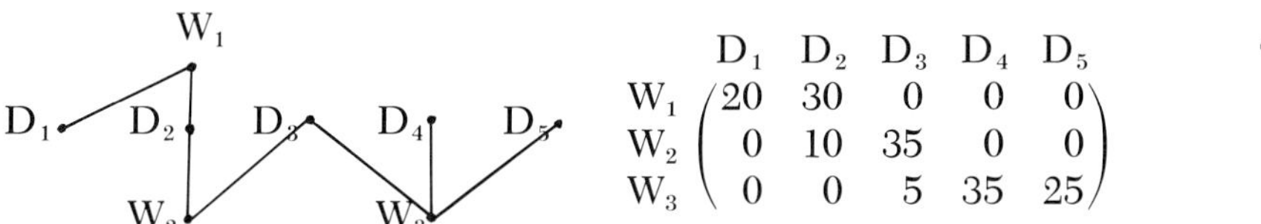

$$
\begin{array}{c} \\ W_1 \\ W_2 \\ W_3 \end{array}
\begin{array}{c} \begin{array}{ccccc} D_1 & D_2 & D_3 & D_4 & D_5 \end{array} \\ \begin{pmatrix} 20 & 30 & 0 & 0 & 0 \\ 0 & 10 & 35 & 0 & 0 \\ 0 & 0 & 5 & 35 & 25 \end{pmatrix} \end{array}
$$

We consider, now, the missing arcs; there are eight of these. The arc W_1D_3 closes the loop $W_1D_3W_2D_2W_1$; accordingly, we calculate

$$c_{13} - c_{23} + c_{22} - c_{12} = 5 - 7 + 9 - 6 = 1$$

Since this is positive, we conclude that the change would not be profitable. We consider other arcs: the arc W_2D_5, finally, is seen to bring a favorable change. In fact, W_2D_5 closes the loop $W_2D_5W_3D_3W_2$. We calculate

$$c_{25} - c_{35} + c_{33} - c_{23} = 3 - 6 + 8 - 7 = -2$$

which is negative. Now, $x_{35} = 25$, and $x_{23} = 35$. As $x_{35} < x_{23}$, we remove the arc W_3D_5; we obtain

$$
\begin{array}{c} \\ W_1 \\ W_2 \\ W_3 \end{array}
\begin{array}{c} \begin{array}{ccccc} D_1 & D_2 & D_3 & D_4 & D_5 \end{array} \\ \begin{pmatrix} 20 & 30 & 0 & 0 & 0 \\ 0 & 10 & 10 & 0 & 25 \\ 0 & 0 & 30 & 35 & 0 \end{pmatrix} \end{array}
$$

For this new scheme, we consider the missing arcs. Arc W_3D_1, will close the loop $W_3D_1W_1D_2W_2D_3W_3$. We calculate

$$c_{31} - c_{11} + c_{12} - c_{22} + c_{23} - c_{33} = 2 - 2 + 6 - 9 + 7 - 8 = -4$$

which is negative. We have $x_{11} = 20$, $x_{22} = 10$, and $x_{33} = 30$. As x_{22} is the smallest of these, we shall add the arc W_3D_1 and remove W_2D_2. We now have

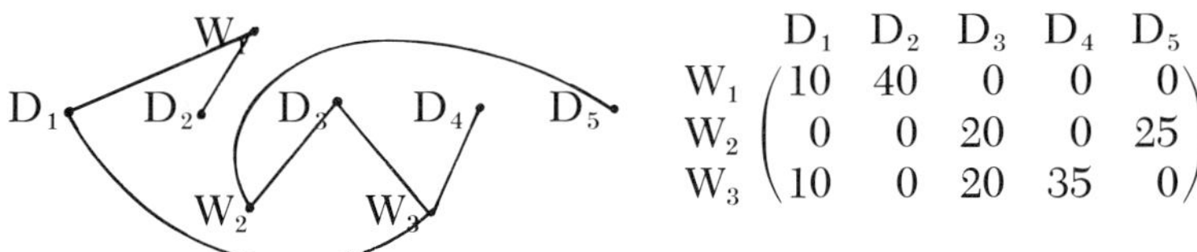

$$
\begin{array}{c} \\ W_1 \\ W_2 \\ W_3 \end{array}
\begin{array}{c} \begin{array}{ccccc} D_1 & D_2 & D_3 & D_4 & D_5 \end{array} \\ \begin{pmatrix} 10 & 40 & 0 & 0 & 0 \\ 0 & 0 & 20 & 0 & 25 \\ 10 & 0 & 20 & 35 & 0 \end{pmatrix} \end{array}
$$

For this scheme, the arc W_1D_3 closes the loop $W_1D_3W_3D_1W_1$. We have

$$c_{13} - c_{33} + c_{31} - c_{11} = 5 - 8 + 2 - 2 = -3$$

478 We have $x_{33} = 20$ and $x_{11} = 10$. Accordingly, we adjoin W_1D_3 and remove W_1D_1, obtaining

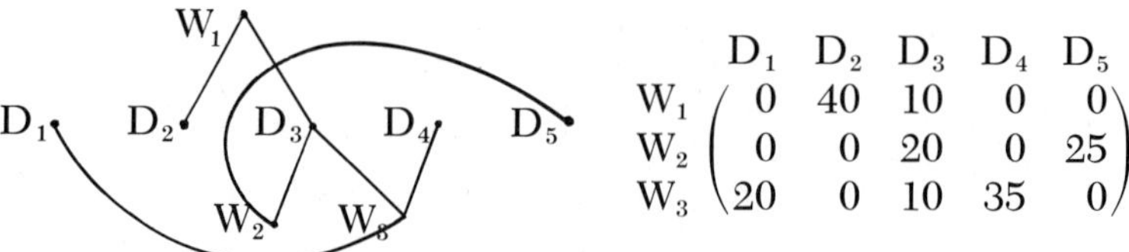

$$\begin{array}{c} \\ W_1 \\ W_2 \\ W_3 \end{array} \begin{array}{c} \begin{array}{ccccc} D_1 & D_2 & D_3 & D_4 & D_5 \end{array} \\ \begin{pmatrix} 0 & 40 & 10 & 0 & 0 \\ 0 & 0 & 20 & 0 & 25 \\ 20 & 0 & 10 & 35 & 0 \end{pmatrix} \end{array}$$

The arc W_3D_2 closes the loop $W_3D_2W_1D_3W_3$. We have

$$c_{32} - c_{12} + c_{13} - c_{33} = 3 - 6 + 5 - 8 = -6$$

We have, also, $x_{12} = 40$ and $x_{33} = 10$. Therefore, we add the arc W_3D_2 and remove W_3D_3. This gives us

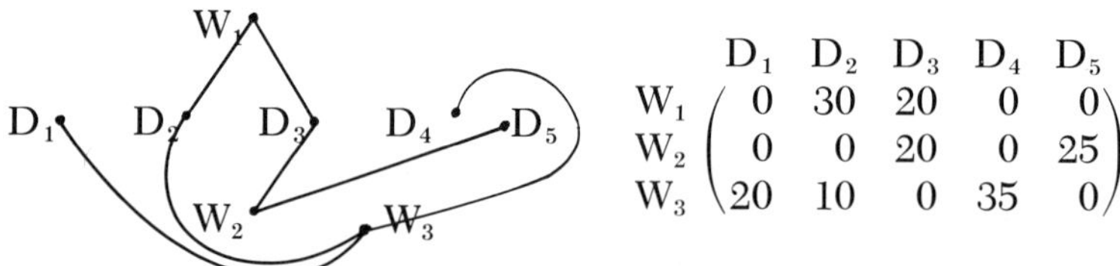

$$\begin{array}{c} \\ W_1 \\ W_2 \\ W_3 \end{array} \begin{array}{c} \begin{array}{ccccc} D_1 & D_2 & D_3 & D_4 & D_5 \end{array} \\ \begin{pmatrix} 0 & 30 & 20 & 0 & 0 \\ 0 & 0 & 20 & 0 & 25 \\ 20 & 10 & 0 & 35 & 0 \end{pmatrix} \end{array}$$

For this scheme, W_1D_1 closes the loop $W_1D_1W_3D_2W_1$. We calculate

$$c_{11} - c_{31} + c_{32} - c_{12} = 2 - 2 + 3 - 6 = -3$$

We have $x_{31} = 20$ and $x_{12} = 30$, so we add the arc W_1D_1 and remove W_3D_1. We have, now,

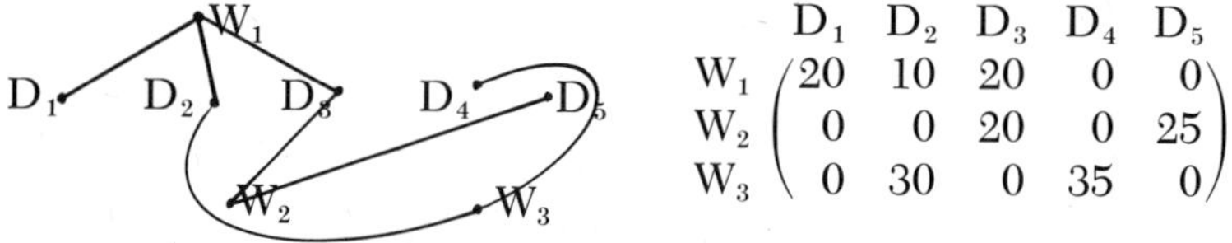

$$\begin{array}{c} \\ W_1 \\ W_2 \\ W_3 \end{array} \begin{array}{c} \begin{array}{ccccc} D_1 & D_2 & D_3 & D_4 & D_5 \end{array} \\ \begin{pmatrix} 20 & 10 & 20 & 0 & 0 \\ 0 & 0 & 20 & 0 & 25 \\ 0 & 30 & 0 & 35 & 0 \end{pmatrix} \end{array}$$

Here, W_1D_4 closes the loop $W_1D_4W_3D_2W_1$. We compute

$$c_{14} - c_{34} + c_{32} - c_{12} = 3 - 4 + 3 - 6 = -4$$

We have $x_{34} = 35$ and $x_{12} = 10$. We therefore add W_1D_4 and delete W_1D_2, which gives us

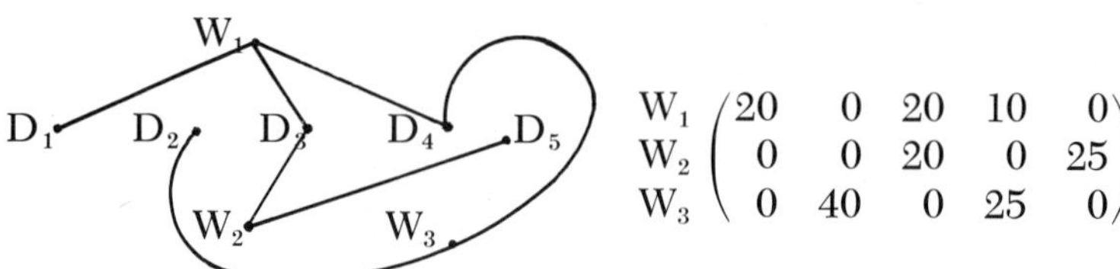

$$\begin{array}{c} W_1 \\ W_2 \\ W_3 \end{array} \begin{pmatrix} 20 & 0 & 20 & 10 & 0 \\ 0 & 0 & 20 & 0 & 25 \\ 0 & 40 & 0 & 25 & 0 \end{pmatrix}$$

Now, the arc W_3D_1 closes the loop $W_3D_1W_1D_4W_3$. We calculate

$$c_{31} - c_{11} + c_{14} - c_{34} = 2 - 2 + 3 - 4 = -1$$

We have $x_{11} = 20$, $x_{34} = 25$. Adding W_3D_1 and deleting W_1D_1, we obtain

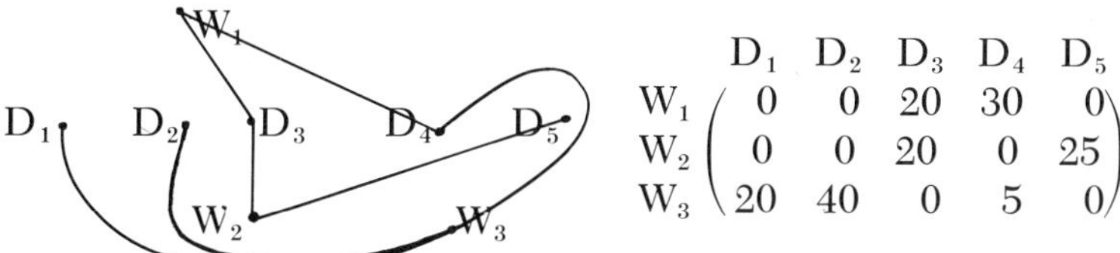

$$\begin{array}{c} \\ W_1 \\ W_2 \\ W_3 \end{array} \begin{array}{c} \begin{array}{ccccc} D_1 & D_2 & D_3 & D_4 & D_5 \end{array} \\ \begin{pmatrix} 0 & 0 & 20 & 30 & 0 \\ 0 & 0 & 20 & 0 & 25 \\ 20 & 40 & 0 & 5 & 0 \end{pmatrix} \end{array}$$

This scheme is optimal: none of the missing arcs can give an improvement. The total cost, as may be verified directly, is 585, which compares most favorably with the cost of 885, given by the first b.f.p. obtained.

13. ASSIGNMENT PROBLEMS

Example VIII.12.4 shows the strength of the flow-type algorithm developed by T. C. Koopmans. It should be pointed out that this example has 15 variables: using the simplex algorithm, its tableaux would have eight rows and nine columns each. Clearly, there is a very real economy of time and effort in using this method. In fact, transportation problems involving some 20 sources and 1000 destinations (and therefore some 20,000 variables) are easily handled in this way with the aid of medium-sized computers.

In case of degeneracy, of course, the Koopmans algorithm fails because, for degenerate programs, adding a new arc to the graph of a b.f.p. will not necessarily close a loop. If the program is not greatly degenerate, a perturbation of the program is usually sufficient to solve it; it is enough to put a small flow—an amount ϵ—on a missing arc that will not close a loop. (Generally, the arc chosen is the one with smallest unit costs among those that do not close a loop.) If, on the other hand, the program is very degenerate—and there are some problems that by their very nature give rise to very degenerate programs—then substantially different techniques are necessary. The *assignment problem* is one such.

Let us assume that a company has several positions to be filled and an equal number of job applicants; the only problem is to determine which of the applicants should be given which position. Thanks to aptitude tests, or, perhaps, to some study of the applicants' backgrounds, it is estimated that (for each value of i and j) the company will derive a "utility" equal to a_{ij} if the ith applicant is assigned to the jth job. The assignment of applicants to jobs is to be made, then, in such a way as to maximize the sum of these expected utilities.

To describe this problem mathematically, we give a definition.

VIII.13.1 Definition. An $n \times n$ matrix $X = (x_{ij})$ is said to be a *permutation matrix* if all its entries are either 0 or 1, and, moreover, there is exactly one 1 in each row and one 1 in each column.

Thus, the following 5×5 matrices are permutation matrices:

(a)
$$\begin{pmatrix} 1 & 0 & 0 & 0 & 0 \\ 0 & 1 & 0 & 0 & 0 \\ 0 & 0 & 1 & 0 & 0 \\ 0 & 0 & 0 & 1 & 0 \\ 0 & 0 & 0 & 0 & 1 \end{pmatrix}$$

(b)
$$\begin{pmatrix} 1 & 0 & 0 & 0 & 0 \\ 0 & 0 & 0 & 1 & 0 \\ 0 & 1 & 0 & 0 & 0 \\ 0 & 0 & 0 & 0 & 1 \\ 0 & 0 & 1 & 0 & 0 \end{pmatrix}$$

(c)
$$\begin{pmatrix} 0 & 1 & 0 & 0 & 0 \\ 1 & 0 & 0 & 0 & 0 \\ 0 & 0 & 0 & 1 & 0 \\ 0 & 0 & 0 & 0 & 1 \\ 0 & 0 & 1 & 0 & 0 \end{pmatrix}$$

It is clear that we can think of each permutation matrix X as an assignment: the ith applicant is assigned to the jth job if $x_{ij} = 1$. Definition VIII.13.1 guarantees that no person is assigned to two jobs and that no two persons are assigned to the same job.

We can now express the assignment problem, mathematically, in the form

8.13.1 Maximize

$$\sum_{j=1}^{n} \sum_{i=1}^{n} a_{ij} x_{ij}$$

Subject to

8.13.2 $\quad X$ is a permutation matrix.

Now, this is not, strictly speaking, a linear programming problem. In fact, the constraint set is finite rather than convex; this type of problem is called a *combinatorial* problem. It would not seem, at first glance, that the assignment problem could be solved by linear programming methods. On the other hand, we should remember that the solution to a linear program (if it exists) can always be found by considering a finite set of points, the extreme points of the constraint set. Hence the assignment problem reduces to a linear program if we can only choose a constraint set whose extreme points are precisely the permutation matrices.

VIII.13.2 Definition. An $n \times n$ matrix $X = (x_{ij})$ is said to be *doubly stochastic* if $x_{ij} \geq 0$ and, moreover,

$$\sum_{j=1}^{n} x_{ij} = 1 \qquad \text{for each } i$$

and

$$\sum_{i=1}^{n} x_{ij} = 1 \qquad \text{for each } j$$

In essence, a doubly stochastic matrix is a non-negative matrix whose row sums and column sums are all equal to 1. It is clear that an $n \times n$ matrix can be thought of as representing a point in n^2-dimensional space. If so, the set of doubly stochastic matrices forms a convex set, since it is defined solely by linear equations and inequalities. It is easy to see that a permutation matrix is doubly stochastic; the relation between doubly stochastic matrices and permutation matrices is given by the following theorem.

VIII.13.3 Theorem. The extreme points of the set of doubly stochastic matrices are precisely the permutation matrices.

We will not give a proof of Theorem VIII.13.3; it may be obtained by an adaptation of the proof of VIII.12.1. We can now restate the assignment problem in the form

8.13.3 Maximize

$$\sum_{i=1}^{n} \sum_{j=1}^{n} a_{ij} x_{ij}$$

Subject to

8.13.4 $$\sum_{j=1}^{n} x_{ij} = 1 \qquad \text{for each } i$$

8.13.5 $$\sum_{i=1}^{n} x_{ij} = 1 \qquad \text{for each } j$$

8.13.6 $$x_{ij} \geq 0$$

What we have done is to replace the set of permutation matrices by the set of doubly stochastic matrices. The program (8.13.3) to

(8.13.6) is clearly feasible and bounded. It must have a maximum, which is attained at one of the extreme points of the constraint set. This extreme point will be the optimal assignment.

It may be seen that program (8.13.3) to (8.13.6) is a transportation-type problem, with availabilities and requirements all equal to 1. Unfortunately, it cannot be handled by the Koopmans algorithm because it is extremely degenerate. In fact, the constraints (8.13.4) and (8.13.5) are $2n$ equations, one of which is redundant. Thus, a b.f.p. should have $2n-1$ non-zero variables. But we know that each b.f.p. has exactly n non-zero variables (one in each row of the matrix). This degeneracy is enough to make the Koopmans algorithm useless.

A common procedure in case of degeneracy is to solve, not the given (primal) linear program, but its dual.

Let us look, then, for the dual of (8.13.3) to (8.13.6). In Section VIII.11, the dual problem was defined, but only for programs with inequality constraints. This program has equation constraints. As was pointed out earlier, each equation can be replaced by two opposite inequalities; in general, we replace $\alpha = \beta$ by $\alpha \leq \beta$ and $-\alpha \leq -\beta$. Thus, two inequalities arise, in which the coefficients are negatives of each other. The dual program will have one variable for each of these inequalities; these two variables appear, in each of the constraints, with coefficients that are negatives of each other. This being so, we can deal with the difference of the two variables. The two variables are non-negative, but this places no such restriction on their difference. We have, as a rule for duality: *in duality, the variable corresponding to an equation constraint is not restricted to be non-negative.*

Using this rule, we can see that the dual of program (8.13.3) to (8.13.6) is

8.13.7 Minimize

$$\sum_{i=1}^{n} u_i + \sum_{j=1}^{n} v_j$$

Subject to

8.13.8
$$u_i + v_j \geq a_{ij}$$

The dual program attempts to put "shadow prices," u_i on the job applicants, and v_j on the several jobs. The number u_i represents, in some way, a "natural wage" for the ith applicant, while the w_j represents, perhaps, the value of the equipment used for the jth job.

Let us suppose that X^* is an optimal assignment, while $(\mathbf{u}^*, \mathbf{v}^*)$ is a solution to the dual program. We know from duality that this is equivalent to the condition

8.13.9
$$\sum \sum a_{ij} x_{ij}^* = \sum u_i^* + \sum v_j^*$$

and it is not difficult to see that a necessary condition is that,

8.13.10 $$\text{if } x_{ij}^* = 1, \text{ then } u_i^* + v_j^* = a_{ij}$$

This condition may be interpreted as saying that an employee's salary, plus the value of the job's equipment, should be no greater than his utility to the firm in that job.

Condition (8.13.10) is also sufficient: if there is a permutation matrix X^* such that (8.13.10) holds, then (8.13.9) will also hold, and X^* will be the optimal assignment. Our approach to the assignment problem will be along these lines. We shall look for vectors (**u**,**v**) satisfying (8.13.8) and decrease the objective function (8.13.7), until a permutation matrix X^*, satisfying (8.13.10) is found.

We shall assume that the entries in the matrix $A = (a_{ij})$ are all integers. This is not a great loss in generality, since, whatever the numbers a_{ij} may be, they can be approximated as closely as desired by rational numbers. In turn, these rational numbers may all be multiplied by their least common denominator, thus obtaining an obviously equivalent assignment problem. The following theorem, which is intuitively reasonable, is given without proof.

VIII.13.4 Theorem. Let $B = (b_{ij})$ be an $n \times n$ matrix all of whose entries are 0 or 1. A necessary and sufficient condition for the existence of a permutation matrix $X \leq B$ is that every set of k rows have non-zero entries in at least k columns.

Our method of solution of the assignment problem is as follows. A vector (**u**,**v**) satisfying (8.13.8) is found. This is most easily done by letting $v_j = 0$, and letting u_i be equal to the largest entry in the ith row. An auxiliary matrix $B = (b_{ij})$ is then formed, defined by

8.13.11 $$b_{ij} = \begin{cases} 1 & \text{if } u_i + v_j = a_{ij} \\ 0 & \text{if } u_i + v_j > a_{ij} \end{cases}$$

Note that, by (8.13.10), we can have $x_{ij}^* = 1$ only if $b_{ij} = 1$. That is, $x_{ij}^* \leq b_{ij}$. If there is a permutation matrix $X \leq B$, we know from the foregoing discussion that this is the optimal assignment. If not, then, by Theorem VIII.13.4, there is a set of k rows that have 1's in only $k-1$ or fewer columns. Let us say that these k rows form the set I, and the corresponding $k-1$ (or fewer) columns form the set J. We then define a new vector (**u**′,**v**′) by

8.13.12 $$u_i' = \begin{cases} u_i - 1 & \text{for } i \text{ in } I \\ u_i & \text{for other } i \end{cases}$$

8.13.13 $$v_j' = \begin{cases} v_j + 1 & \text{for } j \text{ in } J \\ v_j & \text{for other } j \end{cases}$$

Consider this vector. We notice, first of all, that $(\mathbf{u}',\mathbf{v}')$ satisfies (8.13.8). In fact, suppose $b_{ij}=0$. In this case, $a_{ij} < u_i + v_j$. Now, a_{ij}, u_i and v_j are all integers, and so

$$a_{ij} \leq u_i + v_j - 1$$

Now, by (8.13.12), $u_i' \geq u_i - 1$, and by (8.13.13), $v_j' \geq v_j$. Hence $a_{ij} \leq u_i' + v_j'$.

Suppose, on the other hand, that $b_{ij} = 1$. In this case, there are two possibilities. If i is in I, then j is in J, so $u_i' = u_i - 1$, $v_j' = v_j + 1$, and so $a_{ij} = u_i' + v_j'$. If i is not in I, then $u_i' = u_i$, $v_j' \geq v_j$, and so $a_{ij} \leq u_i' + v_j'$.

In any case, we see that $(\mathbf{u}', \mathbf{v}')$ satisfies (8.13.7). Consider, next, the objective function (8.13.8). We see that

$$\sum u_i' = \sum u_i - k$$

but

$$\sum v_j' \leq \sum v_j + k - 1$$

since there are k rows in I, but at most $k-1$ columns in J. At each step, the objective function (8.13.7) decreases by at least 1. It follows that the minimum of the program (8.13.7) and (8.13.8) will be reached in a finite number of steps; this will give us the optimal assignment. We conclude with an example of this algorithm.

VIII.13.5 Example. Find the optimal assignment corresponding to the matrix

$$A = \begin{pmatrix} 14 & 18 & 14 & 12 & 19 & 13 & 14 \\ 20 & 17 & 18 & 15 & 12 & 14 & 14 \\ 19 & 18 & 13 & 18 & 17 & 13 & 12 \\ 12 & 20 & 12 & 16 & 13 & 18 & 18 \\ 11 & 16 & 17 & 16 & 11 & 18 & 14 \\ 16 & 14 & 16 & 18 & 20 & 15 & 13 \\ 17 & 12 & 15 & 18 & 13 & 18 & 19 \end{pmatrix}$$

Starting by setting $v_j = 0$ and $\mathbf{u} = (19,20,19,20,18,20,19)$, we obtain the following auxiliary matrix:

$$B = \begin{pmatrix} 0 & 0 & 0 & 0 & 1 & 0 & 0 \\ 1 & 0 & 0 & 0 & 0 & 0 & 0 \\ 1 & 0 & 0 & 0 & 0 & 0 & 0 \\ 0 & 1 & 0 & 0 & 0 & 0 & 0 \\ 0 & 0 & 0 & 0 & 0 & 1 & 0 \\ 0 & 0 & 0 & 0 & 1 & 0 & 0 \\ 0 & 0 & 0 & 0 & 0 & 0 & 1 \end{pmatrix} \quad \overset{\mathbf{u}}{\begin{pmatrix} 19 \\ 20 \\ 19 \\ 20 \\ 18 \\ 20 \\ 19 \end{pmatrix}}$$

$$\mathbf{v} = (0 \quad 0 \quad 0 \quad 0 \quad 0 \quad 0 \quad 0)$$

We note that the first, second, third and sixth rows of the matrix B have 1's only in the first and fifth columns. We therefore decrease u_1, u_2, u_3, u_6, and increase v_1, v_5:

$$B = \begin{pmatrix} 0 & 1 & 0 & 0 & 1 & 0 & 0 \\ 1 & 0 & 0 & 0 & 0 & 0 & 0 \\ 1 & 1 & 0 & 1 & 0 & 0 & 0 \\ 0 & 1 & 0 & 0 & 0 & 0 & 0 \\ 0 & 0 & 0 & 0 & 0 & 1 & 0 \\ 0 & 0 & 0 & 0 & 1 & 0 & 0 \\ 0 & 0 & 0 & 0 & 0 & 0 & 1 \end{pmatrix} \overset{\mathbf{u}}{\begin{pmatrix} 18 \\ 19 \\ 18 \\ 20 \\ 18 \\ 19 \\ 19 \end{pmatrix}}$$

$$\mathbf{v} = (1 \quad 0 \quad 0 \quad 0 \quad 1 \quad 0 \quad 0)$$

This time, we note that first, second, third, fourth, and sixth rows of B have 1's only in the first, second, fourth, and fifth columns. We decrease u_1, u_2, u_3, u_4, and u_6, and increase v_1, v_2, v_4, and v_5:

$$B = \begin{pmatrix} 0 & 1 & 0 & 0 & 1 & 0 & 0 \\ 1 & 0 & 1 & 0 & 0 & 0 & 0 \\ 1 & 1 & 0 & 1 & 0 & 0 & 0 \\ 0 & 1 & 0 & 0 & 0 & 0 & 0 \\ 0 & 0 & 0 & 0 & 0 & 1 & 0 \\ 0 & 0 & 0 & 0 & 1 & 0 & 0 \\ 0 & 0 & 0 & 0 & 0 & 0 & 1 \end{pmatrix} \overset{\mathbf{u}}{\begin{pmatrix} 17 \\ 18 \\ 17 \\ 19 \\ 18 \\ 18 \\ 19 \end{pmatrix}}$$

$$\mathbf{v} = (2 \quad 1 \quad 0 \quad 1 \quad 2 \quad 0 \quad 0)$$

For this scheme, we note that the first, fourth, and sixth rows of B have 1's only in the second and fifth columns. We will therefore decrease u_1, u_4, and u_6, and increase v_2 and v_5, to obtain the scheme

$$B = \begin{pmatrix} 0 & 1 & 0 & 0 & 1^* & 0 & 0 \\ 1 & 0 & 1^* & 0 & 0 & 0 & 0 \\ 1^* & 0 & 0 & 1 & 0 & 0 & 0 \\ 0 & 1^* & 0 & 0 & 0 & 1 & 1 \\ 0 & 0 & 0 & 0 & 0 & 1^* & 0 \\ 0 & 0 & 0 & 1^* & 1 & 0 & 0 \\ 0 & 0 & 0 & 0 & 0 & 0 & 1^* \end{pmatrix} \overset{\mathbf{u}}{\begin{pmatrix} 16 \\ 18 \\ 17 \\ 18 \\ 18 \\ 17 \\ 19 \end{pmatrix}}$$

$$\mathbf{v} = (2 \quad 2 \quad 0 \quad 1 \quad 3 \quad 0 \quad 0)$$

We check, and find that this matrix B does, in fact, contain a permutation matrix; this is shown by the starred elements. The optimal assignment is shown by the starred elements in the matrix A that follows:

$$\begin{pmatrix} 14 & 18 & 14 & 12 & 19^* & 13 & 14 \\ 20 & 17 & 18^* & 15 & 12 & 14 & 14 \\ 19^* & 18 & 13 & 18 & 17 & 13 & 12 \\ 12 & 20^* & 12 & 16 & 13 & 18 & 18 \\ 11 & 16 & 17 & 16 & 11 & 18^* & 14 \\ 16 & 14 & 16 & 18^* & 20 & 15 & 13 \\ 17 & 12 & 15 & 18 & 13 & 18 & 19^* \end{pmatrix}$$

It may be checked directly that this is, indeed, the optimal assignment. In fact, we have a value of 131 for the objective functions of both programs; duality assures that these are the solutions.

We point out that this problem has 49 variables, constrained by 13 inequalities. Each simplex tableau would have 14 rows and 37 columns (counting the column of constant terms and the objective function row). If we tried, alternatively, to enumerate all the possible assignments, we would find that there are $7! = 5040$ such assignments. In either case, it is clear that this algorithm affords us a very substantial economy of time and labor.

Many types of problem can be solved by methods similar to those considered in this section. One such is the flow problem, which is that of maximizing the total flow through a network whose arcs and nodes have finite capacities. It is not, unfortunately, possible to consider all such problems here.

One that does not seem soluble by these methods is the so-called "traveling salesman" problem, which consists, essentially, of finding the shortest route through a network, touching all the nodes. Heuristically, it describes the problem faced by a salesman who wishes to visit several cities and is looking for the shortest (or cheapest) route through all of them. As such, it does not seem greatly different from the assignment problem; it cannot, however, be reduced to a linear program without the introduction of a very large number of constraints. No solution has yet been given in general (though some particular problems of this type have been solved).

PROBLEMS ON TRANSPORTATION AND ASSIGNMENT

1. Solve the following transportation problems. In each case, the matrix $C = (c_{ij})$ is the cost matrix, while $\mathbf{a} = (a_1, \ldots, a_m)^t$ and $\mathbf{b} = (b_1, \ldots, b_n)$ are the availability and requirement vectors, respectively. (In cases, in which the total availability is greater than the total requirement, introduce a *fictitious destination* to receive the remaining goods, with 0 costs from each warehouse to this destination.)

(a)
$$C = \begin{pmatrix} 3 & 1 & 4 \\ 1 & 5 & 9 \\ 2 & 6 & 5 \end{pmatrix} \quad \mathbf{a} = \begin{pmatrix} 100 \\ 150 \\ 120 \end{pmatrix}$$
$$\mathbf{b} = (140 \quad 90 \quad 140)$$

(b)
$$C = \begin{pmatrix} 3 & 5 & 8 & 9 \\ 7 & 9 & 3 & 2 \\ 3 & 8 & 4 & 6 \end{pmatrix} \quad \mathbf{a} = \begin{pmatrix} 50 \\ 70 \\ 40 \end{pmatrix}$$
$$\mathbf{b} = (30 \quad 45 \quad 30 \quad 55)$$

(c)
$$C = \begin{pmatrix} 1 & 4 & 5 & 1 & 2 \\ 2 & 6 & 1 & 3 & 5 \\ 1 & 3 & 0 & 1 & 1 \end{pmatrix} \quad \mathbf{a} = \begin{pmatrix} 80 \\ 80 \\ 40 \end{pmatrix}$$
$$\mathbf{b} = (50 \quad 50 \quad 20 \quad 35 \quad 45)$$

(d)
$$C = \begin{pmatrix} 1 & 4 & 1 & 5 & 2 \\ 2 & 6 & 3 & 4 & 1 \\ 5 & 8 & 5 & 6 & 4 \end{pmatrix} \quad \mathbf{a} = \begin{pmatrix} 90 \\ 90 \\ 30 \end{pmatrix}$$
$$\mathbf{b} = (40 \quad 60 \quad 25 \quad 35 \quad 40)$$

(e)
$$C = \begin{pmatrix} 2 & 6 & 5 & 3 & 5 \\ 2 & 7 & 1 & 8 & 2 \\ 1 & 4 & 1 & 4 & 2 \\ 1 & 7 & 3 & 2 & 3 \end{pmatrix} \quad \mathbf{a} = \begin{pmatrix} 50 \\ 50 \\ 50 \\ 50 \end{pmatrix}$$
$$\mathbf{b} = (40 \quad 40 \quad 40 \quad 40 \quad 40)$$

2. Formulate the duals of the foregoing problems (a) through (e). What heuristic interpretation can be given to them?

3. Solve the following assignment problems. In each problem, each row represents a candidate, and each column represents a position (If there are more candidates than positions, introduce *fictitious positions*, representing no job, with all entries 0 in the corresponding columns.) What salaries should be paid to the employees in each case?

(a)
$$\begin{pmatrix} 7 & 2 & 8 & 4 & 3 \\ 5 & 1 & 6 & 3 & 3 \\ 6 & 3 & 7 & 4 & 5 \\ 5 & 2 & 5 & 2 & 4 \\ 3 & 0 & 4 & 3 & 1 \end{pmatrix}$$

(b) $\begin{pmatrix} 1 & 4 & 1 & 5 & 9 & 2 & 6 \\ 5 & 3 & 5 & 8 & 9 & 7 & 9 \\ 3 & 2 & 3 & 8 & 4 & 6 & 2 \\ 6 & 4 & 3 & 3 & 8 & 3 & 2 \\ 7 & 9 & 3 & 7 & 5 & 1 & 0 \\ 6 & 0 & 3 & 9 & 9 & 2 & 7 \\ 1 & 8 & 2 & 8 & 5 & 3 & 4 \end{pmatrix}$

(c) $\begin{pmatrix} 1 & 2 & 8 & 3 & 5 \\ 2 & 6 & 7 & 5 & 2 \\ 4 & 9 & 6 & 3 & 3 \\ 2 & 4 & 8 & 5 & 5 \\ 1 & 3 & 5 & 4 & 1 \\ 3 & 8 & 8 & 5 & 2 \end{pmatrix}$

(d) $\begin{pmatrix} 2 & 6 & 5 & 8 & 4 & 3 \\ 1 & 3 & 5 & 7 & 9 & 2 \\ 2 & 5 & 7 & 8 & 1 & 4 \\ 3 & 6 & 9 & 9 & 2 & 5 \\ 1 & 5 & 4 & 6 & 2 & 5 \\ 4 & 7 & 8 & 3 & 1 & 6 \\ 2 & 3 & 6 & 3 & 1 & 0 \end{pmatrix}$

CHAPTER IX

ADVANCED PROBABILITY

1. DISCRETE PROBABILITY SPACES

In Chapter IV we studied the elementary notions of probability: sample spaces, random variables, and expected values. We shall consider here several extensions and generalizations of these notions.

A slight generalization of finite probability space is a *discrete* probability space. For such spaces, there is an infinite sequence (rather than a finite number) of possible outcomes. As a trivial example, we may consider the experiment of throwing a dart at a dart board, until it hits the bull's-eye. The number of throws required for this may be any positive integer.

We would like to generalize the ideas of Section IV.2. We have, now, a sequence $O_1, O_2, O_3, \ldots$ of possible outcomes, to which we assign positive numbers $P(O_i)$, their probabilities. Condition (4.2.1) must also be generalized, and it is here that our first difficulty arises. We are told that the sum of all the terms $P(O_i)$ must be equal to 1, but what, exactly, does this mean? The question is whether the sum of infinitely many positive numbers can be finite. A more fundamental question concerns the meaning of such a sum. We know how to add two numbers; induction allows us to extend this definition so as to add three, four, or any finite collection of numbers. We cannot, however, use induction to obtain the sum of an infinite collection.

An example may be useful at this point.

IX.1.1 Example. The manager of a steel mill has to fulfill a contract, and finds that he needs 1000 more tons of raw steel. He orders production of 1000 T, and this order is carried out. In this production, however, 100 T are used to prime the mill, and as a result, the manager finds that he is still 100 T short of fulfilling the contract. He therefore orders production of 100 T more tons. Production of these, however, requires 10 T to prime the mills, leaving a shortage of 10 T. He orders 10 T production. This time, 1 T is used for priming. The process continues through several more steps; each time, the manager orders just enough steel to fulfill the contract, and finds that one tenth of this amount is consumed for priming. Finally, he despairs of ever fulfilling the contract; no matter how often he orders, there is still a shortage. He therefore resigns, and retires to a quieter life.

For the benefit of those readers who may feel some sympathy for our hypothetical steel mill manager, we hasten to declare that the contract could indeed be fulfilled. We should, however, study the reasoning involved to understand the manager's error, which is based on the belief that the sum of infinitely many terms must be infinite.

Let us suppose that we are given an infinite sequence of numbers

$$a_1, a_2, a_3, \ldots, a_n, \ldots.$$

We would like (if possible) to find the sum of all these terms. We would then write, symbolically,

9.1.1 $$S = \sum_{i=1}^{\infty} a_i$$

and say that S is the sum of the *infinite series* of the a_i.

Consider the so-called n^{th} *partial sum*

9.1.2 $$S_n = \sum_{i=1}^{n} a_i = a_1 + a_2 + \cdots + a_n$$

The partial sum S_n is not the same as the sum S, of course. It seems reasonable, however, to expect that, if S exists (and is finite) then the partial sums S_n should be close to S, especially if n is large (i.e., the sum of very many of the terms should be nearly equal to the sum of all the terms). We define this as a limit.

IX.1.2 Definition. If the partial sums S_n, defined by (9.1.2), tend to a limit S as n increases, then this limit S is the sum of the infinite series (9.1.1). We say then that the series *converges* to S. If the partial sums S_n do not tend to a limit, we say that the series *diverges*.

As in Chapters V and VI, we will not worry too much about giving the definition of a limit. We can define it, as we did in Chapter V, by saying that a limit is the result of a trend.

IX.1.3 Example. Geometric Series. The reader is doubtless familiar with geometric sequences, such as

$$a, ar, ar^2, ar^3, \ldots, ar^n, \ldots.$$

defined by the equation

9.1.3 $$t_i = ar^{i-1}$$

This is known as a geometric sequence, with initial term a, and ratio r. It is not difficult to evaluate the partial sums; if $r \neq 1$, we will have

$$S_n = a + ar + \ldots + ar^{n-1} = \frac{a - ar^n}{1 - r}$$

as may be verified by a direct division of the right-hand side. It is not difficult to see that, if $-1 < r < 1$, then the power r^n will tend to 0 as n increases. It follows that, for large values of n, the terms S_n will approach the limit

$$S = \frac{a}{1 - r}$$

which is therefore the sum of the series:

9.1.4 $$\sum_{i=1}^{\infty} ar^{i-1} = \frac{a}{1 - r}$$

On the other hand, it is easy to see that, if $r \leq -1$, or $r \geq 1$, then the partial sums S_n do not tend to a limit. We conclude that the geometric series converges, with sum (9.1.4), for $-1 < r < 1$, but diverges otherwise.

Returning now to Example IX.1.1, it is easy to see that the total amount of steel which must be produced is the sum of a geometric series with initial term $a = 1000$, and ratio $r = 0.1$. Applying (9.1.4), we see that the series has the sum

$$S = \frac{1000}{1 - 1/10} = \frac{10000}{9}$$

Thus the total amount to be produced is only 1111 1/9 T, not an infinite amount as had been feared. Of this, 111 1/9 will go for priming, leaving a net production of 1000 T.

Now that we have the necessary mathematical background, we proceed to the study of discrete probability spaces and discrete random variables.

IX.1.4 Definition. A discrete probability space is a sequence of possible outcomes

$$X = \{O_1, O_2, O_3, \ldots\}$$

together with a function P which assigns a non-negative number to each of the outcomes O_i, in such a way that

9.1.5 $$\sum_{i=1}^{\infty} P(O_i) = 1$$

Generally speaking, events and probabilities of events are defined for discrete spaces just as for finite spaces.

IX.1.5 Example. An experiment can have as its outcome any positive integer k, with probabilities

$$P(k) = \frac{1}{2^k} \quad \text{for } k = 1, 2, \ldots$$

What is the probability of obtaining an even number?

We should, first of all, check that (9.1.5) holds. In fact,

$$\sum_{k=1}^{\infty} P(k) = \sum_{k=1}^{\infty} \left(\frac{1}{2}\right)^k$$

Now, this is a geometric series, with first term

$$a = \frac{1}{2}$$

and ratio

$$r = \frac{1}{2}$$

Applying the formula for such series, we find

$$S = \frac{a}{1-r} = \frac{\frac{1}{2}}{1 - \frac{1}{2}} = 1$$

and so (9.1.5) does hold.

Now, let E be the event "even number." Thus,

$$E = \{2,4,6,8, \ldots\}$$

and

$$P(E) = \frac{1}{2^2} + \frac{1}{2^4} + \frac{1}{2^6} + \ldots$$

Now, this is also a geometric series, with first term

$$a = \frac{1}{4}$$

and ratio

$$r = \frac{1}{4}$$

so that, for this series,

$$\frac{a}{1-r} = \frac{\frac{1}{4}}{1 - \frac{1}{4}} = \frac{1}{3}$$

Therefore,

$$P(E) = \frac{1}{3}$$

2. DISCRETE RANDOM VARIABLES

In Section 1 of this chapter, we defined a discrete probability space: it is one which has an infinite sequence of possible outcomes. A *discrete random variable* is defined similarly: it is a random variable with a sequence $x_1, x_2, x_3, \ldots$ of possible values. These values have probabilities $f(x_i)$; the numbers $f(x_i)$ are non-negative and satisfy the equation

9.2.1 $$\sum_{i=1}^{\infty} f(x_i) = 1$$

The expectation value of such a random variable is defined by

9.2.2 $$E[X] = \sum_{i=1}^{\infty} x_i f(x_i),$$

assuming that this series converges. Now, the fact that (9.2.1) converges will not guarantee the convergence of (9.2.2). Thus, the expected value of a discrete random variable need not exist; even if the mean exists, the variance need not exist. We will not discuss these possibilities except to state that if the mean exists then Theorem IV.10.2 (The Law of Large Numbers) will hold. If, additionally, the variance exists, then Theorem IV.10.1 (Chebyshev's Inequality) will hold. (In case the mean exists, but not the variance, the proof of IV.10.2 is considerably more complicated.)

The most interesting discrete random variables are integer-valued variables: variables whose possible values are the positive integers (possibly including zero). We give two examples.

IX.2.1 Example. The Geometric Distribution. A simple experiment is to be repeated until a success is obtained. If the probability of success on any trial is p, what is the expected number of trials, X, necessary for this?

Is is clear that this is a discrete integral-valued random variable. To determine its distribution, note that, for X to have the value k, we must have failures on the first $k-1$ trials, followed by a success on the kth trial. Since the trials are independent, we will have

9.2.3
$$f(k) = pq^{k-1} \quad k = 1,2,3, \ldots .$$

where $q = 1 - p$.
The distribution (9.2.3) is known as a *geometric distribution.* It is not difficult to check (by the use of the formula for geometric series) that $\Sigma f(k) = 1$.

The mean of X is given by

$$E[X] = \sum_{k=1}^{\infty} kpq^{k-1}$$

We must evaluate this series. The nth partial sum is

$$S_n = p + 2pq + 3pq^2 + \cdots + npq^{n-1}$$

Then

$$qS_n = pq + 2pq^2 + \cdots + (n-1)pq^{n-1} + npq^n$$

and so, subtracting,

$$(1-q)S_n = p + pq + pq^2 + \cdots + pq^{n-1} - npq^n$$

Using the fact that $1 - q = p$, and the formula for geometric series,

$$pS_n = p\,\frac{1-q^n}{1-q} - npq^n$$

and so

$$S_n = \frac{1 - q^n}{p} - nq^n$$

Now, as n increases, the product nq^n will approach 0, as does q^n. We conclude that

$$\sum_{k=1}^{\infty} kpq^{k-1} = \frac{1}{p}$$

and so

$$E[X] = \frac{1}{p}$$

Thus X has mean $1/p$. Similar considerations will show that the variance of X is q/p^2.

IX.2.2 Example. A man, looking for a job, visits employment agencies. He feels that, at each agency, he has a 0.4 probability of finding a job, and that these probabilities are independent. What is the expected number of agencies which he must visit to find a job? If he visits four agencies per day, what is the probability that he will find a job during the first day? During the second day? What is the probability that he will find one during the third day, given that he has not found one during the first two days?

If we let X be the number of agencies visited, it is clear that X is a geometric random variable, with parameter $p = 0.4$. The expected value of X is $1/p$, or 2.5.

The probability that he finds a job during the first day (visiting, as he does, four agencies) will be

$$Pr\ (X \leq 4) = 0.4 + 0.4(0.6) + 0.4(0.6)^2 + 0.4(0.6)^3 = 0.8704.$$

The probability of finding a job in the second day is

$$Pr\ (5 \leq X \leq 8) = 0.4(0.6)^4 + 0.4(0.6)^5 + 0.4(0.6)^6 + 0.4(0.6)^7 = 0.1128.$$

The probability of finding a job during the third day, given that he has not found one during the first two days, is

$$\frac{Pr(9 \leq X \leq 12)}{Pr(x \geq 9)} = \frac{0.4(0.6)^8 + 0.4(0.6)^9 + 0.4(0.6)^{10} + 0.4(0.6)^{11}}{1 - 0.8704 - 0.1128} = 0.8704$$

so that the probability of finding a job on the third day is no better than on the first day. This is, of course, due to the independence which we assumed. Generally, the geometric random variable has this property: it lacks "memory," in the sense that the job-hunter's past failures do not affect his present probability of finding a job within a day.

IX.2.3 Example. The *Poisson distribution* is a limiting case of the binomial distribution, obtained by assuming that n (the number of trials) is large (say of the order of 100 or larger) whereas p (the probability of success) is very small (say 0.01 or less) so that np is not large.

We have

$$B(n, k; p) = \frac{n!}{k!\,(n-k)!}\,p^k q^{n-k}$$

Let us assume k is a small integer. Then, since n is large, we will have

$$\frac{n!}{(n-k)!} = n(n-1)\ldots(n-k+1)$$

which is approximately equal to n^k.

Thus

$$B(n, k; p) \cong \frac{n^k p^k}{k!}\,q^{n-k}$$

Writing $np = a$, this gives us

$$B(n,k;p) \cong \frac{a^k}{k!}\left(1-\frac{a}{n}\right)^n q^{-k}$$

Assuming p is small, q^{-k} will be quite close to 1, and so

$$B(n,k;p) \cong \frac{a^k}{k!}\left(1-\frac{a}{n}\right)^n$$

Consider, now, the expression

$$h = \left(1-\frac{a}{n}\right)^n$$

It may be shown that, for large n, we have

$$h \cong e^{-a}$$

We obtain thus the approximation

$$B(n,k;p) \cong \frac{a^k e^{-a}}{k!}$$

The discrete random variable, X, with non-negative integral values, and distribution

9.2.4 $$f(k) = \frac{a^k e^{-a}}{k!} \quad k = 0,1,2, \ldots$$

is known as a *Poisson variable with parameter a.* It is not too difficult to see that its mean and variance are both equal to the parameter a. In fact, treating it as a limiting case of the binomial distribution, we will have $\mu = np = a$ and $\sigma^2 = npq = a$ (since q approaches 1 as a limit). The *Poisson distribution* is also known as the *distribution of rare events;* it appears as the distribution of the number of earthquakes in a period of time, as the distribution of the number of eggs lain by an insect, and so on.

IX.2.4 Example. A biscuit company produces cookies by putting a quantity of raisins in the dough and mixing thoroughly before baking. How many raisins should be put into the dough to guarantee that 80 per cent of the cookies have at least two raisins?

Let us assume that there is enough dough for m cookies, and that n raisins are put into the mixture, where both m and n are large. If so, we can consider, for each of the cookies, the "simple experiment" which is a success if a given raisin goes into the cookie, and a failure otherwise. It is clear that this experiment has probability $1/m$ of success, since the raisin might go with equal probability into any one of the m cookies. The number of raisins in a given cookie will then be a random variable equal to the number of successes in n trials (one for each raisin). Since n is large and $p = 1/m$ is small, this is approximately a Poisson distribution with parameter $a = n/m$. The probability that a given cookie will have no raisins at all is then given by (9.2.4):

$$f(0) = \frac{a^0 e^{-a}}{0!} = e^{-a}$$

and the probability that it will have exactly one raisin is given by

$$f(1) = \frac{a^1 e^{-a}}{1!} = ae^{-a}$$

Now, we want the sum of these probabilities to be 0.2, so that 80 per cent of the cookies will have at least 2 raisins. Thus

$$(a + 1)e^{-a} = 0.2$$

This is a transcendental equation, which cannot be solved exactly. It is not difficult to check, however (by the use of tables of the exponential function) that $a = 3$ is an approximate solution. Since we have $n = am$, we conclude that the number of raisins put into the mixture should be at least 3 times the number of cookies which are to be made from the mixture.

IX.2.5 Example. Cars arrive at a toll gate, during a certain period, randomly and independently of each other, at an average rate of one car every two minutes. What is the probability that, during a given one-minute period, no cars arrive at the toll gate? At least two cars?

We have here another application of the Poisson distribution. In effect, the total number of cars passing the gate will be large, but as the one-minute interval is small, the probability that any will arrive during the interval will be very small. From independence, and past considerations, it will follow that we have here a Poisson process. The number of cars arriving in the period has a Poisson distribution with mean $a = 1/2$ (since the average rate is one car every two minutes). Then

$$f(0) = \frac{a^0 e^{-a}}{0!} = e^{-1/2} = 0.61$$

so there is a 61 per cent probability that no cars will arrive.

For the probability that at least two cars arrive, we consider the complementary event: at most one car arrives. Then

$$Pr(X \leq 1) = Pr(X = 0 \text{ or } X = 1)$$

Now

$$f(1) = \frac{a^1 e^{-a}}{1!} = \frac{1}{2} e^{-1/2} = 0.30$$

and so

$$Pr(X \leq 1) = 0.61 + 0.30 = 0.91.$$

Thus the probability that two or more cars arrive (during the one-minute period) is 0.09.

IX.2.6 Example. The Petersburg Paradox. Consider the following game. A player tosses a fair coin until it comes up heads. If it comes up heads for the first time on the kth trial (where $k = 1, 2, \ldots$) the player receives the amount 2^k. How much should the player pay for the privilege of playing this game?

By comparing with the geometric distribution (Example IX.2.1) we see that the probability that the game ends on the kth trial (since $p = q = 1/2$) will be 2^{-k}. Thus the player's winnings, X, can have the values 2^k for $k = 1,2,3, \ldots$, with the probabilities

$$f(2^k) = 2^{-k}$$

Its expected value is

$$E[X] = \sum_{k=1}^{\infty} 2^k \cdot 2^{-k}$$

But it is easy to see that each term in this sum is equal to 1. It follows that X does not have a mean, since the partial sums of the series can be made to increase without bound. The criterion of maximizing expected profit would then suggest that the player be willing to pay any amount, however large, for the privilege of playing the game (since he would expect to win back much more). This absurdity is known as the *Petersburg Paradox.*

PROBLEMS ON DISCRETE RANDOM VARIABLES

1. It may be proved that, if X and Y are independent Poisson variables with parameter a and b, respectively, then their sum $Z = X + Y$ is a Poisson variable with parameter $a + b$. Verify this by letting $a = 1$, $b = 2$, and computing the probabilities

$$P(Z = k)$$

for $k = 0, 1, 2,$ and 3.

2. A mass of cookie dough is mixed with raisins averaging two raisins per cookie. What is the probability that a cookie will have no raisins? exactly one raisin?

3. A machine produces parts to meet specifications; the probability that a part will fit the specifications is 0.98. A lot of 50 parts is taken. What is the probability that it will have 3 or more defectives?

4. A certain city suffers an average of 3 earthquakes per year. What is the probability that it will have 3 or more earthquakes in one six-month period? (Assume a Poisson distribution.)

5. A florist has a very perishable type of flower which he buys for \$2 and sells for \$5. If the demand for this flower is a Poisson variable with parameter 3, how many flowers should the florist buy so as to maximize his expected profits?

6. The random variable X has positive integer values, with the probabilities

$$f(n) = \frac{1}{n(n+1)} \qquad n = 1,2,3, \ldots$$

(a) Prove that

$$\sum_{n=1}^{\infty} f(n) = 1.$$

Hint: Rewrite $f(n)$ as the difference of two fractions.

(b) Find the probability that $X \geq 5$. Given that $X \geq 5$, what is the probability that $X = 6$?

7. Modify Problem 5, above, to the case of a geometric distribution with parameter $p = .4$.

3. MARKOV CHAINS

A Markov chain is a probabilistic process by which a system changes from one of several possible states to another at intervals of time. Examples of Markov processes are the manner in which the length of a waiting line changes, the way in which a gambler's fortune varies, or the economic growth of a country. The characteristic property of such processes is that past history is unimportant; i.e., the future behavior of the system depends (up to a point) on its present state, but is independent of its state at any previous moment of time. Thus, knowledge of the present state gives as much information (for the purposes of prediction) as knowledge of the system's entire history.

We shall assume here that the system has only a finite number of possible states and that the *transitions* (changes from one state to another) can happen only at discrete intervals of time (say, every minute or every hour).

Let us suppose that a Markov process has n states. It is clear from our discussion that the process is entirely described by giving the n^2 probabilities of transition from one state to another: these n^2 numbers can be arranged to form an nth order matrix

$$A = (a_{ij}) \quad i,j = 1,2, \ldots, n$$

where a_{ij} is the conditional probability that the system will be in state j at time $t+1$, given that it is in state i at time t. The matrix A satisfies the conditions

9.3.1 $$a_{ij} \geq 0 \quad \text{for all } i,j$$

9.3.2 $$\sum_{j=1}^{n} a_{ij} = 1 \quad \text{for each } i = 1, \ldots, n$$

The reasons for conditions (9.3.1) and (9.3.2) should be clear. Any matrix that satisfies these conditions is known as a *stochastic matrix*.

(Compare this with the definition of a *doubly* stochastic matrix in Chapter VIII.)

IX.3.1 Example. Two gamblers, G_1 and G_2, have a total of n units in their possession. They carry out a sequence of independent trials of a simple experiment with probability p of success. If the experiment is a success, G_2 pays 1 unit to G_1; if a failure, G_1 pays 1 unit to G_2. This process terminates if either of the two gamblers is ever wiped out (i.e., if he has no units left).

This is a Markov process with $n+1$ states, the ith state (for $i=0, 1, \ldots, n$) coming when G_1 has i units. It is easy to see that, for $1 \leq i \leq n-1$, the system has a probability p of moving to state $i+1$, and q of moving to state $i-1$. For $i=0$ or n, however, there are no further changes: the process is said to have an *absorbing barrier* at states 0 and n. We have, thus:

$$
\begin{aligned}
a_{i,i+1} &= p && \text{for } i=1, \ldots, n-1 \\
a_{i,i-1} &= q && \text{for } i=1, \ldots, n-1 \\
a_{00} &= 1 && \\
a_{nn} &= 1 && \\
a_{ij} &= 0 && \text{for all other } (i,j)
\end{aligned}
$$

For $n=5$, the transition matrix has the form

$$
A = \begin{pmatrix}
1 & 0 & 0 & 0 & 0 & 0 \\
q & 0 & p & 0 & 0 & 0 \\
0 & q & 0 & p & 0 & 0 \\
0 & 0 & q & 0 & p & 0 \\
0 & 0 & 0 & q & 0 & p \\
0 & 0 & 0 & 0 & 0 & 1
\end{pmatrix}
$$

IX.3.2 Example. Consider the following process: an urn contains n balls, k of which are red, and $n-k$, black. At each step, there is a probability k/n that one of the red balls will become black, and an independent probability $(n-k)/n$ that one of the black balls will become red.

Let the kth state hold when the urn has k red balls. We see that we have $n+1$ possible states. To compute the transition probabilities, note that we pass from state k to $k+1$ if one of the black balls "mutates" (i.e., becomes red) but none of the red balls mutates. Therefore

$$
a_{k,k+1} = \frac{n-k}{n}\left(1-\frac{k}{n}\right) = \left(\frac{n-k}{n}\right)^2
$$

 Similarly, we find

$$a_{k,k-1} = \left(\frac{k}{n}\right)^2$$

and, finally,

$$a_{kk} = 1 - \left(\frac{n-k}{n}\right)^2 - \left(\frac{k}{n}\right)^2$$

We thus have:

$$a_{k,k+1} = \left(\frac{n-k}{n}\right)^2$$

$$a_{k,k-1} = \left(\frac{k}{n}\right)^2$$

$$a_{kk} = \frac{2nk - 2k^2}{n^2}$$

$$a_{ij} = 0 \quad \text{otherwise}$$

For $n=5$, the transition matrix would have the form

$$A = \begin{pmatrix} 0 & 1 & 0 & 0 & 0 & 0 \\ 1/25 & 8/25 & 16/25 & 0 & 0 & 0 \\ 0 & 4/25 & 12/25 & 9/25 & 0 & 0 \\ 0 & 0 & 9/25 & 12/25 & 4/25 & 0 \\ 0 & 0 & 0 & 16/25 & 8/25 & 1/25 \\ 0 & 0 & 0 & 0 & 1 & 0 \end{pmatrix}$$

In dealing with Markov processes it is often of interest to predict the state of the system, not in the time period that follows immediately, but at some later period. It becomes natural to look for "*s*-period" transition probabilities, $a_{ij}^{(s)}$: we define $a_{ij}^{(s)}$ as the probability that the system will be in state j at time $t+s$ given that it is in state i at time t. Then

$$A^{(s)} = (a_{ij}^{(s)})$$

is the *s-period transition matrix.*

It is clear that the matrices $A^{(s)}$ depend on the matrix A. Let us see how they are to be computed. What, for instance, is $a_{ij}^{(2)}$? We compute this by considering all the possible ways in which the system can reach state j at time $t+2$, given that it is in state i at time t.

Generally speaking, the system can pass from state i to state k at the intermediate time $t+1$ and then from k to j at time $t+2$. The probability of this is $a_{ik}a_{kj}$. Since k is arbitrary we must add all these terms, obtaining

9.3.3 $$a_{ij}^{(2)} = \sum_{k=1}^{n} a_{ik}a_{kj}$$

But this is simply the rule (studied in Chapter VII) for the multiplication of matrices (in this case, for multiplying A with itself.) We conclude that

$$A^{(2)} = A\ A = A^2$$

and, in general,

9.3.4 $$A^{(s)} = A^s$$

so that the *s-period transition matrix is simply the* sth *power of the single-period transition matrix.*

IX.3.3 Example. The Markov chain with transition matrix

$$A = \begin{pmatrix} q & 1-q \\ 0 & 1 \end{pmatrix}$$

has the 2-period transition matrix

$$A^2 = \begin{pmatrix} q^2 & 1-q^2 \\ 0 & 1 \end{pmatrix}$$

and, in general, it will have the s-period matrix

$$A^s = \begin{pmatrix} q^s & 1-q^s \\ 0 & 1 \end{pmatrix}$$

IX.3.4 Example. The Markov chain with transition matrix

$$A = \begin{pmatrix} 0.3 & 0.6 & 0.1 \\ 0.1 & 0.5 & 0.4 \\ 0.4 & 0.1 & 0.5 \end{pmatrix}$$

has the 2-period matrix

$$A^2 = \begin{pmatrix} 0.19 & 0.49 & 0.32 \\ 0.24 & 0.35 & 0.41 \\ 0.33 & 0.34 & 0.33 \end{pmatrix}$$

and the 3-period matrix

$$A^3 = \begin{pmatrix} 0.234 & 0.391 & 0.375 \\ 0.271 & 0.360 & 0.369 \\ 0.265 & 0.401 & 0.334 \end{pmatrix}$$

4. REGULAR AND ABSORBING MARKOV CHAINS

For most Markov chains, the state in the near future will depend considerably on the present state. On the other hand, it is not inconceivable that (for many such chains at least) the importance of the initial state should vanish over a long period of time. This should be especially so in a Markov chain such as that of Example IX.3.2, in which the system's characteristics seem to push it away from the extremes of many red or many black balls. Example IX.3.4 also seems to behave in this manner: note how the three rows, which are quite dissimilar in A, are nearly equal in A^3. The pattern is even more obvious in the 4-period matrix:

$$A^4 = \begin{pmatrix} 0.259 & 0.374 & 0.367 \\ 0.265 & 0.380 & 0.355 \\ 0.253 & 0.393 & 0.354 \end{pmatrix}$$

Further multiplications would show that for large s, the rows of $A^{(s)}$ are almost equal; in fact it would be seen that they approach the matrix

$$V = \begin{pmatrix} 0.259 & 0.382 & 0.359 \\ 0.259 & 0.382 & 0.359 \\ 0.259 & 0.382 & 0.359 \end{pmatrix}$$

Thus, for large s, the state at time $t+s$ depends only slightly on the state at time t. The vector $\mathrm{v} = (0.259, 0.382, 0.359)$ is a solution of the equation

$$\mathrm{v}A = \mathrm{v}$$

and shows the "steady state" behavior of the system.

We shall show, now, that this convergence of the matrices A^s occurs quite often. Let us assume, first, that all the entries a_{ij} are positive. Let a be the smallest of all these:

$$a = \min_{i\ j} a_{ij}$$

and let $r = 1 - 2a < 1$. Now let $d(s)$ be the largest difference between entries in the same column of the matrix A^s:

$$d(s) = \max_{i,j,k} \{a_{ij}^{(s)} - a_{kj}^{(s)}\}$$

Intuitively, it is easy to see that, if $d(s)$ is small then the rows of A^s are all nearly equal.

We use now the fact that $A^{s+1} = AA^s$ and so

$$a_{ij}^{(s+1)} = \sum_{l=1}^{n} a_{il} a_{lj}^{(s)}$$

We know that A is a stochastic matrix, so

$$\sum_{l=1}^{n} a_{il} = 1$$

Let $a_{pj}^{(s)}$ be the largest entry, and $a_{qj}^{(s)}$ the smallest entry in the jth row of A^s. We have then

$$a_{ij}^{(s+1)} = a_{ip} a_{pj}^{(s)} + \sum_{l \neq p} a_{il} a_{lj}^{(s)}$$

Now $a_{ip} \geq a$, and, for all l, $a_{lj}^{(s)} \geq a_{qj}^{(s)}$. It will follow that

9.4.1 $$a_{ij}^{(s+1)} > a \cdot a_{pj}^{(s)} + (1-a) a_{qj}^{(s)}$$

In a similar way, we will see that:

9.4.2 $$a_{kj}^{(s+1)} \leq (1-a) a_{pj}^{(s)} + a \cdot a_{q}^{(s)}$$

and so

$$a_{kj}^{(s+1)} - a_{ij}^{(s+1)} \leq r(a_{pj}^{(s)} - a_{qj}^{(s)})$$

or

$$a_{kj}^{(s+1)} - a_{ij}^{(s+1)} \leq r\, d(s)$$

Since this holds for all i, j, and k;

9.4.3 $$d(s+1) = \max_{i,j,k} \{a_{kj}^{(s+1)} - a_{ij}^{(s+1)}\} \leq r d(s)$$

We conclude (from (9.4.1) and (9.4.2)), that each entry in the jth column of A^{s+1} is at least as large as the smallest entry in the jth row of A^s, and at least as small as the largest entry in the jth row of A^s. We conclude from (9.4.3) that $d(s) \leq r^s$, and so, as s increases, $d(s)$ tends toward zero, so that, for large s, the rows of $A^{(s)}$ are nearly equal to each other and also nearly equal to the rows of $A^{(s+1)}$ and all higher powers of A. It will follow that the matrices A^s approach a limiting matrix, V, all of whose rows are equal: the rows of V represent the "steady state" condition of the Markov chain.

In general, it is not necessary that all the entries in A be positive; it is sufficient that all the entries in one of the matrices $A^{(s)}$ be positive: convergence can be proved similarly for such matrices.

IX.4.1 Definition. A stochastic matrix A is said to be *regular* if one of its powers, A^s, has all positive entries.

We give the following theorem without further proof:

IX.4.2 Theorem. Let A be a regular matrix. Then

(a) The matrices A^s approach a matrix, V, as s grows.
(b) The rows of V are all equal to the vector $\mathbf{v}$.
(c) $\mathbf{v}A = \mathbf{v}$.

Generally speaking, the easiest way to find the limit matrix V is to solve the equation $\mathbf{v}A = \mathbf{v}$. If A is regular, this equation will have only one solution that satisfies also the stochastic condition

$$\sum v_i = 1$$

IX.4.3 Example. The cigarette company that manufactures brand C starts an aggressive advertising campaign. The results of this campaign are such that, of people smoking brand C in a given week, 80 per cent continue to smoke it the following week; of those smoking any other brands in a given week, 40 per cent are won over to brand C the following week. In the long run, what fraction of the smokers will be smoking brand C?

We have here a Markov chain with transition matrix

$$A = \begin{pmatrix} 0.8 & 0.2 \\ 0.4 & 0.6 \end{pmatrix}$$

Consider the equation

$$\mathbf{v}A = \mathbf{v}$$

or

$$\begin{aligned} 0.8v_1 + 0.4v_2 &= v_1 \\ 0.2v_1 + 0.6v_2 &= v_2 \end{aligned}$$

This has the solution

$$v_2 = 2v_1$$

Since we want $v_1 + v_2 = 1$, we obtain

$$\mathbf{v} = \left(\frac{2}{3}, \frac{1}{3}\right)$$

It follows that

$$V = \begin{pmatrix} \frac{2}{3} & \frac{1}{3} \\ \frac{2}{3} & \frac{1}{3} \end{pmatrix}$$

In the long run the company will have 2/3 of the market. Note that the initial state does not matter; even starting from scratch, the company will, eventually, "nearly" capture its share of the market (though, of course, starting from scratch, it will take longer to do so).

If the transition matrix is not regular, there is still a possibility that the process may behave in this manner (see Example IX.3.3), but generally this will not happen. A special case arises when some of the states in the process form *absorbing barriers*: once the system reaches such a state, there is no further change. Example IX.3.1 has this property, inasmuch as the game terminates as soon as either of the two gamblers is wiped out. It is not too difficult to see, from this example, that, if s is large enough, then the game will "very probably" have terminated before s trials of the experiment; one of the two will have been wiped out. It may be shown (by a process similar to the proof of Theorem IX.4.2) that, as s increases, the matrices A^s approach, as limit, a matrix W, which has non-zero entries only in the first and last columns. The entry w_{i0} then represents the probability that G_1 will eventually be wiped out, given that he starts with i units.

IX.4.4 Example. Let us once again consider the two gamblers of Example IX.3.1, with $p = 0.4$ and $q = 0.6$. What is the probability that G_1 will be wiped out?

We wish here to compute the limit matrix W, which, as mentioned above, will have the form

$$W = \begin{pmatrix} w_{00} & 0 & 0 & 0 & 0 & w_{05} \\ w_{10} & 0 & 0 & 0 & 0 & w_{15} \\ w_{20} & 0 & 0 & 0 & 0 & w_{25} \\ w_{30} & 0 & 0 & 0 & 0 & w_{35} \\ w_{40} & 0 & 0 & 0 & 0 & w_{45} \\ w_{50} & 0 & 0 & 0 & 0 & w_{55} \end{pmatrix}$$

where, for each i, $w_{i0} + w_{i5} = 1$. Here, w_{i0} is G_1's probability of being wiped out, if he starts with i units. It is not difficult to see that $w_{00} = 1$ (if G_1 starts with 0 units, he is already wiped out), while $w_{50} = 0$ (if he starts with 5 units, he has already won).

Since W is the limit of the matrices A^s, we see that it must satisfy

$$AW = W$$

and, in particular, we must have

$$Aw = w$$

or

$$(A - I)w = 0$$

where w is the first column of the matrix W. We have, therefore,

$$\begin{pmatrix} 0 & 0 & 0 & 0 & 0 & 0 \\ 0.6 & -1 & 0.4 & 0 & 0 & 0 \\ 0 & 0.6 & -1 & 0.4 & 0 & 0 \\ 0 & 0 & 0.6 & -1 & 0.4 & 0 \\ 0 & 0 & 0 & 0.6 & -1 & 0.4 \\ 0 & 0 & 0 & 0 & 0 & 0 \end{pmatrix} \begin{pmatrix} 1 \\ w_{10} \\ w_{20} \\ w_{30} \\ w_{40} \\ 0 \end{pmatrix} = \begin{pmatrix} 0 \\ 0 \\ 0 \\ 0 \\ 0 \\ 0 \end{pmatrix}$$

or equivalently,

$$\begin{pmatrix} -1 & 0.4 & 0 & 0 \\ 0.6 & -1 & 0.4 & 0 \\ 0 & 0.6 & -1 & 0.4 \\ 0 & 0 & 0.6 & -1 \end{pmatrix} \begin{pmatrix} w_{10} \\ w_{20} \\ w_{30} \\ w_{40} \end{pmatrix} = \begin{pmatrix} -0.6 \\ 0 \\ 0 \\ 0 \end{pmatrix}$$

(The reader should check that these two matrix equations are indeed equivalent.) This system is easily solved, to give

$$\begin{pmatrix} w_{10} \\ w_{20} \\ w_{30} \\ w_{40} \end{pmatrix} = \begin{pmatrix} 0.924 \\ 0.810 \\ 0.640 \\ 0.384 \end{pmatrix}$$

Thus, G_1's probability of being wiped out is 0.924 if he starts with 1 unit, 0.810 with 2, 0.640 with 3, and 0.384 with 4.

PROBLEMS ON MARKOV CHAINS

1. When a company president finds an item on his desk, he may:

(a) send it to the vice-president, to act on it the following day;

(b) leave it on his own desk for the following day, or

(c) act on it immediately; the probabilities of these three options are 0.3, 0.6, and 0.1 respectively. Similarly, the vice-president has the options of keeping the paper on his own desk, sending it to the president, or acting on it immediately; these choices have probability 0.5, 0.2, and 0.3 respectively.

Represent this as a Markov chain, and form a transition matrix. If the president has the paper, what is the probability that some action will be taken within three days?

2. Three soap brands, I, II, and III, dominate their field. There is a continuous switching among customers, which can be represented probabilistically. If a customer is now using brand I, there is 0.6 probability that he will continue to use it the following week,

0.3 probability that he will switch to brand II, and 0.1 that he will switch to III. If he is using II, there is 0.5 probability that he will continue to use II, 0.4 that he will switch to I, and 0.1 that he will switch to III. If he is using brand III, there is 0.7 probability that he will continue to use it, and 0.3 that he will switch to brand II.

Given that a customer is now using brand II, what is the probability that he will be using brand I, three weeks from now?

3. Using the data of Problem 2, assume that brands I, II, and III are presently holding 50 per cent, 30 per cent and 20 per cent, respectively, of the market. What shares will they hold two weeks from now?

4. Again using the data of Problem 2, what share of the market will the three brands hold in the long run?

5. Customers arrive at random at a service counter that can hold four people. In an interval of time, there is a probability 0.3 that a new customer will arrive (though he will be turned away if there are already four present) and a 0.2 probability that the customer at the head of the line (if there is such a customer) will finish service and leave. The probabilities that more than one person will arrive or that more than one will finish service in such an interval are considered negligible.

(a) Express this situation as a Markov chain, in which the states are the number of persons at the counter.

(b) What is the probability that the counter be idle (i.e., no customers present)?

(c) What is the probability that a customer be turned away?

6. Consider a Markov chain with $m + n$ states $\{S_1, S_2, \ldots, S_{m+n}\}$. Suppose that, for $n + 1 \leq i \leq m + n$, $a_{ii} = 1$, i.e., if the chain reaches one of the last m states, it remains thereafter in that state. Suppose, moreover, that, for $1 \leq i \leq n$,

$$t_i = \sum_{=n+1}^{m+n} a_{ij} > 0$$

i.e., if the chain is in one of the remaining n states, there is a positive probability that it will pass to one of the last m states. Then the states $\{S_{n+1}, \ldots, S_{m+n}\}$ are called *absorbing states*, while $\{S_1, \ldots, S_n\}$ are called *transient states*. Show that, as k approaches infinity, the powers A^k of the transition matrix approach a matrix, W, which is such that $w_{ij} = 0$ for $1 \leq j \leq n$. (Essentially, this means that, with probability 1, the Markov process will eventually reach one of the absorbing states, and remain there. Prove this by showing that, if t is the minimum of the t_j, then the entries in the first n columns of A^k are all smaller than $(1-t)^k$. Since t is positive, these powers all approach 0.)

7. Show that the conclusions of Problem 6 will hold if it is possible to reach one of the absorbing states from each of the transient states in not more than p moves in which p is an integer. (Do this by considering the p-period transition matrix A^p.)

8. With the assumptions of Problem 6, let $B=(b_{ij})$ be the $n\times n$ matrix formed by the first n rows and columns of A, i.e.,

$$b_{ij}=a_{ij}$$

for $i, j=1, \ldots, n$. Then B is called the *transient part* of A. The matrix

$$C=(I-B)^{-1}$$

(which may be shown to exist) is known as the *fundamental* matrix of the chain. Show that, if k is large, C is approximately equal to

$$I+B+B^2+B^3+\ldots+B^k$$

and that the entries c_{ij} represent the expected number of periods which the chain will spend in the jth (transient) state, given that it is in the ith (transient) state.

9. With the assumptions of Problem 6, let d_i be the expected number of periods before the system reaches an absorbing state, given that it is now in state i. Show then that, for $n+1\leq i\leq m+n$,

$$d_i=0$$

whereas, for $1\leq i\leq n$,

$$d_i=1+\sum_{=1}^{m+n} p_{ij}d_j$$

10. Two gamblers, with a total of 5 units between them, play a fair game (i.e., one in which each has probability 0.5 of winning) for stakes of 1 unit each time; the game will terminate whenever either one of the gamblers holds all 5 units. Represent this as a Markov chain.

(a) Given that one gambler holds 2 units now, what is the expected number of times that he will hold 3 units, before the game terminates?

(b) What is the expected duration of the game?

(c) What is the probability that he will eventually be wiped out?

5. CONTINUOUS RANDOM VARIABLES

We have seen that, if an event is certain, its probability must be 1; if possible, it must have probability 0. Through the previous sections of this chapter, we have (rather tacitly) assumed that the converse of this is true: if an event has probability 1, it must be certain; if it has probability 0, it is impossible. In fact, for the type of experiment which we have considered so far, this is so. We shall see now, however, that this need not be so in general.

Let us suppose that we are given a thread and asked to cut a piece 7 inches long. What is the probability that we will do exactly this?

It is not difficult to see that the exact length of the piece of thread which we cut can be thought of as a random variable. Depending on the care which we exert in measuring the thread, this variable will probably be quite close to 7, but measurements are always less than exact, and so there will probably be a small error. It is not impossible that the variable will have the value 7, but some reflection should show us that we should consider this an extremely fortunate occurrence. (Note that we are asking for a thread exactly 7 inches long; it is not sufficient, say, that it be between 6.999 and 7.001 inches long.) We conclude that the probability that the piece of thread be exactly 7 inches long (or, in fact that it have any other pre-assigned length, exactly) must be 0. (In fact, there are so many possible lengths for the thread, that if each of these had a positive probability, the sum of these probabilities would be much greater than 1—infinite, indeed.)

In general, if a random variable can take on every value inside some interval, it is clear that most of these values must have probability 0, even though they are possible. It often happens, moreover, that each of the possible values has probability 0.

Consider the following example. The random variable X can take on any values inside some interval $a \leq X \leq b$. Under certain conditions, it seems reasonable to assume that every value inside this interval is just as likely as any other. (This might happen, say, in the example mentioned above: if we want a piece 7 inches long, and our measuring instrument is accurate to within 1/100 of an inch, we might assume that every value between 6.99 and 7.01 inches is equally likely.) We interpret this by saying that, if c and d are numbers such that

$$a \leq c < d \leq b$$

then

$$P\{c \leq X \leq d\} = \frac{d-c}{b-a}$$

In other words, the probability that the random variable lies inside some sub-interval ($c \leq X \leq d$) is proportional to the length of the sub-interval. (This is known as the *uniform distribution.*) Now, if c and d are close together, so that the sub-interval (c,d) is small, the probability that X lies in this sub-interval will be proportionately small. It will follow, taking a limit as $d \to c$, that

$$P(X = c) = 0;$$

i.e., every single value, thought of as an interval of length 0, must have probability 0.

Since it is not, in general, possible to give the distribution of X by giving the probabilities of each of its possible values, recourse is generally had to the cumulative distribution function

$$F(x) = P(X \leqslant x)$$

IX.5.1 Definition. A *cumulative distribution function* is any function, F, whose domain is the set of all real numbers, satisfying:

9.5.1 $$\lim_{x \to -\infty} F(x) = 0$$

9.5.2 $$\lim_{x \to \infty} F(x) = 1$$

9.5.3 If $x_1 \leqslant x_2$, then $F(x_1) \leqslant F(x_2)$

9.5.4 $$\lim_{\substack{\Delta x \to 0 \\ \Delta x > 0}} F(x + \Delta x) = F(x)$$

Conditions (9.5.1), (9.5.2) and (9.5.3) are easy to understand; they follow directly from the meaning of a cumulative distribution function. Condition (9.5.4) is somewhat more technical; it is generally expressed by saying that F must be continuous on the right. We will not worry about it here.

In general, a cumulative density function is any function which satisfies Definition IX.5.1. There are two important special cases. In one of them, the function has discontinuities (jumps) at each of a sequence of points, and is constant in each of the intervals determined by these points; the random variable is discrete in that case. In another case, the function F is continuous; the random variable is then said to be continuous.

We shall, here, consider continuous random variables. We shall make the additional stipulation that F be the integral of its derivative. (This is not, indeed, a very strong stipulation, except that, quite frequently, there are some points at which the derivative does not exist. Our stipulation says simply that these points are few enough not to make any difference.) If this is so, we may write

9.5.5 $$f(x) = D_x f$$

The function f is then called the *probability density* function of the random variable.

Assuming (as we shall) that the density function f exists, we will have

9.5.6 $$F(x)=\int_{-\infty}^{x} f(t)\,dt$$

and, for any a and b (such that $a \leqslant b$)

9.5.7 $$P(a \leqslant X \leqslant b)=F(b)-F(a)=\int_{a}^{b} f(x)dx$$

The right side of (9.5.6) has the form of an improper integral. This is not, in itself, a great difficulty; moreover, in many cases the function f vanishes outside some interval, so that (9.5.6) may be replaced by a proper integral.

In general, we shall deal, mostly, with the density functions. A density function, as far as we are concerned here, will simply be a function satisfying

9.5.8 $$f(x) \geqslant 0$$

9.5.9 $$\int_{-\infty}^{\infty} f(x)dx=1$$

The intuitive meaning of the density function is simply that, if $f(x)$ is large then the probability that X will be near x is proportionately large. In fact, we have

$$P(x \leqslant X \leqslant x+\Delta x)=F(x+\Delta x)-F(x)$$

If F is differentiable at the point x, we will have the approximation

$$F(x+\Delta x)-F(x) \cong f(x)\,\Delta x$$

Thus, the probability that X lies within a given (short) interval is approximated by the product of the length of the interval and the density f some place within this interval.

IX.5.2 Definition. If x is a continuous random variable with density f, and $U=\phi(x)$ is any function of x, we define the expectation

9.5.10 $$E\,[U]=\int_{-\infty}^{\infty} \phi(x)f(x)\,dx$$

assuming that this (improper) integral exists.

In particular, we define the mean and variance of x:

9.5.11 $$\mu_x = E\,[X] = \int_{-\infty}^{\infty} xf(x)\,dx$$

9.5.12 $$\sigma_x^2 = E\,[(X-\mu_x)^2] = \int_{-\infty}^{\infty} x^2 f(x)\,dx - \mu_x^2$$

assuming, in each case, that these integrals converge. Note how all the properties and definitions given here for continuous variables resemble those given for discrete variables: the sole difference, throughout, is that sums have been replaced by integrals.

IX.5.3 Example. Uniform Distribution. Let $a < b$. We consider here a distribution such that the density, $f(x)$, is a constant, k, for $a \leqslant x \leqslant b$, and zero otherwise. To find the value of the constant k, we write

$$\int_{-\infty}^{\infty} f(x)\,dx = \int_{-\infty}^{a} 0\,dx + \int_{a}^{b} k\,dx + \int_{b}^{\infty} 0\,dx$$

so that

$$\int_{-\infty}^{\infty} f(x)\,dx = k\int_{a}^{b} dx = k(b-a)$$

Now, by (9.5.9),

$$k(b-a) = 1$$

and so

$$k = \frac{1}{b-a}$$

We conclude that the density function is

$$f(x) = \begin{cases} \dfrac{1}{b-a} & \text{for } a \leqslant x \leqslant b \\ 0 & \text{for } x < a \text{ or } b < x \end{cases}$$

We are also interested in the cumulative distribution function F. It is not too difficult to see that, if $x \leqslant a$,

$$\int_{-\infty}^{x} f(t)\,dt = 0$$

If $a \leqslant x \leqslant b$, we have

$$\int_{-\infty}^{x} f(t)\,dt = \int_{-\infty}^{a} 0\,dt + \int_{a}^{x} \frac{1}{b-a}\,dt = \frac{x-a}{b-a}$$

and, if $x \geqslant b$,

$$\int_{-\infty}^{x} f(t)dt = \int_{-\infty}^{a} 0\;dt + \int_{a}^{b} \frac{1}{b-a}\,dt + \int_{b}^{x} 0\;dt = 1$$

Thus

$$F(x) = \begin{cases} 0 & \text{if } x \leqslant a \\ \dfrac{x-a}{b-a} & \text{if } a \leqslant x \leqslant b \\ 1 & \text{if } b \leqslant x \end{cases}$$

We may also compute the mean and variance of this variable. We have

$$\int_{-\infty}^{\infty} x\,f(x)\;dx = \int_{a}^{b} \frac{x}{b-a}\,dx = \frac{x^2}{2(b-a)}\bigg|_{a}^{b} = \frac{b+a}{2}$$

and

$$\int_{-\infty}^{\infty} x^2\,f(x)\;dx = \int_{a}^{b} \frac{x^2}{b-a}\,dx = \frac{x^3}{3(b-a)}\bigg|_{a}^{b} = \frac{b^2+ab+a^2}{3}$$

Thus

$$\mu_x = \frac{b+a}{2}$$

and

$$\sigma_x^2 = \frac{b^2+ab+a^2}{3} - \frac{b^2+2ab+a^2}{4} = \frac{(b-a)^2}{12}$$

that is, the mean is the mid-point of the interval; the standard deviation is the length of the interval divided by $\sqrt{12}$.

IX.5.4 Example. Exponential Distribution. We consider here a random variable which has only positive values. For these, it has a density $f(x) = ke^{-\alpha x}$, where α is a parameter, and k is a constant which must be determined. To do this, we have

$$\int_{-\infty}^{\infty} f(x)\;dx = \int_{-\infty}^{0} 0\;dx + \int_{0}^{\infty} k\;e^{-\alpha x}\;dx$$

The first integral vanishes; the second will give us

$$\int_0^\infty ke^{-\alpha x}\,dx = \frac{-k\,e^{-\alpha x}}{\alpha}\Bigg|_0^\infty$$

This is an improper integral; it may be seen, however, that, if α is positive, then,

$$\lim_{x\to\infty} e^{-\alpha x} = 0$$

and so

$$\frac{-k\,e^{-\alpha x}}{\alpha}\Bigg|_0^\infty = 0 - \left(\frac{-k\cdot 1}{\alpha}\right) = \frac{k}{\alpha}$$

By (9.5.9) this must equal 1. Thus $k = \alpha$. The density function is then

$$f(x) = \begin{cases} 0 & \text{for } x \leqslant 0 \\ \alpha e^{-\alpha x} & \text{for } x > 0 \end{cases}$$

Integrating, we obtain the cumulative distribution funtion:

$$F(x) = \begin{cases} 0 & \text{if } x \leqslant 0 \\ 1 - e^{-\alpha x} & \text{if } x > 0 \end{cases}$$

To find the mean and variance, we have

$$\int_{-\infty}^\infty xf(x)\,dx = \alpha\int_0^\infty x\,e^{-\alpha x}\,dx$$

This integral is almost the same as that of Example VI.5.1. We now have

$$\alpha\int_0^\infty x\,e^{-\alpha x}\,dx = \left(-x\,e^{-\alpha x} - \frac{e^{-\alpha x}}{\alpha}\right)\Bigg|_0^\infty = \frac{1}{\alpha}$$

In a similar way, the expression

$$\int_{-\infty}^\infty x^2\,f(x)\,dx = \alpha\int_0^\infty x^2\,e^{-\alpha x}\,dx$$

may be evaluated (integration by parts may be used, or else we may merely look for this form in a table of integrals). In any case, we will find

$$\alpha\int_0^\infty x^2\,e^{-\alpha x}\,dx = -\left(x^2 + \frac{2x}{\alpha} + \frac{2}{\alpha^2}\right)e^{-\alpha x}\Bigg|_0^\infty$$

Here, the exponential term decreases much more rapidly than the polynomial increases; thus, at the upper limit, the function vanishes. We have then,

$$\int_0^\infty x^2 f(x)\, dx \doteq \frac{2}{\alpha^2}$$

Thus, the mean and variance are given by

$$\mu_x = \frac{1}{\alpha}$$

$$\sigma_x^2 = \frac{2}{\alpha^2} - \mu_x^2 = \frac{1}{\alpha^2}$$

IX.5.5 Example. The life-time, in days, of a certain machine is a random variable, x, with density

$$f(x) = \begin{cases} 0.002\, e^{-0.002\, x} & \text{if } x \geq 0 \\ 0 & \text{if } x < 0 \end{cases}$$

What is the machine's expected life? What is the probability that it will fail during the first 20 days? Given that it has not failed during the first 1000 days, what is the probability that it will fail during the next 20 days?

As we saw above, the expected value of an exponential random variable, with parameter α, is $1/\alpha$. In this case, $\alpha = 0.002$, and so the machine has an expected life-time of 500 days.

The probability that the machine will fail during the first 20 days is, of course,

$$P(X \leq 20) = F(20) = 1 - e^{-0.04} = 0.0392$$

or slightly less than 4 per cent.

The probability that it will fail before 1020 days, given that it has not failed before 1000 days, is

$$P(X \leq 1020 \mid X \geq 1000) = \frac{Pr(1000 \leq X \leq 1020)}{Pr(X \geq 1000)} = \frac{F(1020) - F(1000)}{1 - F(1000)}$$

$$= \frac{(1 - e^{-2.04}) - (1 - e^{-2})}{e^{-2}} = 1 - e^{-0.04}$$

$$= 0.0392$$

and this is the same as the probability of failure during the first twenty days. This shows that, like the geometric random variable, the exponential random variable is "memoryless": no matter how old this machine may be, the probability of failure within the next 20 days (or within any other fixed period of time) will always be the same. (As such, it is clearly not realistic in some cases, but may well be so in others.)

IX.5.6 Example. An adjustable machine cuts pieces of salami of varying lengths. If the machine is set to cut pieces a inches long, then the actual length of a piece will be a random variable with uniform distribution in the interval $a-0.05 \leqslant X \leqslant a+0.05$. A man buys four pieces of salami. If the machine has been set at $a=6$, what is the probability that the four pieces cut will all be at least 6 inches long? What setting should be used, if we want a 90 per cent probability that all four pieces will be at least 6 inches long?

If we set $a=6$, then the length of a piece of salami will have uniform distribution in the interval $5.95 \leqslant X \leqslant 6.05$. The cumulative distribution is then

$$F(x)=\begin{cases} 0 & x \leqslant 5.95 \\ \dfrac{x-5.95}{0.1} & 5.95 \leqslant x \leqslant 6.05 \\ 1 & x \geqslant 6.05 \end{cases}$$

where the 0.1 is the length of the interval from 5.95 to 6.05. Thus, we see that $F(6)=1/2$. The probability that any one piece be over 6 inches long is then $1-F(6)=1/2$, and the probability that all four will be at least 6 inches long is $(1/2)^4=1/16$.

For an arbitrary a, the probability that all four pieces be at least 6 inches long will be $(1-F(6))^4$. We want this to be 0.90, and so we set $1-F(6)=\sqrt[4]{0.90}=0.974$, which gives $F(6)=0.026$.

The distribution function is, of course,

$$F(x)=\begin{cases} 0 & x \leq a-.05 \\ \dfrac{(x-a+0.05)}{0.1} & a-.05 \leq x \leq a+.05 \\ 1 & x \geq a+.05 \end{cases}$$

and we see that we must have

$$10(6-a+0.05)=0.026$$

or

$$a=6.0474.$$

IX.5.7 Example. The Normal Distribution. Consider the function

$$\phi(x)=ke^{-x^2/2}$$

As was mentioned in Chapter VI, it may be shown, through some mathematics which is beyond our scope, that

$$\int_{-\infty}^{\infty} e^{-x^2/2}\, dx = \sqrt{2\pi}.$$

It follows that the function ϕ will be a density function if we set $k = 1/\sqrt{2\pi}$. We have, thus, a random variable x with density

$$\phi(x) = \frac{1}{\sqrt{2\pi}} e^{-x^2/2}$$

and cumulative distribution

$$\Phi(x) = \frac{1}{\sqrt{2\pi}} \int_{-\infty}^{x} e^{-t^2/2} dt.$$

(It is not possible to express the function $\Phi(x)$ in any simple way; i.e., $\phi(x)$ does not have an elementary function as its anti-derivative.)

To find the mean of x, we evaluate the integral

$$\frac{1}{\sqrt{2\pi}} \int_{-\infty}^{\infty} x\, e^{-x^2/2}\, dx$$

by making the substitution $y = x^2/2$.

This gives us

$$dy = x\, dx$$

and so

$$\int xe^{-x^2/2} dx = \int e^{-y} dy = -e^{-y}$$

Thus,

$$\frac{1}{\sqrt{2\pi}} \int_{-\infty}^{\infty} xe^{-x^2/2} dx = -\frac{1}{\sqrt{2\pi}} e^{-x^2/2} \Big|_{-\infty}^{\infty}$$

It may be seen that this improper integral vanishes; we conclude that

$$\mu_x = 0$$

Similarly, the integral

$$\frac{1}{\sqrt{2\pi}} \int_{-\infty}^{\infty} x^2 e^{-x^2/2} dx$$

 may be integrated by parts, setting

$$u = x \qquad dv = xe^{-x^2/2}\,dx$$

and

$$du = dx \qquad v = -e^{-x^2/2}$$

Thus

$$\int_{-\infty}^{\infty} x^2 e^{-x^2/2}\,dx = -xe^{-x^2/2}\Big|_{-\infty}^{\infty} + \int_{-\infty}^{\infty} e^{-x^2/2}\,dx$$

Now, the first term on the right vanishes, while the second is (as we mentioned before) equal to $\sqrt{2\pi}$. We have, then,

$$\frac{1}{\sqrt{2\pi}} \int x^2 e^{-x^2/2}\,dx = 1$$

and so

$$\sigma_x^2 = 1$$

The variable x with this distribution is known as a *normal random variable,* and is extremely important for reasons which will be studied below. The graph of the function ϕ is the familiar bell-shaped curve shown in Figure IX.5.1.

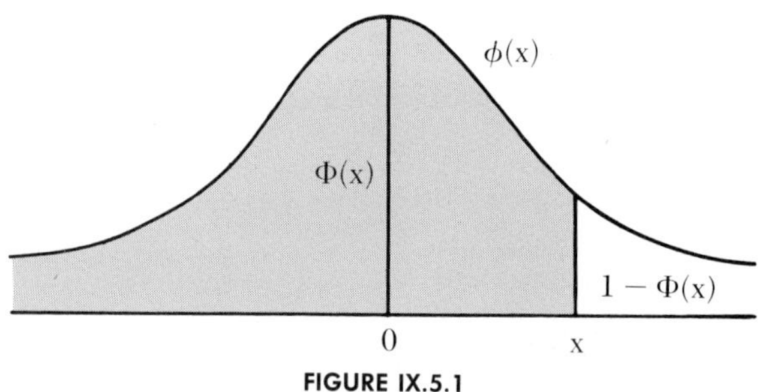

FIGURE IX.5.1

6. TRANSFORMATION OF CONTINUOUS RANDOM VARIABLES

If X is a random variable with known distribution, and U is a function of X, it is generally of interest to find the distribution of U. This can normally be done by restating events which concern the variable U in terms of the variable X. Thus, if $U = X^2$, then the event

$$U \leq 4$$

may be restated as

$$X^2 \leq 4$$

or

$$-2 \leq X \leq 2$$

Thus,

$$P(U \leq 4) = P(-2 \leq X \leq 2)$$

and, if X and U have cumulative distributions F and G respectively, we will have

$$G(4) = F(2) - F(-2)$$

Thus the cumulative distribution of U can be obtained from that of X. Differentiation will then yield the density of U.

IX.6.1 Example. Suppose X has the distribution of Example IX.5.7. Find the distribution of

$$Y = e^X$$

The exponential function is always positive; thus Y can only have positive values. For $y > 0$, we see that the event

$$Y \leq y$$

is the same as

$$X \leq \ln y.$$

Thus Y has distribution G, where

$$G(y) = \Phi(\ln y)$$

Differentiating, we have

$$\frac{dG}{dy} = \frac{d\Phi}{dx}\,\frac{dx}{dy}$$

(where $x = \ln y$). Since ϕ is the derivative of Φ, this gives us the density

$$g(y) = \frac{1}{y}\phi(\ln y)$$

or

$$g(y) = \frac{1}{y\sqrt{2\pi}} \exp\left\{-\frac{1}{2} (\ln y)^2\right\}$$

(where $\exp\{t\}$ means e^t). This distribution is known as the *log-normal distribution,* since the logarithm of Y has a normal distribution.

In case we have the relation

$$U = \alpha X + \beta$$

where α and β are constants ($\alpha > 0$), the event

$$U \leq u$$

can be written

$$\alpha X + \beta \leq u$$

or, since $\alpha > 0$,

$$X \leq \frac{u - \beta}{\alpha}$$

Thus the cumulative distribution functions F and G (of X and U respectively) are related by

$$G(u) = F\left(\frac{u - \beta}{\alpha}\right)$$

Thus

$$\frac{dG}{du} = \frac{dF}{dx}\frac{dx}{du}$$

where $x = (u - \beta)/\alpha$, and so $\dfrac{dx}{du} = \dfrac{1}{\alpha}$. Consequently, the densities f and g are related by

9.6.1 $$g(u) = \frac{1}{\alpha} f\left(\frac{u - \beta}{\alpha}\right)$$

IX.6.2 *Example.* Let X have the exponential density

$$f(x) = \begin{cases} e^{-x} & \text{for } x > 0 \\ 0 & \text{for } x \leq 0 \end{cases}$$

and let

$$U = \alpha X + \beta$$

Then

$$g(u)=\begin{cases}\dfrac{1}{\alpha}e^{-(u-\beta)/\alpha} & \text{if } \dfrac{u-\beta}{\alpha}>0\\[2ex] 0 & \text{if } \dfrac{u-\beta}{\alpha}\le 0\end{cases}$$

If we set $\gamma = 1/\alpha$, this reduces to

$$g(u)=\begin{cases}\gamma e^{-\gamma(u-\beta)} & \text{if } u>\beta\\ 0 & \text{if } u\le\beta\end{cases}$$

IX.6.3 Example. A machine cuts circular pieces of rubber. Because of random variations, the radius (in inches) of the piece cut is a random variable, X, with uniform distribution in the interval $2.9 \le X \le 3.1$. Let Y be the area of the piece of rubber cut. What is the distribution of Y? Find the expected value of Y.

It is easy to see that

$$Y = \pi X^2.$$

We know that X has the distribution

$$F(x)=\begin{cases}0 & x\le 2.9\\ 5x-14.5 & 2.9\le x\le 3.1\\ 1 & x\ge 3.1\end{cases}$$

and, therefore, Y has the distribution

$$G(y)=\begin{cases}0 & y\le 8.41\pi\\ 5\sqrt{y/\pi}-14.5 & 8.41\pi\le y\le 9.61\pi\\ 1 & y\le 9.61\pi\end{cases}$$

Differentiating this, we obtain the density

$$g(y)=D_yG=\begin{cases}0 & y\le 8.41\pi\\ \dfrac{5}{2\sqrt{\pi y}} & 8.41\pi\le y\le 9.61\pi\\ 0 & y\le 9.61\pi\end{cases}$$

so that the distribution of Y is not quite uniform. To find its mean, we integrate:

$$E[Y]=\int_{8.41\pi}^{9.61\pi} yg(y)\,dy=\int_{8.41\pi}^{9.61\pi}\frac{5\sqrt{y}}{2\sqrt{\pi}}\,dy=9.003\pi,$$

so that the average area is 9.003π square inches. The average radius is 3 in., and we see that the average area is slightly more than π times the square of the average radius.

PROBLEMS ON TRANSFORMATION OF CONTINUOUS RANDOM VARIABLES

1. Let X be a continuous random variable with uniform distribution in the interval $0 \leq X \leq 1$. What are the mean and variance of X? Find the cumulative distribution and density of the variable $Y = X^2$. What are its mean and variance?

2. The variable X has a density function f, given by

$$f(x) = k(x - x^2)$$

for $0 \leq x \leq 1$, and $f(x) = 0$ otherwise. What is the value of k? Find the probability that $0 \leq X \leq 1/4$. What are the mean and variance of X?

3. Let X have the normal distribution with density

$$\phi(x) = \frac{1}{\sqrt{2\pi}} e^{-x^2/2}$$

and let $Y = X^2$. Show that the distribution of Y is given by the density

$$f(y) = \frac{1}{\sqrt{2\pi y}} e^{-y/2}$$

4. The random variable X has the exponential density

$$f(x) = 3e^{-3x}$$

for $x > 0$, and $f(x) = 0$ for $x \leq 0$. What is the probability that $1 \leq X \leq 2$? Find the density of the variable $Y = 3X + 2$.

5. Let X have the exponential density

$$f(x) = \alpha e^{-\alpha x}$$

for $x > 0$, and $f(x) = 0$ for $x \leq 0$. For what value of the parameter α will the event $1 \leq X \leq 2$ have the greatest probability?

7. THE NORMAL DISTRIBUTION

In a previous section (Example IX.5.7) we studied the random variable X with density

9.7.1 $$\phi(x) = \frac{1}{\sqrt{2\pi}} e^{-x^2/2}$$

and cumulative distribution function

9.7.2 $$\Phi(x) = \frac{1}{\sqrt{2\pi}} \int_{-\infty}^{x} e^{-t^2/2} dt$$

Both of these functions, ϕ and Φ, have been carefully tabulated. In Appendix 8, we have a table which gives values of these functions for positive values of the argument, x. For negative values of x, these functions can be obtained by noting that, in the right side of (9.7.1), x appears only as a square. Since $(-x)^2 = x^2$, we conclude that

9.7.3 $$\phi(-x) = \phi(x)$$

Application of (9.7.3) makes it clear that

$$\frac{1}{\sqrt{2\pi}} \int_{-\infty}^{-x} e^{-t^2/2} dt = \frac{1}{\sqrt{2\pi}} \int_{x}^{\infty} e^{-t^2/2} dt$$

The left side of this equation is $\Phi(-x)$. The right side is clearly the probability that $X \geq x$. In other words, it is $1 - \Phi(x)$. Thus,

9.7.4 $$\Phi(-x) = 1 - \Phi(x)$$

Equations (9.7.3) and (9.7.4) can be used to find the values of ϕ and Φ for negative x.

Thus, for example, suppose that we wish to find $\phi(-1.5)$. From the table, we find

$$\phi(1.5) = 0.1295$$

Therefore,

$$\phi(-1.5) = \phi(1.5) = 0.1295$$

Similarly, suppose we wish to find $\Phi(-0.8)$. From the table, we find

$$\Phi(0.8) = 0.7881$$

and so

$$\Phi(-0.8) = 1 - \Phi(0.8) = 0.2119$$

Let us suppose we set

$$Y = \alpha X + \beta$$

Then, according to (9.6.1), Y will have the density

$$f(y)=\frac{1}{\alpha}\,\phi\left(\frac{y-\beta}{\alpha}\right)$$

$$f(y)=\frac{1}{\alpha\sqrt{2\pi}}\exp\left\{-\frac{(y-\beta)^2}{2\alpha^2}\right\}$$

We saw above that $\mu_x=0$ and $\sigma_x^2=1$. It follows that

$$\mu_y=\alpha\mu_x+\beta=\beta$$

and

$$\sigma_y^2=\alpha^2\,\sigma_x^2=\alpha^2$$

Thus β and α are, respectively, the mean and standard deviation of the variable Y. This being so, we generally replace β by μ, and α by σ. In that case, the density of Y is

9.7.5 $$f(y)=\frac{1}{\sigma}\,\phi\left(\frac{y-\mu}{\sigma}\right)$$

We have, moreover, for the cumulative distribution function,

9.7.6 $$F(y)=\Phi\left(\frac{y-\mu}{\sigma}\right)$$

We shall say that the variable X, with distribution given by (9.7.1) and (9.7.2), is a *normalized normal variable.* The variable Y, whose distribution is given by (9.7.5) and (9.7.6), will be called *a normal variable with mean μ and variance σ^2.* We will use the notation

X has $N\,(0,1)$
Y has $N\,(\mu,\sigma^2)$

to represent this.

For large values of n, the binomial distribution

9.7.7 $$B(n,k;p)=\binom{n}{k}p^kq^{n-k}$$

is difficult to evaluate because it involves the three factorials $n!,k!$, and $(n-k)!$

It may be proved that, under the assumption that np is close to k (and we will, in general, be interested in values of k close to np, since other values are very improbable), this can be approximated by

9.7.8 $$B(n,k;p)\cong\frac{1}{\sqrt{2\pi npq}}\exp\left\{\frac{(k-np)^2}{2npq}\right\}$$

which looks remarkably like a normal density function: it is, indeed, the function (9.7.5), with $y = k$, $\mu = np$, $\sigma^2 = npq$.

We conclude that, if p is not too small, and n is large, the binomial distribution can be approximated by the normal distribution, with mean np and variance npq. (Note that these are the mean and variance of the binomial distribution as well.)

IX.7.1 Example. A fair die is tossed 100 times. What is the probability of obtaining exactly 18 ones?

We shall here use the normal approximation, with $n = 100$, $p = 1/6$, $q = 5/6$. Here the number of ones thrown has approximately $N(100/6, 500/36)$. Using (9.7.5) and (9.7.6), we have

$$B(100,18;1/6) \cong \frac{1}{\sqrt{500/36}}\,\phi\left(\frac{18 - \dfrac{100}{6}}{\sqrt{500/36}}\right)$$

This can be simplified to give

$$B(100,18;1/6) \cong 0.268\,\phi(0.36)$$

From the table, we find $\phi(0.36) = 0.3739$. We conclude that the probability of exactly 18 ones is approximately $0.268(0.374) = 0.100$.

Quite often, when applying the normal approximation, we are interested not so much in the probability of obtaining exactly a certain number of successes, but, rather, in the probability that the number of successes lies between two given numbers. This could be done, of course, by computing the probability of each successive number. Unfortunately, this process might be exceedingly long. (Suppose, for example that we were asked to find the probability of obtaining between 5000 and 6000 heads in 10,000 tosses of a fair coin.) It becomes simpler, in these cases, to use the fact that, if f is the derivative of F, and, moreover, f does not change too quickly, then

$$f(k) \cong F\left(k + \frac{1}{2}\right) - F\left(k - \frac{1}{2}\right)$$

by the rectangular approximation to the area under the distribution curve, and so the probability of obtaining between m_1 and m_2 successes (inclusively) can be approximated by

$$\sum_{k=m_1}^{m_2} f(k) = F\left(m_2 + \frac{1}{2}\right) - F\left(m_1 - \frac{1}{2}\right)$$

Thus the cumulative distribution function avoids the need for long sums. (If $m_2 - m_1$ is large, the 1/2 may be dropped without causing a great error.)

IX.7.2 Example. A salesman estimates that, at each call he makes, he has a 0.4 probability of making a sale. If he makes 600 calls, what is the probability that he will make between 220 and 300 sales? That he will make between 235 and 247 sales?

In this case, we have $np = 240$, and $npq = 144$. Thus the number of sales is approximately normally distributed with $\mu = 240$, and $\sigma = 12$. We have

$$\begin{aligned} P(220 \leq Y \leq 300) &= F(300) - F(220) \\ &= \Phi\left(\frac{300-240}{12}\right) - \Phi\left(\frac{220-240}{12}\right) \\ &= \Phi(5) - \Phi(-1.67) \end{aligned}$$

From the tables, we find

$$\Phi(5) = 1; \ \Phi(1.67) = 0.9525$$

Thus

$$\begin{aligned} P(220 \leq Y \leq 300) &= 1 - (1 - 0.9525) \\ &= 0.9525 \end{aligned}$$

In the second case, the difference $247 - 235 = 12$ is not large enough to disregard the halves; thus

$$\begin{aligned} P(235 \leq Y \leq 247) &= F(247.5) - F(234.5) \\ &= \Phi\left(\frac{247.5-240}{12}\right) - \Phi\left(\frac{234.5-240}{12}\right) \\ &= \Phi(0.63) - \Phi(-0.46) \end{aligned}$$

From the table, we have

$$\Phi(0.63) = 0.7357; \ \Phi(0.46) = 0.6772$$

Thus

$$\begin{aligned} P(235 \leq Y \leq 247) &= 0.7357 - (1 - 0.6772) \\ &= 0.4129 \end{aligned}$$

We have seen that the normal distribution can be used as an approximation of the binomial distribution, when the number of trials is large. In fact, this is only a special case of a very general theorem, which we give below. We shall say that a sequence of random variables $Y_1, Y_2, Y_3, \ldots,$ with cumulative distribution functions $F_1, F_2, F_3, \ldots$ *converges stochastically* to the random variable X with distribution F if, for every point of continuity, a, of the function F,

$$\lim_{n\to\infty} F_n(a) = F(a)$$

Essentially, this means that, for large n, the variables Y_n and X have "almost the same" distribution. We have then the following very important theorem.

IX.7.3 Theorem (The Central Limit Theorem). Let $X_1, X_2, X_3, \ldots$ be a sequence of independent, identically distributed random variables with mean μ and variance σ^2.

Let

$$Y_n = \frac{1}{\sigma\sqrt{n}}\left(\sum_{i=1}^{n} X_i - n\mu\right)$$

Then the sequence $Y_1, Y_2, Y_3, \ldots$ converges stochastically to the random variable X with $N(0,1)$.

We omit the proof of the Central Limit Theorem, as it is beyond the scope of our mathematics. In essence, it means that, for large n, the sum $X_1 + X_2 + \ldots + X_n$ will have approximately a normal distribution with mean $n\mu$ and variance $n\sigma^2$. In fact, the conditions of the theorem can be considerably weakened; e.g., it is not necessary that the X_i be identically distributed. The following theorem is also of importance. (We must omit its proof as well.)

IX.7.4 Theorem. Let X and Y be normal random variables. Then $X + Y$ is also a normal variable.

Theorems IX.7.3 and IX.7.4 together imply that, when a random variable is the sum of a large number of independent variables, it will (under certain very weak conditions) have a normal distribution. Thus, any effect which can be thought of as being the result of very many small, independent, superimposed causes will tend to have a normal distribution. Examples are the strength of a rope (thought of as the sum of the strengths of very many strands in the rope), a person's intelligence quotient (thought of as the resultant of aptitudes in several different fields), or the height of a man. So often does this distribution appear, in fact, that it is generally recognized as the most important of all probability distributions. This is the reason for its meticulous tabulation.

IX.7.5 Example. The life, T, in hours of the bulbs produced by a company has approximately a normal distribution with mean 500 and variance 400. What is the probability that a given bulb will last less than 485 hours? Less than 510 hours?

Since $20^2 = 400$, we are looking here for

$$F(485) = \Phi\left(\frac{485 - 500}{20}\right) = \Phi(-0.75)$$

From the table, we find

$$\Phi(0.75) = 0.7734$$

Thus

$$F(485) = \Phi(-0.75) = 1 - 0.7734$$
$$F(485) = 0.2266$$

For the second question, we have

$$F(510) = \Phi\left(\frac{510 - 500}{20}\right) = \Phi(0.5)$$

Thus

$$F(510) = 0.6915$$

IX.7.6 Example. An aptitude test is calibrated so that scores on the test will be normally distributed with mean 500 and standard deviation 100. What is the probability that an applicant will obtain a score between 400 and 600? A score greater than 700?

The probability of a score between 400 and 600 is

$$F(600) - F(400) = \Phi\left(\frac{600 - 500}{100}\right) - \Phi\left(\frac{400 - 500}{100}\right)$$
$$= \Phi(1) - \Phi(-1)$$

From the table, we find

$$\Phi(1) = 0.8413$$

Thus,

$$\Phi(1) - \Phi(-1) = 0.8413 - (1 - 0.8413)$$
$$= 0.6826$$

The probability of a score under 700 is

$$F(700) = \Phi\left(\frac{700 - 500}{100}\right) = \Phi(2)$$
$$= 0.9773$$

Thus the probability of a score above 700 will be $1 - 0.9773 = 0.0227$.

IX.7.7 Example. A company screens its job applicants by requiring them to take the test described in Example IX.7.6. If it has 100 applicants, what is the probability that at least 12 of them will receive scores of 600 or over?

Let us first compute the probability that one applicant will obtain 600 or over. This is

$$1 - F(600) = 1 - \Phi\left(\frac{600 - 500}{100}\right) = 1 - \Phi(1)$$
$$= 0.1587$$

We are now interested in the probability of obtaining 12 or more successes in 100 trials of a simple experiment with probability of success $p = .1587$. This is approximated as a normal distribution with mean

$$\mu = np = 15.87$$

and variance

$$\sigma^2 = npq = 13.34$$

so that $\sigma = \sqrt{13.34} = 3.65$. The probability of 12 or more sucesses is then

$$G(100) - G(11.5) = 1 - \Phi\left(\frac{11.5 - 15.87}{3.65}\right)$$
$$= 1 - \Phi(-1.20)$$
$$= 1 - (1 - 0.8849)$$
$$= 0.8849$$

IX.7.8 Example. We say that a person taking a test is in the kth percentile if the probability of obtaining a lower score than that person is $k/100$. In the test described in Example IX.7.6, what score must an applicant obtain to be in the 60th percentile? In the 85th? In the 99th?

We are looking, first, for a value y, such that

$$F(y) = 0.6$$

Since Y has $N(500, 100^2)$, this means that Y must satisfy

$$\Phi\left(\frac{y - 500}{100}\right) = 0.6$$

From the table, we find

$$\Phi(0.25) = 0.5987$$

as the closest value to 0.6. Thus we must have

$$\frac{y - 500}{100} = 0.25$$

or

$$y = 525$$

For the 85th percentile, we want

$$\Phi\left(\frac{y-500}{100}\right) = 0.85$$

From the table, we find

$$\Phi(1.04) = 0.8508$$

Thus

$$\frac{y-500}{100} = 1.04$$

or

$$y = 604$$

For the 99th percentile, we find

$$\Phi(2.33) = 0.9901$$

and so

$$\frac{y-500}{100} = 2.33$$

or

$$y = 733$$

Thus: a score of 525 represents the 60th percentile; 604, the 85th percentile; and 733, the 99th percentile.

IX.7.9 Example. The amount of coffee (in fluid ounces) poured by an electric urn is a normal random variable whose mean is the setting on a dial, and whose standard deviation is 0.15. If the dial is set to 8, what is the probability that the urn will pour between 7.8 and 8.3 ounces?

We want here

$$\begin{aligned} F(8.3) - F(7.8) &= \Phi\left(\frac{8.3-8}{0.15}\right) - \Phi\left(\frac{7.8-8}{0.15}\right) \\ &= \Phi(2) - \Phi(-1.67) \\ &= 0.9773 - (1 - 0.9525) \\ &= 0.9298 \end{aligned}$$

IX.7.10 Example. With the same coffee urn as in Example IX.7.9, how high should the dial be set so that 90 per cent of the servings from the urn will contain at least 8 ounces?

In this case we wish to determine μ so that

$$F(8) = 0.1.$$

Now

$$F(8) = \Phi\left(\frac{8-\mu}{0.15}\right)$$

and, from the table,

$$\Phi(1.28) = 0.8997$$

so that

$$\Phi(-1.28) = 0.1003$$

Hence

$$\frac{8-\mu}{0.15} = -1.28$$

and so

$$\mu = 8.192$$

Thus, the urn should be set to pour approximately 8.2 ounces.

PROBLEMS ON THE NORMAL DISTRIBUTION

1. The scores on a test are normally distributed with mean 100 and standard deviation 20. What is the probability of obtaining a score between 70 and 95? between 90 and 110? above 150?

2. Two hundred applicants take the test described in Problem 1. What is the probability that at least 35 of them will score above 120?

3. One hundred applicants take the test described in Problem 1. What is the probability that at least 2 of them will score over 140? That at least 1 will score over 150? (Use a Poisson approximation.)

4. Given the same test as in Problem 1, what score must an applicant obtain to be in the 75th percentile? in the 90th percentile? in the 99th percentile?

5. A machine produces bolts. The diameter of each bolt, in inches, is a random variable with the distribution $N(0.2, 0.0001)$. What proportion of the bolts will have a diameter between 0.195 and 0.205 inches? between 0.2 and 0.21? smaller than 0.208?

6. The amount of liquid poured by a bottling machine is a normally distributed random variable whose mean can be controlled, and whose standard deviation (in fluid ounces) is 0.03. How large

should the mean be in order to guarantee that at least 95 per cent of the bottles receive 11.99 fluid ounces or more? In this case, how many of the bottles will receive less than 11.97 fluid ounces? How many will receive more than 12.05 fluid ounces?

7. The diameter, in inches, of the ball bearings produced by a certain machine has the distribution $N(0.3, 0.000016)$. What fraction of the bearings will have diameters between 0.3 and 0.305 inches? smaller than 0.298 in.?

8. It is desired to construct casings for the bearings described in Problem 7. How large should the casing be, if it is desired that at least 95 per cent of the bearings fit into the casing?

APPENDIX

1. THE SOLUTION OF EQUATIONS

From the beginning of this book, we have been dealing with the solution of equations, mainly linear in nature, but some also of higher degree—and some, indeed, transcendental (i.e., not algebraic). However, our principal purpose (in Chapter I, especially) was to show how *systems of simultaneous equations* may be solved; it was generally assumed that the student would know how to solve single equations with one unknown. This is usually the content (or part of the content) of high school algebra courses. We give here, for reference, the principal rules for the solution of such equations.

The general rule for equations is, briefly:

A.1.1 If an operation is performed on one side of an equation, it must also be performed on the other. This will preserve the equation if the operation is uniquely defined.

In effect, rule (A.1.1) states that the same quantity may be added or subtracted from both sides of an equation. Both sides may be multiplied by the same number or raised to the same power. On the other hand, some care must be exercised when dividing both sides: we must remember that division is not defined when the divisor is zero. Similarly, taking square roots is not permissible, unless some care is exercised, since a positive number has two square roots, (and a negative number has none, at least among the real numbers).

An important question is whether, after transforming an equation according to (A.1.1), the solutions of the transformed equation are also solutions of the original equation. The answer is that this will be so if the operations performed can be uniquely inverted (reversed). Addition and subtraction can be uniquely reversed, but multiplication cannot always be reversed (it might be multiplication by 0), and taking squares cannot be uniquely reversed. Any time that such an operation is performed, there is a definite danger that so-called *extraneous roots* (roots that were not solutions of the original equation) will be introduced.

In brief, if an operation is well defined and has a well-defined inverse, there is no problem. If it is well defined, but not invertible, it may introduce extraneous roots. If it is not well defined, some solutions may be lost.

As an example of these difficulties, consider the problem of dividing by zero: Suppose $x = y$. Then:

(1) $$x = y$$

(2) $$x^2 = xy$$

(3) $$x^2 - y^2 = xy - y^2$$

(4) $$(x + y)(x - y) = y(x - y)$$

(5) $$x + y = y$$

(6) $$2y = y$$

(7) $$y = 0$$

We pass from equation (1), here, to equation (7), by a sequence of steps that seem valid. Note, however, that the step from (4) to (5) is not a well-defined operation: it involves division by zero. As a result, equation (7) has only one root, whereas equation (1) has an infinity of roots, most of which have been lost.

As an example of the introduction of extraneous roots, consider the following: Suppose $x^2 = -1$. Then

(1) $$x^2 = -1$$

(2) $$(x^2)^2 = (-1)^2$$

(3) $$x^4 = 1$$

(4) $$x = \sqrt[4]{1}$$

(5) $$x = \pm 1$$

Here, the step from equation (1) to equation (2) is valid, but is not invertible: the square root operation in not well defined. As a result no roots are lost (there were none to begin with), but new, extraneous roots have been introduced. Whenever something of this sort is done, it is always necessary to check each of the solutions obtained by introducing them into the original equation; this rids us of the extraneous solutions.

Let us return, now to the solution of equations. The simplest is the general *linear* equation:

A.1.2 $$ax + b = c$$

To solve this, we subtract b from both sides:

$$ax + b - b = c - b$$

or

$$ax = c - b$$

and divide both sides by a:

$$\frac{ax}{a} = \frac{c - b}{a}$$

or

A.1.3 $$x = \frac{c - b}{a}$$

Note that this last step is permissible only when a is different from 0.

The next simplest equation is the *quadratic*, or second-degree equation,

A.1.4 $$ax^2 + bx + c = 0$$

in which $a \neq 0$.

This equation is solved by the method of *completing the square.* We divide both sides by a, to obtain

$$x^2 + \frac{b}{a}x + \frac{c}{a} = 0$$

and subtract c/a:

$$x^2 + \frac{b}{a}x = -\frac{c}{a}$$

Next, we add the term $b^2/4a^2$ to both sides:

$$x^2 + \frac{b}{a}x + \frac{b^2}{4a^2} = \frac{b^2}{4a^2} - \frac{c}{a}$$

This last equation may be rewritten in the form

$$\left(x + \frac{b}{2a}\right)^2 = \frac{b^2 - 4ac}{4a^2}$$

and we can take the square roots of both sides, being careful not to lose any roots:

$$x + \frac{b}{2a} = \pm\sqrt{\frac{b^2 - 4ac}{4a^2}}$$

or

$$x + \frac{b}{2a} = \frac{\pm\sqrt{b^2 - 4ac}}{2a}$$

We finally subtract $b/2a$ from both sides of this equation, obtaining the well-known quadratic formula

A.1.5 $$x = \frac{-b \pm \sqrt{b^2 - 4ac}}{2a}$$

Note that this equation has two solutions, corresponding to the positive and negative signs in front of the radical. There are three cases, depending on the sign of the *discriminant*

$$D = b^2 - 4ac$$

If $D > 0$, it has a positive and a negative square root, and so (A.1.5) gives two *real and distinct* values for x. If $D = 0$ then (A.1.5) gives only one value for x; we say that (A.1.4) has *real and coincident* roots. If $D < 0$, its square root is imaginary; in this case, (A.1.4) has two *conjugate-imaginary* roots; the roots are *not* real.

The next equations to be considered should logically, be the *cubic*, or third-degree, followed by the *biquadratic*, or fourth-degree equation. In actuality, the solution of these equations, though technically possible, is seldom treated because of its difficulty. We give brief outlines of the methods.

For the general cubic

A.1.6 $$x^3 + ax^2 + bx + c = 0$$

the change of variables

A.1.7 $$x = y - \frac{a}{3}$$

will give a *deficient* cubic

A.1.8 $$y^3 + py + q = 0$$

(where p, q, are constants depending on a, b, and c). The further change of variables

A.1.9 $$y = z - \frac{p}{3z}$$

gives the equation

$$z^3 + q - \frac{p^3}{27z^3} = 0$$

and multiplication by z^3 gives us

A.1.10 $$z^6 + qz^3 - \frac{p^3}{27} = 0$$

This can then be treated as a quadratic equation in $w = z^3$, which can be solved by using (A.1.5). Knowledge of w will give us z, and, in turn, y and x. Unfortunately, this method has the disadvantage that the solutions of (A.1.10) may all be imaginary. More precisely, (A.1.5) has three solutions, of which one at least is real. The other two may be either real or imaginary. In the case in which (A.1.5) has three real solutions, it happens that (A.1.10) has only imaginary solutions. For the reader who is willing to work with imaginary numbers, this is not a great drawback. Most people, however, prefer to avoid this.

For the general biquadratic

A.1.11 $$x^4 + ax^3 + bx^2 + cx + d = 0$$

the change of variables

A.1.12 $$x = y - \frac{a}{4}$$

will give a *deficient* biquadratic

A.1.13 $$y^4 + py^2 + qy + r = 0$$

This can be rewritten

$$y^4 + 2\lambda y^2 + \lambda^2 + (p - 2\lambda)y^2 + qy + (r - \lambda^2) = 0$$

Subtracting the last three terms from both sides, we obtain

A.1.14 $$(y^2 + \lambda)^2 = (2\lambda - p)y^2 - qy + (\lambda^2 - r)$$

It is now necessary to choose the parameter λ so that the right side of this equation is a perfect square. This will happen if λ satisfies the equation

A.1.15 $$4(2\lambda - p)(\lambda^2 - r) = q^2$$

The equation (A.1.15), which is known as the *resolvent* of (A.1.13), is a cubic, or third-degree equation. It can be solved by the method just considered. If and when this has been done, its solution (any one of its solutions) can be introduced into (A.1.14) to give something of the form

$$(y^2 + \lambda)^2 = (Ay + B)^2$$

or

$$y^2 + \lambda = \pm (Ay + B) \tag{A.1.16}$$

This gives two quadratic equations, each of which can be solved by formula (A.1.5). Note that, in all, there are four solutions, two from each quadratic.

We have seen that the cubic equation can be solved by an auxiliary quadratic equation, and that the biquadratic can be solved through an auxiliary cubic. The question naturally arises whether the fifth-degree equation cannot, perhaps, be solved through a fourth-degree resolvent, and so on. In fact, this question was pursued for several centuries by mathematicians—a pursuit that came to an unsuccessful end when Abel proved that *the general equation of the fifth-degree (or higher) cannot be solved by means of radicals alone.*

In general, it is best to solve such equations by approximation. The usual method for solving, say

$$P(x) = 0 \tag{A.1.17}$$

where $P(x)$ is a polynomial, is to find two values, a and b, such that $P(a) < 0$ but $P(b) > 0$. It will follow that a root of (A.1.17) lies between a and b. A third value, c, is taken, between a and b. If $P(c) > 0$, then a root will lie between a and c, while if $P(c) < 0$, then the root will lie between c and b. In this manner, smaller and smaller intervals are found, so that the root may be obtained with any degree of accuracy desired. Such methods are known as *numerical methods.*

Next in order of difficulty, after the polynomial equations, are the *rational* equations. A *rational function* is defined as the quotient of two polynomials; a rational equation is one in which only rational functions appear. By addition, subtraction, and simplification, such an equation can always be written in the form

$$\frac{P(x)}{Q(x)} = 0 \tag{A.1.18}$$

Multiplication by $Q(x)$ will reduce this to the polynomial equation $P(x) = 0$, so that the solution of (A.1.18) depends on that of (A.1.17). It should be pointed out, however, that multiplication by $Q(x)$ might introduce extraneous roots. The general rule is that *the solutions of (A.1.18) are all those solutions of (A.1.17) that are not solutions of* $Q(x) = 0$.

As an example, consider the equation

$$\frac{x^2 - 1}{x - 1} = 0$$

Multiplication by the denominator gives

$$x^2 - 1 = 0$$

which has the solutions

$$x = \pm 1$$

Now, if we substitute the value $x = -1$, in the original equation, we see that we have a solution. If, however, we substitute $x = 1$, the left side of the original equation takes the form 0/0, which is meaningless. Thus, $x = 1$ is an extraneous root.

We consider, next, equations with *surds*, or radicals. The usual approach to these equations consists in raising them to powers in order to eliminate the radicals. In so doing, one must guard against extraneous roots. Consider the equation

$$\sqrt{2x + 1} = x - 7$$

in which the radical is taken to represent the *positive* square root of $2x + 1$. Squaring both sides, we obtain

$$2x + 1 = x^2 - 14x + 49$$

which, after adding and subtracting some terms, gives the quadratic

$$x^2 - 16x + 48 = 0$$

The use of (A.1.5) will give the two solutions $x = 4$ and $x = 12$. Substitution of $x = 12$ gives

$$\sqrt{25} = 5$$

a correct statement. However, $x = 4$ gives

$$\sqrt{9} = -3$$

which is false since we have agreed that we want only a positive square root. Thus, $x = 12$ is the only solution; $x = 4$ is extraneous.

In this fashion, irrational equations (equations with surds) may be changed into polynomial equations. It is then a question of solving these, which, as we have seen, may be easy or difficult depending on the degree and form.

The last types of equation that we consider are the *transcendental* equations. These are equations that include the exponential, logarithmic, and trigonometric functions. Unless of especially simple form, they cannot be solved analytically. The equation

$$e^x = 3$$

has the obvious solution $x = \ln 3$. But the slightly more complicated

$$xe^x = 3$$

has no such simple solution; it must be solved numerically. In general, transcendental equations must be solved (approximately) by numerical methods.

2. THE PRINCIPLE OF INDUCTION

It is very often necessary to prove that some statement is true for all the natural numbers: 1, 2, 3, ... Since it is not always possible to prove the statement for all of them simultaneously, and since it is definitely impossible to prove it for all the numbers, one at a time, recourse is often had to the *principle of mathematical induction.* We give the principle in one form (there are several other possible forms):

A.2.1 Let T be a subset of the natural numbers, and suppose that T satisfies conditions (a) and (b):

(a) $1 \in T$

(b) If $n \in T$, then $n+1 \in T$

Then T contains all the natural numbers.

The principle (A.2.1) is a basic part of the foundations of all mathematics. It is, in fact, generally used as part of the definition of our number system. Heuristically, we can see the justification for this principle. Suppose T satisfies the two conditions (a) and (b). Then by (a), it must contain the number 1. Since T contains 1, 2 must, by (b), also belong to T. But then 3 must also belong to T, and so on. It will follow that all the natural numbers (i.e., positive integers) belong to T.

Let us suppose, then, that we wish to prove the truth of a statement for all natural numbers. We can let T be the set of all natural numbers for which this statement is true. We must then prove that T has properties (a) and (b) of the principle (A.2.1). To prove (a), we must show that the statement is true for the natural number 1. To prove (b), we must show that whenever the statement is true for the natural number n, it is also true for the next natural number, $n + 1$.

Thus, a *proof by induction* will consist of two steps. First, prove for 1. Second, prove that truth for n implies truth for $n+1$. This effectively proves it for all.

Let us suppose that we have an endless succession of light bulbs (having a first bulb, but no last bulb) which we wish to light. It is obviously impossible to light them all one at a time, since we simply do not have the time to do this. Suppose, however, that there is a mechanism connecting each one to the next, in such a way that whenever a bulb is lit, the next bulb is necessarily lit. Then lighting the first bulb will cause the second one to light, the second will light the third, and so on. To guarantee that all the bulbs are lit, it is sufficient to check: (a) that the first bulb is lit; (b) that the connections between bulbs have the desired property—that, if a bulb is lit, the next is also lit.

To see how this principle is used in practice, consider the problem of finding the sum of the first n squares. We wish to prove that:

A.2.2 $$1^2 + 2^2 + \ldots + n^2 = \frac{n(n+1)(2n+1)}{6}$$

for all values of n.

We prove this by the induction method. Consider, then, the case $n = 1$. In this case, the left side of (A.2.2) is simply 1^2, or 1. The right side is

$$\frac{1(1+1)(2+1)}{6}$$

which is also equal to 1. Thus, (A.2.2) is true when $n = 1$.

We must now prove that, if (A.2.2) is true for one value of n, it is also true for the next larger value. If we replace n by $n + 1$, (A.2.2) will take the form

A.2.3 $$1^2+2^2+\ldots+n^2+(n+1)^2 = \frac{(n+1)[(n+1)+1][2(n+1)+1]}{6}$$

We must prove that, if (A.2.2) is true for a value of n, then (A.2.3) will also be true for that same value of n.

To do this, we add the term $(n+1)^2$ to both sides of (A.2.2). This gives us

A.2.4 $$1^2+2^2+\ldots+n^2+(n+1)^2 = \frac{n(n+1)(2n+1)}{6} + (n+1)^2$$

It is now easy to check that (A.2.3) and (A.2.4) are the same equation. In fact, their left sides are clearly equal; it is simply a matter of expanding their right sides to show that these, also, are equal.

We have now shown that (A.2.2) is true when $n = 1$, and moreover, that if it is true for one value of n, then it is also true for the next value, $n + 1$. By the principle of induction, this means that the statement is true for all natural numbers. The proof is complete.

3. EXPONENTS AND LOGARITHMS

Throughout much of this book, great use has been made of the *exponential* and *logarithmic* functions. We give here some of the elements of exponents and logarithms.

If a is any number, we shall write

$$a^2 = a \cdot a$$
$$a^3 = a \cdot a \cdot a$$

and so on; in general, if m is a positive integer, a^m is the product of m terms, each equal to a:

$$a^m = \underbrace{a \cdot a \cdot \ldots \cdot a}_{m \text{ terms}}$$

A stricter mathematical definition would be the *inductive* definition

A.3.1 $$a^1 = a$$

A.3.2 $$a^{m+1} = a^m \cdot a$$

Note how this inductive definition works: equation (A.3.1) defines a^m for $m=1$; (A.3.2) defines a^{m+1} when a^m is known. According to the principle of induction this defines the power a^m for all positive integers m.

There are three important rules for working with exponents. The first deals with multiplying two powers of the same number. We have

$$a^m \cdot a^n = \underbrace{a \cdot a \cdot \ldots \cdot a}_{m \text{ terms}} \cdot \underbrace{a \cdot a \cdot \ldots \cdot a}_{n \text{ terms}}$$

It may be seen that the right side of this equation has $m+n$ terms, and so we conclude that

A.3.3 $$a^m \cdot a^n = a^{m+n}$$

The second rule deals with "a power of a power." We have

$$(a^m)^n = \underbrace{a^m \cdot a^m \cdot \ldots \cdot a^m}_{n \text{ terms}}$$

Each of the n terms on the right side of this equation can itself be expanded; doing this, we will have

$$(a^m)^n = \left.\begin{array}{l} (a \cdot a \cdot \ldots \cdot a) \\ \cdot (a \cdot a \cdot \ldots \cdot a) \\ \cdots\cdots\cdots\cdots \\ \underbrace{\cdot (a \cdot a \cdot \ldots \cdot a)}_{m \text{ terms}} \end{array}\right\} n \text{ terms}$$

Now, the right side of this has m terms in each of n rows. There are mn terms all told, and so

A.3.4 $$(a^m)^n = a^{mn}$$

The third rule deals with obtaining a power of the product of two numbers, a and b.

$$(ab)^m = \underbrace{ab \cdot ab \cdot \ldots \cdot ab}_{m \text{ times}}$$

The right side of this equation may be rearranged to give us

$$(ab)^m = \underbrace{a \cdot a \ldots a}_{m \text{ terms}} \cdot \underbrace{b \cdot b \ldots b}_{m \text{ terms}}$$

and so

A.3.5 $$(ab)^m = a^m b^m$$

Note that this last can only be obtained because *multiplication of numbers is commutative*, i.e., $ab = ba$ for any numbers. If this were not so, the rearrangement of terms would not be possible. If we deal, say, with matrices (Chapter VII) we will find that laws analogous to (A.3.3) and (A.3.4) hold, but the analogue of (A.3.5) does not hold.

For the reader who is not entirely convinced by the proofs of (A.3.3), (A.3.4), and (A.3.5), a more formal proof, depending on the definitions (A.3.1) and (A.3.2) and on the principle of induction (A.2.1), is available. Suppose that we wish to prove (A.3.3). We do this by induction on n. Letting $n = 1$, we have

$$a^m \cdot a^1 = a^m \cdot a$$

since $a^1 = a$ by definition. But also by definition, the right side of this is equal to a^{m+1}. Thus,

$$a^m \cdot a^1 = a^{m+1}$$

Note that this is true for *all* values of m.

Proceeding by induction, let us assume that (A.3.3) is true for *all* values of m and for a *particular* value of n. Now we will have

$$a^m \cdot a^{n+1} = a^m (a^n \cdot a)$$

by the definition of a^{n+1}. The associative law for multiplication gives us

$$a^m (a^n \cdot a) = (a^m \cdot a^n) \cdot a$$

By the induction hypothesis, $a^m\ a^n = a^{m+n}$, and so

$$(a^m \cdot a^n) a = a^{m+n} \cdot a$$

and, by definition, the right side here is a^{m+n+1}. Thus,

$$a^m \cdot a^{n+1} = a^{m+n+1}$$

By the principle of induction, this proves (A.3.3) for all values of m and n (among the positive integers). Similar proofs may be obtained for both (A.3.4) and (A.3.5).

We have, thus, defined the meaning of the expression a^m, when m is a positive integer. We would like, now, to define the analogous expressions a^r, when r is negative, fractional, or even, for that matter, irrational. We define these by *extension*, i.e., in such a way as to preserve the properties (A.3.3) and (A.3.4).

Consider, for example, the expression a^0. It is, clearly, meaningless to talk about the product of 0 terms. Let us however, use rule (A.3.3), with $n = 0$. We have

$$a^m \cdot a^0 = a^{m+0} = a^m$$

and so (if a is not zero) we see that a natural definition is

A.3.6 $$a^0 = 1$$

(Note, however, that the expression 0^0 is meaningless.)

Similarly, we can find a meaning for negative integer exponents. We have, indeed,

$$a^m \cdot a^{-m} = a^{m-m} = a^0 = 1$$

and it follows that we can define

A.3.7 $$a^{-m} = \frac{1}{a^m}$$

Let us now consider fractional exponents. If p and q are integers, we see by rule (A.3.4) that

$$(a^{p/q})^q = a^{pq/q} = a^p$$

Thus, $a^{p/q}$ should be defined so that its qth power is equal to a^p. But this is precisely what is known as a qth root. We define

A.3.8 $$a^{p/q} = \sqrt[q]{a^p}$$

The definitions (A.3.7) and (A.3.8) allow us to deal with all rational exponents. We will have

$$a^{-1/2} = \frac{1}{\sqrt{a}}$$

$$a^{1.3} = \sqrt[10]{a^{13}}$$

and so on. There remains only the case of irrational exponents to be disposed of.

For irrational exponents, the method of definition depends on continuity. In effect, we know that every irrational number can be approximated as closely as desired by rational numbers. If r is any real number, we know that there is a sequence $q_1, q_2, \ldots$ of rational numbers that *converges* to r. Then the power a^r is defined as the *limit* of the sequence:

$$a^{q_1}, a^{q_2}, \ldots \rightarrow a^r$$

To give an example, we know that the irrational $\sqrt{2}$ is the limit of the sequence

$$1, 1.4, 1.41, 1.414, 1.4142, \ldots$$

(represented, of course, by the decimal expansion of $\sqrt{2}$). This will mean that the power $a^{\sqrt{2}}$ can be defined as the limit of the sequence

$$a, a^{1.4}, a^{1.41}, a^{1.414}, a^{1.4142}, \ldots$$

In other words, a term of this sequence, such as $a^{1.414}$, is a good approximation to the desired $a^{\sqrt{2}}$ —just as 1.414 is a good approximation to $\sqrt{2}$.

Next we consider *logarithms*. Briefly speaking, a logarithm is the same as an exponent. If

A.3.9 $$a^y = x$$

we say that y is the logarithm of x, with base a. The notation for this is

A.3.10 $$y = \log_a x$$

Statements (A.3.9) and (A.3.10) are equivalent. From (A.3.9), we say that x is the *exponential* of y (with base a); we see that the exponential and logarithmic functions are (by definition) mutually inverse. We shall now develop rules for logarithms. These shall all be natural consequences of the rules (A.3.3) and (A.3.4) for exponents.

Let us suppose we have

$$x = a^y; \; z = a^w$$

or, equivalently,

$$y = \log_a x; \; w = \log_a z$$

Now, by (A.3.3),

$$xz = a^y \cdot a^w = a^{y+w}$$

and so

$$\log_a (xz) = y + w$$

or

A.3.11 $$\log_a (xz) = \log_a x + \log_a z$$

Similarly, we have, by (A.3.7)

$$\frac{1}{x} = \frac{1}{a^y} = a^{-y}$$

or

A.3.12 $$\log_a \left(\frac{1}{x}\right) = -\log_a x$$

Properties (A.3.11) and (A.3.12) can be combined to give us

A.3.13 $$\log_a \left(\frac{x}{z}\right) = \log_a x - \log_a z$$

Once again, assume $y = \log_a x$. We have, by (A.3.4),

$$x^r = (a^y)^r = a^{ry}$$

and so

A.3.14 $$\log_a x^r = r \log_a x$$

Rules (A.3.11) to (A.3.14) explain the usefulness of logarithms. In fact, we see that multiplications and divisions can be replaced by the simpler operations of addition and subtraction while the very difficult operation of raising to a power reduces to a multiplication.

One last property of logarithms interests us now, and this deals with a *change in base.* Let us suppose once again that

$$x = a^y$$

and, moreover,

$$a = b^c$$

We have, then,

$$x = (b^c)^y = b^{cy}$$

and therefore

A.3.15 $$\log_b x = (\log_b a)(\log_a x)$$

For a change in base, it is sufficient to multiply the logarithm using the old base, times the logarithm of the old base with respect to the new base. Equivalently, we have

A.3.16 $$\log_a x = \frac{\log_b x}{\log_b a}$$

so that it is sufficient to divide the logarithm of the variable by the logarithm of the new base.

As base, any positive number other than 1 may be used. Negative numbers are not used as base, since many of their powers are imaginary. The numbers 0 and 1 are useless, since all their powers are equal. Numbers smaller than 1 may be used as base, but they have the disadvantage that, with such bases, the logarithm of a number decreases as the number increases. With base 1/2, for example,

$$\log_{1/2} 2 = -1$$

$$\log_{1/2} 8 = -3$$

and so on. As a practical matter, the base should be a number greater than 1. In actual practice, only three bases have any use at all (and one of those is only used by computers). These three bases are the numbers 10, e, and 2.

The most commonly used logarithms are logarithms to the base 10. Their main advantage is the well-known fact that, with our numerical system, multiplication by 10 or any integral power of 10 can be accomplished by shifting a decimal point, with no change of digits. The importance of this can best be seen if we consider that, for example,

$$\log 2352 = \log 1000 + \log 2.352$$

But $\log_{10} 1000 = 3$, and so

$$\log_{10} 2352 = \log_{10} 2.352 + 3$$

Similarly,

$$\log_{10} 0.0158 = \log_{10} 1.58 - 2$$

and so on. It is sufficient to tabulate logarithms for numbers from 1 to 10; logarithms for smaller or larger numbers can easily be obtained from them.

The advantages of the base e, while not inconsiderable, are of a more theoretical nature. Logarithms to the base e are known as *natural* logarithms. They are also known as *Napierian* logarithms, after Napier, who was one of the first men to study logarithms. We use the notations

$$\log_e x, \ \ln x$$

interchangeably, to denote these logarithms. Such logarithms are also commonly tabulated. They may be obtained from the common (base 10) logarithms, however, by using (A.3.15). This gives us,

$$\ln x = (\ln 10)(\log_{10} x)$$

Since the natural logarithm of 10 is approximately 2.3026, we have

A.3.17 $$\ln x = 2.3026 \log_{10} x$$

One last number, 2, is in current use as base for logarithms. Its use is mainly confined to computers, which can work best with 2 as basis for all their calculations; there is also some use for these logarithms in the computer-related science of information theory. Should the reader ever need these logarithms, they can be obtained from (A.3.15) with the formula

A.3.18 $$\log_2 x = 3.3219 \log_{10} x$$

4. THE SUMMATION SYMBOL

Throughout this book, we have used the summation symbol, Σ. We will explain its use simply in this appendix.

Let us suppose we have an *indexed* set of numbers, with indices among the integers:

$$a_1, a_2, a_3, \ldots$$

Let k and l be integers such that $k < l$. Then the symbol

$$\sum_{j=k}^{l} a_j$$

will be taken to mean the sum of all the numbers a_j, as j takes on all integral values from k to l, inclusive:

A.4.1 $$\sum_{j=k}^{l} a_j = a_k + a_{k+1} + \ldots + a_{l-1} + a_l$$

For example,

$$\sum_{j=5}^{9} b_j = b_5 + b_6 + b_7 + b_8 + b_9$$

$$\sum_{i=0}^{5} a_i = a_0 + a_1 + a_2 + a_3 + a_4 + a_5$$

Sometimes, of course, the indexed number, a_i, may be defined as a function, $f(i)$, of the index. Technically, this is no difference at all. We have, then, for example,

$$\sum_{j=1}^{4} \frac{1}{j} = \frac{1}{1} + \frac{1}{2} + \frac{1}{3} + \frac{1}{4}$$

$$\sum_{i=11}^{13} i^2 = 11^2 + 12^2 + 13^2$$

Strictly speaking, it is not even necessary that the indices be among the integers; they can come from any set at all. If Ω is such a set, the notation

$$\sum_{\omega \in \Omega} a_\omega$$

represents the sum of the terms a_ω, for all ω in the index set Ω.

As an example, we can let Ω be the set of people inside an elevator, and a_ω represent the weight, in pounds, of the individual ω. The inequality

$$\sum_{\omega \in \Omega} a_\omega \leq 1600$$

is then symbolic notation for a sign that may appear inside this elevator.

There are certain rules for working with summations, following from the well-known rules for sums and products of numbers. Thus

$$\sum_{j=k}^{l} c\,a_j = c\,a_k + c\,a_{k+1} + \ldots + c\,a_l$$

$$= c(a_k + a_{k+1} + \ldots + a_l)$$

so

A.4.2 $$\sum_{j=k}^{l} ca_j = c \sum_{j=k}^{l} a_j$$

i.e., a constant factor (one not dependent on the index j) can be "factored out" of a sum.

Similarly,

$$\sum_{j=k}^{l} (a_j + b_j) = (a_k + b_k) + \ldots + (a_l + b_l)$$

$$= (a_k + a_{k+1} + \ldots + a_l) + (b_k + b_{k+1} \;+ \ldots + b_l)$$

and so

A.4.3 $$\sum_{j=k}^{l} (a_j + b_j) = \sum_{j=k}^{l} a_j + \sum_{j=k}^{l} b_j$$

i.e., a sum can be "split" into two sums in this manner.

Double sums are also encountered on occasion. Given a doubly indexed set of numbers a_{ij}, we have the notation

A.4.4 $$\sum_{i=m}^{n} \sum_{j=k}^{l} a_{ij}$$

to represent the sum of the terms a_{ij}, as i and j are allowed, independently, to take all values between m and n, and between k and l, respectively. These double sums can also be rewritten as *iterated sums*:

A.4.5 $$\sum_{i=m}^{n} \left\{ \sum_{j=k}^{1} a_{ij} \right\}$$

To evaluate this expression, we must first compute the sum

$$b_i = \sum_{j=k}^{l} a_{ij}$$

for each i, and then the sum

$$\sum_{i=m}^{n} b_i$$

It is not difficult to see that (A.4.4) and (A.4.5) are equal to each other, and also to the other iterated sum

$$\sum_{j=k}^{l} \left\{ \sum_{i=m}^{n} a_{ij} \right\}$$

in which the sum is taken first over i and then over j.

Similarly, triple and, more generally, n-tuple sums may be defined. These can always be evaluated as iterated sums. Thus

$$\sum_{i \in I} \sum_{j \in J} \sum_{k \in K} a_{ijk} = \sum_{i \in I} \left\{ \sum_{j \in J} \left[\sum_{k \in K} a_{ijk} \right] \right\}$$

and so on.

5. TABLE OF COMMON LOGARITHMS

$$y = \log_{10} x$$

											Average differences								
x	0	1	2	3	4	5	6	7	8	9	1	2	3	4	5	6	7	8	9
10	0000	0043	0086	0128	0170	0212	0253	0294	0334	0374	4	8	12	17	21	25	29	33	37
11	0414	0453	0492	0531	0569	0607	0645	0682	0719	0755	4	8	11	15	19	23	26	30	34
12	0792	0828	0864	0899	0934	0969	1004	1038	1072	1106	3	7	10	14	17	21	24	28	31
13	1139	1173	1206	1239	1271	1303	1335	1367	1399	1430	3	6	10	13	16	19	23	26	29
14	1461	1492	1523	1553	1584	1614	1644	1673	1703	1732	3	6	9	12	15	18	21	24	27
15	1761	1790	1818	1847	1875	1903	1931	1959	1987	2014	3	6	8	11	14	17	20	22	25
16	2041	2068	2095	2122	2148	2175	2201	2227	2253	2279	3	5	8	11	13	16	18	21	24
17	2304	2330	2355	2380	2405	2430	2455	2480	2504	2529	2	5	7	10	12	15	17	20	22
18	2553	2577	2601	2625	2648	2672	2695	2718	2742	2765	2	5	7	9	12	14	16	19	21
19	2788	2810	2833	2856	2878	2900	2923	2945	2967	2989	2	4	7	9	11	13	16	18	20
20	3010	3032	3054	3075	3096	3118	3139	3160	3181	3201	2	4	6	8	11	13	15	17	19
21	3222	3243	3263	3284	3304	3324	3345	3365	3385	3404	2	4	6	8	10	12	14	16	18
22	3424	3444	3464	3483	3502	3522	3541	3560	3579	3598	2	4	6	8	10	12	14	15	17
23	3617	3636	3655	3674	3692	3711	3729	3747	3766	3784	2	4	6	7	9	11	13	15	17
24	3802	3820	3838	3856	3874	3892	3909	3927	3945	3962	2	4	5	7	9	11	12	14	16
25	3979	3997	4014	4031	4048	4065	4082	4099	4116	4133	2	3	5	7	9	10	12	14	15
26	4150	4166	4183	4200	4216	4232	4249	4265	4281	4298	2	3	5	7	8	10	11	13	15
27	4314	4330	4346	4362	4378	4393	4409	4425	4440	4456	2	3	5	6	8	9	11	13	14
28	4472	4487	4502	4518	4533	4548	4564	4579	4594	4609	2	3	5	6	8	9	11	12	14
29	4624	4639	4654	4669	4683	4698	4713	4728	4742	4757	1	3	4	6	7	9	10	12	13
30	4771	4786	4800	4814	4829	4843	4857	4871	4886	4900	1	3	4	6	7	9	10	11	13
31	4914	4928	4942	4955	4969	4983	4997	5011	5024	5038	1	3	4	6	7	8	10	11	12
32	5051	5065	5079	5092	5105	5119	5132	5145	5159	5172	1	3	4	5	7	8	9	11	12
33	5185	5198	5211	5224	5237	5250	5263	5276	5289	5302	1	3	4	5	6	8	9	10	12
34	5315	5328	5340	5353	5366	5378	5391	5403	5416	5428	1	3	4	5	6	8	9	10	11
35	5441	5453	5465	5478	5490	5502	5514	5527	5539	5551	1	2	4	5	6	7	9	10	11
36	5563	5575	5587	5599	5611	5623	5635	5647	5658	5670	1	2	4	5	6	7	8	10	11
37	5682	5694	5705	5717	5729	5740	5752	5763	5775	5786	1	2	3	5	6	7	8	9	10
38	5798	5809	5821	5832	5843	5855	5866	5877	5888	5899	1	2	3	5	6	7	8	9	10
39	5911	5922	5933	5944	5955	5966	5977	5988	5999	6010	1	2	3	4	5	7	8	9	10
40	6021	6031	6042	6053	6064	6075	6085	6096	6107	6117	1	2	3	4	5	6	8	9	10
41	6128	6138	6149	6160	6170	6180	6191	6201	6212	6222	1	2	3	4	5	6	7	8	9
42	6232	6243	6253	6263	6274	6284	6294	6304	6314	6325	1	2	3	4	5	6	7	8	9
43	6335	6345	6355	6365	6375	6385	6395	6405	6415	6425	1	2	3	4	5	6	7	8	9
44	6435	6444	6454	6464	6474	6484	6493	6503	6513	6522	1	2	3	4	5	6	7	8	9
45	6532	6542	6551	6561	6571	6580	6590	6599	6609	6618	1	2	3	4	5	6	7	8	9
46	6628	6637	6646	6656	6665	6675	6684	6693	6702	6712	1	2	3	4	5	6	7	7	8
47	6721	6730	6739	6749	6758	6767	6776	6785	6794	6803	1	2	3	4	5	5	6	7	8
48	6812	6821	6830	6839	6848	6857	6866	6875	6884	6893	1	2	3	4	5	5	6	7	8
49	6902	6911	6920	6928	6937	6946	6955	6964	6972	6981	1	2	3	4	4	5	6	7	8
50	6990	6998	7007	7016	7024	7033	7042	7050	7059	7067	1	2	3	3	4	5	6	7	8
51	7076	7084	7093	7101	7110	7118	7126	7135	7143	7152	1	2	3	3	4	5	6	7	8
52	7160	7168	7177	7185	7193	7202	7210	7218	7226	7235	1	2	2	3	4	5	6	7	7
53	7243	7251	7259	7267	7275	7284	7292	7300	7308	7316	1	2	2	3	4	5	6	6	7
54	7324	7332	7340	7348	7356	7364	7372	7380	7388	7396	1	2	2	3	4	5	6	6	7
x	0	1	2	3	4	5	6	7	8	9	1	2	3	4	5	6	7	8	9

5. TABLE OF COMMON LOGARITHMS *(Continued)*

$$y = \log_{10} x$$

x	0	1	2	3	4	5	6	7	8	9	Average differences 1	2	3	4	5	6	7	8	9
55	7404	7412	7419	7427	7435	7443	7451	7459	7466	7474	1	2	2	3	4	5	5	6	7
56	7482	7490	7497	7505	7513	7520	7528	7536	7543	7551	1	2	2	3	4	5	5	6	7
57	7559	7566	7574	7582	7589	7597	7604	7612	7619	7627	1	2	2	3	4	5	5	6	7
58	7634	7642	7649	7657	7664	7672	7679	7686	7694	7701	1	1	2	3	4	4	5	6	7
59	7709	7716	7723	7731	7738	7745	7752	7760	7767	7774	1	1	2	3	4	4	5	6	7
60	7782	7789	7796	7803	7810	7818	7825	7832	7839	7846	1	1	2	3	4	4	5	6	6
61	7853	7860	7868	7875	7882	7889	7896	7903	7910	7917	1	1	2	3	4	4	5	6	6
62	7924	7931	7938	7945	7952	7959	7966	7973	7980	7987	1	1	2	3	3	4	5	6	6
63	7993	8000	8007	8014	8021	8028	8035	8041	8048	8055	1	1	2	3	3	4	5	5	6
64	8062	8069	8075	8082	8089	8096	8102	8109	8116	8122	1	1	2	3	3	4	5	5	6
65	8129	8136	8142	8149	8156	8162	8169	8176	8182	8189	1	1	2	3	3	4	5	5	6
66	8195	8202	8209	8215	8222	8228	8235	8241	8248	8254	1	1	2	3	3	4	5	5	6
67	8261	8267	8274	8280	8287	8293	8299	8306	8312	8319	1	1	2	3	3	4	5	5	6
68	8325	8331	8338	8344	8351	8357	8363	8370	8376	8382	1	1	2	3	3	4	4	5	6
69	8388	8395	8401	8407	8414	8420	8426	8432	8439	8445	1	1	2	2	3	4	4	5	6
70	8451	8457	8463	8470	8476	8482	8488	8494	8500	8506	1	1	2	2	3	4	4	5	6
71	8513	8519	8525	8531	8537	8543	8549	8555	8561	8567	1	1	2	2	3	4	4	5	5
72	8573	8579	8585	8591	8597	8603	8609	8615	8621	8627	1	1	2	2	3	4	4	5	5
73	8633	8639	8645	8651	8657	8663	8669	8675	8681	8686	1	1	2	2	3	4	4	5	5
74	8692	8698	8704	8710	8716	8722	8727	8733	8739	8745	1	1	2	2	3	4	4	5	5
75	8751	8756	8762	8768	8774	8779	8785	8791	8797	8802	1	1	2	2	3	3	4	5	5
76	8808	8814	8820	8825	8831	8837	8842	8848	8854	8859	1	1	2	2	3	3	4	5	5
77	8865	8871	8876	8882	8887	8893	8899	8904	8910	8915	1	1	2	2	3	3	4	4	5
78	8921	8927	8932	8938	8943	8949	8954	8960	8965	8971	1	1	2	2	3	3	4	4	5
79	8976	8982	8987	8993	8998	9004	9009	9015	9020	9025	1	1	2	2	3	3	4	4	5
80	9031	9036	9042	9047	9053	9058	9063	9069	9074	9079	1	1	2	2	3	3	4	4	5
81	9085	9090	9096	9101	9106	9112	9117	9122	9128	9133	1	1	2	2	3	3	4	4	5
82	9138	9143	9149	9154	9159	9165	9170	9175	9180	9186	1	1	2	2	3	3	4	4	5
83	9191	9196	9201	9206	9212	9217	9222	9227	9232	9238	1	1	2	2	3	3	4	4	5
84	9243	9248	9253	9258	9263	9269	9274	9279	9284	9289	1	1	2	2	3	3	4	4	5
85	9294	9299	9304	9309	9315	9320	9325	9330	9335	9340	1	1	2	2	3	3	4	4	5
86	9345	9350	9355	9360	9365	9370	9375	9380	9385	9390	1	1	2	2	3	3	4	4	5
87	9395	9400	9405	9410	9415	9420	9425	9430	9435	9440	0	1	1	2	2	3	3	4	4
88	9445	9450	9455	9460	9465	9469	9474	9479	9484	9489	0	1	1	2	2	3	3	4	4
89	9494	9499	9504	9509	9513	9518	9523	9528	9533	9538	0	1	1	2	2	3	3	4	4
90	9542	9547	9552	9557	9562	9566	9571	9576	9581	9586	0	1	1	2	2	3	3	4	4
91	9590	9595	9600	9605	9609	9614	9619	9624	9628	9633	0	1	1	2	2	3	3	4	4
92	9638	9643	9647	9652	9657	9661	9666	9671	9675	9680	0	1	1	2	2	3	3	4	4
93	9685	9689	9694	9699	9703	9708	9713	9717	9722	9727	0	1	1	2	2	3	3	4	4
94	9731	9736	9741	9745	9750	9754	9759	9763	9768	9773	0	1	1	2	2	3	3	4	4
95	9777	9782	9786	9791	9795	9800	9805	9809	9814	9818	0	1	1	2	2	3	3	4	4
96	9823	9827	9832	9836	9841	9845	9850	9854	9859	9863	0	1	1	2	2	3	3	4	4
97	9868	9872	9877	9881	9886	9890	9894	9899	9903	9908	0	1	1	2	2	3	3	4	4
98	9912	9917	9921	9926	9930	9934	9939	9943	9948	9952	0	1	1	2	2	3	3	4	4
99	9956	9961	9965	9969	9974	9978	9983	9987	9991	9996	0	1	1	2	2	3	3	3	4
x	0	1	2	3	4	5	6	7	8	9	1	2	3	4	5	6	7	8	9

6. TABLE OF NATURAL (NAPERIAN) LOGARITHMS

ln x

x	0	0.01	0.02	0.03	0.04	0.05	0.06	0.07	0.08	0.09
1.0	0.0000	0.0100	0.0198	0.0296	0.0392	0.0488	0.0583	0.0677	0.0770	0.0862
1.1	0.0953	0.1044	0.1133	0.1222	0.1310	0.1398	0.1484	0.1570	0.1655	0.1740
1.2	0.1823	0.1906	0.1989	0.2070	0.2151	0.2231	0.2311	0.2390	0.2469	0.2546
1.3	0.2624	0.2700	0.2776	0.2852	0.2927	0.3001	0.3075	0.3148	0.3221	0.3293
1.4	0.3365	0.3436	0.3507	0.3577	0.3646	0.3716	0.3784	0.3853	0.3920	0.3988
1.5	0.4055	0.4121	0.4187	0.4253	0.4318	0.4383	0.4447	0.4511	0.4574	0.4637
1.6	0.4700	0.4762	0.4824	0.4886	0.4947	0.5008	0.5068	0.5128	0.5188	0.5247
1.7	0.5306	0.5365	0.5423	0.5481	0.5539	0.5596	0.5653	0.5710	0.5766	0.5822
1.8	0.5878	0.5933	0.5988	0.6043	0.6098	0.6152	0.6206	0.6259	0.6313	0.6366
1.9	0.6419	0.6471	0.6523	0.6575	0.6627	0.6678	0.6729	0.6780	0.6831	0.6881
2.0	0.6931	0.6981	0.7031	0.7080	0.7129	0.7178	0.7227	0.7275	0.7324	0.7372
2.1	0.7419	0.7467	0.7514	0.7561	0.7608	0.7655	0.7701	0.7747	0.7793	0.7839
2.2	0.7885	0.7930	0.7975	0.8020	0.8065	0.8109	0.8154	0.8198	0.8242	0.8286
2.3	0.8329	0.8372	0.8416	0.8459	0.8502	0.8544	0.8587	0.8629	0.8671	0.8713
2.4	0.8755	0.8796	0.8838	0.8879	0.8920	0.8961	0.9002	0.9042	0.9083	0.9123
2.5	0.9163	0.9203	0.9243	0.9282	0.9322	0.9361	0.9400	0.9439	0.9478	0.9517
2.6	0.9555	0.9594	0.9632	0.9670	0.9708	0.9746	0.9783	0.9821	0.9858	0.9895
2.7	0.9933	0.9969	1.0006	1.0043	1.0080	1.0116	1.0152	1.0188	1.0225	1.0260
2.8	1.0296	1.0332	1.0367	1.0403	1.0438	1.0473	1.0508	1.0543	1.0578	1.0613
2.9	1.0647	1.0682	1.0716	1.0750	1.0784	1.0818	1.0852	1.0886	1.0919	1.0953
3.0	1.0986	1.1019	1.1053	1.1086	1.1119	1.1151	1.1184	1.1217	1.1249	1.1282
3.1	1.1314	1.1346	1.1378	1.1410	1.1442	1.1474	1.1506	1.1537	1.1569	1.1600
3.2	1.1632	1.1663	1.1694	1.1725	1.1756	1.1787	1.1817	1.1848	1.1878	1.1909
3.3	1.1939	1.1969	1.2000	1.2030	1.2060	1.2090	1.2119	1.2149	1.2179	1.2208
3.4	1.2238	1.2267	1.2296	1.2326	1.2355	1.2384	1.2413	1.2442	1.2470	1.2499
3.5	1.2528	1.2556	1.2585	1.2613	1.2641	1.2669	1.2698	1.2726	1.2754	1.2782
3.6	1.2809	1.2837	1.2865	1.2892	1.2920	1.2947	1.2975	1.3002	1.3029	1.3056
3.7	1.3083	1.3110	1.3137	1.3164	1.3191	1.3218	1.3244	1.3271	1.3297	1.3324
3.8	1.3350	1.3376	1.3403	1.3429	1.3455	1.3481	1.3507	1.3533	1.3558	1.3584
3.9	1.3610	1.3635	1.3661	1.3686	1.3712	1.3737	1.3762	1.3788	1.3813	1.3838
4.0	1.3863	1.3888	1.3913	1.3938	1.3962	1.3987	1.4012	1.4036	1.4061	1.4085
4.1	1.4110	1.4134	1.4159	1.4183	1.4207	1.4231	1.4255	1.4279	1.4303	1.4327
4.2	1.4351	1.4375	1.4398	1.4422	1.4446	1.4469	1.4493	1.4516	1.4540	1.4563
4.3	1.4586	1.4609	1.4633	1.4656	1.4679	1.4702	1.4725	1.4748	1.4770	1.4793
4.4	1.4816	1.4839	1.4861	1.4884	1.4907	1.4929	1.4951	1.4974	1.4996	1.5019
4.5	1.5041	1.5063	1.5085	1.5107	1.5129	1.5151	1.5173	1.5195	1.5217	1.5239
4.6	1.5261	1.5282	1.5304	1.5326	1.5347	1.5369	1.5390	1.5412	1.5433	1.5454
4.7	1.5476	1.5497	1.5518	1.5539	1.5560	1.5581	1.5602	1.5623	1.5644	1.5665
4.8	1.5686	1.5707	1.5728	1.5748	1.5769	1.5790	1.5810	1.5831	1.5851	1.5872
4.9	1.5892	1.5913	1.5933	1.5953	1.5974	1.5994	1.6014	1.6034	1.6054	1.6074
5.0	1.6094	1.6114	1.6134	1.6154	1.6174	1.6194	1.6214	1.6233	1.6253	1.6273
5.1	1.6292	1.6312	1.6332	1.6351	1.6371	1.6390	1.6409	1.6429	1.6448	1.6467
5.2	1.6487	1.6506	1.6525	1.6544	1.6563	1.6582	1.6601	1.6620	1.6639	1.6658
5.3	1.6677	1.6696	1.6715	1.6734	1.6752	1.6771	1.6790	1.6808	1.6827	1.6845
5.4	1.6864	1.6882	1.6901	1.6919	1.6938	1.6956	1.6974	1.6993	1.7011	1.7029
x	0	0.01	0.02	0.03	0.04	0.05	0.06	0.07	0.08	0.09

6. TABLE OF NATURAL (NAPERIAN) LOGARITHMS *(Continued)*

ln x

x	0	0.01	0.02	0.03	0.04	0.05	0.06	0.07	0.08	0.09
5.5	1.7047	1.7066	1.7084	1.7102	1.7120	1.7138	1.7156	1.7174	1.7192	1.7210
5.6	1.7228	1.7246	1.7263	1.7281	1.7299	1.7317	1.7334	1.7352	1.7370	1.7387
5.7	1.7405	1.7422	1.7440	1.7457	1.7475	1.7492	1.7509	1.7527	1.7544	1.7561
5.8	1.7579	1.7596	1.7613	1.7630	1.7647	1.7664	1.7681	1.7699	1.7716	1.7733
5.9	1.7750	1.7766	1.7783	1.7800	1.7817	1.7834	1.7851	1.7867	1.7884	1.7901
6.0	1.7918	1.7934	1.7951	1.7967	1.7984	1.8001	1.8017	1.8034	1.8050	1.8066
6.1	1.8083	1.8099	1.8116	1.8132	1.8148	1.8165	1.8181	1.8197	1.8213	1.8229
6.2	1.8245	1.8262	1.8278	1.8294	1.8310	1.8326	1.8342	1.8358	1.8374	1.8390
6.3	1.8405	1.8421	1.8437	1.8453	1.8469	1.8485	1.8500	1.8516	1.8532	1.8547
6.4	1.8563	1.8579	1.8594	1.8610	1.8625	1.8641	1.8656	1.8672	1.8687	1.8703
6.5	1.8718	1.8733	1.8749	1.8764	1.8779	1.8795	1.8810	1.8825	1.8840	1.8856
6.6	1.8871	1.8886	1.8901	1.8916	1.8931	1.8946	1.8961	1.8976	1.8991	1.9006
6.7	1.9021	1.9036	1.9051	1.9066	1.9081	1.9095	1.9110	1.9125	1.9140	1.9155
6.8	1.9169	1.9184	1.9199	1.9213	1.9228	1.9242	1.9257	1.9272	1.9286	1.9301
6.9	1.9315	1.9330	1.9344	1.9359	1.9373	1.9387	1.9402	1.9416	1.9430	1.9445
7.0	1.9459	1.9473	1.9488	1.9502	1.9516	1.9530	1.9544	1.9559	1.9573	1.9587
7.1	1.9601	1.9615	1.9629	1.9643	1.9657	1.9671	1.9685	1.9699	1.9713	1.9727
7.2	1.9741	1.9755	1.9769	1.9782	1.9796	1.9810	1.9824	1.9838	1.9851	1.9865
7.3	1.9879	1.9892	1.9906	1.9920	1.9933	1.9947	1.9961	1.9974	1.9988	2.0001
7.4	2.0015	2.0028	2.0042	2.0055	2.0069	2.0082	2.0096	2.0109	2.0122	2.0136
7.5	2.0149	2.0162	2.0176	2.0189	2.0202	2.0215	2.0229	2.0242	2.0255	2.0268
7.6	2.0281	2.0295	2.0308	2.0321	2.0334	2.0347	2.0360	2.0373	2.0386	2.0399
7.7	2.0412	2.0425	2.0438	2.0451	2.0464	2.0477	2.0490	2.0503	2.0516	2.0528
7.8	2.0541	2.0554	2.0567	2.0580	2.0592	2.0605	2.0618	2.0631	2.0643	2.0656
7.9	2.0669	2.0681	2.0694	2.0707	2.0719	2.0732	2.0744	2.0757	2.0769	2.0782
8.0	2.0794	2.0807	2.0819	2.0832	2.0844	2.0857	2.0869	2.0882	2.0894	2.0906
8.1	2.0919	2.0931	2.0943	2.0956	2.0968	2.0980	2.0992	2.1005	2.1017	2.1029
8.2	2.1041	2.1054	2.1066	2.1078	2.1090	2.1102	2.1114	2.1126	2.1138	2.1150
8.3	2.1163	2.1175	2.1187	2.1199	2.1211	2.1223	2.1235	2.1247	2.1258	2.1270
8.4	2.1282	2.1294	2.1306	2.1318	2.1330	2.1342	2.1353	2.1365	2.1377	2.1389
8.5	2.1401	2.1412	2.1424	2.1436	2.1448	2.1459	2.1471	2.1483	2.1494	2.1506
8.6	2.1518	2.1529	2.1541	2.1552	2.1564	2.1576	2.1587	2.1599	2.1610	2.1622
8.7	2.1633	2.1645	2.1656	2.1668	2.1679	2.1691	2.1702	2.1713	2.1725	2.1736
8.8	2.1748	2.1759	2.1770	2.1782	2.1793	2.1804	2.1815	2.1827	2.1838	2.1849
8.9	2.1861	2.1872	2.1883	2.1894	2.1905	2.1917	2.1928	2.1939	2.1950	2.1961
9.0	2.1972	2.1983	2.1994	2.2006	2.2017	2.2028	2.2039	2.2050	2.2061	2.2072
9.1	2.2083	2.2094	2.2105	2.2116	2.2127	2.2138	2.2148	2.2159	2.2170	2.2181
9.2	2.2192	2.2203	2.2214	2.2225	2.2235	2.2246	2.2257	2.2268	2.2279	2.2289
9.3	2.2300	2.2311	2.2322	2.2332	2.2343	2.2354	2.2364	2.2375	2.2386	2.2396
9.4	2.2407	2.2418	2.2428	2.2439	2.2450	2.2460	2.2471	2.2481	2.2492	2.2502
9.5	2.2513	2.2523	2.2534	2.2544	2.2555	2.2565	2.2576	2.2586	2.2597	2.2607
9.6	2.2618	2.2628	2.2638	2.2649	2.2659	2.2670	2.2680	2.2690	2.2701	2.2711
9.7	2.2721	2.2732	2.2742	2.2752	2.2762	2.2773	2.2783	2.2793	2.2803	2.2814
9.8	2.2824	2.2834	2.2844	2.2854	2.2865	2.2875	2.2885	2.2895	2.2905	2.2915
9.9	2.2925	2.2935	2.2946	2.2956	2.2966	2.2976	2.2986	2.2996	2.3006	2.3016
x	0	0.01	0.02	0.03	0.04	0.05	0.06	0.07	0.08	0.09

7. TABLE OF EXPONENTIAL VALUES

e^x

The bracket gives the multiplying exponent.

Examples: exp (0.0) = 0.1000 × 10^1 = 1.000, exp (5.0) = 0.1484 × 10^3

x	0	0.01	0.02	0.03	0.04	0.05	0.06	0.07	0.08	0.09
0.0	0.1000(E+1)	0.1010(E+1)	0.1020(E+1)	0.1030(E+1)	0.1041(E+1)	0.1051(E+1)	0.1062(E+1)	0.1073(E+1)	0.1083(E+1)	0.1094(E+1)
0.1	0.1105(E+1)	0.1116(E+1)	0.1127(E+1)	0.1139(E+1)	0.1150(E+1)	0.1162(E+1)	0.1174(E+1)	0.1185(E+1)	0.1197(E+1)	0.1209(E+1)
0.2	0.1221(E+1)	0.1234(E+1)	0.1246(E+1)	0.1259(E+1)	0.1271(E+1)	0.1284(E+1)	0.1297(E+1)	0.1310(E+1)	0.1323(E+1)	0.1336(E+1)
0.3	0.1350(E+1)	0.1363(E+1)	0.1377(E+1)	0.1391(E+1)	0.1405(E+1)	0.1419(E+1)	0.1433(E+1)	0.1448(E+1)	0.1462(E+1)	0.1477(E+1)
0.4	0.1492(E+1)	0.1507(E+1)	0.1522(E+1)	0.1537(E+1)	0.1553(E+1)	0.1568(E+1)	0.1584(E+1)	0.1600(E+1)	0.1616(E+1)	0.1632(E+1)
0.5	0.1649(E+1)	0.1665(E+1)	0.1682(E+1)	0.1699(E+1)	0.1716(E+1)	0.1733(E+1)	0.1751(E+1)	0.1768(E+1)	0.1786(E+1)	0.1804(E+1)
0.6	0.1822(E+1)	0.1840(E+1)	0.1859(E+1)	0.1878(E+1)	0.1896(E+1)	0.1916(E+1)	0.1935(E+1)	0.1954(E+1)	0.1974(E+1)	0.1994(E+1)
0.7	0.2014(E+1)	0.2034(E+1)	0.2054(E+1)	0.2075(E+1)	0.2096(E+1)	0.2117(E+1)	0.2138(E+1)	0.2160(E+1)	0.2181(E+1)	0.2203(E+1)
0.8	0.2226(E+1)	0.2248(E+1)	0.2270(E+1)	0.2293(E+1)	0.2316(E+1)	0.2340(E+1)	0.2363(E+1)	0.2387(E+1)	0.2411(E+1)	0.2435(E+1)
0.9	0.2460(E+1)	0.2484(E+1)	0.2509(E+1)	0.2535(E+1)	0.2560(E+1)	0.2586(E+1)	0.2612(E+1)	0.2638(E+1)	0.2664(E+1)	0.2691(E+1)
1.0	0.2718(E+1)	0.2746(E+1)	0.2773(E+1)	0.2801(E+1)	0.2829(E+1)	0.2858(E+1)	0.2886(E+1)	0.2915(E+1)	0.2945(E+1)	0.2974(E+1)
1.1	0.3004(E+1)	0.3034(E+1)	0.3065(E+1)	0.3096(E+1)	0.3127(E+1)	0.3158(E+1)	0.3190(E+1)	0.3222(E+1)	0.3254(E+1)	0.3287(E+1)
1.2	0.3320(E+1)	0.3353(E+1)	0.3387(E+1)	0.3421(E+1)	0.3456(E+1)	0.3490(E+1)	0.3525(E+1)	0.3561(E+1)	0.3597(E+1)	0.3633(E+1)
1.3	0.3669(E+1)	0.3706(E+1)	0.3743(E+1)	0.3781(E+1)	0.3819(E+1)	0.3857(E+1)	0.3896(E+1)	0.3935(E+1)	0.3975(E+1)	0.4015(E+1)
1.4	0.4055(E+1)	0.4096(E+1)	0.4137(E+1)	0.4179(E+1)	0.4221(E+1)	0.4263(E+1)	0.4306(E+1)	0.4349(E+1)	0.4393(E+1)	0.4437(E+1)
1.5	0.4482(E+1)	0.4527(E+1)	0.4572(E+1)	0.4618(E+1)	0.4665(E+1)	0.4711(E+1)	0.4759(E+1)	0.4807(E+1)	0.4855(E+1)	0.4904(E+1)
1.6	0.4953(E+1)	0.5003(E+1)	0.5053(E+1)	0.5104(E+1)	0.5155(E+1)	0.5207(E+1)	0.5259(E+1)	0.5312(E+1)	0.5366(E+1)	0.5419(E+1)
1.7	0.5474(E+1)	0.5529(E+1)	0.5585(E+1)	0.5641(E+1)	0.5697(E+1)	0.5755(E+1)	0.5812(E+1)	0.5871(E+1)	0.5930(E+1)	0.5989(E+1)
1.8	0.6050(E+1)	0.6110(E+1)	0.6172(E+1)	0.6234(E+1)	0.6297(E+1)	0.6360(E+1)	0.6424(E+1)	0.6488(E+1)	0.6554(E+1)	0.6619(E+1)
1.9	0.6686(E+1)	0.6753(E+1)	0.6821(E+1)	0.6890(E+1)	0.6959(E+1)	0.7029(E+1)	0.7099(E+1)	0.7171(E+1)	0.7243(E+1)	0.7316(E+1)
2.0	0.7389(E+1)	0.7463(E+1)	0.7538(E+1)	0.7614(E+1)	0.7691(E+1)	0.7768(E+1)	0.7846(E+1)	0.7925(E+1)	0.8004(E+1)	0.8085(E+1)
2.1	0.8166(E+1)	0.8248(E+1)	0.8331(E+1)	0.8415(E+1)	0.8499(E+1)	0.8585(E+1)	0.8671(E+1)	0.8758(E+1)	0.8846(E+1)	0.8935(E+1)
2.2	0.9025(E+1)	0.9116(E+1)	0.9207(E+1)	0.9300(E+1)	0.9393(E+1)	0.9488(E+1)	0.9583(E+1)	0.9679(E+1)	0.9777(E+1)	0.9875(E+1)
2.3	0.9974(E+1)	0.1007(E+2)	0.1018(E+2)	0.1028(E+2)	0.1038(E+2)	0.1049(E+2)	0.1059(E+2)	0.1070(E+2)	0.1080(E+2)	0.1091(E+2)
2.4	0.1102(E+2)	0.1113(E+2)	0.1125(E+2)	0.1136(E+2)	0.1147(E+2)	0.1159(E+2)	0.1170(E+2)	0.1182(E+2)	0.1194(E+2)	0.1206(E+2)
2.5	0.1218(E+2)	0.1230(E+2)	0.1243(E+2)	0.1255(E+2)	0.1268(E+2)	0.1281(E+2)	0.1294(E+2)	0.1307(E+2)	0.1320(E+2)	0.1333(E+2)
2.6	0.1346(E+2)	0.1360(E+2)	0.1374(E+2)	0.1387(E+2)	0.1401(E+2)	0.1415(E+2)	0.1430(E+2)	0.1444(E+2)	0.1459(E+2)	0.1473(E+2)
2.7	0.1488(E+2)	0.1503(E+2)	0.1518(E+2)	0.1533(E+2)	0.1549(E+2)	0.1564(E+2)	0.1580(E+2)	0.1596(E+2)	0.1612(E+2)	0.1628(E+2)
2.8	0.1644(E+2)	0.1661(E+2)	0.1678(E+2)	0.1695(E+2)	0.1712(E+2)	0.1729(E+2)	0.1746(E+2)	0.1764(E+2)	0.1781(E+2)	0.1799(E+2)
2.9	0.1817(E+2)	0.1836(E+2)	0.1854(E+2)	0.1873(E+2)	0.1892(E+2)	0.1911(E+2)	0.1930(E+2)	0.1949(E+2)	0.1969(E+2)	0.1989(E+2)
3.0	0.2009(E+2)	0.2029(E+2)	0.2049(E+2)	0.2070(E+2)	0.2091(E+2)	0.2112(E+2)	0.2133(E+2)	0.2154(E+2)	0.2176(E+2)	0.2198(E+2)
3.1	0.2220(E+2)	0.2242(E+2)	0.2265(E+2)	0.2287(E+2)	0.2310(E+2)	0.2334(E+2)	0.2357(E+2)	0.2381(E+2)	0.2405(E+2)	0.2429(E+2)
3.2	0.2453(E+2)	0.2478(E+2)	0.2503(E+2)	0.2528(E+2)	0.2553(E+2)	0.2579(E+2)	0.2605(E+2)	0.2631(E+2)	0.2658(E+2)	0.2684(E+2)
3.3	0.2711(E+2)	0.2739(E+2)	0.2766(E+2)	0.2794(E+2)	0.2822(E+2)	0.2850(E+2)	0.2879(E+2)	0.2908(E+2)	0.2937(E+2)	0.2967(E+2)
3.4	0.2996(E+2)	0.3027(E+2)	0.3057(E+2)	0.3088(E+2)	0.3119(E+2)	0.3150(E+2)	0.3182(E+2)	0.3214(E+2)	0.3246(E+2)	0.3279(E+2)
3.5	0.3312(E+2)	0.3345(E+2)	0.3378(E+2)	0.3412(E+2)	0.3447(E+2)	0.3481(E+2)	0.3516(E+2)	0.3552(E+2)	0.3587(E+2)	0.3623(E+2)
3.6	0.3660(E+2)	0.3697(E+2)	0.3734(E+2)	0.3771(E+2)	0.3809(E+2)	0.3847(E+2)	0.3886(E+2)	0.3925(E+2)	0.3965(E+2)	0.4004(E+2)
3.7	0.4045(E+2)	0.4085(E+2)	0.4126(E+2)	0.4168(E+2)	0.4210(E+2)	0.4252(E+2)	0.4295(E+2)	0.4338(E+2)	0.4382(E+2)	0.4426(E+2)
3.8	0.4470(E+2)	0.4515(E+2)	0.4560(E+2)	0.4606(E+2)	0.4653(E+2)	0.4699(E+2)	0.4747(E+2)	0.4794(E+2)	0.4842(E+2)	0.4891(E+2)
3.9	0.4940(E+2)	0.4990(E+2)	0.5040(E+2)	0.5091(E+2)	0.5142(E+2)	0.5194(E+2)	0.5246(E+2)	0.5298(E+2)	0.5352(E+2)	0.5405(E+2)
4.0	0.5460(E+2)	0.5515(E+2)	0.5570(E+2)	0.5626(E+2)	0.5683(E+2)	0.5740(E+2)	0.5797(E+2)	0.5856(E+2)	0.5915(E+2)	0.5974(E+2)
4.1	0.6034(E+2)	0.6095(E+2)	0.6156(E+2)	0.6218(E+2)	0.6280(E+2)	0.6343(E+2)	0.6407(E+2)	0.6472(E+2)	0.6537(E+2)	0.6602(E+2)
4.2	0.6669(E+2)	0.6736(E+2)	0.6803(E+2)	0.6872(E+2)	0.6941(E+2)	0.7011(E+2)	0.7081(E+2)	0.7152(E+2)	0.7224(E+2)	0.7297(E+2)
4.3	0.7370(E+2)	0.7444(E+2)	0.7519(E+2)	0.7594(E+2)	0.7671(E+2)	0.7748(E+2)	0.7826(E+2)	0.7904(E+2)	0.7984(E+2)	0.8064(E+2)
4.4	0.8145(E+2)	0.8227(E+2)	0.8310(E+2)	0.8393(E+2)	0.8477(E+2)	0.8563(E+2)	0.8649(E+2)	0.8736(E+2)	0.8823(E+2)	0.8912(E+2)
4.5	0.9002(E+2)	0.9092(E+2)	0.9184(E+2)	0.9276(E+2)	0.9369(E+2)	0.9463(E+2)	0.9558(E+2)	0.9654(E+2)	0.9751(E+2)	0.9849(E+2)
4.6	0.9948(E+2)	0.1005(E+3)	0.1015(E+3)	0.1025(E+3)	0.1035(E+3)	0.1046(E+3)	0.1056(E+3)	0.1067(E+3)	0.1078(E+3)	0.1089(E+3)
4.7	0.1099(E+3)	0.1111(E+3)	0.1122(E+3)	0.1133(E+3)	0.1144(E+3)	0.1156(E+3)	0.1167(E+3)	0.1179(E+3)	0.1191(E+3)	0.1203(E+3)
4.8	0.1215(E+3)	0.1227(E+3)	0.1240(E+3)	0.1252(E+3)	0.1265(E+3)	0.1277(E+3)	0.1290(E+3)	0.1303(E+3)	0.1316(E+3)	0.1330(E+3)
4.9	0.1343(E+3)	0.1356(E+3)	0.1370(E+3)	0.1384(E+3)	0.1398(E+3)	0.1412(E+3)	0.1426(E+3)	0.1440(E+3)	0.1455(E+3)	0.1469(E+3)
5.0	0.1484(E+3)	0.1499(E+3)	0.1514(E+3)	0.1529(E+3)	0.1545(E+3)	0.1560(E+3)	0.1576(E+3)	0.1592(E+3)	0.1608(E+3)	0.1624(E+3)
x	0	0.01	0.02	0.03	0.04	0.05	0.06	0.07	0.08	0.09

8. THE NORMAL DISTRIBUTION

These functions are those given by equations (9.7.1) and (9.7.2), and illustrated in Figure IX.5.1.

x	$\phi(x)$	$\Phi(x)$	x	$\phi(x)$	$\Phi(x)$	x	$\phi(x)$	$\Phi(x)$
.00	.3989	.5000	1.40	.1497	.9192	2.80	.0079	.9974
.05	.3984	.5199	1.45	.1394	.9265	2.85	.0069	.9978
.10	.3970	.5398	1.50	.1295	.9332	2.90	.0060	.9981
.15	.3945	.5596	1.55	.1200	.9394	2.95	.0051	.9984
.20	.3910	.5793	1.60	.1109	.9452	3.00	.0044	.9987
.25	.3867	.5987	1.65	.1923	.9505	3.05	.0038	.9989
.30	.3814	.6179	1.70	.0940	.9554	3.10	.0033	.9990
.35	.3752	.6368	1.75	.0863	.9599	3.15	.0028	.9992
.40	.3683	.6554	1.80	.0790	.9641	3.20	.0024	.9993
.45	.3605	.6736	1.85	.0721	.9678	3.25	.0020	.9994
.50	.3521	.6915	1.90	.0656	.9713	3.30	.0017	.9995
.55	.3429	.7088	1.95	.0596	.9744	3.35	.0015	.9996
.60	.3332	.7257	2.00	.0540	.9773	3.40	.0012	.9997
.65	.3230	.7422	2.05	.0488	.9798	3.45	.0010	.9997
.70	.3123	.7580	2.10	.0440	.9821	3.50	.0009	.9998
.75	.3011	.7734	2.15	.0396	.9842	3.55	.0007	.9998
.80	.2897	.7881	2.20	.0355	.9861	3.60	.0006	.9998
.85	.2780	.8023	2.25	.0317	.9878	3.65	.0005	.9999
.90	.2661	.8159	2.30	.0283	.9893	3.70	.0004	.9999
.95	.2541	.8289	2.35	.0252	.9906	3.75	.0004	.9999
1.00	.2420	.8413	2.40	.0224	.9918	3.80	.0003	.9999
1.05	.2299	.8531	2.45	.0198	.9929	3.85	.0002	.9999
1.10	.2179	.8643	2.50	.0175	.9938	3.90	.0002	.9999
1.15	.2059	.8749	2.55	.0155	.9946	3.95	.0002	1.0000
1.20	.1942	.8849	2.60	.0136	.9953	4.00	.0001	1.0000
1.25	.1827	.8944	2.65	.0119	.9960			
1.30	.1714	.9032	2.70	.0104	.9965			
1.35	.1604	.9115	2.75	.0091	.9970			

ANSWERS TO SELECTED PROBLEMS

SECTION 1.1

1. (a) {O1, O2, O7, O8, O17, O18}
   (b) {O5, O7, O8, O10, O11, O12, O13, O14, O15, O17, O18}
   (c) {O6, O11, O15, O16}
   (d) ∅
   (e) {O2, O3, O4, O5, O10, O12, O14, O15, O16, O18}

2. (a), (d), (f) and (h) are equal. (c) and (i) are equal. (e) and (g) are equal.

SECTION 1.2

2. (a) {O1, O2, O6, O7, O8, O11, O15, O16, O17, O18}
   (b) ∅
   (c) ∅
   (d) {O1, O2, O3, O4, O6, O9, O11, O15, O16}
   (e) {O1, O2, O3, O4, O5, O7, O8, O10, O12, O14, O15, O16, O17, O18}
   (f) {O3, O4, O5, O6, O9, O10, O11, O12, O13, O14, O15, O16}
   (g) {O2, O5, O10, O12, O14, O15, O18}

3. (a) 149
   (b) 395
   (c) 321
   (d) 125

(e) 105

(f) 128

(g) 200

(h) 500

SECTION 1.3

1. $f/g(x)$ is the population density of x (in inhabitants per square mile).

2. $\{(a,a), (a,x), (a,y), (b,a), (b,x), (b,y), (c,a), (c,x), (c,y), (d,a) (d,x), (d,y)\}$

3. $f \circ g$ is the set of all pairs (a,b), where a is a state, and b is the population of the capital of a.

5. {(O1,Root), (O2,Root), (O3,Root), (O4,Root), (O5,Brown), (O6,Brown), (O7,Brown), (O8,Brown), (O9,Brown), (O10,Brown), (O11,Gordon), (O12,Gordon), (O13,Gordon), (O14,Gordon), (O15,Ernest), (O16,Ernest), (O17,Ernest), (O18,Ernest)}

6. (a) $f+h(x) = x^2 - x + \sqrt[3]{x}$

   (b) $g/f(x) = (3x+4)/(x^2 - x)$

   (c) $f \circ h(x) = \sqrt[3]{x^2} - \sqrt[3]{x}$

   (d) $f - 2g + h \circ f(x) = x^2 - 7x - 8 + \sqrt[3]{x^2 - x}$

   (e) $(f+g) \circ (h-g)\ (x) = (\sqrt[3]{x} - 3x - 2)^2$

   (f) $(h-g) \circ (f+g)\ (x) = \sqrt[3]{x^2 + 2x + 4} - 3x^2 - 6x - 16.$

SECTION 1.4

1. 30 boys

2. 19.

3. 168.

SECTION 1.5

1. 84.

2. $10!/3! = 604,800$.

3. 1764.

4. 1596, 1617.

5. 64.

6. 120.

7. 15.

8. 1,048,576.

10. Hint: note that both sides of the equation give the total number of subsets in a set with n elements.

SECTION 1.6

1. (a) $p =$ the grass is green

$q =$ the weather is hot

$r =$ it has rained

$$(p \wedge q) \wedge \sim r$$

(b) $p =$ the people are happy

$q =$ the weather is bad

$$p \wedge q$$

(c) $p =$ the nation is prosperous

$q =$ there is war

$$p \vee \sim q\ .$$

(d) $p =$ we are together

$q =$ all are good friends

$$p \vee q$$

(e) $p =$ we are arming

$q =$ the enemy will attack us

$$\sim p \wedge \sim q$$

SECTION 1.7

1. (a)

p	T	T	F	F
q	T	F	T	F
compound	T	F	F	T

(b)

p	T	T	F	F
q	T	F	T	F
compound	F	T	F	T

(c)

p	T	T	T	T	F	F	F	F
q	T	T	F	F	T	T	F	F
r	T	F	T	F	T	F	T	F
compound	T	T	T	T	T	T	F	F

(d)

p	T	T	T	T	F	F	F	F
q	T	T	F	F	T	T	F	F
r	T	F	T	F	T	F	T	F
compound	T	F	T	T	F	F	T	T

(e)

p	T	T	F	F
q	T	F	T	F
compound	F	F	F	F

2. (a) true whenever p and q are both F.

(b) true unless p and q are both T.

(c) always true.

(d) always false.

SECTION 1.8

2. (d), (e), (f), (g), and (h) are all true.

SECTION 1.9

1. (a)

(b)

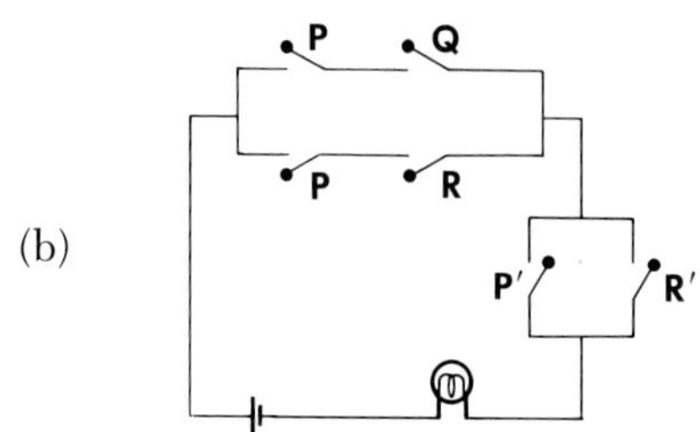

(c)

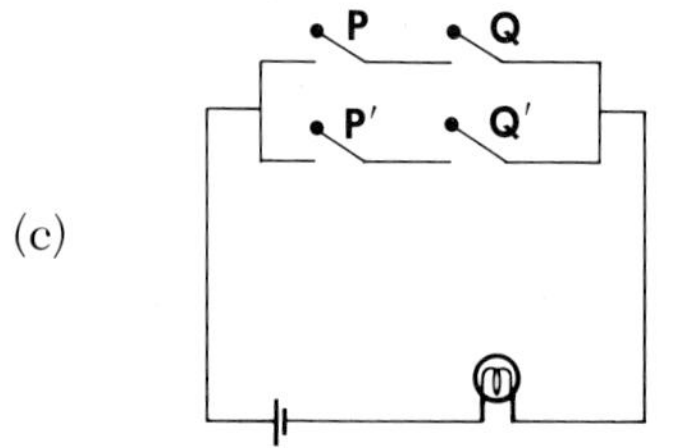

(d)

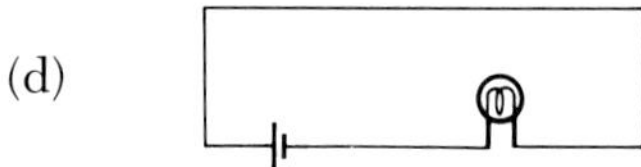

(e)

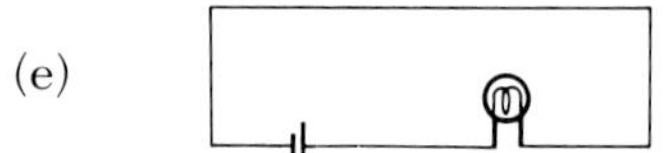

2. $(p \vee q) \wedge ((\sim p \wedge \sim q) \vee ((p \vee r) \wedge \sim q) \vee (\sim r \wedge q))$

3.

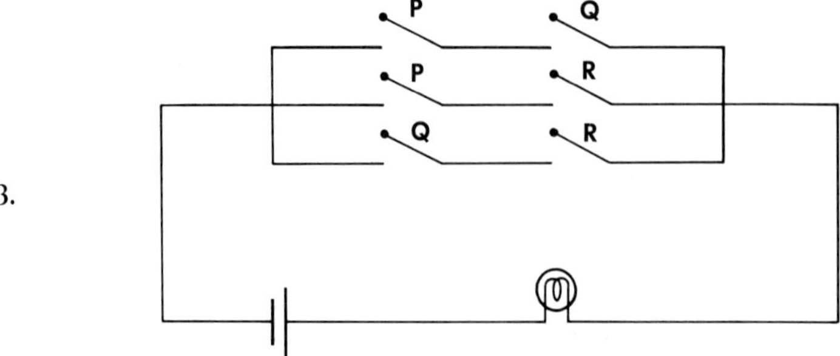

4.

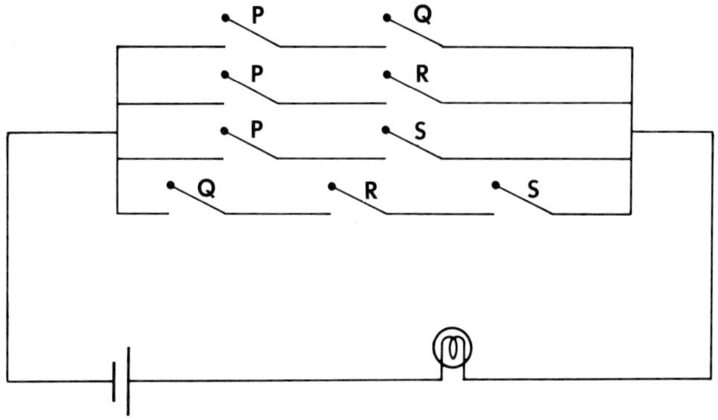

5.

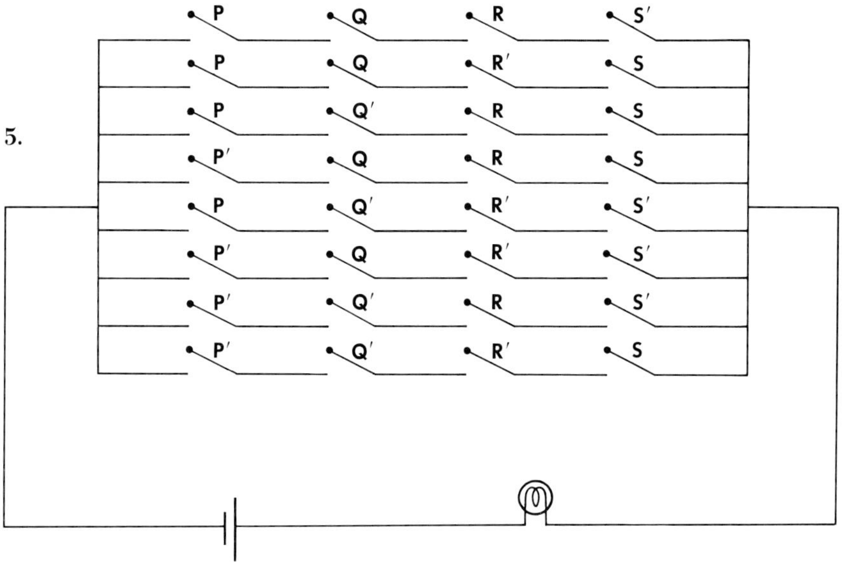

SECTION 1.10

1. 6701

2. 1873

3. 11010111, 1011011011, 1010010010, 10101000110.

4. 45, 53, 62, 17.

5.

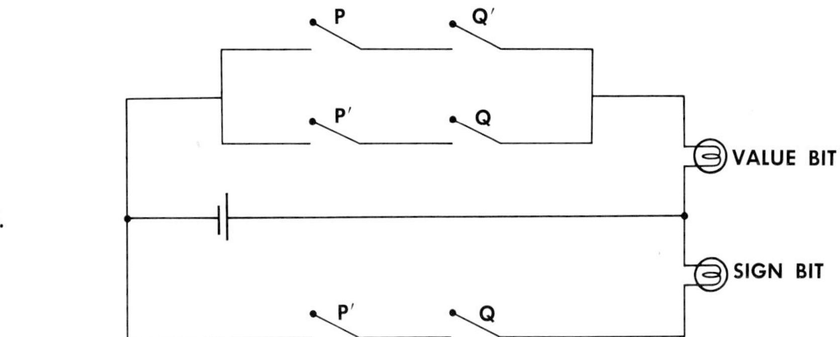

6.

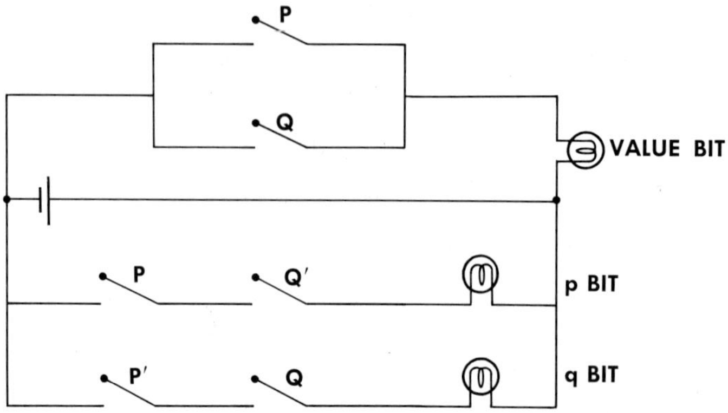

P
Q
VALUE BIT
P
Q′
p BIT
P′
Q
q BIT

7. 326, 1365, 265.

SECTION II.3

1. (a) $2x - y = -1$
(b) $x - y = -2$
(c) $2x + y = 0$
(d) $x = -1$
(e) $y = 4$
(f) $-x + 2y = 8$
(g) $2x - y = 0$
(h) $y = 1$
(i) $x - y = 5$
(j) $x - 4y = 1$
(k) $2x + 4y = 14$
(l) $x = 0$
(m) $x - 2y = -7$
(n) $x - y = 4$
(o) $x - y = 3$
(p) $x + 3y = 13$
(q) $2x - y = 10$
(r) $x + 2y = 4$
(s) $-2x + 3y = -6$
(t) $-4x - y = 4$
(u) $y = 3x + 6$
(v) $y + x = 5$
(w) $y = 2$
(x) $2x + y = -6$
(y) $2x + y = 4$

2. (a) $m = -1/3$, $x_0 = 6$, $y_0 = 2$
(b) $m = 3/2$, $x_0 = 5/3$, $y_0 = -5/2$
(c) $m = -1/4$, $x_0 = 12$, $y_0 = 3$
(d) $m = 1/2$, $x_0 = 3$, $y_0 = -3/2$
(e) $m = 1/4$, $x_0 = 16$, $y_0 = -4$

3. (a) 1 (b) $\sqrt{10}$ (c) $\sqrt{13}$ (d) $\sqrt{26}$ (e) $\sqrt{73}$

4. (a) (7/2, 5/2), (3, 1/2), and (3/2, 2).
(b) $2x + 7y = 17$
$7x + 2y = 22$
$x - y = 1$
(c) $x + 4y = 9$
$4x + y = 12$
$x - y = 1$

5. $s = 5000p - 5000$.

6. $d = 9000 - 4000p$.

7. $c = 30x + 8000$.

8. Fixed costs are $900; variable costs are $50 per ton.

SECTION II.4

2. 200 units

3. 180 tons

4. 2500 units per year.

SECTION II.5, PART 1.

1. (a) $x = 1$, $y = 2$.
   (b) $x = 10$, $y = 1$.
   (c) $x = 11/7$, $y = 9/7$.
   (d) $x = 3$, $y = 2$.
   (e) x arbitrary, $y = 2 - 1/2x$.
   (f) $x = 3$, $y = 1$.
   (g) $x = -1$, $y = -1$.
   (h) no solution.
   (i) $x = 2$, $y = 3$.
   (j) $x = 6$, $y = 3$.
   (k) x arbitrary, $y = 5/2 - 3/2\,x$.
   (l) no solution.
   (m) $x = 1$, $y = 2$.
   (n) $x = -1$, $y = 1$.
   (o) no solution.
   (p)' no solution.
   (q) $x = 2$, $y = 1$.
   (r) $x = 0$, $y = 3$.

2. The points are (a), (8/3, 5/3), and (b), (13/5, 8/5).

3. 6 ounces of the 40% solution, and 14 ounces of the 90%.

4. 17½ pounds of peanuts, 7½ pounds of cashews.

5. 6 pounds of A, 1 pound of B.

6. $33.33 per ton.

SECTION II.5, PART 2.

1. The extreme points are:
   (a) (6/5, 27/5), (0,6), (3,0), and (0,0).
   (b) (9, 2), (39/5, 14/5), (5,0), (0,0), and (10, 0).
   (c) (12/5, 44/5), (16/3, 0), and (20, 0).
   (d) (9, 2), (25/3, 10/3), and (12, 0).
   (e) (8/5, 16/5), (0, 4), (10/3, 4/3), (2, 0), and (0, 0).

2. Let x = quantity of expensive mixture.

   y = quantity of cheaper mixture.

   z = quantity of unmixed peanuts.

   Then

$$\begin{aligned} 5x + y \quad &\leqq 6000 \\ x + y + z &\leqq 4000 \\ x \quad &\geqq 0 \\ y \quad &\geqq 0 \\ z &\geqq 0. \end{aligned}$$

3.

$$\begin{aligned} x + \quad y &\geq 6 \\ 200x + 50y &\geq 800 \\ 1000x + 2800y &\geq 8000 \\ x, \quad y &\geq 0 \end{aligned}$$

4.

$$\begin{aligned} 50x + 40y &\leq 10{,}000 \\ 20x + 35y &\leq 8{,}000 \\ x, \quad y &\geq 0 \end{aligned}$$

SECTION II.6

1. (a) $x = 1$, $y = 1$, $z = -1$.

   (b) $x = 1$, $y = -1$, $z = 2$.

   (c) no solution.

   (d) x arbitrary, $y = 3x - 11$, $z = 5/2x - 17/2$.

   (e) $x = 3$, $y = 1$, $z = 1$.

   (f) $x = 1$, $y = -4$, $z = 1$.

2. Let x, y, and z be the number of units of A, B, and C respectively. Then

$$\begin{aligned} 0 \leq z &\leq 10/3 \\ x &= 3 - 2z/5 \\ y &= 4 - 6z/5 \end{aligned}$$

3. $x = 41/19$, $y = 28/19$, $z = 40/19$.

4.

$$\begin{aligned} 3x + \quad y + 2z &\leq 40 \\ x + 5y + 6z &\leq 40 \\ 4x + 9y + 5z &\leq 60 \\ x, y, z &\geq 0 \end{aligned}$$

SECTION III.2

2. (a) ellipse

   (b) parabola

   (c) hyperbola

   (d) parabola (degenerate)

   (e) hyperbola

   (f) ellipse (degenerate)

(g) hyperbola

(h) parabola

(i) hyperbola

(j) parabola

SECTION III.3

1. (a) Center $(2, -1)$, radius 1.
   (b) Center $(4, -6)$, radius $\sqrt{42}$.
   (c) Center $(-1, 0)$, radius 3.
   (d) Center $(6, -2)$, radius $\sqrt{24}$.
   (e) Center $(0, -8)$, radius 4.
2. (a) $x^2 + y^2 - 6x - 10y - 15 = 0$.
   (b) $x^2 + y^2 + 2x - 4y - 11 = 0$.
   (c) $x^2 + y^2 + 2x - 8y - 19 = 0$.
   (d) $x^2 + y^2 + 6y - 91 = 0$.
   (e) $x^2 + y^2 - 8x - 4y + 19 = 0$.
   (f) $x^2 + y^2 - 6x + 2y + 1 = 0$.
   (g) $x^2 + y^2 - 20x - 10y + 100 = 0$.
   (h) $x^2 + y^2 + 2x + 2y - 47 = 0$.
3. This is a circle with center $(-5, -100)$ and radius 150. The maximum amount of table grapes is 111.8 pounds; the maximum amount of wine is 49.9 gallons.
4. This is a circle with center $(-1500, -1000)$ and radius 4500. The highest possible price is (approximately) $2890.

SECTION III.4

1. (a) Center $(3, -3)$. Major axis, parallel to x-axis, $\sqrt{172}$ units. Minor axis, $\sqrt{43}$ units. Focal length, $\sqrt{129}$ units.
   (b) Center $(-1, 2)$. Major axis, parallel to y-axis, 3 units. Minor axis, 2 units. Focal length, $\sqrt{5}$ units.
   (c) Center $(3, 0)$. Major axis, parallel to x-axis, 8 units. Minor axis, 2 units. Focal length, $\sqrt{60}$ units.
   (d) Center $(1, -1)$. Major axis, parallel to y-axis, 10 units. Minor axis, 4 units. Focal length, $\sqrt{84}$ units.
   (e) Center $(3, -2)$. Major axis, parallel to y-axis, 28 units. Minor axis, 16 units. Focal length, $\sqrt{528}$ units.
2. (a) $\sqrt{3/4}$.
   (b) $\sqrt{5/9}$.

(c) $\sqrt{15/16}$.

(d) $\sqrt{21/25}$.

(e) $\sqrt{33/49}$.

3. The ratio is the eccentricity e. The line $x = -a^2/f$ is also a directrix.

5. The curve is an ellipse with center $(-10, -5)$, major axis $\frac{400}{3}\sqrt{3}$ parallel to the x-axis, and minor axis $100\sqrt{3}$.

6. The curve is an ellipse with its center at $(150, -200)$, and axes of 600 and 1600 units, respectively, together with the line $y = 600$, which is tangent to the ellipse at $(150, 600)$.

SECTION III.5

1. (a) Vertex $(14/3, -4)$. Focus $(19/6, -4)$. Directrix $x = 37/6$.

   (b) Vertex $(-2/5, -64/5)$. Focus $(-2/5, -51/4)$. Directrix $y = -257/20$.

   (c) Vertex $(-5/12, -1)$. Focus $(31/12, -1)$. Directrix $x = -41/12$.

   (d) Vertex $(3/28, -849/56)$. Focus $(3/28, -106/7)$. Directrix $y = -425/28$.

   (e) Vertex $(-6, 2)$. Focus $(-11/2, 2)$. Directrix $x = -13/2$.

2. The relation of revenue, r, to rent, x, is

$$r - 15{,}625 = -\frac{1}{5}(x - 125)^2$$

which shows that the maximum of r is 15,625, obtained by setting a rent of \$125 per month.

3. The break-even point is 18.5 bushels; the maximum amount which can be produced at a profit is 681 bushels.

SECTION III.6

1. (a) Center $(-2, 1)$. Asymptotes $2(x+2) = \pm(y-1)$

   (b) Center $(-2/3, -2)$. Asymptotes $4y + 8 = \pm(3x+2)$.

   (c) Center $(4, 3)$. Asymptotes $2y - 6 = \pm(x-4)$.

   (d) Center $(1, 1)$. Asymptotes $5x - 5 = \pm 2(y-1)$.

   (e) Center $(1, 1)$. Asymptotes $5x - 5 = \pm 2(y-1)$.

2. (a) $4x^2 - y^2 + 16x + 2y + 40 = 0$.

   (b) $9x^2 - 16y^2 + 12x - 64y - 145 = 0$.

   (c) $x^2 - 4y^2 - 8x + 24y - 104 = 0$.

   (d) $25x^2 - 4y^2 - 50x + 8y + 113 = 0$.

   (e) $25x^2 - 4y^2 - 50x + 8y + 171 = 0$.

3. (a) Asymptotes $x = -4$, $y = 2$.
   (b) Asymptotes $x = 2$, $y = -4$.
   (c) Asymptotes $x = -8$, $y = 2$.
   (d) Asymptotes $x = 4$, $y = -3$.
   (e) Asymptotes $x = 0$, $y = 6$.

5. The ratio is equal to the eccentricity, e.

6. This is a hyperbola with center at $(80, -100)$. The asymptotes are given by $10(x-80) = \pm(3y+300)$. More exactly, it is that part of the left branch of the hyperbola which lies in the positive quadrant of the plane.

7. The total amount bought, y, will be given by

$$y = 5 + \frac{2500}{x}$$

where x is the price. The graph of this curve is a hyperbola with asymptotes $x = 0$ and $y = 5$.

SECTION III.7

1. (a) no solutions.
   (b) $\left(\frac{17+\sqrt{73}}{3}, \frac{-11-\sqrt{73}}{2}\right)$, $\left(\frac{17-\sqrt{73}}{3}, \frac{-11+\sqrt{73}}{2}\right)$
   (c) $(5, 4)$ and $(17/5, 8/5)$.
   (d) $(2, -9)$ and $(-2, -13)$.
   (e) $(2, 2)$ and $(-7/9, -13/6)$.
   (f) $(2, 3)$ and $(22/5, -21/5)$.
   (g) $(2, 3)$ and $(786/185, -4406/185)$.
   (h) $(1, 1)$ and $(1/5, 13/5)$.
   (i) $(3, 2)$ and $(383/61, 420/61)$.
   (j) no solutions.

3. The equilbrium price is $x = 7.20$. Supply and demand are then both equal to 0.8.

4. Approximately 5.8 dozen.

5. The break-even point is 40 miles. The maximum amount that can be produced at a profit is 44 miles.

SECTION III.8

2. \$12,155.06.

3. $100\,(2^{5/3})$, or approximately 317.

4. Approximately $\frac{5\sqrt{2}}{2}$, or 3.54 defectives per hour.

5. 108,000/13, or approximately 8308.

6. $\frac{\log 7 - \log 2}{\log 3 - \log 2}$, or about \$3.16.

SECTION IV.2

1. $P(E) = 1/2$, $P(S) = 1/3$, $P(E \cap S) = 1/6$, $P(E - S) = 1/3$, $P(E \cup S) = 2/3$, $P(S - E) = 1/6$.

2. (a) 1/2
   (b) 3/4
   (c) 1/4
   (d) 1/8
   (e) 7/8
   (f) 1/4
   (g) 1/2

3. (a) 1/13
   (b) 1/52
   (c) 4/13

4. (a) 5/36
   (b) 15/36
   (c) 6/36

SECTION IV.4, PART 1.

1. 1/2
2. 3/4
3. 211/243, 1/243
4. 203/396
5. $29!/(20!\ 30^9)$, or .185
6. 0.777
7. 244/495
8. $(0.95)^{10}$, or .599
9. 4/9

SECTION IV.4, PART 2.

1. 26/33
2. 3/14
3. (a) 7/178
   (b) 9/22
4. 1/26
5. 24/41

SECTION IV.6

1. (a) 924/4096 (b) 794/4096 (c) 2508/4096

2. (a) 3125/15,552
   (b) 1453/23,328

3. 75%

4. .085

5. 11097/15625

6. 938/969

7. 21/46

8. 27/28

9. (a) 3679/3876
   (b) 1144/1615
   (c) 284/285

SECTION IV.10

1. mean 50/3, variance 125/9

2. 0

3. 3 units

4. .01

7. $\sigma_U^2 = \sigma_X^2 + \sigma_Z^2,\ \sigma_V^2 = \sigma_Y^2 + \sigma_Z^2,\ \sigma_{UV} = \sigma_Z^2$

$$\rho_{UV} = \frac{\sigma_Z^2}{\sqrt{(\sigma_X^2 + \sigma_Z^2)(\sigma_Y^2 + \sigma_Z^2)}}$$

8. .978

9. .398

10. .303

SECTION V.1

1. 1

3. $-2/x^3$

5. $2x - 2$

7. $2x$

9. $4x - y - 4 = 0$

11. $x + y + 2 = 0$

13. $2x + y - 3 = 0$

15. $\dfrac{50}{\sqrt{x}} - 0.02x$

SECTION V.2

1. $15x^2 + 12x$

2. $42x^6 - 15x^4$

5. $x - \dfrac{4}{x^3}$

7. $\frac{1}{3}x^{-2/3} - \frac{1}{3}x^{-4/3}$

9. $\frac{1}{2\sqrt{x}} - \frac{1}{x\sqrt{x}}$

11. $2x + 2$

13. $3x^2 - 6x$

15. $\frac{3\sqrt{x}}{2}$

17. $12x - y - 21 = 0$

19. $21x - y - 45 = 0$

21. $x = 0$ or $x = 2/3$

23. $R = 15x - 0.01x^2$
Marginal price $= -0.01$

SECTION V.3

1. $\frac{20x^3 - 9x^2}{10y^4}$

3. $\sqrt{x/y}$

5. $\frac{2x - 2}{3y^2 + 1}$

7. $\frac{6x}{8y}$

9. $3x + 4y - 25 = 0$

11. $3x + 2y - 30 = 0$

15. $\frac{3x^2 + \frac{1}{2}x^{-1/2}}{2y + 1}$

17. $-\sqrt{y/x}$

SECTION V.4

1. $\frac{x}{\sqrt{1 + x^2}}$

3. $\frac{x}{(1 - x^2)^{3/2}}$

5. $12(x^2-2x)(x^3-3x^2+5)^3$

7. $\dfrac{3x-1}{\sqrt{3x^2-2x+1}}$

9. $\dfrac{1}{3x^{2/3}(1-x^{1/3})^2}$

11. $\dfrac{1+3(1-x^2)}{2\sqrt{x-(1-x)^3}}$

13. $\dfrac{1}{2(1-x)^{3/2}}-\dfrac{1}{2(1+x)^{3/2}}$

15. $-\dfrac{1}{2}\left(x+\dfrac{1}{1-x}\right)^{-3/2}\left[1+\dfrac{1}{(1-x)^2}\right]$

17. $\dfrac{2}{9}(x^{2/3}-x^{-2/3})^{-2/3}(x^{-1/3}+x^{-5/3})$

19. $\dfrac{2x^2-2x+1}{(1-2x)^2}$

21. $(x-2)(4-t^{-1/2})$

23. $4xt$

25. $\dfrac{8t-3}{2\sqrt{x}}$

27. $\dfrac{3t}{1+\sqrt{x}}$

29. $\dfrac{24t}{x^4\sqrt{1+4t^2}}$

31. $\dfrac{1-2t}{(1-x)^2}$

33. $6x$

35. $\dfrac{2t}{\sqrt{u}}$

37. $\dfrac{-9ux^2}{t^2\sqrt{x^3+4}}$

39. $\dfrac{6u}{\sqrt{t}}$

41. $-\dfrac{x+10}{\sqrt{6400-20x-x^2}}, -25+\dfrac{6400x-30x^2-2x}{\sqrt{6400x^2-20x^3-x^4}}$

43. $7-\dfrac{25x}{\sqrt{19600-10x^2}}$

SECTION V.5

1. $\dfrac{1+2x^2}{\sqrt{1+x^2}}$

3. $\dfrac{15y^2-11y+10}{\sqrt{y^2-5}}$

5. $\dfrac{12t^6-16t^3-15}{2\sqrt{(t^3-3)(3t^4+5t)}}$

7. $(u^2-1)^4(26u^4-6u^2)$

9. $\dfrac{25y^2-60y}{3(y^2-3y)^{2/3}}$

11. $5t^4-12t^3+6t^2$

13. $\dfrac{5x^2-9x+2}{2\sqrt{2x(x-1)(x-2)}\,\sqrt{2x}}$

15. $\dfrac{2}{(1-t)^2}$

17. $\dfrac{2x}{(x^2+1)^{3/2}(x^2-1)^{1/2}}$

19. $\dfrac{3u^2}{\sqrt{2}}$

21. $\dfrac{2-3x^2}{\sqrt{2x}\,(x^2+2)^2}$

23. $\dfrac{-4t+t^{1/2}+3t^{5/2}}{6(t^3-t)^{4/3}}$

25. $\dfrac{6z^2-14z+9}{z^4(2z-3)^2}$

27. $1000+\dfrac{300x^2-2x^3}{\sqrt{200x^3-x^4}}$

29. $\dfrac{100{,}000x-10x^2-10{,}000\sqrt{200x-x^2}}{\sqrt{10{,}000-2x}\,\sqrt{200x-x^2}\,(1000+\sqrt{200x-x^2})^2}$

SECTION V.6

1. $6x^5-15x^4+8x^3-3,\ 30x^4-60x^3+24x^2$

3. $12x^3 - 6x^2 + 12x - 1,\ 36x^2 - 12x + 12$

5. $15(x^3+3x)^{14}(3x^2+3),\ 210(x^3+3x)^{13}(3x^2+3)^2 + 90x(x^3+3x)^{14}$

7. $\frac{1}{3}x^{-2/3} - \frac{1}{3}x^{-4/3},\ -\frac{2}{9}x^{-5/3} + \frac{4}{9}x^{-7/3}$

9. $\frac{2x-3x^3}{\sqrt{1-x^2}},\ \frac{6x^4-9x^2+2}{(1-x^2)^{3/2}}$

11. $3x\sqrt{x^2+1},\ \frac{3x^2}{\sqrt{x^2+1}} + 3\sqrt{x^2+1}$

13. $\frac{1}{(x+1)^2},\ \frac{-2}{(x+1)^3}$

15. $-\frac{2x+y}{x+2y},\ \frac{6x^2+8xy+y^2}{(x+2y)^3}$

17. $\frac{-2xy}{x^2+2y},\ \frac{6x^4y+8x^2y^2-8y^3}{(x^2+2y)^3}$

19. $\frac{2x-x^2}{(1-x)^2},\ \frac{2+x-x^2}{(1-x)^3}$

21. $\frac{1}{3}(x+1000)^{-2/3},\ -\frac{2}{9}(x+1000)^{-5/3}$

SECTION V.7

1. 20 by 20

3. $24\sqrt{2}$ by $18\sqrt{2}$

5. 12 by 8

7. 4 by 4 by 2

9. 16 by 16 by 4

11. 8 by 16 by 4

13. $r=5,\ h=10$

15. $r=\frac{48}{8+3\pi},\ h=\frac{48+12\pi}{8+3\pi}$

17. $r=\sqrt[3]{\frac{10{,}000}{14\pi}},\ h=\frac{1000}{\pi r^2}$

19. $285

21. \$540

23. \$400, \$100

25. $\dfrac{A-b}{2(B+a)}$

27. $\text{tax} = \dfrac{1}{2}(A-b)$; price increase $= \dfrac{B(A-b)}{4(B+a)}$

SECTION V.8

1. $(2x-1)e^{x^2-x}$

3. $\dfrac{1-e^x}{2\sqrt{x-e^x}}$

5. $e^x - e^{-x}$

7. $e^{-x}(2x-x^2)$

9. $\dfrac{-e^{1/x}}{x^2}$

13. $e^{-0.01t}$

15. $D_x y = -\dfrac{x}{\sigma^2}e^{-x^2/2\sigma^2}$, $D_x^2 y = \dfrac{x^2-\sigma^2}{\sigma^4}e^{-x^2/2\sigma^2}$

SECTION V.9

1. $2x \ln x + x$

3. $\dfrac{-1}{x(\ln x)^2}$

5. $\dfrac{2 \ln x}{x}$

7. $e^{-x}\left(\dfrac{1}{x} - \ln x\right)$

9. $\ln b \, e^{x \ln b}$

11. $\dfrac{e^x}{1+e^x}$

13. $\ln x$

15. $\dfrac{3}{5}x^{-2/5}$

17. $\frac{1}{3}(2x-1)(x^2-x-6)^{-2/3}$

19. $\dfrac{5x^4-28x^3+48x^2-32x}{3[(x^2-3x+4)(x^3-4x^2)]^{2/3}}$

21. $\dfrac{7x^6+12x^5+30x^4+44x^3+42x^2+36x+24}{3[(x^2+2x+2)(x^2+4)(x^2+3)]^{2/3}}$

SECTION V.10

1. $6 \cos 3x$

3. $2x \sin x^2$

5. $\dfrac{-\cos x}{\sqrt{1-\sin x}}$

7. $\cos \frac{1}{x} + \frac{1}{x} \sin \frac{1}{x}$

9. $\dfrac{(\ln x+\cos x)[e^x \sin(2x+1)+2e^x \cos(2x+1)]-e^x \sin(2x+1)\left(\frac{1}{x}-\sin x\right)}{(\ln x+\cos x)^2}$

11. $\dfrac{-3 \sin x \cos x}{(\cos^2 x+1)^{1/2}(\sin^2 x+1)^{3/2}}$

13. $\dfrac{e^x \cos y+\sin y}{e^x \sin y - x \cos y}$

15. 1

17. $\sin 2x$

19. He should produce 19.629 lb. of table grapes and 47.964 gal. of wine.

SECTION V.11

1. $2xy-2y^2,\ x^2-4xy$

3. $1/y,\ -x/y^2$

5. $\dfrac{x}{\sqrt{x^2+y^2}},\ \dfrac{y}{\sqrt{x^2+y^2}}$

7. $\dfrac{2x}{x^2+y^2},\ \dfrac{2y}{x^2+y^2}$

9. $\dfrac{-1}{y}e^{-x/y},\ \dfrac{x}{y^2}e^{-x/y}$

11. $-e^{-xy}(y\,dx + x\,dy)$

13. $\dfrac{2(x\,dx + y\,dy)}{x^2 + y^2}$

15. $(2x + y)\,dx + (x + 2y)\,dy$

17. $e^{-x/y}\left(\dfrac{x\,dy - y\,dx}{y^2}\right)$

19. $\min\left(-\frac{2}{3} + \frac{1}{6}\sqrt{52}, -\frac{10}{3} - \frac{1}{6}\sqrt{52}\right)$

21. none

23. approx. 1673 lb. beef and 1775 lb. pork.

SECTION V.12

1. $bx = ay$ for min; $x = 0$ for max

3. $x = 2y$ for min; $x = 0$ for max

5. $x = 0$ for min; $y = 0$ for max

7. $\max\left(\frac{\sqrt{2}}{2}, \frac{\sqrt{2}}{2}, 0\right)$, $\min\left(-\frac{\sqrt{2}}{2}, -\frac{\sqrt{2}}{2}, 0\right)$

9. $\max\left(\pm\frac{\sqrt{2}}{2}, \pm\frac{\sqrt{2}}{2}\right)$, $\min\left(\pm\frac{\sqrt{2}}{2}, \mp\frac{\sqrt{2}}{2}\right)$

11. $\max\left(\pm\frac{\sqrt{3}}{3}, \pm\frac{\sqrt{3}}{3}, \frac{\sqrt{3}}{3}\right)$ or $\left(\pm\frac{\sqrt{3}}{3}, \mp\frac{\sqrt{3}}{3}, -\frac{\sqrt{3}}{3}\right)$

    $\min\left(\pm\frac{\sqrt{3}}{3}, \pm\frac{\sqrt{3}}{3}, -\frac{\sqrt{3}}{3}\right)$ or $\left(\pm\frac{\sqrt{3}}{3}, \mp\frac{\sqrt{3}}{3}, \frac{\sqrt{3}}{3}\right)$

13. cube

15. $\sqrt{5}\,h = (1 + \sqrt{5})\,r$, $\sec\alpha = 3/2$

17. \$89, \$94

19. $x = 8$, $y = 4$, $z = 3$

SECTION VI.1

1. $\displaystyle\int_0^1 (2x - 3x^2)\,dx$

3. $\displaystyle\int_0^2 \sqrt{3 + x}\,dx$

5. $\int_0^{10} \sqrt{x}\,dx$

7. $\int_0^2 (5+2x)^3\,dx$

9. $\int_1^2 x^3 dx$

11. 1/2

13. 19/3

15. $(b^3 - a^3)/3$

17. $(b^{k+1} - a^{k+1})/(k+1)$

SECTION VI.2

1. 8/3

3. 17/6

5. 27/24

7. 6/5

9. 9/2

11. 46/15

13. $2 \ln 2$

15. $1 + \ln 2$

17. $1/\sqrt{2}$

19. $21,566

SECTION VI.3

1. $(2\sqrt{2} - 1)/3$

3. $(3\sqrt[3]{4} - 3)/4$

5. 3/20

7. $(9\sqrt[3]{9} - 9)/2$

9. 2/5

11. 0

13. $5\sqrt[3]{2}/2$

15. $2\sqrt{2} - 2$

17. $\ln \frac{1+e}{2}$

19. 1/2

21. 1

23. $\ln (4/5)$

25. $(e^3 - 1)/3$

27. 1

29. 5/12

31. 0

33. 3/4

35. 0

37. $500.48

SECTION VI.4

1. 4/15

3. 5787/10

5. 9/140

7. 64/15

9. 5/24

11. 8/9

13. $81\pi/16$

15. $\pi/4$

17. $-38/3$

19. 4000 ln 10

SECTION VI.5

1. $(4\sqrt{2} + 4)/15$

3. $(8\sqrt{2} - 10)/3$

5. $(14\sqrt{2} - 16)/15$

7. $-\pi$

9. $\pi/2$

11. 0

13. $(e^2 - 1)/4$

15. $(e^{-\pi/2} + 1)/5$

17. $\pi^3 - 6\pi$

19. 8/7

21. $2(\ln 2)^2 - 4 \ln 2 + 2$

23. approx. \$38,338

25. approx. 70

SECTION VI.6

1. 48/5

3. 8

5. 9/2

7. 8/3

9. 1/12

11. $3 \ln 3 + 2$

13. $4\sqrt{50}/3$, 0, 80/3

15. $25 - 5 \ln 2$, 10/3

SECTION VI.7

1. Divergent

3. Divergent

5. 0

7. 0

9. 2

11. -1

17. 4000 lb.

SECTION VI.8

1. Exact value 2

3. Exact value $2\sqrt{2} - 2 \approx 0.828428$

SECTION VII.2

1. (a) $(5, 13, -3)$.

(b) $\begin{pmatrix} 13 \\ -6 \end{pmatrix}$

(c) $(-6, -14, 20, 6)$.

(d) $\begin{pmatrix} 1 \\ 13 \\ 10 \end{pmatrix}$

(e) $(-9, +28)$

(f) $(-5, 29, -15)$.

(g) $(8, 12, 8)$

(h) $\begin{pmatrix} 6 \\ 14 \end{pmatrix}$

(i) $\begin{pmatrix} 12 \\ 13 \\ -5 \\ -8 \end{pmatrix}$

(j) $(-6, -1, 31)$.

2. (a) $x = 3$, $y = 5$.

(b) $x = -2$, $y = 1$.

(c) $x = 11/5$, $y = 6/5$.

(d) $x = -2$, $y = 3$.

(e) $x = -1$, $y = 2$.

(f) $x = 1$, $y = 5$.

(g) $x = 3$, $y = 1$.

(h) no solution.

(i) $x = -1$, $y = 2$.

(j) $x = 102/43$, $y = 25/43$, $z = 115/43$.

(k) $x = 1$, $y = 1$, $z = -3$.

3. (a) yes (b) yes (c) no (d) no (e) yes.

4. (a) $\begin{pmatrix} 1 & 3 \\ 11 & 6 \end{pmatrix}$

(b) $\begin{pmatrix} 6 & 43 \\ 32 & -9 \\ 29 & -29 \end{pmatrix}$

(c) $\begin{pmatrix} 17 & 14 & -5 & 24 \\ 34 & 17 & -8 & -4 \\ -27 & 23 & 0 & 7 \end{pmatrix}$

(d) $\begin{pmatrix} 2 & -21 \\ 12 & -8 \\ 15 & -10 \\ 13 & 15 \end{pmatrix}$

(e) $\begin{pmatrix} 2 & -1 & -1 \\ 12 & -15 & 19 \\ -15 & 11 & 13 \end{pmatrix}$

5. (a)
$$4.3(A + B) = \begin{pmatrix} 0 & 240.8 & 73.1 \\ 55.9 & 0 & 301 \\ 103.2 & 120.4 & 0 \end{pmatrix}$$

(b)
$$4.9A + 5.5B = \begin{pmatrix} 0 & 289.4 & 89.3 \\ 68.5 & 0 & 365.8 \\ 125.4 & 144.4 & 0 \end{pmatrix}$$

(c)
$$0.35A + 0.42B = \begin{pmatrix} 0 & 21.35 & 6.65 \\ 51.1 & 0 & 27.16 \\ 9.31 & 10.64 & 0 \end{pmatrix}$$

6. $\begin{pmatrix} 19 & 77 & 56 & 29 \\ 73 & 38 & 22 & 31 \\ 92 & 12 & 21 & 9 \end{pmatrix}$

7. At the two plants,

$$A + B = \begin{pmatrix} 59 & 60 \\ 103 & 31 \end{pmatrix}$$

At Denver,

$$2B = \begin{pmatrix} 64 & 58 \\ 110 & 28 \end{pmatrix}$$

At all three,

$$A + 3B = \begin{pmatrix} 123 & 118 \\ 213 & 59 \end{pmatrix}$$

SECTION VII.5

1. (a) 40

(b) -6

(c) -10

(d) 19

(e) $\begin{pmatrix} -16 & -2 & 29 \\ 6 & -34 & 34 \end{pmatrix}$

(f) $\begin{pmatrix} 19 & -4 \\ 2 & 4 \\ 6 & -3 \end{pmatrix}$

(g) $\begin{pmatrix} 14 & -7 & -3 \\ 36 & 26 & -2 \\ 29 & 5 & 7 \end{pmatrix}$

(h) $\begin{pmatrix} 8 & 19 & -11 \\ 17 & 9 & 21 \\ -16 & -28 & 16 \end{pmatrix}$

(i) $\begin{pmatrix} 14 & 36 & 29 \\ -7 & 26 & 5 \\ -3 & -2 & -7 \end{pmatrix}$

(j) $\begin{pmatrix} 0 & 0 & 0 \\ 0 & 0 & 0 \\ 0 & 0 & 0 \end{pmatrix}$

(k) $\begin{pmatrix} 7 \\ 8 \\ -1 \end{pmatrix}$

(l) $\begin{pmatrix} 25 & 32 \\ 11 & -4 \end{pmatrix}$

(m) $\begin{pmatrix} -10 & 30 & 14 \\ 3 & 19 & 7 \\ 4 & 32 & 12 \end{pmatrix}$

(n) $\begin{pmatrix} 7 & 46 & 54 & 55 \\ -7 & 37 & 0 & 4 \end{pmatrix}$

(o) $\begin{pmatrix} 19 & 2 \\ 26 & 6 \end{pmatrix}$

(p) $\begin{pmatrix} 26 & 54 \\ 7 & 17 \end{pmatrix}$

(q) $\begin{pmatrix} 29 & 6 \\ 38 & 20 \end{pmatrix}$

(r) $\begin{pmatrix} 6 & 34 \\ 2 & 16 \\ -5 & -32 \end{pmatrix}$

(s) 6

(t) 19.

2. (a) $\begin{pmatrix} -7 & -21 & -7 \\ 14 & 42 & 14 \end{pmatrix}$

(b) $\begin{pmatrix} 48 & 82 \\ -16 & -4 \end{pmatrix}$

(c) $\begin{pmatrix} -55 & -36 & 95 \\ -46 & -46 & 108 \\ 233 & 81 & -229 \end{pmatrix}$

(d) $\begin{pmatrix} -6 & -164 & -200 \\ -23 & -266 & -336 \end{pmatrix}$

(e) 99.

3. (a) $x = 2, y = 1.$

(b) $x = 1/2, y = 0.$

(c) $x = -2, y = 1.$

(d) $x = 5, y = -1, z = -3, w = 2.$

(e) $x = -7, y = -7, z = 5, w = 6.$

(f) $x = -1, y = -17, z = 1, w = 7.$

4. $(78, 43, 21, 8)\,(70, 120, 50, 110)^t = 12{,}550$

5. Approximately 10,796,569

6. b_{ii} is the number of persons (in the set) that P_i knows.

7.

$$A^2 = \begin{pmatrix} 2 & 1 & 1 & 1 & 0 \\ 1 & 3 & 0 & 1 & 0 \\ 1 & 0 & 1 & 1 & 0 \\ 1 & 1 & 1 & 2 & 0 \\ 0 & 0 & 0 & 0 & 0 \end{pmatrix}$$

8. The expected payoffs from buying stock A, B, or C are given by the vector (4700, 7000, 4000). Stock B is the best investment.

9. The vector of payrolls is $p = sA$, where s is the salary vector and A is the matrix of employees.

$$p = (15100, 6550, 8800, 18475, 6775).$$

10. $x/12 - y/12$ of ground steak,
$7y/48 - x/48$ of hamburger.

In this case it is 40/3 pounds of ground steak and 50/3 pounds of hamburger.

11. The units are: T of fuel, T-mi. of transportation.

$$A = \begin{pmatrix} 0 & 100 \\ 0.001 & 0 \end{pmatrix} \qquad x = (1139, 138889)$$

12. 55/9¢ per lb. of fuel, 200/9¢ per T-mi. of transportation.

SECTION VII.7

1. (a) $x = 2$, $y = 0$.
   (b) $x = -27$, $y = 17$.
   (c) $x = 31$, $y = 20$, $z = -47$.
   (d) $x = 1$, $y = 3$, $z = 5$.
   (e) $x = -7$, $y = -4$, $z = 2$, $w = -5$.
   (f) $x = -1$, $y = 1$, $z = 4$.
   (g) $x = -5/2\,z + 5/2$, $y = 9/4\,z + 1/4$, $w = -6z + 3$.
   (h) $x = -137/14$, $y = 39/14$, $z = 25/2$, $w = 9/14$.
   (i) $x = 1$, $y = 1$, $z = 1$, $w = 1$.
   (j) $x = -6$, $y = 7$, $z = 3$, $w = 4$.
   (k) $x = -25/44$, $y = -127/44$, $z = 61/44$, $w = -133/44$.
   (l) $x = 1$, $y = 2$, $z = 1$, $w = 3$.
   (m) no solution.
   (n) no solution.
   (o) w arbitrary, $x = -1/20w + 21/20$, $y = -2/5w - 8/5$, $z = 31/40w + 89/40$.

2. The mixture should contain 8 grams of A, 6 of B, and 2 of C. This weighs exactly 16 grams.

3. The general solution is

$$\begin{aligned} 40/3 \le\ & z \le 15 \\ x &= 60 - 4z \\ y &= 3z - 40. \end{aligned}$$

4. It is not possible to make the mixture cost 56¢ or 64¢ per gallon. For a cost of 60¢/gal., we must have $x = 20/3$, $y = 0$, and $z = 40/3$.

5. 200 pounds of X, 50 pounds of Y, and 1550 pounds of Z.

SECTION VII.8

1. (a) $\begin{pmatrix} -1/28 & 5/28 \\ 6/28 & -2/28 \end{pmatrix}$

(b) $\begin{pmatrix} -2 & 1 \\ 7/4 & -3/4 \end{pmatrix}$

(c) $\begin{pmatrix} 5/11 & 3/11 \\ -2/11 & 1/11 \end{pmatrix}$

(d) $\begin{pmatrix} -5 & 6 \\ 1 & -1 \end{pmatrix}$

(e) $\begin{pmatrix} 2 & -3/2 \\ 1 & -1/2 \end{pmatrix}$

(f) $\begin{pmatrix} -23 & -13 & 12 \\ 60 & 34 & -31 \\ -2 & -1 & 1 \end{pmatrix}$

(g) $\begin{pmatrix} 14/99 & -53/198 & 2/9 \\ 1/9 & 1/9 & -1/9 \\ -1/11 & 3/22 & 0 \end{pmatrix}$

(h) not invertible.

(i) $\begin{pmatrix} 389/2268 & -34/2268 & -344/2268 & -13/2268 \\ -412/2268 & 380/2268 & 376/2268 & 212/2268 \\ 455/2268 & -238/2268 & -140/2268 & -91/2268 \\ -47/2268 & 214/2268 & 164/2268 & -185/2268 \end{pmatrix}$

(j) not invertible.

2. The amounts which can be made are:

$$\begin{aligned} &-5a/3 + b + c \text{ of } X \\ &59a/12 - 15b/4 + c/4 \text{ of } Y \\ &-9a/4 + 15b/4 - c/4 \text{ of } Z. \end{aligned}$$

3. One pound of steel, net, requires gross production of 1.042 lb. of steel, 0.164 lb. of fuel, and 0.199 T-mi. of transportation. One pound of fuel, net, requires 0.215 lb. of steel, 1.052 lb. of fuel, and 0.096 T-mi. of transportation. One T-mi. of transportation requires 0.116 lb. of steel, 0.323 lb. of fuel, and 1.04 T-mi. of transportation.

4. Prices are 8.91¢ per lb. of steel, 8.149¢ per lb. of fuel, and 6.587¢ per T-mi. of transportation.

5. The technological matrix is

$$A = \begin{pmatrix} 0 & 3 & 1 & 0 & 0 & 0 & 6 & 3 \\ 0 & 0 & 0 & 2 & 0 & 0 & 2 & 0 \\ 0 & 0 & 0 & 1 & 1 & 0 & 1 & 2 \\ 0 & 0 & 0 & 0 & 0 & 2 & 2 & 0 \\ 0 & 0 & 0 & 0 & 0 & 3 & 2 & 2 \\ 0 & 0 & 0 & 0 & 0 & 0 & 0 & 0 \\ 0 & 0 & 0 & 0 & 0 & 0 & 0 & 0 \\ 0 & 0 & 0 & 0 & 0 & 0 & 0 & 0 \end{pmatrix}$$

which gives us

$$(I-A)^{-1}=\begin{pmatrix} 1 & 3 & 1 & 7 & 1 & 17 & 29 & 7 \\ 0 & 1 & 0 & 2 & 0 & 4 & 6 & 0 \\ 0 & 0 & 1 & 1 & 1 & 5 & 5 & 4 \\ 0 & 0 & 0 & 1 & 0 & 2 & 2 & 0 \\ 0 & 0 & 0 & 0 & 1 & 3 & 2 & 2 \\ 0 & 0 & 0 & 0 & 0 & 1 & 0 & 0 \\ 0 & 0 & 0 & 0 & 0 & 0 & 1 & 0 \\ 0 & 0 & 0 & 0 & 0 & 0 & 0 & 1 \end{pmatrix}$$

so that one scaffolding requires 17 rods, 29 bolts, and 7 coils of wire.

SECTION VIII.1

1. (a) $x=0$, $y=4$, value 20.

(b) $x=4$, $y=0$, value 8.

(c) $x=3/5$, $y=9/5$, value 27/5.

(d) $x=12$, $y=0$, value 36.

2. 70/3 of A, 40/3 of B, cost \$6.17.

3. 0, 640 lb., 360 lb. Profit \$404.

4. 2000 lb., 4000/3 lb. Profit \$2333.33.

SECTION VIII.3

1. (a)
$$\begin{aligned} -1/8\,s-1/8\,t-3/4\,u+7/2&=-x\\ 9/4\,s+1/4\,t+1/2\,u\quad\;&=-y\\ 11/8\,s+3/8\,t+1/4\,u-1/2&=-z. \end{aligned}$$

(b)
$$\begin{aligned} 7s-2t-23&=-x\\ -3s+t+10&=-y\\ -5s+t+11&=-z. \end{aligned}$$

(c)
$$\begin{aligned} 5r+22s+2t+11&=-x\\ -20r-89s-7t-42&=-y\\ 3r+13s+t+6&=-z. \end{aligned}$$

(d)
$$\begin{aligned} -0.8s-0.5t+0.7r-8.9&=-x\\ 1.2s+t-0.8r+13.6&=-y\\ 0.2s+0.5t-0.3r+5.1&=-z. \end{aligned}$$

(e)
$$\begin{aligned} -r-2t-12s+9&=-x\\ 2r+3t+21s-13&=-y\\ -r-t-7s+8&=-z \end{aligned}$$

(f)
$$\begin{aligned} 5/7r+2/7s-1/7t-19/7&=-x\\ 16/7r+5/7s+1/7t-23/7&=-y\\ 62/7r+8/7s+3/7t+1/7&=-z. \end{aligned}$$

(g)
$$\begin{aligned} -\frac{3}{8}s-\frac{7}{8}t+\frac{5}{8}u+1&=-x\\ s+2t-u\quad&=-y\\ \frac{1}{8}s-\frac{3}{8}t+\frac{1}{8}u-1&=-z. \end{aligned}$$

SECTION VIII.10

1. $x = 40/7$, $y = 64/35$, $z = 53/35$, $w = 129/7$.

2. $x = 4/5$, $y = 0$, $z = 44/5$, $w = 48/5$.

3. $x = 0$, $y = 10/3$, $z = 25/3$, $w = 80/3$.

4. $x = 5$, $y = 4$, $z = 3$, $w = 42$.

5. 135/34 of A, 175/34 of C. Cost \$5.52.

6. 5 of A, 15 of C. Revenues \$155.

7. 25/2 chairs, 5/2 tables, profit \$187.50.

8. 4273 lb. of A, 21,818 lb. of C, 2909 lb. of D. Revenues, \$18,391.

9. 50 of A, 175/2 of B. Revenue \$1025.

10. 485 T. of A, 50 T. of C. Cost \$2725.

SECTION VIII.11

1. (a) Primal solution, $x = 0$, $y = 5$, $z = 10$.
Dual solution, $r = 24/7$, $s = 0$, $t = 1/7$.
Value 90.

(b) Primal solution, $x = 0$, $y = 25/3$, $z = 0$.
Dual solution, $r = 0$, $s = 2/3$, $t = 0$.
Value 50/3.

(c) Primal solution, $x = 45/4$, $y = 25/4$, $z = 0$.
Dual solution, $r = 1/8$, $s = 0$, $t = 5/8$.
Value 115/4.

(d) Primal solution, $x = 5$, $y = 15$, $z = 0$.
Dual solution, $r = 0$, $s = 3$, $t = 5$.
Value 45.

2. (5) $r = 31/4$, $s = 0$, $t = 5/34$.
(6) $r = 7/2$, $s = 3/4$, $t = 0$.
(7) $r = 25/4$, $s = 5/4$. $t = 0$.
(8) $r = 215/88$, $s = 375/88$, $t = 45/8$.
(9) $r = 3/2$, $s = 0$, $t - 11/2$.
(10) $r = 50$, $s = 0$, $t = 225$.

SECTION VIII.13

1. (a) $\begin{pmatrix} 0 & 80 & 20 \\ 140 & 10 & 0 \\ 0 & 0 & 120 \end{pmatrix}$ cost 950

(b) $\begin{pmatrix} 5 & 45 & 0 & 0 \\ 0 & 0 & 15 & 55 \\ 0 & 25 & 15 & 0 \end{pmatrix}$ cost 530

(c) $\begin{pmatrix} 0 & 50 & 0 & 25 & 5 \\ 50 & 0 & 20 & 10 & 0 \\ 0 & 0 & 0 & 0 & 40 \end{pmatrix}$ cost 425

(d) $\begin{pmatrix} 5 & 60 & 25 & 0 & 0 \\ 35 & 0 & 0 & 15 & 40 \\ 0 & 0 & 0 & 20 & 0 \end{pmatrix}$ cost 560

(e) $\begin{pmatrix} 40 & 10 & 0 & 0 & 0 \\ 0 & 0 & 20 & 0 & 30 \\ 0 & 30 & 20 & 0 & 0 \\ 0 & 0 & 0 & 40 & 10 \end{pmatrix}$ cost 470

3. (a) (1,1), (2,3), (3,5), (4,2), (5,4). Value 23. Salaries (dual variables) 4, 2, 3, 2, 1.

(b) (1,5), (2,7), (3,6), (4,3), (5,1), (6,4), (7,2). Salaries 4, 6, 3, 3, 4, 4, 3.

(c) (1,3), (2,4), (3,1), (4,5), (6,2). Value 30. Salaries 2, 1, 3, 2, 0, 2.

(d) (1,2), (2,5), (3,4), (4,3), (5,6), (6,1). Value 41. Salaries 1, 6, 1, 3, 1, 2, 0.

SECTION IX.2

2. Probability (no raisins) $= e^{-2} = 0.14$
Probability (1 raisin) $= 2e^{-2} = 0.27$

3. Probability (3 or more defectives) $= 0.07$

4. Probability (3 or more earthquakes) $= 0.19$

5. The expected profits, for 1, 2, 3, 4, or 5 flowers bought, are \$2.75, \$4.75, \$5.65, \$5.45, and \$4.40, respectively. The maximum is \$5.65, for 3 flowers.

6. (a) Clearly, we have $f(n) = \frac{1}{n} - \frac{1}{n+1}$. It will follow that $P(X \le n) = \frac{n}{n+1}$.

(b) $P(X \ge 5) = 0.2$
$P(X = 6) \mid X \ge 5) = 5/42$.

7. The maximum expectation is \$3, obtained by buying either 1 or 2 flowers.

SECTION IX.4

1. $A = \begin{pmatrix} .5 & .2 & .3 \\ .3 & .6 & .1 \\ 0 & 0 & 1.0 \end{pmatrix}$, .391

2. .424

3. (.396, .372, .232)

4. (3/8, 3/8, 1/4)

5. (a) $\begin{pmatrix} 0.7 & 0.3 & 0 & 0 & 0 \\ 0.14 & 0.62 & 0.24 & 0 & 0 \\ 0 & 0.14 & 0.62 & 0.24 & 0 \\ 0 & 0 & 0 & 0.2 & 0.8 \end{pmatrix}$

   (b) 160/301

   (c) 245/3483

8. Note that $(I-B)(I+B+B^2+\ldots+B^K)=I-B^{K+1}$. Then note that, for large values of K, all entries of B^{K+1} will be small.

10. $\begin{pmatrix} 1 & 0 & 0 & 0 & 0 & 0 \\ 1/2 & 0 & 1/2 & 0 & 0 & 0 \\ 0 & 1/2 & 0 & 1/2 & 0 & 0 \\ 0 & 0 & 1/2 & 0 & 1/2 & 0 \\ 0 & 0 & 0 & 1/2 & 0 & 1/2 \\ 0 & 0 & 0 & 0 & 0 & 1 \end{pmatrix}$

   (a) 8/5

   (b) 5 plays

   (c) 3/5

SECTION IX.6

1. Y has the cumulative distribution

$$F(y)=\begin{cases} 0 & \text{for } y \le 0 \\ \sqrt{y} & 0 \le y \le 1 \\ 1 & y \ge 1 \end{cases}$$

   X has mean 1/2 and variance 1/12. Y has mean 1/3 and variance 5/36.

2. The value of k is 6. $P(0 \le X \le 1/4) = 5/32$. The mean and variance of X are 0.5 and 0.05, respectively.

4. $P(1 \le X \le 2) = 0.0475$.
   Y has the density

$$g(y)=\begin{cases} e^{2-y} & \text{for } y \ge 2 \\ 0 & y < 2 \end{cases}$$

5. We will generally have $P(1 \le X \le 2) = e^{-\alpha} - e^{-2\alpha}$, which is greatest for $\alpha = \ln 2$.

SECTION IX.7

1. $P(70 \le X \le 95) = 0.3345$
   $P(90 \le X \le 110) = 0.3830$
   $P(X \ge 150) = 0.0062$

2. 0.2967

3. $P(\text{at least 2 over } 140) = 0.662$
   $P(\text{at least one over } 150) = 0.462$

4. 75th percentile = 113.5
   90th percentile = 125.6
   99th percentile = 146.6

5. $P(0.195 \leq X \leq 0.205) = 0.3830$
   $P(0.2 \leq X \leq 0.21) = 0.3413$
   $P(X \leq 0.208) = 0.7881$

6. The mean must be set at 12.04. Then, 1 per cent of the bottles will have less than 11.97 fl. oz., and 37 per cent will have more than 12.05 fl. oz.

7. $P(0.3 \leq X \leq 0.305) = 0.3944$
   $P(X \leq 0.298) = 0.3085.$

8. 0.3066 inches in diameter.

INDEX

Absolute distribution, 203
Absolute probability, 172, 178
Absolute value, 225
Absorbing states, 507, 509
Abstraction, 354–355
Addition
  circuit for, 50–51
  in binary notation, 49
  of functions, 15
  of matrices, 356–357
Additive identity, 358
Additive inverse, 358
Adjacent basic points, 432
Adjacent extreme points, 423, 432–433
Altitude, of triangle, 71, 89
Antidifferentiation, 309
Antiderivative formulas, 310, 314
Apollonius, 53, 113
A posteriori probability, 187
Approximate integration, 349
A priori probability, 187
Archimedes, 53, 307
Area, 301–303, 334–337, 341
Aristotle, 30
Assignment problem, 479–486
Associative laws
  for set operations, 10
  for matrix addition, 358
  for matrix multiplication, 371
  for scalar multiplication, 358
Asymptotes, 138, 142
Augmented matrix, 390
Average, 206
Axes
  coordinate, in plane, 55–56
    in three dimensions, 99
  of an ellipse, 129
  of a hyperbola, 138
  of a parabola, 133

Base (of logarithms), 164, 548
  change in, 549
Basic feasible point, 432
Basic points, 432
Basic variables, 427
Bayes' formula, 187–188
Bernoulli, Jakob, 192
Bernoulli coefficient, 192, 193
Bernoulli trials, 192
Bi-conditional, 38
Binary notation, 48–51
  addition in, 49
  decimal-modified, 52
  multiplication in, 49
Binary operation, 355
Binomial distribution, 191–192, 211, 526–527
  negative, 198
Biquadratic equation, 150, 539–540
Boole, George, 30
Break-even point, 75

Calculus, fundamental theorem of, 309, 348
Cartesian product, 11–12
  number of elements of, 23
Cayley, Arthur, 366
Center
  of a circle, 124
  of a hyperbola, 138
  of an ellipse, 129
Central limit theorem, 529
Certain event, 172
Chain rule, 236, 314
Characteristic property, 2
Chebyshev's inequality, 213–214, 494

Circle, 144–124
  radius, 124
  center, 124
  common chord, 151–152
Circuits, 40–51
  for addition, 50–51
  for multiplication, 50
  for subtraction, 51
  parallel, 41
  series, 41
Coefficient(s), 65
  Bernoulli, 192, 193
Coincident lines, 81, 84
Column, 355
  vector, 356
Combination, 27
  linear, 81–82
Commutative laws
  for matrix addition, 358
  for set operations, 10
Complement, 7
  relative, 8
Completing the square, 537–538
Components of vector, 356
Composition of functions, 16–17, 236
Compound experiment, 180–181, 186
Conditional, 36–37
  consequence, 37
  hypothesis, 37
Conditional probability, 177–178
  Bayes' formula for, 187–188
Conic section, 113
  classification, 121–123
  degenerate, 116
Conjugate axis, 138
Conjugate hyperbola, 138
Conjunction, 31–32, 41
Consequence, 37
Constraints, 287–289, 415
  equation, 287–289, 447–450
    duality for, 482
  non-negativity of, 97–98
Consumer's surplus, 338
Continuous random variables, 511–534
  density of, 512–513
  exponential, 515–516
  log-normal, 521–522
  normal, 518–519, 524–534
  uniform, 514–515
Convergence, of series, 490–491
  of simplex algorithm, 441–446
Convex sets, 415–416
  extreme points of, 94, 418
Coordinate axes, in plane, 55, 56
  in three or more dimensions, 99
Coordinate planes, in space, 99
Coordinates, in plane, 55
  in three or more dimensions, 100–101
  on line, 54
Correlation, 218–219
Covariance, 218
Cubic equation, 538–539
Dantzig, G. B., 414
Decimal logarithm, 549–550
Decimal-modified binary form, 52
Decimal notation, 47–51
Deficient biquadratic equation, 539
Degeneracy, of conic section, 116
  of linear program, 442, 445, 450, 472, 479, 482
Demand law, 85, 337–338
Density, probability, 512–513
Derivative, 222, 223
  formulas, 227, 236, 242, 264, 267, 273, 274
  higher order, 247
  operator, 247
  partial, 279
Descartes, 53
Determinant, 379
Deviation, standard, 209
Differentiable function, 223
Differential, 233, 242, 280
  equation, 261, 274
Differentiation, 229
Dimensions of matrix, 355
Directrix, 132, 133, 146
Discrete probability spaces, 492
Discrete random variable, 493
Discounted future earnings, 330–331
Disjoint, events, 174
  sets, 19, 174
    pairwise, 19
Disjunction, 31–32, 41
Dispersion, 209
Distance, focal, 129
  between points, in plane, 58
    on line, 54
Distribution, 199
  absolute, 203
  binomial, 191–192, 211
  conditional, 203–204
  continuous, 512–520
  exponential, 515–516
  geometric, 494–495
  joint, 202
  negative binomial, 198
  normal, 518–520, 524–534
  Poisson, 496–497
  uniform, 514–515
Distribution function (cumulative), 199, 512
Distributive laws
  for matrix multiplication, 371
  for scalar multiplication, 358
  for set operations, 10
Divergence, of series, 490
Division, of functions, 15
  of matrices, 374
Domain, 14
Doubly stochastic matrix, 481
Drawings, with replacement, 194
  without replacement, 194
Duality, 456–469
  for equation constraints, 482

Eccentricity, 114, 132, 146
Edge, 423–424
Elasticity, 268
Element, 1
  of a set, 1, 2
Elimination, process of, 82
Ellipse, 114, 128
  axes, 129
  center, 129
  directrix, 132
  eccentricity, 132
  focal distance, 129
  focus, 129
Empty set, 4
Entry, 355
Equality, of matrices, 355, 356
  of sets, 3
Equations, solution of, 536–542
    biquadratic, 539–540
    cubic, 538–539
    higher degree, 540
    linear, 536–537
    quadratic, 538–539
    rational, 540–541
    transcendental, 541
    with surds, 541–542
  first-degree, 63, 536–537
  systems of, 80–89, 147–153
Equilibrium price, 85
Euclid, 53
Event, 172
  certain, 172
  disjoint, 174
  impossible, 172
  independent, 179
Expected value, 206, 207, 493, 513
Experiment, 169–170
  compound, 180–181, 186
  simple, 191
Exponent, 157, 544–547
Exponential distribution, 515–516
Exponential function, 157–158, 261–264
Extraneous root, 536, 541
Extreme points, adjacent, 423
  of a convex set, 94, 418

Factorial (of a number), 25
Failure, 191
Fair coin, 170
Feasible points, 415
  basic, 432
Fechner, G. T., 166
Fermat, Pierre de, 169
First-degree equation, 63, 536–537
Fixed costs, 69
Focus, 129, 133, 146
Function(s), 13–14
  addition of, 15–16
  composition of, 16–17
  division of, 15–16
  domain of, 14

Function(s) (*Continued*)
  multiplication of, 15–16
  range of, 14
  subtraction of, 15–16
Fundamental theorem of calculus, 309, 348

Geometric distribution, 494–495
Geometric series, 491
Graph, of exponential function, 157–158
  of first-degree equation, 61, 64
  of inequality, 76, 79
  of logarithmic function, 165
  of quadratic equation, 113
  of relation or equation, 56
Growth problem, 159, 162–163, 261–264

Half-plane, 78
Homogeneity law, 367–368
Hyperbola, 115, 137
  asymptotes, 138, 142
  center, 138
  conjugate axis, 138
  conjugate hyperbola, 138
  directrix, 146
  eccentricity, 146
  focus, 146
  transverse axis, 138
Hyper-plane, 110
Hypothesis, 37

Identity, additive, 358
  function, 18
  multiplicative, 375
Image, 13–14
Implicit function theorem, 282
Impossible event, 172
Improper integral, 343–344
Inconsistent system, 81, 92, 446
Independent, events, 179
  random variables, 204, 209–210, 529
  Induction, 542–543
  definition by, 544
  proof by, 542–543
Inequality, 71–72
  Chebyshev's, 213–214, 494
  loose, 74, 78
  strict, 74, 78
  systems of, 90, 107, 413–414
Inequality signs, 72–73
Infeasible linear program, 446
Integral, 303, 308–309
Integration, by parts, 328–329
  by substitution, 320–321
Intensity vector, 381–382
Intercepts, 67–68
Intersection (of sets), 7
  associative law, 10

Intersection (of sets) (*Continued*)
  commutative law, 10
  distributive law, 10
Inverse, additive, 358
  matrix, 376
    computation of, 379, 405–408
  multiplicative, 376
Invertible matrix, 376

Joint distribution, 202
Joint probability, 179

Lagrange multipliers, 289–291
Law, associative, commutative, and distributive
    for matrix addition, 358
    for matrix multiplication, 371
    for scalar multiplication, 358
    for set operations, 10
  demand and supply, 85, 337–339
  homogeneity, 367–368
  of large numbers, 214, 494
  Weber-Fechner, 165–166
Leibnitz, Gottfried, W., 232
Limit(s), 222, 303, 490, 506
  of integration, 303
Linear, algebra, 355
  combination, 81–82
  equations
    definition, 63
    solution, 536–537
    systems, 80, 101
  programs, 413–488
    assignment problem, 479–486
    constraint set, 415
    degenerate, 442, 450, 482
    dual, 457–459
    fundamental theorem of, 465
    infeasible, 446
    objective function, 415
    perturbation, 450–451
    primal, 459
    solution of, 415, 426–441
    transportation problem, 470–479
    unbounded, 420, 443
    value, 415
Logarithm, 164–165, 266–267, 547–550
  base, 548
  binary, 550
  change in base, 549
  decimal, 549–550
  natural (Naperian), 266, 550
Logarithmic derivative, 268
Logarithmic differentiation, 269
Log-normal distribution, 521–522

Mapping, 14
Marginal quantities, 222, 253, 256, 278–279, 311
Markov chains
  definition, 500
  fundamental matrix, 510
  regular, 506
  transition matrix, 500, 502
Matrix (matrices), 355–412
  addition of, 356
  augmented, 390
  determinant of, 379
  dimensions of, 355
  doubly stochastic, 481
  fundamental, 510
  inverse, 376, 379, 405–411
  invertible, 376
  multiplication of, 369
  n^{th} order, 356
  permutation, 480
  regular, 506
  scalar, 376
  scalar multiplication of, 357
  singular, 376
  square, 356
  stochastic, 500
    absorbing, 501
  technological matrix, 381
  transition, 500
  transposition, 365
  unit, 375
Maximum, 251–253, 282–283, 289–291, 414–415
Mean, 209, 514
Median, of triangle, 71, 89
Minimum, 251–253, 282–283, 289–291, 415
Multiplication, circuit for, 50
  in binary notation, 49
  of functions, 15
  of matrices, 368–369
  of vectors, 366
  scalar, 357
Multiplicative, indentity, 375
  inverse, 376

n^{th} Order matrix, 356
Napier, John, 550
Natural (Naperian) logarithms, 266–267, 550
Negation, 30–33
Negative binomial distribution, 198
Negative numbers, 72–73
Network, 470
Non-basic variables, 427
Non-degeneracy, 442
Non-negativity constraints, 97–98
Normal distribution, 519–520, 524–534
  approximation to binomial, 526–527
  central limit theorem, 529
Northwest corner rule, 474–476
Notation, binary, 48–51
  decimal, 47–51
  decimal-modified binary, 52
Numbers, positive and negative, 72–73
  rational, 2

Numbers (*Continued*)
real, 2
system of, 354

Objective function, 415
Ordered n-tuples, 12
Ordered pair, 11–13
Ordinate, 55
Origin, in line, 54
in plane, 55
in three or more dimensions, 99
Orthogonal vectors, 368
Outcome, 169–170

Parabola, 57, 114, 132
directrix, 133
focus, 133
vertex, 133
Parallel lines, 59, 65, 84
Parameter, 67, 132
Pascal, Blaise, 169, 181
Permutation, 24, 26
matrix, 480
Perpendicular lines, 60
Perturbation, 450–451
Petersburg Paradox, 498–499
Pivot, 428
rules for choosing, 434, 438–439
step (transformation), 428
Point-slope formula, 61
Poisson distribution, 496–497
Positive numbers, 72–73
Price, shadow, 467
vector, 366, 387
Primal linear program, 459
Probability,
absolute, 172
a posteriori, 187
a priori, 187
conditional, 177–178
Bayes' formula, 187–188
joint, 179
subjective, 171
Probability function, 199, 492
Probability space, continuous, 511
discrete, 492
finite, 172
Producers' surplus, 339
Product transformation curve, 126
Proposition, 30–50
bi-conditional, 38
conditional, 36–37
conjunction, 31–32, 41
disjunction, 31–32, 41
negation, 30–32, 43
tautology, 34
truth-table of, 33, 37, 38, 43
truth-value of, 30
Pure competition, 339
Pythagoras, 53

Quadratic equation, 113, 537–538
system of, 150

Radian measure, 272
Radius, 124
Random variable, 199
continuous, 511
correlation, 218–219
covariance, 218
discrete, 493
expected value, 206–207, 493, 513
independent, 204
mean, 209
normal, 518–520, 524–534
standard deviation, 209
variance, 109, 494, 514
Range, 14
Rational equation, 540
Rational function, 540
Rational number, 2
Regular matrix, 506
Relation, 12, 13
Resolvent, 539–540
Right-handed system, 100
Root, extraneous, 536
Roster, 2
Row, 355
operations, 388–390

Scalar, matrix, 376
multiplication, 357
product, 366
Series, 489–490
convergent, 490
divergent, 490
geometric, 491
partial sums, 490
Set, 1–30
Cartesian product, 11–12
complement, 7–8
disjoint, 19
element of, 1–2
empty, 4
equality of, 3
intersection of, 7
number of elements, 19–23
subset of, 3, 23
union of, 6
universal, 3
Shadow prices, 467
Simple experiment, 191
Simplex algorithm, 426–453
basic feasible point, 432
basic point, 432
convergence, 441–447
non-degeneracy, 442
pivot, 428
rules for choosing, 434, 438–439
simplex tableau, 426
slack variables, 425

Simplex algorithm (*Continued*)
  Stage I, 438–441, 445–446
  Stage II, 433–438, 442–443
Simplex tableau, 426
Simpson's rule, 349–352
Sine function, 272–274
Singular matrix, 376
Slack variables, 425
Slope, 58–59
  infinite, 59, 63
  of tangent line, 221, 233
  point-slope formula, 61
  slope-intercept formula, 68
Solutions, infinity of, 81
  of equation, 81, 535–540
  of linear program, 415
  of system of equations, 81–89, 102–105, 149–152, 387–403
Square, completing, 537–538
  matrix, 356
Standard deviation, 209
Stochastic matrix, 500
  absorbing, 507
  doubly, 481
  regular, 506
Subjective probability, 171
Subset, 3, 23, 27
Success, 191
Subtraction, circuit for, 51
  of functions, 15
  of matrices, 359
Summation symbol, 550–553
  double, 552
  iterated, 552
  triple, 553
Supply and demand, 85, 337–339
Supply law, 85, 338–339
Surd, 541
Sylvester, James, 366
System(s)
  inconsistent, 81, 92
  left- and right-handed, 100
  $m \times n$, 110–111
  of equations, 80, 101, 147
    solution of, 81–89, 102–105, 147–156
  of inequalities, 90–98, 107–108
  of numbers, 354

Tautology, 34
Technological matrix, 381–382
Thales, 53
Theorem, central limit, 529
  of calculus, fundamental, 309, 348
  implicit function, 282
Transcendental equations, 541, 542
Transient state, 509
Transition matrix, 500
Transportation problem, 470–479
Transposition of matrices, 364–365
Truth-table, 33, 37–38
Truth-value, 30, 33
Tucker, A. W., 414, 426
Two-point formula, 61

Unbounded linear program, 420, 443
Uniform distribution, 514–515
Union (of sets), 6
  associative law, 10
  commutative law, 10
  distributive law, 10
  number of elements, 19–21
Unit matrix, 375
Unitary operation, 364
Universal set, 3
Unrestricted variables, 415, 482

Value, absolute, 225
  expected, 206–207, 493, 513
  of linear program, 415
  truth–, 30
Variable costs, 69
Variables, basic, 432
  non-basic, 432
  random, 199, 493, 511
  slack, 425
  unrestricted, 415, 482
Variance, 209, 211, 514
Vector, 356
  column, 356
  components of, 356
  intensity, 381
  multiplication of, 366
  orthogonal, 368
  price, 366, 387
  row, 356
Venn diagram, 6
Vertex, of parabola, 133
  of polyhedron, 94, 418

Weber, E. H., 166
Weber-Fechner law, 165–166